A NEW SHORT TEXTBOOK OF

MICROBIAL AND PARASITIC INFECTION

NEW SHORT TEXTBOOK SERIES

Some titles in the series

A New Short Textbook of Anaesthetics, Intensive Care and Pain Relief
Jose Ponte
David W. Green

A New Short Textbook of Surgery
Leonard Cotton
Kevin Lafferty

A NEW SHORT TEXTBOOK OF
MICROBIAL AND PARASITIC INFECTION

B. I. DUERDEN
BSc, MD, MRCPath

Professor of Medical Microbiology, University of Sheffield;
Honorary Consultant in Microbiology, Sheffield Health Authority

T. M. S. REID
BMedBiol, MB, ChB, MRCPath

Consultant Bacteriologist, City Hospital, Aberdeen;
Honorary Clinical Senior Lecturer in Bacteriology, University of Aberdeen

J. M. JEWSBURY
PhD, DIC, ARCS

Senior Lecturer in Medical Parasitology, Liverpool School of Tropical Medicine

D. C. TURK
DM, MRCP, FRCPath

formerly Consultant Microbiologist, Regional Public Health Laboratory, Sheffield

Edward Arnold
A division of Hodder & Stoughton
LONDON MELBOURNE AUCKLAND

© 1987 B.I. Duerden, T.M.S. Reid, J.M. Jewsbury and D.C. Turk

First published as *A Short Textbook of Medical Microbiology* 1965
First published as *A New Short Textbook of Microbial and Parasitic Infection* 1987
Second Impression 1988
Reprinted 1991

British Library Cataloguing in Publication Data

A New short textbook of microbial and parasitic infection.
1. Medical microbiology   2. Parasitic diseases
I. Duerden, B.I.
616'.01 QR46

ISBN 0 340 40178 8

All rights reserved. No part of this publication may be reproduced or transmitted in any form or by any means, electronically or mechanically, including photocopying, recording or any information storage or retrieval system, without either prior permission in writing from the publisher or a licence permitting restricted copying. In the United Kingdom such licences are issued by the Copyright Licensing Agency: 90 Tottenham Court Road, London W1P 9HE.

Whilst the advice and information in this book is believed to be true and accurate at the date of going to press, neither the author nor the publisher can accept any legal responsibility or liability for any errors or omissions that may be made.

Typeset in 10/12pt Palatino by Multiplex medway limited.
Printed and bound in Great Britain for Edward Arnold, a division of Hodder and Stoughton Limited, Mill Road, Dunton Green, Sevenoaks, Kent TN13 2YA by Athenaeum Press Ltd, Newcastle upon Tyne.

Contents

	Page
Authors' Preface	ix

PART I MICROBES, WORMS AND MAN

1 **Perspective** — 3
Introduction—Microbiology and medicine—Historical perspective

2 **Biological Background** — 10
Major groups of micro-organisms—Their physiology

3 **Ecology** — 26

4 **Epidemiology—the Transmission of Pathogens** — 31

5 **Pathogenic Mechanisms** — 37

6 **Host Defences and Immunity** — 46
Non-specific defences—Specific immunity—Some possible results of infection—Hypersensitivity reactions—Immunological tolerance and auto-immunity—Fever

PART II ORGANISMS THAT INFECT MAN

7 **Bacteria** — 67
Taxonomy and nomenclature—Major criteria for classification and identification—Systematic classification of bacteria—Gram-positive cocci—Gram-negative cocci—Gram-positive bacilli—Gram-negative bacilli—Acid-fast bacilli—Branching bacteria—Spirochaetes—Mycoplasmas

8 **Rickettsiae, Coxiella and Chlamydiae** — 111

9 **Viruses** — 117
Picornaviruses—Reoviruses—Rotaviruses and gastro-enteritis—Arboviruses—Arenaviruses—Marburg and Ebola viruses—Myxoviruses—Rhabdoviruses—Coronaviruses—Retroviruses—Parvoviruses—Papovaviruses—Adenoviruses—Herpesviruses—Poxviruses—Hepatitis viruses—Chemotherapy of virus infections

10 **Fungi** — 137
Superficial mycoses—Deep mycoses—Respiratory allergies due to fungi—Antifungal drugs

11 **Protozoa** — 143
Amoebiasis—Naegleria and Acanthamoeba infection—Giardiasis—Cryptosporidiosis—Trichomoniasis—Trypanosomiasis—Leishmaniasis—Malaria—Toxoplasmosis—Babesiosis—Pneumocystis infection—Balantidiosis—Antiprotozoan drugs

12 **Helminths (Worms)** — 153
Trematodes (flukes)—Cestodes (tapeworms)—Nematodes (roundworms)

13 **Laboratory Investigation of Infection (and Skin Testing)** — 160
Using and understanding the laboratory—Laboratory procedures for

bacteria; for viruses; for fungi; for protozoa and helminths—Immunodiagnosis

PART III INFECTIONS OF MAN

14 **Mouth and Upper Respiratory Tract** 189
Normal microbial flora—Oral infections—Upper respiratory tract infections—Otitis media, mastoiditis and sinusitis—Acute epiglottitis

15 **Lower Respiratory Tract** 199
Bacterial and virus infections of the airways—Pneumonia—Tuberculosis—Helminth infections—Examination of sputum and other laboratory procedures

16 **Central Nervous System** 212
Meningitis—Encephalitis, meningo-encephalitis and myelitis—Brain abscesses–CSF shunt infection—Laboratory diagnosis of CNS infections

17 **Blood and Lymphoid Tissue** 230
Some bacteraemic illnesses—Viraemia—Infections of the heart—Blood cultures—Protozoan infections of blood—Lymphoid tissue infections—Pyrexia of unknown origin

18 **Gastro-intestinal Tract** 249
Normal flora—Bacterial enteritis—Virus gastro-enteritis—Protozoan infections—Water-borne infections—Bacterial food-poisoning—Antibiotic-associated diarrhoea—Diarrhoea in systemic infections—Helminth infections—Examination of faeces

19 **Liver and Biliary Tract** 264
Virus hepatitis—Bacterial infections of the liver—Biliary tract infections—Protozoan infections of the liver—Helminth infections

20 **Genito-urinary Tract** 272
Urinary tract infections—Genital tract infections

21 **Soft Tissue and Eye Infections** 285
Bacterial infections of soft tissues—Helminth infections of soft tissues—Flies and their larvae—Eye infections

22 **Bone and Joint Infections** 299
Osteomyelitis—Septic arthritis—Immune-mediated arthritis—Bone and joint effects of protozoan and helminth infections

23 **Skin Infections** 305
Normal flora—Bacterial infections—Virus infections—Fungal infections—Protozoan infections—Ectoparasites

24 **The Compromised Host** 313
Immune deficiency states—Other compromising factors—Reactivation of latent infections

25 **Congenital and Neonatal Infections** 320

PART IV PREVENTION AND TREATMENT OF INFECTION

26 **Principles of Infection Control; Sterilization and Disinfection** 333

27 **Infection Control in Hospital and in the Community** 345
Hospital problems—Protection of health-care staff—Community (public health) problems

28 **Immunization** 362
Active immunization—Passive immunization—Vaccines—Immunization against particular diseases due to bacteria—Immunization against particular diseases due to viruses—Immunization programmes

29 **Antibacterial Drugs** 373
Perspective—Use and abuse—Drug resistance—Laboratory procedures—The drugs themselves—Some special problems

APPENDICES

		Page
A	Glossary of Technical Terms	401
B	Meanings of Some Abbreviations	406
	Index	408

Authors' Preface

To remain useful, a textbook must keep up with advances in knowledge and with changes in teaching patterns. *A Short Textbook of Medical Microbiology* was well received for over twenty years, largely because its five editions allowed it to adjust to changing emphases in the teaching of microbiology to medical students; but recent years have seen a more rapid and largely commendable shift in those emphases, at least in British medical schools. The spotlight has been less on the organisms and more on what they do to their human hosts. Our change in title to *Microbial and Parasitic Infection* is in keeping with this shift and also reflects major alterations in the contents of the book.

We have retained much that was in its predecessor, with appropriate revision and a good deal of condensation, because medical students still need a basic knowledge of the properties of man's microbial and other parasites, as well as of the various other infection-related topics with which we have dealt previously. However, since the practical concern of most of our readers will be not with laboratory matters but with clinical problems, a large part of the new book is devoted to descriptions of infections in relation to the major systems of the body and as they affect particular groups of patients. The introduction of the word 'Parasitic' into our title marks the welcome addition of Dr John Jewsbury to our team. His special knowledge and expertise have enabled us to deal more competently with protozoan infections and to include helminth infection for the first time.

Like its predecessor this book is primarily intended for medical students, providing them with a short, readable account of man's relationships with his various parasites, and so with a basis for further reading in the subject and for an informed approach to the investigation, prevention and treatment of infective disease. However, we shall be happy if it also resembles its predecessor in proving useful to other readers — notably paramedical workers and qualified doctors, even including trainee microbiologists. As before we have avoided details of laboratory methods, assuming that our readers have opportunities to learn all that they need to know about these in practical classrooms or from other sources. Nor have we catered for readers who want practical instructions about the use of sterilizers, antiseptics, immunizing agents or antimicrobial drugs, though we trust that what they read here will enable them to understand the principles of the use of all these.

Suggestions for further reading are given at the ends of most chapters, but two larger textbooks are relevant to so many chapters that they are mentioned here instead. The first volume of *Medical Microbiology*, 13th edition, edited by J.P. Duguid and others (Churchill Livingstone, Edinburgh and London, 1978) is entitled 'Microbial Infections' and contains a great deal that is of interest to medical students and clinical doctors. (Its second volume, 'The Practice of Medical Microbiology', contains more technical laboratory information.) The clinical pictures, epidemiology and management of diseases due to micro-organisms are clearly and authoritatively described in *Infectious Diseases*, 3rd edition, by A.B. Christie (Churchill Livingstone, 1980).

Dr. I.A. Porter, a co-author of all five editions of *A Short Textbook of Medical Microbiology*, has now retired from the team and we thank him for his contributions over the years.

In connection with the new book we are particularly grateful to Hazel Storer for her secretarial skill in preparing the typescript, to Ian Geary FIMLS for the drawings on pages 70 and 212, and to our publishers for their sustained interest and unfailing helpfulness.

B.I. Duerden, T.M.S. Reid
J.M. Jewsbury and D.C. Turk.

PART I

Microbes, Worms and Man

1
Perspective

Introduction

Medical students often first encounter microbiology merely as a small part of the general training in preclinical and paraclinical sciences which provides the foundation for their entry into clinical medicine. At that stage they rarely appreciate that infections with the organisms about which they are hearing intrude into every aspect of medicine and are an inevitable major component of medical practice. Infective diseases, which may be thought of as battles between man and his microbial and helminth parasites, are the world's major cause of illness (morbidity) and death (mortality). The patterns of infection that command medical attention have a wide spectrum—from the classical infectious diseases such as typhoid, malaria and tuberculosis, which continue to be the outstanding medical problems of the less affluent parts of the world, to the often puzzling opportunist infections that can complicate transplant surgery and cancer chemotherapy in places where medical services are most advanced.

In this book our approach to the study of infective diseases is first to provide a sound basis of knowledge about organisms that cause infection, their interactions with hosts in producing disease, and the laboratory tests available for detecting their presence and assessing their significance. We can then discuss clinical infections, their clinical and laboratory diagnosis, their management and their prevention. So Part I, after this introductory chapter, outlines the chief biological properties and divergences of the major groups of organisms with which we are concerned, and then deals briefly with the normal inter-relations of microbes and men (ecology). The great majority of microbes, even those that live in or on man, do us no harm—indeed, many of them are beneficial; but in the medical context we are clearly concerned mainly with those that can produce disease (pathogens). The next three chapters therefore concentrate on these—initially on how they spread from person to person (epidemiology), how they establish themselves in and do damage to the host (pathogenesis) and how the host responds to their challenge (immunology). In Part II a series of chapters deals in turn with each of the major groups of organisms. Our aim in these chapters has been to give such information about the biology and laboratory behaviour of the organisms as is relevant to the understanding and diagnosis of the infections that they cause—with particular reference to properties that affect the collection of appropriate specimens for laboratory investigation and the speed and reliability with which the laboratory can derive useful results from these specimens. Having established a basis of knowledge about the organisms, we proceed in Part III to describe the kinds of infection that occur in particular tracts or regions of the body, their investigation, management and prevention—including in our descriptions the spectra of infection seen in hospital and in community practice, in developed countries and in those with less sophisticated medical services. General aspects of antimicrobial therapy and the prevention of infective diseases are brought together in Part IV.

Throughout the book we shall make repeated reference to the need for precise definition of the meanings of words. The uninitiated reader will find the first of our two appendices (Appendix A, Glossary of Technical Terms) helpful in this respect—for example, it makes clear the distinction between the terms 'infective diseases' and 'infectious diseases', both of which we have used in the opening paragraph of this chapter.

Microbiology and Medicine

Microbiology, largely a paramedical subject during its childhood and adolescence, has matured into a wide-ranging science with medical microbiology as just one of its many subdivisions. The medical microbiologist has always been interested chiefly in micro-organisms that are parasites of man and cause disease. He has tended to regard other organisms as unimportant, and has avoided them by using cultural conditions more favourable to the growth of parasites. In his attempts at classification he has been preoccupied with the problem of recognizing pathogens; and as ability to produce disease is not consistently linked with other microbial properties, he has used a different set of criteria for the subdivision of each group of organisms. Many of his tests have been, and still are, empirical and scientifically crude—e.g. he talks of bacteria producing 'acid' and 'gas' from a sugar or an alcohol without reference to the nature of the acid or gas or the mechanisms of their production. Yet by such means he has gone a long way towards unravelling the problems of human microbial disease and has provided clinicians with much valuable information.

In sharp contrast, the academic microbiologist is interested in micro-organisms for their own sake; to him the question whether they can cause disease is merely one aspect of their biology. He wants to know the minutiae of microbial structure made available by electron microscopy and the details of microbial metabolism, for which he needs media of defined chemical composition and apparatus that maintains continuous control of that composition even while metabolism is going on. He investigates microbial reproduction and genetics, and attempts to evolve classifications based upon the properties of the organisms themselves rather than upon what they do to other creatures.

Moreover, many biologists, physiologists and biochemists are not primarily interested in the organisms themselves but use bacteria or fungi as relatively simple and manageable models in which to study processes that also occur in the cells of more complex organisms. Such workers have added much to our knowledge of micro-organisms; so have biochemists who use them either as sources of interesting compounds or as tools with which to carry out chemical manipulations, and genetic engineers who have programmed them to synthesize useful materials, including therapeutic agents such as insulin and interferon.

These various disciplines overlap and learn from each other, and medical microbiologists must keep abreast of the conceptual and technical advances made by such colleagues in order to apply them in medicine; but medical students and clinicians are not under any such obligation, and they need to know only a little of the technology of medical microbiology. What matters to them is the help that medical microbiology can give in the understanding, investigation, treatment and (best of all) prevention of infective diseases of man.

Historical Perspective

Man has always lived in an environment that abounds with minute living organisms, and he has always carried them in countless billions around and within his person. However, the science of microbiology is little more than a century old, having its origins in the work of Louis Pasteur in the 1850s, though that great breakthrough was preceded by many centuries of speculation and investigation.

The concept of contagion

From the most ancient writings we learn that plagues and pestilences were well-recognized features of human existence. Man attributed them to all sorts of real and hypothetical factors in his environment, such as divine anger, cosmic influences, witchcraft, the seasons of the year or bad air. It was appreciated at an early date that the introduction of a sick person into a community could result in spread of the disease to the local population. The Biblical book of *Leviticus* indicates that the methods of spread of certain skin and venereal diseases were known in the days of Moses, and that their victims were accordingly excluded from contact with their fellow men. In his *De Contagione* published in 1546, Fracastorius struck a remarkably modern note with the statement that diseases could be spread by direct contact between individuals, by the agency of inanimate objects such as clothing and personal possessions (which he called *fomites*—a three-syllable Latin word originally meaning 'kindling wood') or through the air. He suggested that this spread involved the passage of small infective particles (*seminaria*) from an affected person to others, but since he could not demonstrate the existence of these particles, his theory received little attention.

In the late eighteenth and early nineteenth centuries the theory of contagion was again propounded vigorously by certain medical men. In 1795 Gordon of Aberdeen showed that an epidemic of puerperal fever 'seized such women only as were visited or delivered by a practitioner or taken care of by nurses who had previously attended patients affected with the disease.' He recommended washing and the changing of clothes to prevent carriage of contagion from one puerperal woman to another. Almost 50 years later Semmelweiss showed that the spread of puerperal infection by students and practitioners who commonly went straight from the post-mortem room to the maternity wards could be reduced by cleanliness and the washing of their hands in a solution of chloride of lime. Both these men believed that the carriage of 'something' from one patient to another was responsible for the development of puerperal fever.

Early observers of micro-organisms

In 1671 Kircher reported the presence of little worms in the blood of patients with plague, and claimed that they were responsible for the disease. However, it is likely that what he saw were aggregations of red blood cells, and that the honour of being the first observer of micro-organisms belongs to Antony Leeuwenhoek, a linen-draper of Delft in Holland. His hobby was the making of simple but ingenious microscopes, and with these he was able in 1674 to observe minute living creatures in rain, sea and pond water, and in various other fluids. He communicated his findings in a series of letters to the Royal Society in London, but neither he nor his contemporaries appear to have realized the significance of his observations. During the eighteenth century several workers suggested that small creatures such as he had described might be responsible for various dieases, but their ideas were not accepted for lack of any factual support.

By the early nineteenth century improvements in microscope design had made possible the beginning of systematic description of micro-organisms. In 1838 Ehrenberg, in his work on *Infusoria* (the small creatures found in infusions), introduced such terms as **bacterium, vibrio, spirillum** and **spirochaete.** Meanwhile, in 1835, Agostino Bassi had described the fungus (later named *Botrytis bassiana*) which caused muscardine, a disease of silkworms, and he had suggested that this disease was transmitted by contact or by infection of food. This, the first reliable report of a disease caused by a transmissible parasitic micro-organism, was followed in 1839 by Schoenlein's description of the fungus that causes the human disease favus. In 1850 Rayer and Davaine reported the presence of rod-shaped organisms in the blood of animals that had died of anthrax, and Davaine later showed that this disease could be transmitted by inoculation of blood containing such rods but not of blood from which they were absent. During this era other claims to have found microbial causes of disease were put forward with inadequate experimental backing, and in 1840 Henle pointed out that a micro-organism causing a disease should be present in every case and should produce a similar disease in animals into which it was inoculated—criteria which were later expanded into 'Koch's postulates' (p. 37).

The theory of spontaneous generation

Up to the seventeenth century philosophers and scientists had generally accepted that at least some animals could develop entirely from non-living materials. Thus putrefying meat was believed to give rise to maggots and the mud of the Nile to snakes, whereas corn and a linen cloth stored in a jar were considered suitable ingredients for the production of mice! However, in 1688 Redi showed that putrefying meat did not produce maggots if flies were kept away from it, and thereby convinced many that the theory of spontaneous generation was inaccurate, at least in relation to flies and larger creatures. It survived in relation to microscopic creatures for another 200 years, and in the latter half of the eighteenth century it was the subject of a fierce controversy, with the Italian abbot Spallanzani and the Irish priest Needham as the central figures. Needham's claim that micro-organisms reappeared in infusions which had previously been heated to kill all living creatures were countered by Spallanzani's demonstration that this did not occur if the heating was vigorous enough and if air was subsequently excluded from the container. According to Needham, this was because excessive heat destroyed a 'vegetative force' that was necessary for the generation of organisms. In 1854 Schroeder and Dusch showed that the growth of micro-organisms which took place in previously heated infusion if air was allowed to enter the container could be prevented by first passing the air through a cotton-wool filter—in other words, that the important component of the air was particulate. However, their later findings were erratic and confusing because the amount of heat that they used was inadequate to sterilize all of the fluids tested.

Pasteur, Lister and Koch

To the French chemist Louis Pasteur belongs the credit both for terminating the dispute about spontaneous generation and for establishing beyond doubt the role of micro-organisms in transmissible diseases. He entered the field of microbiology at a point far removed from medicine—the study of fermentation. This phenomenon had been known to man from very early times but was without an explanation until 1837, when Schwann and Cagniard-Latour discovered independently that the yeasts always associated with alcoholic fermentation of sugar solutions were living organisms. Their belief that these organisms actually caused the fermentations was disputed by Liebig and others who upheld purely chemical explanations. In a series of brilliant experiments and papers between 1855 and 1860 Pasteur showed conclusively that lactic and butyric acid fermentations were the work of bacteria, and that the fermentations involved in the production of beer and wines were the work of yeasts. He also showed that there was a relationship between the type of micro-organism involved and the type of fermentation produced. He then proceeded to destroy the theory of spontaneous generation (though its supporters were slow to admit defeat) by showing that living micro-organisms were invariably derived from exactly similar living organisms. In the course of this work he learnt a great deal about the scrupulous care needed in dealing with bacteria and fungi, and about their differing nutritional requirements, and so he laid the foundations of modern microbiological technique. But he was far more than a careful technician. His brilliance lay in his ability to see the far-reaching significance of his discoveries. From problems of preparation and preservation of wine, beer and vinegar he went on to rescue the silkworm industry from the scourge of an infectious disease called pebrine, to show farmers how the spread of anthrax among their animals could be prevented, and then to discover how to immunize these animals against anthrax, fowls against chicken cholera and finally man against rabies.

Meanwhile, news of Pasteur's work on fermentation reached Joseph Lister, the Professor of Surgery in Glasgow. At that time virtually all wounds suppurated and the mortality following surgery was fearful. Failure of wounds to produce 'laudable pus' was in fact considered a bad sign—quite rightly, as we can see today, for all wounds were infected and lack of suppuration frequently meant absence of resistance to infection on the part of the patient. Lister concluded that if micro-organisms caused fermentation they might also cause suppuration of wounds, and that in their absence wounds might heal cleanly and without risk to the patients' lives. So he introduced his **antiseptic technique,** which he first described in 1867. By washing wounds with carbolic acid, spraying this substance into the air of the operating theatre and applying protective dressings to keep fresh organisms from entering the wounds, he achieved a striking reduction of postoperative sepsis and mortality. Because carbolic acid was toxic to patients and to their attendants, it was far from being the perfect answer to the surgeon's problems, and today more emphasis is placed upon preventing the introduction of organisms into wounds (**asepsis**) than upon their destruction, but Lister's procedure proved their importance and prepared the way for modern surgery.

In 1870 a young German general practitioner, Robert Koch, began to follow up the work of Davaine on anthrax. He was able to grow in artificial culture the rod-shaped organisms seen in the blood of animals suffering from this disease, and to reproduce the disease by injecting his cultures into animals. He also showed that the rods could turn into rounded spores resistant to adverse conditions, and then back into rods. During the last quarter of the nineteenth century Koch and his bacteriological pupils identified the causative organisms of tuberculosis, cholera, typhoid, diphtheria and many other major diseases of man and animals, and began to establish a systematic classification of bacteria. This work was made possible by technical advances for which Koch himself was largely responsible, including the use of aniline dyes for staining micro-organisms, of oil-immersed microscope objectives for examining them and of media solidified with agar for growing them.

Immunology

Sometimes slowly and sometimes more rapidly, immunology has grown during the past century from being primarily concerned with immunity against infectious diseases to being a wide-ranging science that has made valuable contributions to the development of many branches of biology and medicine. Some of its important contributions have been in the field of prevention or treatment of infectious diseases by artificial induction of immunity–i.e. **immunization.** Such work began long ago. Immunization against smallpox (variola), by inoculation with material from a lesion of a patient (**variolation**), had been practised for centuries in the East before its introduction into Britain in 1721. It was a hazardous procedure, but smallpox was a widespread and terrible disease. In 1796 Jenner discovered that protection against smallpox could be achieved much more safely by inoculation with material from a lesion of cowpox, a natural disease of cattle. This process became known as **vaccination**, from the Latin *vacca*, a cow. We now know that its success was due to the close relationships between the viruses of cowpox and smallpox. The final stages of this story are outlined on p. 333.

Almost a century after Jenner's discovery, Pasteur found that fowls inoculated with an old laboratory culture of the organism of chicken cholera developed only a mild illness and were subsequently resistant to infection with fresh cultures of the organism. Then he discovered that sheep could be protected against anthrax by inoculation with cultures of anthrax bacilli attenuated (i.e. rendered harmless) by growing them at 42°C. This work provided a basis for a rational approach to the prevention of microbial diseases, and it was found possible to attenuate many other pathogenic organisms. Then in 1890, following the discovery by Roux and Yersin that the symptoms of diphtheria were mainly due to the release of a soluble poison (**toxin**) from the bacteria, Behring showed that guinea-pigs could be protected against this disease by injections of diphtheria **toxoid** (toxin treated to make it harmless). With Kitasato he similarly immunized animals against tetanus. The sera of such immunized animals were shown to neutralize the appropriate toxins specifically, and also to give protection against the appropriate diseases to other animals into which they were injected. Within a few years the treatment of human diphtheria was greatly advanced by the introduction of effective **antitoxic sera** (animal sera that neutralized diphtheria toxin) which could be injected into human beings with reasonable safety.

Meanwhile Jules Bordet demonstrated that two factors were required for bacterial lysis—a heat-stable substance which he called 'sensitizer' (**antibody**) and a heat-labile 'alexine' now referred to as **complement**. In 1884 Metchnikoff had shown that motile cells which he called **phagocytes** are important in the development of immunity to infection. This set the scene for a fierce battle of words between those who believed that immunity is essentially 'humoral' and the 'cellular' school of thought. As we shall see in Chapter 6, there was no need for a battle; both types of mechanism are important, and they are closely inter-related. At the turn of the century the concept that all immune reactions were not necessarily beneficial to the host found support in the work of Arthus, Sanarelli, Shwartzman and von Pirquet who studied serum sickness and the nature of tuberculin hypersenstivity.

The branch of immunology known as **diagnostic serology** can be said to have started in 1896, when Gruber and Durham demonstrated the clumping of cholera organisms by specific antiserum—a means of definitive identification of these organisms. Later in the same year Widal described his diagnostic test for typhoid, which consisted of showing that the patient's serum would clump (agglutinate) typhoid bacilli. Some of the innumerable later developments of diagnostic serology are discussed in later chapters, notably in Chapter 13.

The present century has seen enormous advances in the understanding of the mechanisms underlying immunological phenomena and in the application of that new knowledge. The theoretical aspects of immunology, in particular the mechanism of antibody formation, interested Paul Ehrlich who in 1896 propounded his 'side chain theory', embodying the concept that antibody-forming cells were triggered by antigen binding to specific receptors on their surface membranes. Only in recent times has the presence of such receptors on immunocompetent cells been verified and the theory substantiated.

In the 1930s the advent of modern techniques such as ultracentrifugation and electrophoresis enabled Heidelberger and Tiselius and also Kabat to identify the slow-moving γ-globulins as the components of serum in which antibody activity resided. The detailed elucidation of the structure and function of immunoglobulins by Porter, Edelman and others soon followed. The role of the thymus was first appreciated as a result of work by Miller, which led to the description of T and B lymphocytes and their functions, discussed in Chapter 6.

Virology

In the very early days of bacteriology it became apparent that some undoubtedly infectious diseases had no detectable bacterial causes. Pasteur, for example, demonstrated the infectivity of rabies and the possibility of preventing it by immunization, but he could not find its aetiological agent. He suggested that this might be because it was very small. In 1892 the Russian botanist Ivanovsky transmitted tobacco mosaic disease to healthy plants by means of a bacterium-free filtrate of sap from affected plants—work which was corroborated by the Dutch bacteriologist Beijerink. Six years later Loeffler and Frosch showed that foot-and-mouth disease of cattle was also transmissible by means of a bacterium-free filtrate. From that time onwards it was generally accepted that some diseases are due to living agents even smaller than bacteria. For many years their further study was hampered by lack of suitable techniques. By light microscopy it was possible to detect single particles of some of the larger viruses ('elementary bodies') and others could be seen as cytoplasmic or intra-nuclear aggregates of particles ('inclusion bodies') inside infected cells. A certain amount could be learned by transmission experiments in living animal hosts. But whereas from the time of Koch almost any hospital laboratory could provide a diagnostic bacteriological service, medical virology remained a subject for research workers because it lacked any equivalent to bacteriology's routine 'microscopy and culture' approach. The overcoming of these problems in the 1950s by the development of electron-microscopy and tissue-culture techniques opened the way to the provision of a routine diagnostic virological service. Today this is generally available and is being steadily developed. As more effective antivirus drugs become available, rapid diagnosis of virus infections becomes increasingly important and useful.

Antimicrobial drugs

Naturally occurring compounds have been used with success in the treatment of infections for several centuries—at least since the first recorded use of an extract of cinchona bark (quinine) for malaria in 1619; but it was in the opening years of the twentieth century that Paul Ehrlich began the search for synthetic substances specifically designed to attack harmful microbes (his 'magic bullets'). His arsenical compounds were effective against a limited number of such organisms, notably those causing syphilis and trypanosomiasis. In the late 1920s, while studying the produce of fungal moulds, including *Penicillium notatum*, which were capable of inhibiting the growth of many bacteria, Fleming discovered an agent he called **penicillin**. However, it was not until 1935, when Domagk introduced the first of the **sulphonamides** (prontosil) and later when the potential of Fleming's discovery was exploited by the work of Florey and Chain, that major advances were made. By now virtually all bacterial, fungal, protozoan and worm infections and even a few due to viruses have come within the reach of effective drug treatment. Many diseases that were virtually untreatable and commonly fatal a mere 40 years ago can today be treated with almost invariable success. Medical microbiology and parasitology have been transformed by these developments, since their contributions have become far more relevant to patient care and are more urgently needed; but the transformation has included the whole of medical practice and indeed the way of life and life expectation of us all. In almost any branch of medicine a doctor frequently has to decide whether to use an antimicrobial drug, and, if so, which to choose from the increasingly and confusingly wide selection available to him. A firm grasp of basic microbiological facts and principles provides the best foundation for his decisions.

Suggestions for Further Reading

Vallery-Radot, R. (1923). *The Life of Pasteur.* Constable, London

de Kruif, P. (first published 1926, currently available as paperback), *Microbe Hunters*, Pocket Books Inc, New York.

Brock, T.D. (translator and editor). (1961). *Milestones in Microbiology.* Prentice-Hall, London.

Foster, W.D. (1970). *A History of Bacteriology.* Heinemann, London—for the period 1840-1940.

Burnet, Sir Macfarlane (1968). *Changing Patterns: An Atypical Autobiography.* Heinemann, Melbourne and London—for developments in bacteriology, virology and immunology as seen and influenced by an eminent worker in these fields.

Reid, R. (1974). *Microbes and Men.* British Broadcasting Corporation.

Biological Background

Infections in man may be caused by a wide range of living creatures. Many of them are micro-organisms (microbes), so small that individuals can be seen only by various forms of microscopy. The study of these is known as **microbiology**. However, helminths (worms) are much larger and more complex, and in many cases are easily visible to the naked eye. Whereas the term parasites, as defined on p. 26, includes virtually all of the organisms with which we are concerned in this book, **parasitology** has by a somewhat confusing convention come to mean the study of parasitic eukaryotes (see below), notably protozoa and helminths; and it is also conventional to use for these organisms the collective term parasites in much the same way as we use the term micro-organisms inclusively for the smaller creatures. Protozoa, which are eukaryotic micro-organisms, are included in both disciplines.

Major Groups of Micro-organisms

Micro-organisms of medical importance are found within five groups: (1) bacteria; (2) rickettsiae and chlamydiae; (3) viruses; (4) fungi; (5) protozoa. The cells of fungi and protozoa are essentially similar in structure to those of higher plants and animals, and these organisms are therefore known as **eukaryotic**, whereas bacteria, rickettsiae and chlamydiae are known as **prokaryotic** because their cells have a much simpler nuclear structure, do not have nuclear or other internal dividing membranes and have in their walls a mucopeptide substance not found in eukaryotic cells (see below). Viruses are even simpler in structure and cannot be described as living cells.

Bacteria

These are cellular (usually unicellular) organisms. A typical bacterial cell is able to carry out many different metabolic activities and to increase its size and reproduce itself by fission. Individual cells are of the order of 0.5–1 μm broad by 0.5–8 μm long (1 μm = a micrometre, a thousandth of a millimetre). The shape of a bacterial cell is determined by its rigid but permeable **cell wall,** which also prevents it from swelling up and bursting under the influence of its high internal osmotic pressure. The main structural component of this wall is **mucopeptide** (or **peptidoglycan**), which consists of chains of alternating molecules of N-acetylglucosamine and N-acetylmuramic acid cross-linked by peptide chains. It is responsible for the rigidity of the cell wall. Long before anything was known about mucopeptide, Gram devised the differential staining method still widely used by bacteriologists (pp. 71–2). This divides bacteria into **gram-positive** (blue-staining) and **gram-negative** (red-staining). It is now clear that the principal difference between these two groups is that mucopeptide constitutes 50–90% of the walls of gram-positive bacteria but only 5–10% of those of gram-negative bacteria, which have a thicker outer layer composed of a phospholipid–polysaccharide–protein complex. Polymers of glycerol phosphate or ribitol phosphate known as **teichoic acids** occur in the cell walls of gram-positive but not of gram-negative bacteria. Within the cell wall is the **protoplast,** which is mainly semi-solid **cytoplasm** surrounded by a thin elastic semi-permeable **cytoplasmic membrane**—a complex structure which is of great importance in determining what substances can enter or leave the cell and is the site of most of its enzymic activities. In some bacteria the cytoplasmic membrane forms highly convoluted, invaginated membranous organelles called **mesosomes,** which appear to be the sites of specialized metabolic activity and are prominent during cell-wall synthesis and during sporulation. Among the structures to be found within the cytoplasm are many granular **ribosomes,** which contain much of the cell's ribonucleic acid (RNA), and a **chromosome** or **nuclear body** (sometimes more than one) consisting of a long double-stranded deoxyribonucleic acid (DNA) molecule in the form of a much twisted and contorted ring; other DNA may be present in the form of small extra-chromosomal portions called **plasmids.** Both chromosomes and plasmids are anchored to membrane attachment sites that control their replication. Some bacteria form **capsules,** usually composed of polysaccharide, outside their cell walls. Some have fine whip-like organelles of locomotion called **flagella** (singular **flagellum**) protruding from their surfaces. Some have numerous shorter hair-like protrusions called **fimbriae** or **pili;** most of these are apparently organelles of adhesion enabling the bacteria to attach themselves to surfaces such as those of host cells, and may be important in the production of disease, but some have a special role in bacterial conjugation (**sex-fimbriae,** p. 20). A few bacterial species form non-reproductive **spores,** which develop intracellularly and have thick walls, greatly reduced metabolic activity and greatly increased resistance to adverse conditions. (For more detail about these and other morphological features of bacteria see pp. 69–71.)

Bacteria reproduce by **binary fission,** in which one cell enlarges and then divides into two approximately equal parts. This division is preceded by simple replication of the nuclear ring, without the polarized mitosis that is part of the reproductive process of most nucleated cells, and therefore without the possibility of segregation and re-assortment of chromosomal genes.

Bacteria are divisible into large groups that differ in the shapes of their bacterial cells. Cells that are spherical, or nearly so, are called **cocci.** These commonly occur grouped together. If in pairs, they are often referred to as diplococci. Repeated division in the same plane produces chains; division in two or three planes at right angles produces regular packets of four, eight or more; and division without any definite orientation produces irregular clusters. The **bacilli** or rods are elongated cylindrical forms, straight or slightly curved, with ends that are rounded, square, pointed or sometimes swollen to form clubs. Certain bacteria which resemble bacilli but are more definitely curved are known as **vibrios** or comma bacilli. The **spirochaetes** are corkscrew-like spirals. **Actinomycetes** and other **higher bacteria** (so called because they are thought to represent a more advanced state of evolution, although they are still prokaryotes) resemble the fungi (eukaryotes) in forming branched filaments. The **mycoplasmas** are an exception to the rule that bacterial cells are encased in rigid cell walls. Their cells are essentially protoplasts, and are smaller in size than those of other bacteria (c. 0.25 μm diameter). They can grow and reproduce like other bacteria, but their lack of cell wall means that they are of variable shape and can survive only in roughly isotonic conditions. They have much in common with the mutant forms of normal bacteria known as **L-forms** (p. 71).

Rickettsiae, Coxiella burneti and Chlamydiae

These organisms should be regarded as bacteria in that they contain both RNA and DNA, have muramic acid in their outer coats, reproduce by binary fission and are susceptible to the action of antibacterial drugs that have no effect on viruses. On the other hand, with diameters of only 0.2–0.5 μm they are nearer in size to viruses than to bacteria, and they are also (with only one known exception) unable to reproduce except inside the cells of the host organisms.

Rickettsiae and coxiella are pleomorphic, forming cocci, bacilli and filaments. By contrast chlamydiae are spherical and have an unusual type of predominantly intracellular developmental cycle. The infective forms (**elementary bodies**), about 0.3 μm in diameter, are phagocytosed by host cells and develop within them into larger forms (**reticulate bodies**) up to 2 μm in diameter. More reticulate bodies are formed by binary fission during the next 20 hours or so, but by about 40 hours these have become reorganized into large numbers of elementary bodies. Rupture of the host cells 48–72 hours after infection releases the elementary bodies, which can then infect new host cells. Intracellular clusters of chlamydiae can be seen as basophilic inclusion bodies in Giemsa-stained smears, whereas those formed by viruses (p. 22) are acidophilic.

Viruses

Though some viruses are similar in size to the organisms just described, most are smaller—some very much smaller—than any other known living organisms, and are too small to be seen with an ordinary light microscope unless they form inclusion bodies (p. 22). Since viruses have no metabolism of their own and cannot reproduce themselves, it is arguable whether they should be described as living organisms. For this reason, viruses which are able to invade host cells and to replicate there are often described as **active** rather than alive, and those which have lost the ability to do these things are then described as **inactivated** rather than dead. The virus particle is called a **virion,** not a cell. At its simplest, as in the viruses of poliomyelitis, this is a mere 25–30 nm in diameter (1 nm = a nanometre, a thousandth of a micrometre) and consists only of a nucleic acid core, the **genome,** packed within a protein coat, the **capsid,** which protects the genome during transmission between host cells. At the other end of the range, the virions of poxviruses measure about 200 × 300 nm and are chemically and structurally a good deal more complex, though they are still developments of the same basic plan. The nucleic acid found in a virus of any given type is either RNA or DNA, but not both as in bacteria and other cellular organisms. Viruses increase in number not by fission but by **replication** (p. 21) inside bacterial, plant or animal host cells, which they have converted into virus-production units.

Increased knowledge of virus structure has made possible their logical classification, and although there is not complete agreement as to the best system, the one on which Table 9.1 (p. 118) is based is widely accepted. The characters used in delineating the main groups include:

(1) The nature of the **nucleic acid** in the genome—either RNA or DNA. The nucleic acid contains the genetic information for replication of the virus.

(2) The **symmetry** of the capsid. This is determined by the shapes and mutual attractions of the units, called **capsomeres**, that make up the capsid. In some virus groups these protein 'building bricks' produce a capsid that is an icosahedron—a hollow near-spherical structure with 20 identical triangular faces. Such viruses are said to show **cubic** symmetry, and their classification is based upon their numbers of capsomeres. In other groups the capsomeres are in a spiral arrangement, forming a hollow cylinder; they are said to show **helical** symmetry. Some groups have more elaborate structures and are described as complex—e.g. poxviruses are brick-shaped. Since analysis of capsid structure depends on interpretation of fine details in two-dimensional electronmicrographs, it is understandable that even among virus groups that are of medical interest there are some of which the capsid designs are as yet incompletely understood or unknown. The complex tadpole-like design of many bacteriophages is described below.

(3) The presence of an **envelope.** In some virus groups the **nucleocapsid** (nucleic acid core + capsid) is surrounded by a loose membranous envelope consisting of lipids, proteins and carbohydrates. Some of these components closely resemble those of host cells, and are believed to be derived from host-cell membrane as the virus is liberated from the host cell. Many enveloped viruses are described as ether-sensitive because they are inactivated by treatment with ether (or with other lipid solvents). Presumably this means that lipid components of their envelopes are necessary for their activity. Closely associated with the

envelopes of the myxoviruses are numerous projecting spikes, and these are connected with haemagglutinating activity of these viruses. Viruses that do not have envelopes are described as naked.

(4) Particle **size**. The size of virus particles used to be determined by filtration through membranes of known pore size, by measuring their rate of sedimentation in a high-speed centrifuge, or by density-gradient centrifugation; though the last of these, which involves finding the depth to which the particles can be centrifuged through a fluid that increases in density from top to bottom, is primarily a means of measuring the specific gravity of particles rather than their size. All of these procedures are still useful as means of purifying virus preparations, but for the determination of virus particle size they have been superseded by electron microscopic comparison of the virus particles with particles or structures of known size.

Bacteriophages

Bacterial cultures on solid media sometimes have a 'moth-eaten' appearance, due to the presence of many small areas in which the bacteria have been lysed. This phenomenon was first studied by Twort in 1915. He found that bacterium-free filtrates of a suspension of such a culture of staphylococci could produce similar lysis in other staphylococcal cultures. In 1917 d'Herelle, working with fluid cultures of dysentery bacilli, discovered a similar lytic activity, and concluded that it was due to very small living agents parasitic upon the bacteria. He called them **bacteriophages**, a name which has persisted and is often abbreviated to **phages**.

Bacteriophages are a group of viruses. They are found in many bacterial species, but show a high degree of host specificity, so that any one phage is usually limited not only to a single bacterial species but to certain strains of that species. This specificity is of practical value in the subdivision of species for epidemiological purposes (p. 23). Bacteriophages can be recovered from such natural sources as faeces, sewage and polluted water. Their presence can be demonstrated by applying bacterium-free filtrates of such materials to cultures of susceptible bacteria. They are also carried inside bacteria which have only a latent infection and do not undergo lysis (lysogenic bacteria, p. 23).

Phages differ in size and form, but as a group they have the greatest structural complexity of any viruses. The nucleic acid is DNA in most of those that have been studied, including those that have been the subject of the most intensive investigation, the T phages of *Escherichia coli*. Such a phage is tadpole shaped, its head consisting of a DNA core surrounded by a protein coat that corresponds to the capsid of an animal virus. Taking the T2 phage as an example, the head is about 100 nm long and consists of a short tube, hexagonal in cross-section and about 60 nm across, closed at its ends by caps that are two halves of an icosahedron. From one of these ends emerges a thin hollow tubular tail, with an end-plate and terminal fibres by which the virus becomes attached and adsorbed to the cell wall of its bacterial host. Such adsorption is possible only if the phage, which is non-motile, happens to come into contact with a bacterium of a strain with the appropriate specific receptors on its surface, and if various other conditions are fulfilled. Once adsorbed, the phage digests a small area of bacterial cell wall and then the tail contracts, injecting its DNA into the bacterial body.

Fungi

These are generally larger than bacteria, and are commonly multicellular. Their relatively thick cell walls owe their rigidity not to mucopeptide, as do those of bacteria, but to fibrils of **chitin** embedded in a matrix of protein, mannan or glucan. Inside the cell wall lies a sterol-containing cytoplasmic membrane which is the target for the polyene antifungal agents (p. 142). The **moulds** or **filamentous fungi** grow as tubular branching filaments (**hyphae**) which become interwoven to form a network (**mycelium**). In some families the hyphae are divided into short lengths by cross-walls (**septa**). The yeasts are oval or spherical cells which commonly reproduce by budding. Some yeasts, including the medically important genus *Candida*, show a modified form of budding, in which the buds elongate into filaments (**pseudohyphae**) that remain linked in chains and resemble mould mycelium. Many fungi pathogenic for man are **dimorphic**; some of these are yeast-like when invading tissue but form a mycelium when growing saprophytically in soil or in culture (e.g. *Histoplasma* and *Blastomyces*), whereas in others it is the parasitic form that may be mycelial (e.g. *Candida albicans* and *Malassezia furfur*). Notable exceptions are *Cryptococcus neoformans* which retains its yeast-like morphology in both phases, and the moulds— e.g. *Aspergillus fumigatus*—which are filamentous both in culture and in infected tissues.

Protozoa

These unicellular organisms, mostly much larger than bacteria, show clear differentiation of their protoplasm into nucleus and cytoplasm (i.e. they are eukaryotes). Their reproductive mechanisms vary from simple binary fission, with nuclear replication by mitosis, to complex life-cycles involving sexual and asexual phases and the formation of cysts.

Helminths (Worms)

Helminths of three groups infect man. The *Trematoda* (**flukes**) and the *Cestoda* (**tapeworms**) are included in the *Platyhelminthes* (**flatworms**); they are usually flattened dorso-ventrally and bilaterally symmetrical, with three body layers but no true body cavities, and are usually hermaphrodite. The third group, the *Nematoda* (**roundworms**) are zoologically distinct: they are more tubular but still with bilateral symmetry, they have a simple but complete digestive system with mouth and anus, and the sexes are separate.

Adult **flukes** mostly live in their hosts' intestinal tracts or lungs, but those of some species live in blood vessels. They have simple digestive systems, with no anus. Each adult lays large numbers of eggs, which are excreted by the hosts in faeces, sputum or urine. All flukes have complex life cycles, with snails as first intermediate hosts (p. 28), in which asexual multiplication occurs, and often with second intermediate hosts also. Most of the species that are common parasites of man are acquired by mouth but the most important group (the schistosomes) usually enter the human body through skin.

Adult **tapeworms** are intestinal parasites. They have no digestive systems, but absorb nutrients from the hosts' gut contents through their surfaces. An adult has a small head (**scolex**), with two or four suckers and usually a circle of hooks by which it attaches to the host's intestinal wall; and a 'body' (**strobila**) which is generally long and tape-like and consists of many (maybe several thousand) units. These units, the **proglottides** (singular, **proglottis**), are often incorrectly called segments, but in fact they are formed one by one immediately posterior to the scolex; those near the scolex are immature, those futher down the strobila are sexually mature and those at the far end are gravid (full of eggs) and eventually drop off the end of the strobila and are passed in the host's faeces. The larval stages occur in various organs and tissues of vertebrate intermediate hosts. Depending on the species of tapeworm, man may act as the host of the adult stage, of the larval stage, or of both.

Adult **roundworms** that infect man are found, depending on their species, in his gut, blood, lymphatic vessels or subcutaneous tissues. Adult females produce eggs (enormous numbers each day in some species), or in some species larvae instead of eggs. A common pattern among species that are intestinal parasites is that the eggs are excreted in the host's faeces, undergo maturation outside the body, and are then ingested by fresh human hosts. Some of the species that live in blood or lymphatic vessels or subcutaneous tissues produce larvae that must be ingested by mosquitoes as intermediate hosts, and some need other invertebrates. Around 14 species infect man, and some do so with very high frequencies in particular areas, with up to 80% of the human population infected.

Physiology: Bacteria

Bacteria are of diverse sizes, shapes and structures, and live in widely varied environments. It is not suprising that they also differ widely in the details of their physiology, even though their biochemical mechanisms in general are similar to those of all living creatures, including man.

Metabolic Needs

The bacterial cell is a complex structure. Within its minute confines are included a wide variety of proteins, nucleic acids, polysaccharides, lipids, and their derivatives. Some bacteria are motile, some generate light, but the main activity of bacteria as a whole is reproduction—i.e. the making of new bacteria. This process may go on at an amazing speed. Under optimal conditions some species divide as often as three or four times per hour, which means that a single bacterium, visible only by microscopy, may grow overnight into a colony several millimetres in diameter made up of billions of its progeny. (The word 'grow' in relation to micro-organisms is generally used to mean 'increase in numbers'.) Such a formidable synthetic operation requires an adequate supply of energy and raw materials, and appropriate environmental conditions. The precise needs of a particular organism depend largely upon the range of **enzymes** that it possesses, in consequence of its genetic make-up or **genotype**. Some enzymes are **constitutive**, produced by the organisms in almost all circumstances, and others are **inducible**, produced (after some initial delay) in response to special circumstances, usually the presence of their specific substrates.

Sources of energy

Some micro-organisms are **phototrophs,** able to derive their energy from sunlight. The majority, however, are **chemotrophs,** getting their energy from the oxidation of chemical compounds. Those that are parasites of man or animals are called **chemo-organotrophs** because they utilize organic compounds.

Oxygen

Organisms that grow readily in the presence of air are described as **aerobes.** Some species are **obligate** (or **strict**) **aerobes,** unable to grow in the absence of free oxygen, but others are **facultative** organisms, able to grow in the presence or absence of oxygen, though they often grow more vigorously under aerobic conditions. **Obligate anaerobes** require a highly reduced environment and cannot grow if more than a trace of free oxygen is present. The energy-producing pathways (see below) in anaerobes can operate only at a very low redox potential that cannot be sustained in the presence of free oxygen. Moreover they poison themselves in the presence of oxygen by making peroxides and superoxides, which they cannot destroy since they do not possess effective catalase and superoxide dismutase enzymes. Anaerobes are though to be the most primitive micro-organisms in the evolutionary scale. **Micro-aerophiles** are organisms that grow best in the presence of a little oxygen, though the term is often applied loosely and incorrectly to CO_2-dependent bacteria (see below).

The oxidations on which chemotrophs depend for their energy can be carried out in three different ways: **aerobic respiration,** with free oxygen as the final electron-acceptor in a chain of oxidation–reduction reactions; **anaerobic respiration,** with inorganic compounds (nitrates, sulphates and carbonates) as final electron-acceptors; and **anaerobic fermentation** of a carbohydrate or other organic substance, the electron-acceptor being another molecule of the energy source or some other organic molecule. Various organic acids and the gases CO_2 and H_2 may be formed as end-products of fermentation. (In medical microbiological writings the word fermentation is commonly used in a less precise sense—p. 73). Some bacterial species use only one of these three ways, others are more versatile.

Carbon dioxide

This is probably necessary in small amounts, such as are present in the atmosphere, for the growth of most micro-organisms. A higher concentration, 5–10%, improves the growth of many parasitic species, notably of *Neisseria gonorrhoeae* and many anaerobes, and is usually necessary for the primary isolation of *Brucella abortus* from pathological materials. Some bacteria, such as *Streptococcus milleri,* have an absolute requirement for a higher concentration of CO_2 and are called **CO_2-dependent** or **carboxyphilic**. Free CO_2 can be the sole carbon source for autotrophs.

Raw materials

Some chemotrophic bacteria (called **autotrophs**) can grow in simple inorganic salt solution. At the other end of the scale are the leprosy bacillus and the spirochaete of syphilis, which cannot be grown in non-living media, and rickettsiae, which (like viruses) depend upon their host cells for essential enzymes as well as for raw materials. Between these extremes are **heterotrophs** with innumerable gradations of requirements for organic and inorganic substrates. Among common parasites of man, *Escherichia coli* can grow in a solution containing glucose (carbon and energy source), ammonium sulphate (nitrogen source) and a few other inorganic salts. In contrast, *Haemophilus influenzae* has exacting requirements; as well as a suitable carbohydrate, various minerals and an assortment of amino-acids, purines and vitamins, this species must be supplied with nicotinamide-adenine dinucleotide (NAD) or its phosphate (NADP) as a co-dehydrogenase, and with haemin or some closely related substance for the synthesis of respiratory enzymes (p. 98). Such differences in the nutritional requirements of organisms are of great importance to the medical microbiologist in his choice of culture media.

Temperature

Psychrophiles grow best at low temperatures, some below 0°C. They are important in connection with cold storage of food and blood, but otherwise are not relevant to medical microbiology. **Thermophiles,** found in such situations as hot springs and rotting vegetable matter, are of no medical importance except as sensitizing agents in 'farmer's lung' etc. (p. 141); but the term 'thermophilic' is applied to some medically important species that are unusually tolerant of temperatures over 40°C (e.g. certain *Campylobacter* species—p. 97). Most bacteria, including all parasites of man, are **mesophiles,** with optimal growth temperatures somewhere between 20 and 40°C; they vary considerably in the ranges of temperature over which they will grow. As might be expected, nearly all of man's parasites are best suited by temperatures around 37°C. Many of them will multiply at lower temperatures, down to 20°C or less, but few at more than 45°C. Some will grow only within a narrow temperature range—e.g. *N. gonorrhoeae*, 30–39°C. Unusual among human pathogens is *Yersinia pestis*, the causative organism of bubonic plague. Its optimal growth temperature of about 27°C is probably related to the fact that multiplication in the proventriculus of the rat flea is an important stage in its transmission (p. 95).

Hydrogen ion concentration

Micro-organisms differ widely in their preferences and tolerances concerning the pH of their environment. Most of those of medical importance grow best in slightly alkaline conditions. Artificial culture media must be carefully buffered to prevent the rapid lowering of the pH value by acid metabolic products to a level at which organisms can no longer multiply. Lactobacilli are unusual among the bacterial flora of the human body in preferring an acid medium (pH 4.0). Indeed, they help to protect the adult vagina because they form lactic acid from the glycogen of the mucosa and thereby keep the vaginal secretion too acid for the growth of most other organisms. The medium devised by Sabouraud for the isolation of relatively slow-growing fungi uses their ability to grow at pH 5.4; this is inhibitory to bacteria which would otherwise overgrow them. At the other end of the scale, *Vibrio cholerae* grows best around pH 8.5.

Metabolic Products

Obviously the most important product of bacterial metabolism is more bacteria. This section deals with some others, which for convenience of discussion are somewhat arbitrarily classified under four headings: toxins, extracellular enzymes, pigments and other products. Antibiotics are considered in Chapter 29.

Toxins

The potent toxins that some living bacteria liberate into their environment are called **exotoxins;** they are proteins with enzymic activity, and are heat-labile. One of the most powerful poisons known is the toxin produced by *Clostridium botulinum*, a soil bacterium that sometimes grows in human or animal foods and renders them highly lethal (pp. 40 and 258). Similar exotoxins are released by the causative organisms of tetanus, diphtheria and scarlet fever and produce the characteristic clinical features of these diseases. The bacteria which cause cholera and some types of dysentery multiply in the lumen of the host's intestine and produce exotoxins known as **enterotoxins** that damage the intestinal mucosa (p. 40). Other exotoxins named according to their effects include the haemolysins (red-cell-destroying toxins) and leukocidins (leukocyte-destroying toxins) of streptococci and staphylococci, and the phospholipase C (lecithinase) of *Cl. perfringens* (one of the gas-gangrene bacilli) which hydrolyses the phospholipid lecithin, a constituent of cell membranes. The important exotoxins are discussed more fully in Chapter 5 (p. 40) and later chapters.

Gram-negative bacteria have, outside the mucopeptide structural layer of their cell walls, a thicker phospholipid–polysaccharide–protein layer. The name **endotoxin** is given to complex lipopolysaccharide-containing material derived from this layer (p. 10). It is more heat-stable than exotoxins and, unlike them, is mostly liberated on the death and disintegration of the bacteria, though some organisms shed small amounts of it from their surfaces during life. The endotoxins of different species differ somewhat in composition and effect, but all are weight-for-weight less potent than most exotoxins and much less specific in their effects. When in the human or animal blood-stream, endotoxin can cause fever and coagulation and circulatory disturbances (p. 40), which are mainly consequences of macrophage activation by endotoxin (p. 48). A sensitive test for the presence of endotoxin in body or other fluids is provided by its ability to cause a lysate of amoebocytes (blood cells) from the horseshoe crab *Limulus polyphemus* to form a gel (the Limulus lysate test).

Extracellular enzymes

Some of these enzymes have already been mentioned in their capacity as exotoxins. Others, although not so directly harmful to the host, still contribute to the pathogenicity of the organisms, and may be called **aggressins** (p. 41). The coagulase produced by most pathogenic staphylococci may give some protection against host defences by coating the cocci with fibrin formed from plasma fibrinogen. The pathogenicity of these organisms may also be enhanced by the fact that clumps of cocci in fibrin clots become trapped in capillary blood vessels and multiply there, whereas isolated cocci are removed from circulation by phagocytosis as described on pp. 48–9. Conversely, the streptokinase of haemolytic streptococci probably facilitates their passage through clots by activating plasminogen to the fibrinolytic enzyme plasmin. The hyaluronidases (p. 41) formed by various species are spreading factors that open up connective tissues to bacterial invasion.

Many micro-organisms depend for their survival on enzymes that destroy toxic substances. For example, we have already mentioned catalase, which destroys peroxides (p. 15), and we shall refer repeatedly to another group of enzymes which have acquired great medical importance—**β-lactamases**, by means of which many bacteria can destroy penicillins and cephalosporins.

Other extracellular enzymes are concerned with the nutrition of their producers. Before an organism can use nutrients of high molecular weights, they must be broken down into molecules small enough to pass through its cytoplasmic membrane. Extracellular digestion by excreted enzymes is of particular importance to saprophytic organisms (p. 26).

Pigments

Phototrophic micro-organisms trap the energy of sunlight by means of their pigments, in much the same way as do the blue-green algae and higher plants. All chemotrophs also contain pigments—flavoproteins and cytochromes—which participate as electron donors and acceptors in their respiratory pathways. The pyocyanin of *Pseudomonas aeruginosa* (formerly *pyocyanea*), which gives a characteristic green colour to its cultures and also to pus from infected wounds, may have a respiratory function, whereas the black or brown pigment of *Bacteroides melaninogenicus* is merely a by-product of its metabolism of the haemoglobin in the culture medium. Red, yellow, violet and other pigments are produced by some bacteria, mostly saprophytes of no medical importance, and also by many moulds. Growths of such organisms have been responsible for a number of curious episodes, such as 'bleeding' of statues and strange discolourations of foods. Some of the ringworm fungi can be differentiated by the characteristic colours that they release into media on which they are growing.

Other products

Some of man's essential vitamins are synthesized for him by his intestinal flora—a point which it is sometimes dangerous to forget. Severe deficiencies of vitamins B and K can appear with startling speed in an already malnourished patient if his intestinal bacteria are largely suppressed by antibiotic treatment. In nature the end-products of the metabolism of one species are often the food supplies of another—an important factor in maintaining stable populations. For example, dextran produced by oral streptococci may be broken down for their own use by *Bacteroides* species in dental plaque. A well-known laboratory illustration of the same principle is the satellite growth of *H. influenzae* around colonies of other bacteria which supply it with V factor (NAD; p. 16). Some microbial metabolic products contribute to the nutrition of the hosts; e.g. herbivores, especially ruminants such as cows and sheep, depend upon anaerobic bacteria to break down cellulose and produce absorbable fatty acids and alcohols. It is probable that a similar, though less significant, bacterial contribution to host nutrition occurs in man, when anaerobes in the large intestine scavenge carbohydrates that have reached that level undigested.

However, many products of microbial metabolism are probably of no use to any organisms; indeed, accumulated waste products are often fatal to their producers and to others. The precise nature of the end-products of the metabolism of any particular organism depends in part upon the organism itself and in part upon the substrates available and the conditions of growth. If the substrates and conditions are standardized, analysis of the end-products may help to identify the organism. This principle underlies many of the tests used in medical microbiology. The crude but informative procedure of testing an organism's ability to produce acid and gas from various sugars and alcohols (p. 4) has for many years been used to provide a basis for the classification of gram-negative bacilli and of several other groups of bacteria as well as fungi. Some other tests commonly used for similar purposes are mentioned on p. 91. Gas–liquid chromatography (p. 73) can provide precise identification of products of bacterial metabolism, principally short-chain fatty acids and alcohols produced by anaerobes. These can be detected in cultures, and also in some clinical specimens—a rapid means of detecting the presence of the relevant organisms in the specimens.

Reproduction

Genetics

A bacterium that reproduces by simple binary fission, as described on p.11, gives rise to two identical organisms, and consistent repetition of this process would produce a population of identical organisms. However, changes in the genetic composition (**genotypic variation**) of bacteria can happen in several ways. These include:

(1) **Mutation.** The bacterial chromosome consists of double-stranded DNA, which in many bacteria at least is in the form of a loop. Each intertwined spiral strand is a long sequence of nucleotide units; each unit consists of a deoxyribose component, which is part of the backbone of the strand, and a projecting

nitrogenous base component, which is linked with that of a unit on the other strand. The base is any one of four substances which are present in all natural DNA—the purines adenine and guanine and the pyrimidines cytosine and thymine. The sequence in which these four bases occur down the length of the strand constitutes a code or formula which determines the structure of the cell's enzymes and structural proteins and therefore the properties of the organism. The two strands of the loop are not identical but are complementary. The deoxyribose units are orientated in opposite directions, and the adenine on one strand is always matched by and linked with thymine on the other; guanine is similarly paired with cytosine. When the cell is about to divide, the two strands separate and each acts as a template or guide for the construction of its new partner, so that the two new double-stranded molecules are formed, each identical with and containing one strand of the original molecule (semi-conservative replication). This complex and delicate process is occasionally disturbed by factors that break the strands or cause errors of copying and consequent changes in the code (**mutation**). The altered pattern is then faithfully handed on to later generations, provided that it is compatible with survival.

Spontaneous mutations that cause recognizable changes in properties are relatively rare. Any given 'mistake' will arise in ordinary circumstances only once in several millions of divisions (though its likelihood can be considerably increased by exposing the dividing organisms to ultraviolet or X-irradiation or to one of various 'mutagenic' chemicals). However, bacteria reproduce so fast that mutants are continually appearing, and reliable occurrence rates can be calculated for many specific mutations. The significance of mutation is greatly enhanced by circumstances that favour the mutants. For example, if 100 million bacteria, including one streptomycin-resistant mutant, are added to a suitable streptomycin−containing broth, the broth will soon (within 18–24 hours for most species) be populated entirely by streptomycin-resistant descendants of the one mutant. (Environmental selective mechanisms are similarly important in determining the significance of any other form of genotypic variation).

Note that mutation occurs within a single cell. Each of the other three processes described below involves movement of genetic material from one cell to another.

(2) **Transformation.** Certain bacteria can acquire genetic characteristics by soluble-DNA-mediated transformation. This provided the first clear evidence of the central role of DNA in inheritance when, in 1944, Avery, McLeod and McCarty reported their studies of pneumococcal type-transformation. If pneumococci of one capsular type (p. 79) are grown under defined conditions in the presence of soluble DNA from pneumococci of another type, a minute proportion of the dividing cells take up the 'foreign' DNA, incorporate it into their genetic make-up, and produce progeny that make capsular material appropriate to the type which provided the DNA. Transformation can occur in many bacterial species other than pneumococci, but only when donor and recipient strains are of the same or closely related species.

(3) **Transduction.** Sometimes when a bacterial culture is infected by a virus (called a bacteriophage—p. 23) from another strain, a small minority of the recipient organisms acquire some property of the donor strain and transmit it as stable genetic character to their descendants. This happens because the bacteriophage brings with it some of its previous host's DNA—usually in place of some of its own DNA, so transducing phages are often defective. Since as a rule any bacteriophage has only a narrow host range, transduction is usually between closely related strains. **Lysogenic conversion** is somewhat different, in that the phage does not transfer a property from one strain to another but the phage's own DNA confers new properties upon strains with which it enters into a lysogenic relationship (p. 23). Thus both erythrogenic toxin production by *Streptococcus pyogenes* (p. 77) and toxigenicity of *Corynebacterium diphtheriae* (p. 84) depend on possession of appropriate lysogenic phages.

(4) **Conjugation.** The genetic material of bacteria is not all located in the nuclear chromosome. In some bacteria at least, part of it is in the form of extra-chromosomal DNA units

known as **plasmids**, some of which alternate between being free and being integrated into the chromosome. Free plasmids replicate independently of the chromosomes. They determine possession of properties that are essential for the survival of the bacterium in a favourable environment, but provide a pool of 'optional' characters that may confer advantages in other circumstances. Individual bacterial cells of many different species contain special plasmids, **transfer factors**, which confer on them the ability to form **sex-fimbriae** (p. 11) by which they attach themselves to and conjugate with other bacterial cells that do not already have transfer-factor plasmids. In conjugation, genetic material—the transfer factor with or without one or more other plasmids or part of the chromosome—is transferred from the initiating cell to its partner, which can then pass on some or all of the newly acquired genetic material (and therefore the associated properties) to its progeny, and to other transfer-factor-negative cells by conjugation. Transfer of DNA in this way also involves replication; only one DNA strand is transferred, the other remains in the donor cell and new complementary strands are synthesised in donor and recipient. This form of genetic transfer is a great deal more complex than our brief outline suggests, and there are a number of variants of it. Although transfers occurs most readily between related strains, it is not restricted to donor and recipient strains with close taxonomic relationships This has important medical implications; for example, antibiotic-resistant but harmless organisms in the human or animal intestine can confer antibiotic resistance, by plasmid transfer, on potentially pathogenic but previously antibiotic-sensitive bacteria of other genera which the host happens to ingest. The practical significance of such **transferable** or **infective drug resistance** is discussed in Chapter 29 (pp. 382–4).

Phenotypic variations are changes of appearance or behaviour in response to changes in environmental factors and involve no alteration of genetic structure. The microscopic and colonial appearance of bacteria, their possession of flagella or of capsules and many of

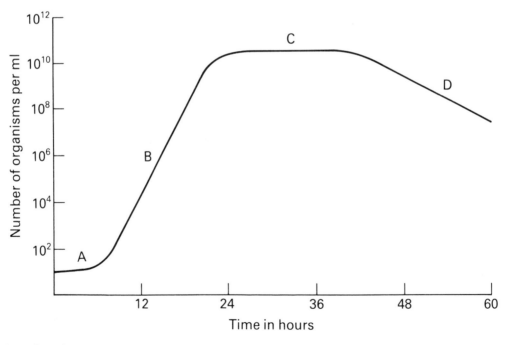

Figure 2.1 *Growth curve of a bacterial culture in a liquid medium, showing lag phase (A), logarithmic phase (B), stationary phase (C) and phase of decline (D).*

their metabolic activities vary according to their circumstances. Particularly clear examples of phenotypic variation are provided by bacteria that show enzymic induction; these inherit potential ability to make certain enzymes but only 'learn' to do so after being exposed for a while to appropriate substrates (p. 15).

Phases of growth

When bacteria are introduced into or on to sterile culture media many factors determine their subsequent rate of multiplication. Each bacterial strain has its own maximal rate which it can achieve under optimal conditions; but it does not necessarily begin to reproduce at this rate immediately.

Figure 2.1 shows a typical growth curve of a bacterial culture in broth. The count of living bacteria is plotted on a logarithmic scale against the time after the inoculation of the broth. The exact shape of the curve depends on the nature of the organism, the size of the inoculum, the age of the culture from which it was taken, the composition of the medium, the conditions of incubation etc.; but four stages of activity can usually be discerned. In the first stage, the **lag phase,** the bacterial cells adapt themselves to their new environment and prepare for division. They increase in size, but there is little increase in numbers. In due course division speeds up and is soon occurring at the maximal rate for the system. The increase in numbers is now exponential, the population doubling at regular intervals. Since such an increase takes the form of a straight line when plotted on a logarithmic scale against time, this stage is called the **logarithmic phase**. The maximal rate of division continues until it is slowed by exhaustion of nutrients or accumulation of toxic metabolites or both. The population increase gradually comes to a standstill, the **stationary phase,** in which the rate of production of new cells by division equals the rate of loss by death. This is succeeded by the **phase of decline,** in which the number of living bacteria slowly decreases and the proportion of dead cells increases. The speed and shape of this decline depends upon the susceptibility of the organisms to their own waste products; some delicate species are extinct within a few days,

whereas others may survive for years. Rapidly multiplying bacteria of the logarithmic phase are particularly susceptible to damage by antiseptics and antibiotics. They are also able to multiply at maximal speed immediately, without a lag phase, if transferred to suitable fresh medium.

A logarithmic rate of multiplication can be maintained indefinitely if the culture medium is repeatedly or continuously renewed. Such **continuous culture** has important industrial applications—e.g. in the manufacture of antibiotics; it also makes possible the study of bacterial metabolism under constant and controllable conditions that can be made to simulate natural conditions in the host.

Physiology: Viruses

Viruses do not reproduce themselves; they are **replicated** by host cells. The essence of this process is that the virus nucleic acid enters a host cell and uses its nucleic acid and protein synthesis mechanisms, diverting them by means of virus messenger RNA to production of virus components. In the laboratory such a take-over can be achieved by nucleic acid alone, free from capsid or envelope. Retroviruses, e.g. human immunodeficiency virus (p. 128), possess a unique enzyme, reverse transcriptase, which constructs a DNA copy of the RNA virus genome. This copy is then integrated into the host cell DNA. The capsid protects the nucleic acid when it is not in the host cell, and in the case of a naked virus makes possible its entry into the fresh host cells and determines the range of hosts into whose cells such entry can be made. (It also largely determines and is the main target for the host's immunological responses to the virus.) Entry of enveloped viruses into host cells is determined by the envelope. The first step towards invasion of fresh host cells is **attachment** and **adsorption** of viruses to their surfaces. For some and possibly for all groups of animal viruses adsorption depends on interaction—chemical in some cases and electrostatic in others—between parts of the virus surface and receptor areas on the surfaces of the host cells; hence the limitation of host range imposed by the outer coat of the virus, and probably also the fact that the various tissues of a host differ in their susceptibility to a given virus. Much remains to be discovered about adsorption mechanisms, but some facts are clearly established. It is known, for example, that adsorption of myxoviruses is a function of projections associated with their envelopes (p. 124) and that their host-cell receptors are mucoproteins, whereas receptors for polioviruses are lipoproteins. Once adsorbed, viruses gain admission to animal cells (**penetration**) by a process resembling phagocytosis if they are naked or, if they have envelopes, by fusion of these with host-cell membranes (in contrast to the self-injecting mechanism of bacteriophages—p. 13). The nucleic acid loses its capsid (**uncoating**) and is released into the cells. At this stage and for some hours afterwards no infective virus can be recovered from the host cells, and the virus is said to be in **eclipse**. During this period the take-over is proceeding, and the cells, acting on instructions given by the invading nucleic acid, are producing enzymes with which to make new viruses. Where these viruses are made and assembled depends upon the group to which the virus belongs. Some are made within the host nucleus and some within the cytoplasm. Some viruses appear to be made entirely in one part of the cell, whereas for others the nucleic acid core is made in one site and capsomeres are transported ready-made from other sites (within the same cell). The degree of disturbance of host cell metabolism also varies, as does its fate. For example, orthomyxoviruses are released from the cell surface by a budding process, acquiring their envelopes as they emerge, and the cell can continue to produce and release them for a long period; on the other hand polioviruses accumulate inside the host cell, kill it and are released in large numbers when it bursts. Such changes leading to lysis and death of the host cell are called **cytopathic effects** (CPE), which may be seen as plaques of damaged cells in cell monolayers seeded with a virus (p. 170). The nature and severity of the disease produced by a virus depends upon the number and type of cells destroyed. This destruction is not always the direct effect of the virus, as virus-infected cells may also be destroyed by host immunological mechanisms (Chapter 6).

Inclusion bodies

These are accumulations of virus material, up to 30 μm in diameter, which are formed within host cells in some virus infections. Such structures were recognized in association with some diseases long before it was possible to isolate the viruses, and many of them were named after their discoverers—e.g. those found in rabies are called **Negri** bodies and those in smallpox and vaccinia were called **Guarnieri** bodies. Their appearances and situations are often sufficiently characteristic to be of diagnostic value. Those of herpes simplex, poliomyelitis, yellow fever and adenovirus infections are found within the nuclei of the infected cells, whereas others, such as those of poxvirus infections and rabies, are cytoplasmic. Most of them contain active viruses and are intracellular 'colonies', but some intranuclear inclusions seem to be merely deposits of material left over from previous virus synthesis.

Latent infection

After an initial acute episode, some viruses can remain latent in host cells for months or years in a form other than replicating infectious virus. They do not kill the cell, or provoke an immunological attack which would do so, but retain the potential for reactivation at a later date in response to various stimuli or changes in the host's immune status. In such reactivated infections there is full expression of the viral genome, with production of infectious virus and of acute manifestations of diease.

In latent herpesvirus infections the DNA can become integrated into the host cell DNA in a manner similar to bacteriophage lysogeny (p. 23).

Persistent infection

Some viruses persist in host cells without being cytopathic, i.e. they do not destroy the infected cells, but they impair the 'specialized' functions of the cell. One feature of such persistent infection is the reduction in expression of virus antigens on the surface of the infected cells; this enables them to survive despite a vigorous immune response. However, small amounts of infectious virus are produced by persistently infected cells and individuals with persistent virus infections who have circulating antibodies to the virus may, in consequence, develop immune complex diseases (p. 61).

Slow virus infections are caused only by atypical viruses. They are characterized by prolonged incubation periods, often lasting many years, during which time a slow chronic degenerative disease process takes effect. Most known slow virus infections affect the central nervous system (p. 223).

Transformation

When virus nucleic acid is inserted into host cell DNA the cell may become transformed into a malignant (cancer) cell which multiplies rapidly, unchecked by normal cell-proliferation controls. Despite much research, direct evidence for the role of viruses in human cancer is limited to the role of Epstein Barr (EB) virus infection in Burkitt's lymphoma (p. 133), but there is also evidence for an association between hepatitis B virus and hepatoma (p. 266) and between genital herpes infection and carcinoma of the cervix (p. 131).

Bacteriophages

Bacteriolysis

If a large excess of phages is added to a susceptible bacterial culture, so that many phages are adsorbed on to each bacterium, the cells may be disrupted because many small areas of their walls are dissolved at once—'lysis from without'. If the infection is not so heavy, the bacteria survive this external assault, but intracellular phage multiplication may then occur, beginning within a few minutes of the injection of the phage DNA. Instead of synthesizing its own DNA, the bacterial cell provides that for new phages, and also supplies them with their protein coats. Within an hour or less many phages have been formed, and these then lyse the cell and are released—'lysis from within'.

As we have already indicated (p. 13), bacteriolysis by phage can be studied in cultures either on solid or in fluid media. A bacterial 'lawn' can be produced by inoculating the whole surface of an agar plate with the bacterial strain to be studied. Before this culture is incubated, fluid prepations of various phages can be spotted on to it in marked positions. The culture is then incubated overnight, and lytic action of bacteriophages is shown by failure of bacterial growth in the areas on to which the preparations were spotted. If the number of effective phage particles is small, each one produces a separate small defect in the bacterial lawn (a **plaque**) roughly analogous to a single bacterial colony; but a large number of effective phage particles will produce confluent lysis of the whole area covered by the original spot. Bacteriolysis in a fluid culture becomes apparent far more rapidly, a broth culture of a susceptible bacterium becoming crystal clear within 30–60 minutes after the addition of a phage preparation; but of course only one phage preparation can be tested for lytic action on any one culture in this way, whereas the lawn procedure on a solid medium allows the testing of many preparations simultaneously on a single plate.

Lysogeny

Infection of a bacterium by phage is not necessarily followed by phage multiplication and bacteriolysis. Sometimes when phage DNA enters a host cell it does not derange its synthetic activities but is integrated into the bacterial chromosome. Such a phage is described as **temperate**, as distinct from **virulent** or **lytic**, and the integrated non-lytic form in which it persists in the cell is called **prophage**. In this form it is reproduced synchronously with the bacterium and handed down to all the progeny of the original cell. This relationship is described as **lysogeny** and the host bacterium is **lysogenic** because in certain circumstances propagation and release of the phage may occur, and it may then cause lysis of other bacterial strains. Such propagation may occur spontaneously, and it can also be induced by ultra-violet light or various chemicals. A lysogenic bacterial strain is resistant to lysis by another preparation of the phage that it is carrying.

Physiology: Fungi

Most fungi that infect man are tolerant of a wide range of temperatures, although growth is usually optimal at 25–35°C. The spectrum of human infections caused by fungi reflects their individual preference. Thus dermatophytes (ringworm fungi) grow well at 28–30°C, temperatures similar to that of the skin surface, whereas those capable of infecting internal organs (e.g. *Candida albicans* and *Aspergillus fumigatus*) grow well at 37°C. Most of these fungi are not demanding in their nutritional requirements. A simple sugar, usually glucose, will suffice as an energy source, nitrate or ammonia as a source of nitrogen and mineral salts for electrolytes and trace elements. These requirements are met by Sabouraud's agar; this contains D-glucose, peptone and water and is widely used in medical microbiology laboratories.

Fungi are predominantly aerobic, but many yeasts can produce alcohol by fermentation (anaerobic metabolism). When fungi are grown under controlled conditions they can produce a range of metabolites of great importance in clinical microbiology—e.g. antibiotics such as penicillins and cephalosporins and even antifungal agents such as griseofulvin and amphotericin B. Many moulds also produce mycotoxins, such as **aflatoxins** which are hepatotoxic and carcinogenic in animals, but their role in human disease is unclear. However, it has long been established that **ergotism** results from eating bread prepared from rye infected with the fungus *Claviceps purpurea* which produces ergot alkaloids.

Reproduction

Virtually all fungi have the potential to reproduce by a process of mitosis, forming asexual spores. These may be **conidia** produced in large numbers by filamentous fungi for dissemination, or **chlamydospores** or **arthrospores** produced in small numbers for survival in extreme conditions. Until recently most fungi pathogenic to man were thought to lack a sexual phase to their life cycle and hence were classified as *fungi imperfecti*. Many have now been mated in laboratory conditions and have formed sexual spores. The modes of sexual reproduction of fungi, and in particular the characteristics of the spores, form the basis of fungal identification and classification

Physiology: Protozoa and Helminths

The diversity of life styles and complexity of life cycles of these parasites make study of their physiology difficult and generalizations almost impossible. There are of course vast physiological differences between free-living forms, encysted stages, migratory stages in tissues, stages found in the mammalian intestine, etc., and in many cases the physiological processes are not known in any detail.

Nutrition

The gut morphology of flukes and nematodes is consistent with digestive and/or absorptive functions, but it is not clear to what extent these occur. Absorption through their external surfaces is for most parasites other than the nematodes an important means of acquiring nutrients, but in only a few species has this been studied in any detail. Simple diffusion and active transport across surface membranes are involved, but the mechanisms may differ considerably from mammalian systems. However, nematodes are covered by a cuticle that is biologically inert and impervious to solutes. Glycogen is the most commonly stored polysaccharide.

Metabolic Pathways

Glycolysis is the most important pathway for energy metabolism, especially in species living in sites with low oxygen tension; a wide variety of end-products is formed. The tricarboxylic acid cycle is present in some species, incomplete in or absent from others, and used in reverse by some helminths, which reduce malate to succinate, which is then excreted. Two CO_2-fixation pathways have been identified in gut parasites. Many parasites take up oxygen when it is available, but its metabolic importance to them is not clear. The amino-acid content of proteins, amino-acid biochemistry and protein synthesis have been studied in only a few species, and little is known about lipid metabolism.

Excretory systems

These range in complexity from the specialized flame-cell system of the flukes and the excretory ampulla found in roundworms to the simple diffusion of waste products from the external surface which probably occurs to some extent at least in most, if not all, parasites and is the likely route for excretion of ammonia, urea and uric acid. The excretory systems have been studied in only a few species of helminths. Most helminths are probably osmo-conformers, so that osmoregulation is of little significance.

Reproduction

The life cycles of these parasites are described in the next chapter. Asexual reproduction features in the life cycles of many parasitic protozoa, all of the flukes and some of the tapeworms. Sexual reproduction occurs in the protozoan group called *Sporozoa* (which includes the malarial parasites) and in all helminths. In those species which use both methods, asexual and sexual reproduction take place alternately.

Asexual reproduction in most protozoa is by simple binary fission, each parent cell undergoing mitotic nuclear division and then dividing into two daughter cells. Multiple fission occurs in some protozoa (e.g. trypanosomes), and malarial parasites multiply asexually by schizogony—multiple division of the nucleus followed by formation of separate portions of cytoplasm to surround each of the daughter nuclei. Asexual reproduction of man's helminth parasites is by internal budding—within the snail intermediate host in the case of the flukes and in the hydatid cyst stage (also in the intermediate host) of the tapeworm *Echinococcus*.

Sexual reproduction of flukes, the great majority of which are hermaphrodite, is probably by self-fertilization. The reproductive system of a fluke consists of several (usually two) testes with associated ducts, a single ovary, paired vitellaria (producing egg-shell precursors) and associated ducts, and a long uterus in which eggs are stored before being passed out. Eggs are usually ovoid, with brownish shells of quinone-tanned protein and outer covers called opercula. Tapeworm reproductive systems are broadly similar to those of flukes; each proglottis contains both male and female components, but the male system usually matures before the female and cross-fertilization between different proglottides of one worm or between worms may occur. Tapeworm eggs are usually round, with a thin membraneous shell which is generally soon lost and an outer thick covering which is not a true shell but an embryophore. Whereas flukes discharge eggs continuously so that they appear in the excreta in fairly constant numbers, tapeworm eggs are usually retained in the uterus of the gravid proglottis when it separates from the strobila and are usually found in this situation in the faeces rather than as free eggs. In roundworms the sexes are separate, with specialized reproductive systems. Females can produce large numbers of eggs, which typically are rounded with three-layered shells that are highly resistant to chemical and environmental hazards and allow long survival of free eggs. Some roundworms produce larvae rather than eggs; these may be at a very early stage of development when passed by the female (as in filarial worms) or relatively advanced (as in *Trichinella* and guinea worm).

Suggestions for Further Reading

Chappel, L.H. 1979. *Physiology of Parasites*. Blackie, Glasgow and London

Mandelstam, J., McQuillan, K. and Dawes, I.W. 1980. *Biochemistry of Bacterial Growth*. 3rd edn. Blackwell, Oxford.

3
Ecology

Distribution and Ways of Life of Micro-organisms

Micro-organisms are virtually ubiquitous; some are to be found in almost every environmental niche. They are present in soil, water and air, in most kinds of inorganic or organic non-living matter, as well as in vast numbers within and on the surfaces of all living creatures. The distribution of any particular species is limited by its growth requirements and by its compatibility with other species.

A vital part in the circle of life is played by the many bacteria and fungi which live by breaking down and recycling the bodies of dead animals and plants. Micro-organisms are at both the beginning and the end of some food chains, which thus become cycles. 'If microscopic beings were to disappear from our globe, the surface of the earth would be encumbered with dead organic matter and corpses of all kinds, animal and vegetable .... Without them, life would become impossible because death would be incomplete.' (Louis Pasteur, 1861) Such organisms that live on dead organic matter are described as **saprophytic**. Some normally saprophytic species can occasionally invade the tissues of living animals and humans, but this is rare. Some writers also use the term 'saprophytic' for organisms that are superficial and harmless parasites, but these are better described as commensals (see below).

A small minority of micro-organisms, including nearly all those of medical importance, are commonly or necessarily **parasitic**—i.e. they live inside or on the surfaces of other living organisms from which they derive benefit. The extreme form of parasitism is demonstrated by viruses, which are totally dependent on host cells for their own multiplication. Bacteria themselves harbour parasitic viruses (bacteriophages—p. 13), and plants and animals act as hosts to large and varied microbial populations. Parasites may be **commensal, symbiotic** or **pathogenic**. A **commensal** (literally, one that shares the table) derives nourishment from its hosts but does nothing in return—a non-paying guest. Many examples are found in the secretions of human skin and mucous membranes. A **symbiont** lives in partnership with its host, receiving nourishment but rendering service in return—a paying guest. Such are the nitrogen-fixing bacteria of the root nodules of leguminous plants, the cellulolytic bacteria that digest plant food in the intestine of herbivores, and the vitamin-synthesizing bacteria of the human intestine. Many bacteria commonly regarded as commensals are in fact beneficial to their host because they make it difficult, by their presence and by their metabolites, for potential pathogens to colonize his surfaces (see below). A **pathogen** does harm to its host.

However, these divisions are less clear-cut than they first seem. The three terms refer to relationships between parasites and hosts, not simply to properties of micro-organisms; the same micro-organism may exhibit different forms of parasitism in different hosts, or even in the same host at different times or in different sites.

Microbial Pathogenicity and Carriage in Man

Pathogenicity is as much an expression of the host's susceptibility as it is of the organism's intrinsic power to cause disease. To use a well-worn analogy, the soil is as important as the seed in determining the outcome of infection. A few bacterial species are virtually always pathogenic to man; they are not found in healthy individuals and their detection is diagnostic of disease. They include the mycobacteria of tuberculosis and leprosy, the spirochaete of syphilis and the small gram-negative bacillus of plague. Even with these, however, the severity of the disease varies between individuals and within the one person there may be long periods, often years, during which the infection is quiescent and unnoticed. Many more species with very clear pathogenic potential do not cause disease in every individual whom they colonize. For some of these there is a well-recognized carrier state (see below), and others are so common in the bacterial flora of healthy people (e.g. pneumococci in the nasopharynx) that their mere presence is unremarkable. Then there are other species that are part of the normal bacterial flora of the body but can cause severe infections either when the local conditions change and they overproliferate or when they gain access to sites not normally colonized by them. For example, the intestinal commensals *Escherichia coli* and *Bacteroides fragilis* are the commonest causes of urinary tract infection and of post-operative peritonitis and wound infection respectively. Many relatively non-pathogenic commensals and some environmental saprophytes may cause **opportunist** infections in severely compromised patients, particularly those with impairment of immunity or other host defence systems.

Carriers of potentially pathogenic microorganisms may be **healthy**, suffering no ill effects but being potential sources of clinical infection in others. Alternatively they may be **incubational** (or **precocious**) carriers who will shortly develop clincial infection, or **convalescent** carriers who have already had it. Duration of the carrier state, and of excretion of the organism, varies widely. Short-term convalescent carriage is common, but with some infecting species carriage persists for long periods, sometimes for life. Typhoid is the classic example of a disease that gives rise to such **chronic carriage**, and this may be difficult to detect because *Salmonella typhi* may be excreted intermittently, especially when it has established itself in the gall bladder. In contrast, carriage of other salmonellae seldom lasts for more than a month or two. Carriage of *Staphylococcus aureus* is of particular concern in hospital outbreaks of wound sepsis. The source of such an outbreak may be a healthy carrier, who has no history of significant staphylococcal infection but harbours the staphylococcus in the nose or on the skin.

Host–parasite Relationships of Protozoa and Helminths

Like other specialized subjects, parasitology (as defined on p. 10) has developed an extensive terminology that confuses those not familiar with it. While we do not need to go into much detail about it in this book, the subject cannot be understood without grasping the concepts of a life cycle and knowing the meaning of some of the associated terms. All protozoan and helminth parasites go through a series of developmental stages, following a regular sequence until a final stage is reached where reproduction takes place and a new series of developmental stages is initiated. The cycle may have several developmental stages or few: at least one stage must occur in a host organism (otherwise the parasite would not be a parasite), and the cycle can proceed only in one direction. There may be several phases of parasite multiplication in the cycle or only one; according to species, these phases of multiplication may be sexual, asexual or both. Once one stage has reached its full development, the parasite cannot mature further until the next stage commences, and in many cases this can occur only in another host species (e.g. filarial worms require both men and mosquitoes) or after the passage of resistant stages (e.g. some nematode eggs, passed in the faeces, require a period of development outside the human body before they can infect another person). If the life cycle includes a single host at only one point in the cycle, it is said to be 'direct';

e.g. the parasites causing amoebiasis in man are passed in the faeces in a resistant cyst stage, human infection is acquired by the faecal–oral route via contaminated food or drink and no other species of host is required for transmission. On the other hand, cycles which have two or more species of host which are infected in sequence as essential links, not alternative hosts, are said to be 'indirect'; e.g. some filarial parasites are transmitted from man to man by blood-sucking mosquitoes, and without the latter the life cycle could not be completed. In this example there are two hosts (man and mosquito) and it is necessary to establish a principle to distinguish between them. The species of host in which the parasite reproduces sexually is called the **definitive host**; all other host species in the life cycle (there may be more than one) are called **intermediate hosts**. It is easy to assume that man is usually the definitive host and an insect vector the intermediate host, but in one of man's most important parasitic infections, malaria, sexual reproduction takes place in the mosquito, which is thus the definitive host. Asexual multiplication may occur either in the intermediate host or (as well as sexual multiplication) in the definitive host. Alternatively there may be no multiplication in the intermediate host but merely a period of maturation (e.g. filarial worms in mosquitoes). Some parasites do not multiply sexually at any time, and in theory it is therefore impossible to state which species is the definitive and which the intermediate host—e.g. the trypanosome parasites causing sleeping sickness, for which convention dictates that man is regarded as the definitive and the tsetse fly as the intermediate host.

Whereas the human host is never dependent on protozoan or helminth parasites, such parasites cannot survive indefinitely without hosts (human or other). However, the degree of this host-dependence varies considerably. Thus the protozoa causing malaria and sleeping sickness cannot survive except in the bodies of humans and of either mosquitoes or tsetse flies respectively. At the other extreme, some nematode parasites are capable of surviving for several generations independently of any host, but sooner or later they must return to a truly parasitic existence for several generations. The majority of parasitic species lie between these two extremes. The non-parasitic stages are termed 'free-living' if they have an active existence independent of a host—e.g. hookworms have free-living stages in soil. Other parasites have stages which can survive outside the host but are not strictly free-living; they are cysts or eggs, which have fairly impervious outer coverings and are thus resistant to adverse environmental conditions. Examples include the cysts of the parasitic amoebae and the eggs of tapeworm and nematode worms. These resistant eggs are usually passed in the host's faeces (since most of the parasites which produce them are found in the lumen or wall of the intestine) and are the stage by which infections are usually transmitted.

As we have seen, the distinction between definitive and intermediate hosts is based on the location of sexual multiplication. A different concept is applied in the use of the term **reservoir** host for any host which can maintain the life cycle of the parasite in the absence of man. The reservoir host is therefore an alternative host to man and is often a wild animal; it is usually affected by the disease, not just an unaffected carrier or transmitter. Reservoir hosts are important epidemiologically in that they can maintain in an area a cycle of transmission and a source of infection to which humans may be exposed.

As with microbial infections, parasitic infection does not necessarily result in disease of clinical significance. The many factors that determine whether it does so include the number of parasitic organisms present (which may itself be a function of the parasite's reproductive pattern and potential), their size, the site(s) which they occupy in the body, their metabolic processes (particularly the nature of any waste products) and the general health and immunological status of the host. Thus a few small parasites with low reproductive potential living in the intestinal lumen of a well-nourished individual are of little or no consequence. On the other hand, an infection with malaria parasites (which have a very high reproductive potential) can be life threatening, especially in a young child or pregnant woman.

Man's Normal Microbial Population

There is evidence that in at least some species an apparently healthy young animal may harbour in its tissues viruses which it derived from its mother *in utero* and which may cause it to develop leukaemia later in life. We do not know whether such transmission involves only leukaemia viruses, or whether anything of the sort occurs in man. With this reservation we can say that the healthy human fetus has no resident microbial population up to the time of its birth. It acquires on its surface or swallows or inhales an assortment of micro-organisms from the mother's birth canal, and these are soon reinforced by contributions from various human and inanimate (and possibly also animal) sources in the newborn infant's immediate environment. Those organisms which find themselves in suitable environments, whether on the outer or inner body surfaces, begin to multiply and to enter into complex competitive relationships with other potential colonizers. Within hours of birth the infant has begun to acquire a resident microbial population—or rather, a number of different populations, since some organisms thrive on the skin, others do better in the mouth or throat or nose, others in the intestine and so on. By degrees—and at speeds that depend on many factors, such as frequency and method of washing, diet and living conditions—the combinations of organisms that have taken up residence in different areas (virtually all of them bacteria) form fairly stable, balanced, interdependent populations and come to resemble those commonly found in such sites in adults (often described as the **normal flora**). The bacterial population of a single human body is of the order of 10^{14}; a skin scale may carry many bacterial micro-colonies, each consisting of many thousands of cells; and there are around 10^{12} bacteria in a gram (wet weight) of colonic contents or faeces. Two points need to be remembered: that the range of organisms detectable in any situation depends on the methods used in looking for them as well as on the actual population, and that throughout life there are fluctuations and marked personal differences in the 'normal' microbial populations of the body, dependent on general health, diet, hormonal activity, age, race and many other factors.

The human body is host to several distinct populations of bacteria. The skin of a normal healthy adult will carry representatives of gram-positive cocci of the genera *Staphylococcus* and *Micrococcus*, gram-positive 'coryneform' or 'diphtheroid' bacilli of the genera *Propionibacterium* and *Corynebacterium*, and yeasts such as *Pityrosporum*. Similar organisms are found in the anterior nares, with *Staph. aureus* being carried there by about 30% of adults. The mouth contains a large and varied population including α-haemolytic (viridans) streptococci, lactobacilli, the micro-aerophilic actinomyces and anaerobic gram-negative bacilli of the genera *Bacteroides* and *Fusobacterium*, and small numbers of yeasts of the genus *Candida*. On the mucosa of the pharynx these are joined by haemophili, pneumococci, commensal neisseriae and *Branhamella*. The stomach and upper small intestine are host to few bacteria, but the large intestinal contents have a vast flora that is predominantly anaerobic. The major component genera are shown in Table 3.1. The gram-positive bifidobacteria and gram-negative bacteroides head the list and outnumber the facultative and aerobic species such as *Escherichia coli* by more than 1000:1. The vagina in post-pubertal women has a distinctive flora, dominated by lactobacilli but also containing smaller numbers of anaerobic gram-positive cocci and bacteroides; other species derived from the skin and the faecal flora may colonise the introitus and lower vagina.

These resident bacteria have an important protective function. By their presence in and occupation of colonization sites they may prevent colonization by pathogens—e.g. commensal enterobacteria inhibit the attachment of some pathogens to the intestinal mucosa. Moreover, their metabolic activity may also protect against infection. For example:

(1) lactobacilli in the vagina produce acid from glycogen and maintain a low pH that is unsuitable for many bacteria (p. 16);
(2) propionibacteria on the skin break down sebum and release fatty acids that inhibit potential pathogens; and
(3) some skin staphylococci produce

antibiotics that are active against other members of the skin flora and some pathogens. The role of bacteriocins (p. 250) in this protective function is not clear.

Disturbances of the normal commensal flora may lead to serious opportunist infection. Antibacterial drugs have two major effects on the normal flora. They may reduce all or parts of the normal flora and allow overgrowth or superinfection with, for example, *Candida albicans* causing thrush in the mouth (p. 137), or *Clostridium difficile* causing pseudomembranous colitis (p. 259) in the large intestine. Also their use may result in selective promotion of drug-resistant strains among the normal flora, which may give rise to subsequent treatment problems. The classical example of such a sequence is the too early initiation of penicillin administration to a patient with a damaged heart valve who is undergoing dental treatment; the aim is to prevent his mouth streptococci from causing infective endocarditis, but if selection occurs pre-operatively he may end up with endocarditis due to penicillin-resistant streptococci, which can be difficult to cure (p. 239). Metabolic and hormonal changes such as those found in diabetes or pregnancy may also disturb the balance of the normal flora.

TABLE 3.1 Bacterial flora of normal faeces

Genus or group	Mean number of bacteria ($\log_{10}$)/g
Bacteroides	10.5
Bifidobacterium	10.5
Enterobacteria	8.0
Streptococci (including enterococci)	7.0
Lactobacilli	6.5
Clostridium	5.0
Veillonella	4.0
Pseudomonas	2.5
Total anaerobes	11
Total aerobes	8
Total	12

'Germ-free' Animals

An animal born by Caesarian section to avoid microbial contamination, and then maintained in a germ-free environment and fed sterilized food, develops no microbial population. Because of the uncertainty about intra-uterine transmission of viruses (see above) the term 'germ-free' may not be an accurate description of such animals, and the alternative term **gnotobiotic** has been coined (see Glossary). They can be exposed to specified bacteria with the knowledge that their responses will not be complicated by previous unknown contacts with these or similar organisms, or by the presence of other bacteria. Study of such highly unnatural situations is helping to elucidate the roles of normal flora, and the mechanism of pathogenicity and host resistance. Many of these animals have been reared in sterile isolators which served as the prototypes for the closed isolator systems made of flexible plastic flim used for the care of patients at special risk of infection. (These have now been superseded by the more practical Trexler-type isolation systems in which a linear flow of sterile air is combined with partial plastic mechanical barriers to protect highly susceptible patients; p. 344.)

Suggestions for Further Reading

Dixon, B. 1976. *Invisible Allies: Microbes and Man's Future*. Temple Smith, London.

4
Epidemiology – the Transmission of Pathogens

Epidemiology is the study of the **time**, the **place** and the **person** in relation to a particular disease (in our case, an infection). The analysis of who gets a particular infection, and where and when, enables us to build up a picture of the natural history of the disease, define the risk groups and devise preventive measures. Influences that may be important in the epidemiology of a disease include:

(1) personal factors such as age and sex, nutritional state or the presence of other diseases;
(2) geographical and climatic factors—e.g. malaria can spread only in a climate suitable for the anopheline mosquito;
(3) social and environmental factors—e.g the standard of housing and the provision of an adequate water supply and sewage system; and
(4) occupational factors that bring certain people into contact with particular infections—e.g. farmers and veterinary surgeons with brucellosis.

The following terms are used to describe behaviour patterns of microbial diseases: **epidemic**, a noun or adjective, indicating a temporary marked increase in frequency of a particular disease in a community; **pandemic**, referring to a world-wide epidemic; and **endemic**, used only as an adjective, describing a disease which is persistently present in a community. An endemic disease may from time to time flare up into an epidemic.

There is a surprising lack of unanimity about the precise meaning of the key word **infection**—in particular, as to whether it implies actual invasion of host tissues or merely the presence of potential invaders. It is best defined as the arrival or presence of potentially pathogenic organisms on the surface or in the tissues of an appropriate host. We can then refer to an **infected** animal or plant—even to a bacterium being infected with bacteriophage—but not to infected inanimate objects. For the latter, when they are carrying potential pathogens, and also when they should be sterile but have ceased to be so, the term **contaminated** is preferable. An **infective** disease is one caused by an infecting organism. An **infectious** (or **communicable**) disease is one that is transmissible from patient to patient by transfer of the causative organism; and an infectious patient is one from whom such a disease can be acquired. Infection is called **clinical** when it is causing overt disease, or **subclinical** when there is little or no impairment of the patient's health. Persistent subclinical infection that may be converted into clinical infection by changing circumstances (such as a reduction in the patient's immunity) is described as **latent**.

These definitions do not eliminate all terminological difficulties, since they include the concept of 'potentially pathogenic' organisms. As we have seen (p. 27), organisms from the normal flora can become pathogenic in various circumstances, and otherwise innocent organisms can become opportunist pathogens. Another commonly used term, **colonization**, can add to the confusion, because it has two meanings: it may refer to the presence of members of the normal microbial flora at the normal sites on the skin, mucosal surfaces, etc.; or it may describe the presence of a known pathogen, e.g. *Staph. aureus*, when it has not yet caused any disease and may or may not cause diseases in the future—a definition which overlaps that of subclinical infection.

Sources of Infection

With few exceptions, the sources of human infections are, directly or indirectly, other human beings or animals—i.e. the animate world. Only a few pathogenic organisms, such as some fungi and possibly some clostridia, have their normal habitat in the inanimate environment; in most cases—e.g. where soil appears to be a source of infection—the organisms have come initially from animal sources, often by faecal contamination, and the inanimate material is merely a vehicle.

The source of pathogenic organisms may be either **exogenous**, when they came from outside the patient, or **endogenous** when they come from the patient's body, usually from his own normal flora. In most of the classical infectious diseases the source is exogenous, and the diseases are characterized by spread from person to person or from animals to man. Endogenous infections are specially important when trauma or lowered local or general resistance makes a patient susceptible to attack by his resident parasites. Normally sterile sites may be contaminated during surgery when the site of operation is heavily colonised—e.g. the peritoneum and tissues of the abdominal wall during intestinal operations. Local resistance is lowered by impaired blood supply, and general resistance by malnutrition, debilitating disease such as diabetes, immunosuppressive chemotherapy or radiotherapy or even other infections such as measles.

Human sources

These may themselves be clinically infected or may be carriers (p. 27). Carriers are important in the spread of infection, especially in epidemics, since they are hard to detect, mix freely with other people, and may disseminate large numbers of organisms over long periods; clinical cases may be less 'successful' distributors because they are likely to be segregated, and may even be removed from circulation permanently by death.

Animal sources

The name **zoonoses** (four syllables) is given to infections normally maintained in animals, which man sometimes acquires accidentally—e.g. brucellosis, leptospirosis, rabies, leishmaniasis.

Environmental sources

Few environmental saprophytes are pathogenic for man unless his immune system is severely compromised. The exceptions are fungi that cause local lesions (mycetoma) after implantation or systemic illness after inhalation, and some sporing bacilli of the genera *Bacillus* and *Clostridium*.

Transmission of Pathogens

Pathogenic micro-organisms may be transmitted from the source to the new victim in a wide variety of ways, ranging from direct person-to-person spread (contagion) to spread by indirect routes involving inanimate objects (fomites, p. 5), food or water, or arthropod vectors.

The pathogenicity of an organism (its ability to cause disease) has, on its side, three components: **transmissibility**, the means by which it can reach a new host; **infectivity**, the ease with which it can infect or colonize the new host; and **ability to damage the host** (to be discussed in Chapter 5). The other side of the study consists of the many factors determining the host's susceptibility, which we shall consider later. An organism's **transmissibility**, and the modes of transmission open to it, are largely determined by its ability or otherwise to survive in adverse conditions. Spores of the tetanus bacillus excreted in animals' faeces survive for many years in soil, and the cholera vibrio survives in sewage-contaminated water, but *Treponema pallidum* (the cause of syphilis) and the gonococcus do not survive exposure to light, air or drying. An organism's **infectivity** depends on the minimum number of live cells capable of colonizing or establishing infection, and is also determined in part by the ability of the organism to penetrate to its site of damage. Infectivity may be expressed quantitatively as the **infective dose** or as the **effective dose** (ID or ED—respectively the smallest number of organisms required to establish infection or to achieve some other specified effect). Because of biological variation, the dose that gives a positive result in 50% of a test population is often measured (ID_{50}, ED_{50}). The usual mode of spread may depend on the ID. For example, salmonellae and shigellae both cause diarrhoeal disease, but the ID of shigellae is only 10^1–10^2 organisms and infection spreads readily from person to person by the faecal–oral route; on the other hand the ID of food-poisoning salmonellae may be 10^5–10^6 organisms and a period of multiplication in contaminated food is usually needed to build up this dose. The influence of microbial invasiveness on mode of transmission is exemplified by the two main types of viral hepatitis, A and B: virus A can penetrate the intestinal mucosa and so can spread by the faecal–oral route, whereas virus B cannot penetrate intact skin or mucosa and its common means of entry into the body is by injection.

Routes of spread of infection

Common routes for the spread of microbial and parasitic disease are indicated below. The list is not exhaustive and some of the infections mentioned can be spread in more than one way. Examples of micro-organisms and parasites spread by different routes are given in Table 4.1

Contact

Pathogenic micro-organisms may be transmitted directly from person to person by touch (important when examining wounds) or by more intimate contact such as kissing or sexual activity. Indirect transmission via fomites, including contaminated surgical instruments or dressings, also occurs and is the reason for the emphasis on aseptic techniques and sterile instruments in surgery. Intermediate vehicles need not be solid. Contact with contaminated skin-disinfectant solutions has spread infection in hospitals, and immersion in appropriately contaminated water can permit the spirochaetes of leptospirosis (p. 109) or the cercariae of schistosomiasis (p. 155) to penetrate intact skin.

Inoculation

Some highly virulent organisms are unable to pentrate the epithelial surfaces of the body, i.e. cannot invade, and must be injected into the tissues or blood. Accidental trauma may achieve this—e.g. a penetrating wound may introduce into the tissues soil containing tetanus spores; but it can also follow deliberate penetration of the skin—e.g. hepatitis B can result from an injection being given from a blood-contaminated syringe. Some micro-organisms cause disease only when inoculated by the bite of an animal (e.g. rabies) or of an insect vector (e.g. malaria, yellow fever, African trypanosomiasis).

Ingestion

Many infections spread directly from person to person by the faecal–oral route. These include many diarrhoeal diseases, such as bacillary dysentery, and several systemic infections that gain entry through the intestinal mucosa, e.g. hepatitis A, poliomyelitis. Drinking water contaminated with faeces is a vehicle for the transmission of typhoid, cholera and *Giardia intestinalis* infection. As we have seen, multiplication in food is an essential stage in the transmission of salmonella food-poisoning; but simple faecal contamination of food, without multiplication, is a common method of spread of dysentery (amoebic and bacillary), hepatitis A and other diseases in communities where standards of hygiene are low. Unpasteurized milk and dairy products are important in the spread of zoonoses such as brucellosis, Q fever and tuberculosis from cattle to man, and can also transmit intestinal pathogens, notably campylobacters.

Faecal material, particularly from patients with diarrhoea, may contaminate toilet seats, or be transmitted by the hands to toilet handles, taps, door handles etc. From these it can find its way on to the hands, and so into the mouths, of other people. Similarly, food handlers who are faecal excretors of pathogens may contaminate food. These dangers emphasize the need for adequate

TABLE 4.1 Examples of the spread of infective agents

Mode of spread	Some infections commonly spread in this way
Fairly direct person to person	
In droplets or droplet nuclei	Most acute respiratory tract infections; tuberculosis; meningitis; measles; rubella; mumps; smallpox.
In saliva	Infectious mononucleosis; rabies (from animals).
By the faecal–oral route	Bacterial diarrhoea; amoebiasis; hepatitis A; poliomyelitis and other enterovirus infections; some roundworm infections.
By direct contact	Staphylococcal and streptococcal infections; sexually transmitted infections; herpes simplex; ringworm (from humans or animals).
In blood or blood products	Hepatitis B; AIDS; syphilis; malaria; S. American trypanosomiasis.
From mother to fetus in the uterus ('vertical spread')	Rubella; toxoplasmosis; cytomegalic inclusion disease; syphilis.
From mother to fetus during birth	Herpes simplex; gonococcal, group B streptococcal and chlamydial infections.
In fomites (p. 5) or dust	Many infections due to bacteria, viruses or fungi capable of survival outside the patient.
Via foods	
Eggs, poultry, meat products	Salmonellosis, campylobacter infections; tapeworm infections.
Unpasteurized milk, dairy products	Salmonellosis, brucellosis, campylobacter infections, Q fever, tuberculosis.
Raw vegetables in some hot countries.	Amoebic and bacillary dysentery, hepatitis A.
In water	
By drinking	Typhoid; cholera; amoebiasis; giardiasis.
By immersion	Leptospirosis; amoebic meningo-encephalitis; schistosomiasis; pseudomonas infections of ears or skin.
Via showers, humidifiers, etc.	Legionellosis.
From soil	
	Tetanus; gas-gangrene; some systemic fungal infections (deep mycoses).
Via arthropod vectors	
Mosquitoes, fleas, ticks, mites, lice	Malaria; yellow fever; virus encephalitides; bubonic plague; rickettsial infections; trachoma; filariasis.

personal hygiene. Those who attend to the toilet needs of small children or incapacitated patients must take particular care.

Aerial Spread

Most acute respiratory infections, the common childhood infectious diseases (mumps, varicella, measles, etc), meningococcal meningitis and tuberculosis are examples of infections spread through the air and acquired by inhalation. 'Coughs and sneezes spread diseases.' Even when talking or breathing quietly we constantly emit from our mouths and noses numerous droplets of moisture containing bacteria and viruses. The largest fall rapidly to the ground, where they dry and the organisms are added to the dust. Small droplets, however, evaporate in the air, leaving their solid contents as droplet nuclei. These remain airborne for long periods and travel considerable distances on air currents, and they are readily inhaled by other people.

Other pathogenic bacteria shed into the air, or suspended as dust particles, may deposit on susceptible exposed sites and establish infection. This applies particularly to staphylococci and streptococci, and in the hospital environment when wound surfaces are exposed, and is the reason why ventilation and air flow are important in operating theatres and treatment rooms. Patients with staphylococcal infections shed many organisms into the environment, but healthy nasal and skin carriers of *Staph. aureus* also continually shed large numbers of these bacteria on skin scales and are important sources of surgical sepsis.

Vertical transmission

Some micro-organisms are able to cross the placenta and infect the fetus if the mother is infected during pregnancy. The infant may die, or be born with the typical disease, or may suffer congenital damage, particularly if infection occurred during the first trimester of pregnancy. Of particular concern in this context are rubella, syphilis, toxoplasmosis and cytomegalovirus infection.

Insect vectors

These have been mentioned above in relation to inoculation of micro-organisms. They also have a vital role in the transmission of several important infections in which a stage of development in the insect is an essential part of the life cycle of the parasite (see below). At a less sophisticated level, flies that feed on both faecal material and any available foodstuff readily transmit pathogens to food via their contaminated feet and mouth parts.

Additional points about routes of infection for protozoa and helminths

Life cycles of these parasites are of two distinct types (p. 28)—those with direct transmission (i.e. with no intermediate host) and those with indirect transmission (i.e. with an intermediate host). Those with direct life cycles are parasites of the intestine. Eggs or cysts with various degrees of resistance to adverse environmental conditions are passed in the faeces. Most require a period outside the body to mature to an infective state, though some develop so rapidly that they may be infective when passed. Eggs of some of these species hatch after being passed, liberating larvae which are infective by skin penetration (e.g. hookworms), but eggs of most species and all cysts must be ingested by their next hosts, and transmission is thus by the faecal–oral route. Infection is usually a result of accidental contamination of food, drink or fingers by small amounts of faecal material.

Several important parasites are transmitted through intermediate hosts, which themselves may become infected by biting and sucking the blood of the definitive hosts (e.g. tsetse flies that transmit sleeping sickness), or by swallowing appropriate forms of the parasites (e.g. pigs, cattle and tapeworm eggs).

When insect vectors of parasitic diseases become infected during a blood meal on an infected host, the parasites develop and mature within the insect, and are then transmitted to a new host (or hosts). Control of the insect vectors is therefore an important method of controlling transmission of such infections. There have been instances of blood parasites being transmitted directly (i.e. without development), either when a blood-sucking insect has been disturbed during a blood meal and has moved immediately to another host to complete the meal, or when infected blood has been used in a transfusion or has been transmitted by contaminated syringes or needles (e.g. those used by drug addicts).

Protozoa and helminths are rarely transmitted by inhalation or contact, but close personal contact may result in transfer of arthropod ectoparasites (e.g. lice, mites, fleas); these may also be vectors of other agents of infection (including viruses and other micro-organisms).

There is some evidence that infection with *Enterobius vermicularis* (p. 157) may occasionally result from inhalation of the eggs, which are very light and easily distributed in the air; by far the more common route of infection is by ingestion of the eggs.

Suggestions for Further Reading

Warren, K.S. and Mahmoud, A.A.F. 1984. *Tropical and Geographical Medicine*. McGraw Hill, New York.

5
Pathogenic Mechanisms

Pathogenicity, a micro-organism's capacity to produce disease, has three components—**transmissibility, infectivity** and **ability to damage the host** (p. 32). Transmissibility has been discussed in Chapter 4, along with some aspects of infectivity—notably the 'infective dose' (p. 33). In this chapter we shall look at pathogenicity as a whole, and in particular at the mechanisms of infectivity and of damage to the host.

Proof of Pathogenicity

There are helpful similarities between our uses of the words 'pathogen' and 'criminal'. Some people are known to the police as specialists in particular forms of crime; others as more versatile wrong-doers; and others as generally law-abiding citizens who are liable to occasional lapses. To be a known criminal is not to be incapable of any other sort of existence; and to be unknown to the police is not necessarily the same as being innocent.

Similarly, some micro-organisms, such as *Corynebacterium diphtheriae* and *Clostridium tetani*, are known to be responsible for characteristic diseases; others, such as *Staphylococcus aureus* and *Streptococcus pyogenes*, can cause many different forms of disease; and others, such as *Str. viridans*, are usually harmless but in special circumstances may become pathogens. Most known pathogenic bacteria are capable of existing as harmless commensals; and undoubtedly there are many species with as yet unsuspected pathogenic activities.

To establish beyond doubt that a given organism causes a given disease may be difficult. It is not enough to show that it is constantly present in an appropriate distribution in each case of the disease, for its presence may be a result rather than the cause of the disease. According to the classical criteria of pathogenicity commonly known as **Koch's postulates**, it should be possible (1) to show that the organism is constantly present; (2) to grow it in artificial culture media; and (3) to reproduce the disease in susceptible animals by administering such cultures to them. However, there are many diseases to which these criteria cannot be applied, but which can be confidently attributed to particular organisms. For example, *Treponema pallidum* cannot be grown in culture and does not produce in animals anything closely resembling syphilis, yet its association with that disease is so constant that nobody doubts its causative role.

The Meaning of Virulence

Even when a particular microbial species is certainly pathogenic, this property is not necessarily, or even usually, shared by all strains of the species. It is common to find bacterial strains that are closely similar and undoubtedly of the same species, yet one is pathogenic to certain hosts and the other is not. The term **virulence** is often used in an attempt to quantitfy pathogenicity, but caution is needed here. It is sometimes convenient to be able to describe a strain as highly virulent, or of reduced virulence, or avirulent; but we can give mathematical expression to virulence only if we define carefully the conditions under which it is measured. The lethal dose (LD) of an organism for a host is a particular example of an effective dose (ED) as defined on p. 33, and we learn something about the virulence of a bacterial strain if, for example, we determine the number of bacteria that, when administered to each of a large batch of closely similar mice, will kill 50% of them (the LD_{50} for those mice). However, even this information is only of value for comparison with the LD_{50} of another strain grown under the same conditions in the same medium for the same length of time and administered in the same way to a batch of mice of the same strain which are strictly comparable with the first batch in regard to age, sex, size, nutrition, past experience of infection and any other features that may be relevant. Such a comparison will not necessarily tell us anything about the relative virulence of the two strains for another host species or even for mice of a different breed or age.

As is clear from what we have said (p. 33 and above) about infective, effective and lethal doses, the outcome of a given infection depends to a large extent on the number of organisms initiating it. In an epidemic, or in an outbreak of common-source illness such as food-poisoning after a party, it is often found that some of those exposed have symptomless infections, some are mildly ill and some are more severely affected. In general such variations are mainly due to differences in the doses of pathogens reaching individuals, though other factors such as host immunity can be important.

The impossibility of comparing virulence without specifying the host is well illustrated by the human and bovine tubercle bacilli (*Mycobacterium tuberculosis* and *Myco. bovis*). Both are pathogenic for man and for guinea-pigs, but cattle and rabbits are far less susceptible to human than to bovine strains. Inoculation of mycobacteria into guinea-pigs provides an indication of their virulence for man; to use rabbits for this purpose would be grossly misleading.

The virulence of a microbial strain for a given host species may decrease progressively when the strain is maintained in laboratory culture or in an unrelated host species; such a strain is described as attenuated. Thus the *Bacille Calmette–Guerin (BCG)* is a bovine tubercle bacillus strain attenuated by prolonged artificial culture. When injected into humans it causes only local lesions, but stimulates the development of immunity effective against natural tuberculosis (p. 365).

Pathogenicity of Bacteria

Bacterial pathogens vary widely in the number and nature of the properties that enable them to infect and damage hosts. For example, virulence may depend on a single recognizable feature as shown by the fact that some strains of *C. diphtheriae*, typical in all respects except that they do not produce the diphtheria toxin, are avirulent. In many other species, however, virulence is determined by complex interaction of several factors.

Nevertheless, we can make a number of generalizations about **characters that bacterial pathogens must have**:

(1) Apart from a few special cases of diseases caused by microbial products (e.g. botulism, which is due to ingestion of the pre-formed toxin in food), a bacterial pathogen **must first colonize the appropriate entry site** on the body and initiate an infection. Some species (e.g. *Vibrio cholerae, Bordetella pertussis*) attach to the surfaces of epithelial cells and multiply there without entering the cells. Others (e.g. shigellae, the causative agents of bacillary dysentery) penetrate the epithelial cells but have little or no tendency to spread further. In such 'hit-and-run' infections microbial multiplication and shedding occur before the specific host defences have been mobilized. Most bacterial pathogens **must penetrate further than the epithelial layers** to produce their ill effects—e.g. *Salmonella typhi* which has to pass from the intestine into the blood to produce typhoid fever. Some bacteria enter host phagocytic cells which cannot digest them, and then travel around the body inside these (p. 42). In many instances such invasions are prevented by host defences, but the risk of generalized infection is much increased when these are impaired—e.g. by underlying disease or treatment with immunosuppressive drugs (p. 318).

Endogenous infections due to the host's commensal organisms appear at first sight to be exceptions to these rules; but such organisms have originally come from outside the host, and their change to pathogenicity usually involves penetration into his tissues or transfer to another part of his body, or results from impairment of host defences.

(2) A pathogen **must be able to multiply in or on the host's tissues.** Put the other way round, this means that the host's tissues must supply appropriate nutrients, atmospheric conditions and temperature for the pathogen's growth. Here we can see in broad outline the facts which determine the host ranges of all pathogens, and indeed of all parasites, but we can fill in very few of the details. Similar factors undoubtedly play a large part also in deciding the distribution of parasites within the body of the indiviudal host, and the sites at which pathogens produce their characteristic lesions. For example, the fact that *Myco. ulcerans* and *Myco. marinum* can multiply only in the temperature range 30–33°C presumably accounts for their ability to produce lesions only in the skin of man and not in deeper tissues; and the distribution of leprosy lesions, due to the related *Myco. leprae*, probably has a similar explanation. Abnormal conditions in the tissues of a host may suit organisms which could not otherwise grow there. Thus the abnormal susceptibility of diabetics to infections probably results from the high glucose content and other chemical pecularities of their tissues; and gas-gangrene, caused by anaerobic organisms of the genus *Clostridium*, occurs in tissues which have lost their blood supply and so their source of oxygen. The role of erythritol in localizing brucella infection in bovine abortion (probably the most clearly understood example of the influence of host tissue constituents) is considered on p. 100. In most cases, however, we do not know what determines the localization of organisms in microbial lesions.

(3) It is self-evident that to be a pathogen a bacterium **must be able to damage the host's tissues**. This can happen in many different ways. Some bacteria may cause local or more remote damage by releasing endotoxins or exotoxins (p. 17). Others produce enzymes that destroy cells or tissue components (p. 17). Tissue damage may also be due to host hypersensitivity reactions to the organism rather than to its toxicity.

(4) In order to be able to do any of the things mentioned so far, a pathogen **must be able to resist and overcome the host's defence mechanisms**. The bacteria must survive first the antibacterial mechanism of the skin and mucosal surfaces and then, after penetration into the body, the internal cellular and humoral defences. Ways of doing so are discussed below, notably on p. 41. Ability to resist the non-specific bactericidal action of host serum (i.e. action directed against many bacterial species and not dependent on prior immunization against the species in question) may be a major factor in determining why some bacterial species or strains are virulent.

Bacterial Virulence Factors

Of the bacterial properties which have been identified that are directly associated with pathogenicity, some allow colonization and infection by the pathogen and others are concerned with damaging the host.

Adhesion and **attachment** to epithelial surfaces is the essential first step in most infections and some bacteria possess special surface structures known as **adhesins**. These include the pilus-like K surface antigens of *Escherichia coli* and the pili of gonococci, although it is not clear whether the fimbriae possessed by shigellae and other enterobacteria facilitate their attachment to the intestinal mucosa. The lipoteichoic acid in the fimbriae of group A haemolytic streptococci enables them to bind to host epithelial cells. The dextran secreted by some oral viridans streptococci (e.g. *Str. mutans*) attaches the cocci to the teeth and promotes the build-up of dental plaque; and similar extracellular material has been demonstrated in association with skin staphylococci. A less specific but possibly important factor in attachment is that the surface charges of bacterial and host cells may be mutually attractive or repellant.

Adhesion does not lead to colonization unless the bacteria are able to **resist various protective antibacterial factors** encountered on the skin and mucosal surfaces, such as the fatty acids in sweat and sebum, the lysozyme in tears and the combination of gastric and intestinal digestive enzymes in the alimentary tract. *Staph. aureus* is able to be the pathogen most commonly associated with skin and wound infections because it is resistant to the bactericidal action of skin secretions. Most bacterial pathogens of the intestinal tract need to be ingested in large numbers for some of them to survive the actions of the host's digestive secretions. They may be protected by the buffering effect of food. Patients with reduced gastric acid secretion are much more susceptible to such infections.

After initiation of infection, pathogenicity of bacteria is usually described in terms of **toxigenicity** and **invasiveness**, although there is considerable overlap between the two.

In some infections, virulence is determined almost exclusively by **toxigenicity**. Botulism (p. 258) is a disease entirely due to bacterial toxin, but it is not an infection—the toxin is ingested in food. Tetanus (p. 290) and diphtheria (p. 192) are infective diseases in which the causative organisms remain localized in their initial sites of infection (in wounds or in the throat respectively) but produce potent toxins (p. 17), and these are responsible for causing the major systemic effects of the diseases in parts of the body remote from the infection sites. Toxin-neutralizing antibodies in the patients's blood can prevent such effects. In scarlet fever the causative species, Str. pyogenes (p. 77), is itself an invasive organism that may cause widespread tissue damage and septicaemia, but the skin rash and some of the systemic symptoms are caused by its erythrogenic toxin. A group of toxins concerned with the pathogenesis of diarrhoeal disease are known as **enterotoxins** (p. 17). The most potent is produced by Vibrio cholerae, a non-invasive organism (p. 255), and its action is to stimulate adenylcyclase activity in the intestinal mucosal cells with consequent loss of fluid and electrolytes and the profuse watery diarrhoea characteristic of cholera. A similar toxin is produced by some strains of Esch. coli that cause diarrhoea (p. 252). Other toxins are responsible for the forms of food-poisoning caused by Cl. perfringens (p. 257) and Staph. aureus (p. 257).

All of these toxins described above are **exotoxins** (p. 17), secreted into their surrounding environment by living bacteria. Most are produced by gram-positive bacteria, though some of the enterotoxins are from gram-negative species. Exotoxins are proteins with specific enzymic activity by which they cause damage. They have the physico-chemical characters of enzymes (i.e. they are denatured and inactivated by heat, by formalin and by various other chemicals, and are active only within a restricted pH range), and they are effective in very small amounts (i.e. they are very potent). Like most proteins, they are antigenic, and they can be neutralized by specific antibodies.

A second group of toxins that contribute to bacterial virulence are known as **endotoxins** (p. 17). These are produced exclusively by gram-negative bacteria and are derived from the lipid-containing part of the gram-negative cell wall (p. 10). Chemically, all endotoxins are high molecular-weight lipopolysaccharides (LPS) and comprise a constant lipid core (lipid A) to which is attached a complex and variable polysaccharide portion. Unlike the exotoxins, endotoxins are not actively secreted by living cells and do not have specific enzymic activity, but they do have profound biological effects on living animals. Their most characteristic effect is pyrogenicity (production of fever); this is a universal effect of the injection of endotoxin or of endotoxin-containing bacteria into man or animals because endotoxin causes the release of interleukin 1 (endogenous pyrogen—p. 63) from phagocytic cells (polymorphs and macrophages). Endotoxins also trigger the coagulation enzyme cascade and may cause widespread (disseminated) intravascular coagulation (DIC); this is sometimes seen at its most dramatic in the bleeding into skin and deep tissues that occurs in overwhelming septicaemia due to gram-negative bacteria, notably the meningococcus (Waterhouse–Friderichsen syndrome). DIC may cause profound circulatory disturbance, but endotoxin itself also causes a fall in blood pressure, reduction in venous return and cardiac output and an increase in heart rate, giving a weak thready pulse in a cold and clammy patient. This is the clinical state of 'shock', known in these circumstances as **septic**, or **bacteraemic**, or **'gram-negative'** or **endotoxic shock**.

The pathogenic mechanisms of primarily invasive bacteria are less clear-cut. Some species penetrate the skin or mucosal surfaces fortuitously, through a wound or small area of damage. Many, however, can penetrate intact epithelium. Of these, some invade between the host cells, disrupting the intercellular attachment and gaining access to the deeper tissues, whereas others stimulate endocytosis in the epithelial cells and are transported across the cells in endocytic vacuoles.

Once within the tissues, these invasive organisms cause damage by various means. Many produce extracellular enzymes, known as **aggressins**, that act upon tissue components. These include proteases, some of which have general proteolytic activity and others more specifically destroy collagen (collagenases), elastin (elastases) or fibrin; lipases, phospholipases and phosphatases; neuraminidase; and hyaluronidase ('spreading factor') which breaks down the hyaluronic acid that forms the basis of much of the extracellular tissue substance. Other aggressins are specially toxic for particular types of cells—e.g. leukocidins or haemolysins which attack white and red cells respectively. These extracellular enzymes differ from the classical exotoxins in having more general enzymic properties, in being produced by the organism in conjunction with other exo-enzymes, and in not being the sole agents responsible for the disease processes. They are antigenic and stimulate the host to produce antibodies (e.g. antistreptolysin, antihyaluronidase), though so far it has not been possible to use them in treatment or prevention of the diseases, as can be done with the toxins of tetanus, diphtheria and other exotoxic diseases. However, the distinction between aggressins and exotoxins is not absolute. For example, the phospholipase (lecithinase) produced by *Cl. perfringens* is largely responsible for the tissue destruction in gas-gangrene due to that species, and can be regarded as a potent exotoxin; polyvalent antitoxins (containing antilecithinase) are sometimes used in the treatment of this condition, though their effectiveness is doubtful.

Invasive bacteria have various ways of overcoming host defences. The polysaccharide capsules of *Str. pneumoniae* and some other pathogenic bacteria protect them against phagocytosis (p. 48), and the invasiveness and virulence of such organisms are directly related to both the amount and the nature of the capsule.

However, the capsular material is antigenic and coating of this capsule with specific antibody (opsonization—p. 53) neutralizes its protective action and renders the cells accessible to phagocytosis; this is important in the natural resolution of the infection.

Some bacteria that are susceptible to phagocytosis are not eliminated because they are resistant to phagocytic killing (p. 49). *Myco. tuberculosis* can persist for long periods within macrophages and the intracellular survival of *Salmonella typhi* and of brucellae is an important feature of infection with these organisms, with important implications for therapy because intracellular bacteria are often inaccessible to antibiotics.

Hypersensitivity

It is customary to regard pathogens as the attackers and host responses as the defensive line; but in some bacterial infections the main manifestations of the disease may result from damage done to the host by his own stimulated immune system—i.e. it is due to his hypersensitivity reactions (pp. 61–2). Much of the tissue destruction in tuberculosis is the result of cell-mediated immunity against the infected cells, as is the nerve damage in leprosy (Type IV hypersensitivity). The circulating immune complexes found in the blood in many infectious diseases may well be responsible for late manifestations of infection—e.g. glomerulonephritis and reactive arthritis (Type III hypersensitivity).

Pathogenicity of Viruses

Most of man's virus pathogens are primarily human parasites, but some have other animal species as their principal hosts. A host is susceptible to infection by any virus that can replicate in its cells, but to cause disease a virus in most cases has to be able to survive on and penetrate a mucous surface of the host, and it will need to be able to avoid or interfere with the host's defence mechanisms, replicate in host tissues, reach susceptible cells and cause damage to them.

Attachment, Penetration and Dissemination

Skin is an effective barrier to the entry of viruses into the body unless they are introduced during an episode in which its integrity is breached—e.g. by a needle or by an insect or animal bite. The more usual first stage of virus infection is attachment of the virus, by means of a surface protein, to a virus-specific receptor on the surface of a susceptible host epithelial cell, commonly located on the mucous membrane of the respiratory, the gastro-intestinal or the genito-urinary tract. Provided that it can run the gauntlet of the local defence mechanisms (p. 46) the virus can then enter the cell, where it is safe from being carried away in the mucus layer. In many instances a phase of local replication then takes place, followed by tissue invasion and spread through the body to the target tissues; or there may be further replication in some central site and further blood-stream dissemination before the final target organs are reached. To produce the lesions of the disease in these target sites the viruses have again to invade cells, and must replicate there sufficiently to cause tissue damage. The incubation periods of virus infections (and of some others) can in some cases be explained in terms of these various cycles of replication that precede development of the final disease pattern. Still further replication may take place before virus particles are shed from the body, and are then available for transmission to other susceptible hosts. However, other viruses—particularly those that cause respiratory or gastro-intestinal tract infections—merely replicate in superficial mucosal cells, without invasion of deeper tissues, and are then shed directly from these sites; such infections characteristically have short incubation periods (under 7 days).

After penetrating the surface epithelium a virus meets phagocyctic cells, and may be carried via the lymphatic system to lymph nodes. Phagocytosis may lead to its destruction, but measles virus, adenoviruses and herpesviruses can replicate inside macrophages and lymphocytes and can use them as vehicles in which to travel round the blood-stream out of reach of humoral antibodies (p. 53) and complement (p. 57).

Invading viruses show marked tissue- and organ-specificity—e.g. for salivary glands (mumps virus, cytomegalovirus), for the breast or kidneys (cytomegalovirus) or for the respiratory tract (measles); replication in these sites leads to excretion of virus in saliva, milk, urine or respiratory tract secretions respectively. Skin rashes and arthritis may result from localization of virus in these sites but are more likely to be due to deposition of immune complexes (p. 61).

Blood-borne invasion of the placenta may lead to virus infection of the fetus, with results that range from negligible damage to abortion, perinatal death or congenital malformation. The nature and extent of fetal damage depends largely on the stage of pregnancy at which the infection occurs, since developing organs (especially eyes, ears, brain and heart) are particularly vulnerable during the first three months of pregnancy and the fetal immune system is not developed at this stage.

Nervous system invasion by viruses can follow various routes. Some blood-borne viruses localize in cerebral vessels and invade adjacent brain tissue, whereas others penetrate the capillaries of the blood–CSF junction in the meninges or the choroid plexus. Rabies virus travels to the central nervous system from the body surface or tissues along peripheral nerves. Herpes simplex and varicella–zoster (chickenpox–shingles) viruses travel up sensory nerves from the surface to sensory ganglia and then, in recurrent herpes or in shingles, return to the surface by the same route.

Cell Damage

Some viruses damage the infected cells by shutting down macromolecular synthesis—i.e. they are cytopathic. Others are less obviously harmful, causing only subtle alterations in cellular function and growth rate; though in the extreme case of oncogenic viruses these alterations may amount to transformation into malignant cells. Still other viruses are non-cytopathic, replicating in the host cells but causing no obvious damage.

The extent to which the host is affected by

virus-inflicted cell damage depends on the numbers of cells involved and their location and functional importance. For example, the shutting down of protein synthesis in the motor neurone cells of the anterior horn of the spinal cord, which is caused by poliovirus infection, results in death of the neurones and consequent paralysis of the muscle groups supplied, which may have dire consequences for the host. Cell damage may occur at the site of virus entry, during generalized spread of the infection or after localization in specific target organs.

Persistent Virus Infections

Some viruses are able to evade or interfere with host defence mechanisms and remain in the body for long periods (p. 42). Some, such as polyoma and adenoviruses, can remain latent in this way for many years without any obvious effect on the host. Others—notably the herpes simplex and varicella–zoster viruses mentioned above in connection with nervous system infections—can also persist in a latent non-infectious form but may be reactivated from time to time to produce symptomatic and transmissible infections; such reactivation is usually associated with impairment of general health or immunity or with other precipitating factors.

Immunosuppression by Viruses

This is described in Chapter 24 (p. 314).

Interference

Host cells infected with one virus may be resistant to infection with a second virus. This phenomenon of interference does not depend on the two viruses being closely related, and in some cases it is not even necessary for the first virus to be active; after inactivation by heating or by exposure to ultra-violet light it may still protect the host cells containing it from subsequent infection by an active virus. Interference is sometimes due to an activity of the host cells—production of **interferons** (pp. 49 and 58). These do not prevent the second virus from entering host cells, but they do prevent it from diverting host-cell ribosomes into making the proteins essential for its replication. In other cases interference is a direct result of the activities of the first virus; it may have destroyed all of the receptor areas on the surfaces of the cells so that further viruses are unable to attach themselves, or it may have taken complete control of the enzyme systems of the cells and directed them to its own reproduction, so that a second virus may enter the cells but finds no available enzymes. However, it is possible for two viruses to exist and multiply together in the same cell—e.g. herpes simplex and vaccinia, in the nucleus and the cytoplasm respectively.

It is easy to see in theory the possible relevance of the interference phenomenon to the prophylaxis of virus diseases, but its possibilities have not yet been widely exploited. Oral administration of live poliovirus vaccine can be rapidly effective in preventing spread of an epidemic, because the vaccine virus establishes itself in the recipient's intestine to the exclusion of wild poliovirus strains (p. 370). Conversely, pre-existing natural enterovirus infections may impair the success of such oral vaccination with live poliovirus.

Hypersensitivity

Much of the damage resulting from virus infections is caused by the host's own responses, mainly type III (immune complex) hypersensitivity reactions (p. 61). Circulating immune complexes are probably formed in most virus infections as part of the normal response that leads to resolution of the infection; and immune complex reactions are frequently involved in the inflammatory processes and the associated cell damage. Persistent virus infection may result in immune-complex deposition in blood vessels, skin, joints, the choroid plexus or renal glomeruli, with consequences—such as rashes, arthritis or glomerulonephritis—that may be serious. Systemic immune-complex reactions may trigger disseminated intravascular coagulation, which is a feature of arbovirus haemorrhagic fevers (p. 122) and of the late stages of yellow fever (p. 122). Type IV T-cell-mediated hypersensitivity reactions (p. 62) are responsible for the rashes of many virus infections (e.g. measles) and for much tissue damage that results from killing of virus-infected cells by the host.

Pathogenicity of Fungi

Many fungal infections of man are superficial in location and present no serious threat to general health, though they may be highly inconvenient. Notable amongst them are the dermatophyte infections (**ringworm**, p. 138) of skin, hair and nails, in which only the dead keratinized outer layer of the skin is invaded by the fungi, though the consequent inflammatory reaction involves the epidermis, dermis and hair follicles; secondary bacterial infection may be superimposed. Keratin destruction and increased epidermal replacement results in scaling; the characteristic 'ring' lesions are the result of elimination of fungus from the central areas by the inflammatory response. Inoculation of an appropriate fungus through the skin may lead to invasion of the dermis and subcutaneous tissue, resulting in a **mycetoma** (p. 287). Widespread systemic fungal infection (**deep mycosis**) may be the result of a primary invasive infection of a healthy person by a virulent fungus such as *Histoplasma* (p. 139); but such a condition can also be caused by a relatively avirulent fungus causing an infection in a patient with impaired general health or with impaired immunity. Such opportunist infections are seen with increasing frequency in neutropenic or immunosuppressed patients; *Candida* or *Aspergillus* species (pp. 139–41) are commonly responsible.

Hypersensitivity to fungi can cause skin rashes remote from the fungal lesions—e.g. in dermatophyte infections. Respiratory allergies to fungi are discussed on pp. 141–2.

Pathogenicity of Protozoa and Helminths

The mechanisms of disease production by these more complex parasites are themselves extremely complex. They include mechanical, physiological and immunological components, and they vary considerably according to the parasitic species concerned. The site occupied by the parasite is important; if this is the usual one, the resulting disease may be relatively minor compared with the effects produced by the same parasite in unusual sites—e.g. a few *Paragonimus* eggs in the brain may cause far more serious effects than a much larger number in their more usual situation in the lungs. The severity of disease caused depends not only on the number of parasites present, but also on the physiological state of the host. Any outside cause resulting in a lowering of general health will predispose to more serious consequences of parasite attack. Given numbers of parasites in the blood or tissues (e.g. in malaria or in trichinosis) generally cause more serious disease than do similar numbers belonging to species that remain in the intestinal lumen (e.g. *Giardia*, *Trichuris*). Those which multiply within the body tend to be far more harmful than those which do not, and those which become greatly enlarged in the host's body are often more harmful than those which are present in much larger numbers but remain small. Although the factors determining the degree of pathogenicity in any given case of infection may be complex and interlocked, it is convenient to discuss the mechanisms of pathogenicity under the following headings: (1) mechanical; (2) traumatic and invasive; (3) physiological and toxic; (4) immunological.

(1) Mechanical

Mechanical blockage can be a major component of the disease caused by some parasitic species—e.g. obstruction of the gut lumen by large numbers of *Ascaris* or by smaller numbers of tapeworms; obstruction of intestinal absorption by large numbers of *Giardia* covering the wall of the small intestine; or obstruction of lymphatics, with resultant elephantiasis, by filarial parasites of the genera *Wuchereria* and *Brugia*. Obstruction clearly depends more on parasite biomass than on numbers; thus two or three large tapeworms may have more serious effects than a much larger number (but a smaller mass) of intestinal protozoa. Increasing mass is also important when the parasites are within the tissues. When eggs of either of the tapeworms *Taenia solium* or *Echinococcus granulosus* reach the human brain and develop into growing cysts, pressure from these as they grow may cause epilepsy. Hydatid (*E. granulosus*) cysts may reach volumes of 1–2 litres, and such masses can cause severe damage to any organs in which they occur.

(2) *Traumatic and invasive*

Amoebic dysentery is an example of a parasitic disease in which tissue destruction—manifested in this case as ulceration of the intestinal wall—is the major component, though the host may also be damaged by toxins produced by the amoebae and by blood loss secondary to the ulceration. The ulceration also provides a portal of entry for bacteria, and the same is true of the skin damage caused by skin-penetrating helminths such as *Strongyloides* and hookworms which can permit entry of tetanus spores, for example.

(3) *Physiological and toxic*

Gut helminths are well placed to help themselves to nutrients required by their hosts, whose diet is then deficient. A special example of this is the pernicious anaemia caused by the tapeworm *Diphyllobothrium latum* which, as well as causing mechanical and other ill effects, takes up vitamin B_{12} from the host's gut and so deprives him of it. Other forms of anaemia can result from blood loss (particularly common in hookworm infection) and from red-cell destruction, notably by malarial parasites. Failure of fat absorption is a major component of the effect of heavy *Giardia* infections mentioned above.

Another type of physiological interference is that caused by parasite toxins. Some parasites produce metabolites that may have profound effects on the host. Absorption of such parasite products by the host from the gut lumen is thought to be responsible for some effects of helminth infections—e.g. oedema in *Clonorchis* infection, irritability and general toxic manifestations associated with the presence of various gut parasites. Malarial parasites are thought to produce a metabolite with a vasoconstrictor action. *Trypanosoma cruzi* causes the most serious effects of Chagas' disease by means of a neurotoxin which it secretes. This affects the autonomic nerves supplying the smooth muscle of organs including the oesophagus, the colon and the heart, which in consequence become grossly enlarged, often fatally. (There may also be an immunological component to this condition—see below.)

(4) *Immunological*

Most protozoan and helminth parasites stimulate an immunological response by the host. These are seldom protective, but as in other forms of infection they may be harmful to the host. An auto-immune reaction (p. 62) may follow attachment of parasite antigen to host cells, and result in destruction of these cells; this is the probable mechanism of the massive red-cell destruction, despite relatively small parasite numbers, that occurs in malarial blackwater fever (p. 242). Allergic reactions can also be important—e.g. if a primary sensitizing leakage of hydatid cyst fluid into the patient's tissues is followed by a further major leak (from cyst rupture or surgery), anaphylaxis may result (p. 271). The effects of the neurotoxin in Chagas' disease, mentioned above, may be supplemented by an auto-immune reaction that causes degeneration of autonomic ganglia in the gut wall and consequent mega-oesophagus and megacolon. In schistosomiasis the immunological reactions of the host—at first humoral and then of the delayed hypersensitivity cellular type (p. 62)—produce inflammatory reactions around the parasite eggs in the host tissues; the infiltration of inflammatory cells—mainly lymphocytes, macrophages and eosinophils—is the precursor of space-occupying granulomas that form around the eggs and subsequently become fibrosed.

Suggestions for Further Reading

Mims, C.A. 1982. *The Pathogenesis of Infectious Diseases.* 2nd edn., Academic Press, London.

Crewe, W. and Haddock, D.R.W. 1985. *Parasites and Human Disease.* Edward Arnold, London.

6

Host Defences and Immunity

The human body is protected against potential pathogens both by **non-specific defences**, effective against wide ranges of organisms, and by **specific immunity**, effective only against particular organisms.

Non-specific Defences

As we have already seen, in order to become a pathogen an organism needs in nearly all cases to gain access to host tissues. Many of the barriers to achieving this are non-specific.

Superficial Defences

One important defence mechanism that is in operation on mucous membranes as well as in the tissues and the blood stream—phagocytosis—is discussed in the next section (Cellular Defences). The other main defences of the body's various surfaces are as follows.

SKIN This is both a mechanical barrier to micro-organisms and, by virtue of the fatty acid content of sweat and sebum, a death-bed for many of them. However, hair follicles and glands form comparatively weak points in the defences, and by multiplying in these *Staph. aureus* many give rise to pustules, boils and carbuncles. This same bacterial species, being resistant to the bactericidal action of the skin secretions, is commonly present on the surface and may be carried into the subcutaneous tissues by anything which pierces or lacerates the skin. The importance of the skin barrier is emphasized by the high frequency of superficial infections when its efficiency is impaired—e.g. by burning or by vitamin deficiency—and when it is by-passed by abrasions, puncture wounds, etc.

CONJUNCTIVAE These are less of a mechanical obstruction, but they are constantly washed by tears and wiped by the movement of the eyelids. Tears contain an enzyme **lysozyme** (also present in most other body fluids and in polymorphonuclear leukocytes) which, acting with specific antibodies and complement (p. 56), can lyse bacteria.

MOUTH Saliva contains antimicrobial substances such as lysozyme and lactoferrin. Other defences of the oral cavity are an intact mucous membrane, crevicular fluid, and mechanical flushing that removes bacteria not adherent to epithelial cells or the teeth.

ALIMENTARY TRACT Most micro-organisms are killed or damaged by gastric secretions, particularly the acid component; notable exceptions are *Mycobacterium tuberculosis*, enteroviruses and protozoan cysts (and helminth eggs). This barrier can be overcome if a large dose of pathogenic organisms is ingested, and a much smaller dose may suffice to establish infection if acid secretion is reduced. Hence neonates, and also patients who have recently undergone some forms of gastric surgery, have an increased susceptibility to infection with salmonellae and other ingested pathogens; and regular use of antacid tablets has been shown to increase susceptibility to brucellae in contaminated unpasteurized milk. Establishment of infection is also hindered by peristalsis and rapid transit through the intestine, and by the need for the potential pathogens to compete with the resident bacterial flora for attachment to epithelial cells.

RESPIRATORY TRACT Breathing involves inhalation of very large volumes of air every day, and therefore of considerable numbers of suspended micro-organisms. Larger particles are filtered out in the nose by impinging on hairs or on the sticky mucous membranes. The large and complex normal bacterial flora of the upper respiratory tract constitutes an important barrier to colonization by extraneous organisms (see p. 29), and interference with the balance of this population by use of antibacterial drugs can have undesirable consequences here, as also on other heavily populated surfaces. Of the particles that reach the lower respiratory tract, only those under 5 μm in diameter can penetrate as far as the alveoli. Some are large or irritant enough to provoke reflex bronchoconstriction and coughing, which expels them. Mucus secretion by the goblet cells and mucus-secreting glands combines with ciliary action of the epithelial cells to provide a remarkably efficient clearance system for other inhaled particles and micro-organisms. These are caught on the sticky mucus blanket (where they are exposed to antibacterial substances such as lysozyme and lactoferrin) and are swept upward by the 'mucociliary escalator', to be coughed out or swallowed and dealt with by the gastric secretions. Toxic agents such as cigarette smoke and also some virus infections inhibit ciliary action and so predispose to lower respiratory tract infections. The clearance mechanism is also defective in bronchi damaged by chronic bronchitis or bronchiectasis; and some patients with congenitally defective ciliary function suffer from recurrent or chronic infections of the respiratory tract, paranasal sinusitis or otitis media. Organisms that succeed in reaching the alveoli are usually disposed of by the alveolar macrophages, which may be activated by signals from sensitized T lymphocytes (p. 58).

VAGINA Here there is no mechanical cleansing system, but in the adult there is an important hormone-dependent chemical defence mechanism, put into effect by the characteristic normal flora as explained on p. 29. As in the upper respiratory tract and elsewhere, disturbance of the normal flora by antibiotic treatment can allow endogenous pathogens to proliferate—notably *Candida albicans*.

Cellular Defences

Micro-organisms that get through the outer defences and begin to multiply within the tissues usually provoke inflammatory reactions. The vigour and pattern of these depend, among other things, on the nature of the organism and on the previous experience of the host. For the first few hours of an acute inflammatory respose there is a phase of increased vascular permeability, with extravasation of plasma and leukocytes—neutrophil polymorphonuclear leukocytes (neutrophils) in the initial stages, followed later by mononuclear cells. Phagocytosis (see below) by these leukocytes, with the assistance of other non-specific local defence factors such as complement and free iron, may result in early elimination of the invading organisms, or in marked reduction of their numbers. During the subsequent phase of decreased vascular permeability, antimicrobial drugs may be prevented from reaching the tissues in effective concentrations. Bacteria that survive the local inflammatory response may spread via the lymph nodes, and if not eliminated there they can reach the blood. Presence of bacteria in the blood is called **bacteraemia**, and may progress to a serious and perhaps fatal generalized infection if the bacteria are not removed by phagocytes in the liver, spleen and bone marrow. (For further consideration of bacteraemia, and its relation to **septicaemia**, see p. 60). Viruses, fungi, protozoa and some of the helminths can also reach the blood-stream, with results indicated in the appropriate chapters.

Pus

This common result of microbial (usually bacterial) infection is a complex mixture of surviving or dead micro-organisms and leukocytes (neutrophils in acute infections), together with remnants of broken-down host cells, all suspended in tissue fluid.

Phagocytosis

The engulfing of micro-organisms by host cells is called **phagocytosis**. An effective mechanism for dealing with the first stages of an infection requires the rapid mobilization of appropriate cells (**phagocytes**). As we have seen, the leukocytes that predominate in early inflammatory exudates are **neutrophils**. These are produced in the bone marrow, and released into the circulation after a two-week maturation process. During this time they develop intracytoplasmic granules that contain enzymes and other factors responsible for mediation of the inflammatory response and for their microbial activity. Neutrophils have a short half-life (6–7 hours) in the circulation. During acute inflammation their ability to adhere to and migrate across the vascular endothelium is enhanced—hence their appearance in the exudate. They are therefore in most circumstances the front-line troops that deal with tissue invasion. When neutrophils encounter immune complexes (p. 61) or biologically active fragments of complement such as C3a or C5a (p. 57) they degranulate, releasing powerful enzymes that cause local tissue damage. This process is extremely important in the pathogenesis of immune complex diseases (p. 61).

The other major group of phagocytes are the monocytes and larger macrophages which constitute the **mononuclear phagocyte** (or **reticulo-endothelial**) **system**. These are found free in the blood, free or fixed in the tissues of the lymph nodes or the spleen, fixed in the liver (Kupffer cells), on the pleural and peritoneal surfaces, and in the alveoli of the lungs. As tissue macrophages they have a life of several weeks. They are strategically placed for filtering organisms out of the lymph and the blood, removing them from the pleural and peritoneal fluid, and dealing with those that reach the alveoli. However, some organisms can live and even multiply inside macrophages (p. 58), using them (like the Trojan horse) to carry infection to other parts of the body.

Phagocytes migrate to the site of infection in response to chemical stimuli (**chemotaxis**). Products of the micro-organisms themselves may be either directly chemotactic or capable of generating chemotactic stimuli by activating the complement system (p. 56).

Phagocytosis begins with **attachment** of the micro-organism to the surface of the phagocyte. This is facilitated by the presence of opsonins (p. 53), which can attach to the surface of the organism and for which there are specific receptors on the surface of the phagocyte. Attachment is followed by **engulfment** by the cell, with formation of an intracellular vacuole, the **phagosome**. The cytoplasmic granules (lysosomes) then fuse with the vacuole, discharging their enzymes into it and forming a **phagolysosome**. The mechanism by which phagocytes kill micro-organisms is not yet fully elucidated, but it involves the enzyme myeloperoxidase, a burst of oxidative metabolism and the generation of hydrogen peroxide and superoxide. The ability of phagocytosing neutrophils and monocytes to produce superoxide can be measured in the nitroblue–tetrazolium (NBT) test (p. 313).

Macrophages attracted to an infection site and activated there have a number of functions other than phagocytosis. They may be activated by lymphokines from sensitized T lymphocytes (p. 58), by immune complexes (p. 61) or through activation of the complement system, especially C3 (p. 57); they in their turn liberate substances that activate biological amplification mechanisms such as the complement, kinin, coagulation and fibrinolytic systems. They also secrete a number of other substances that are important in the processes of acute inflammation—prostaglandins and leukotrienes; thromboplastin, which may be important in the development of disseminated intravascular coagulation (p. 40); and monokines (comparable to lymphokines) including interferon and interleukin 1 (p. 58). The last-named of these acts on the hypothalamus, producing fever, and also controls the release of neutrophils into the blood and the generation of inflammatory-cell exudate at an infection site.

The importance of the cellular defences against infection is clearly illustrated by the numerous infections that occur in patients with quantitative or qualitative defects in their phagocytes, as described on pp. 311–3. Some organisms, however, have their own counter-defence against phagocytosis. The polysaccharide capsules of bacteria such as pneumococci allow them to escape phagocyte attachment except in the presence of specific anticapsular antibodies. Some species—notably *Staph. aureus*—produce leukocidins, which are cell-bound, or soluble toxins that kill phagocytes and are in large measure responsible for the number of dead neutrophils often found in pus. As mentioned above, some organisms are ingested by phagocytes but then survive inside them; *Trypanosoma cruzi* and certain enveloped viruses achieve this by escaping from the phagosomes into the cytoplasm, whereas *Mycobacterium tuberculosis* and *Toxoplasma gondi* can prevent phagosome–lysosome fusion and thus escape digestion. Micro-organisms that can survive inside phagocytes can be killed only by the combined action of macrophages and specifically sensitized cytotoxic T lymphocytes (p. 54).

Interferons

These constitute a group of proteins with broad antiviral activity (within the right host species) and other biological activities. They render host cells resistant to virus infection (p. 43), and are the first line of defence against viruses, since they are detectable within a few hours of the initiation of infection—whereas protective antibodies do not appear for several days. Interferons confer protection on cells near to those that produce them by blocking virus replication, but viruses differ in their susceptibility to such blocking action. Three types of interferon are now recognized: α interferon, produced by leukocytes; β-interferon, produced by fibroblasts; and γ-interferon (Type 2 or immune type), the most important of the three, produced by stimulated T and B lymphocytes (p. 58). γ-interferon enhances the activity of the other two types and of natural killer cells.

Natural killer (NK) cells

These are large granular lymphocytes, distinct from T and B lymphocytes, that can act independently to destroy any 'foreign' cell—e.g. a virus-infected or a cancer cell—without any prior sensitization, but do so far more effectively in the presence of interferon or interleukin 2 from activated T or B lymphocytes.

Specific Immunity

Introduction

The term immunity originally related to protection, and early immunology was concerned with the study of that specific resistance to further infection by a particular micro-organism which follows an initial encounter with that organism or its products, or with closely related substances. However, these protective responses are only part of a more general phenomenon—the initiation of processes, as a result of meeting a foreign substance (an **antigen**—p. 50), which cause the body to react differently to it or to closely related substances on subsequent encounter. These altered reactions may be either more or less vigorous than the original, and they may be either protective or harmful—or in some cases both at once. They may in some circumstances become directed at the body's own cells—'auto-immune' disease. They have a special importance in relation to organ transplantation. Immunology is now a large subject, full of excitement. In this chapter we can give only a brief review of those aspects of it that are most relevant to infection, and we shall of necessity present current beliefs and concepts with little or none of the supporting data and often in a simplified form. For a fuller understanding our readers should consult immunological works such as those quoted at the end of the chapter.

Specific immunity to infection is the resistance of an individual to a particular disease as a result of his ability to defend himself against its causative agent. It depends upon his genetic inheritance, his age, his general health and his past experience of micro-organisms. It may be innate or acquired (naturally or artificially).

Innate immunity

Each individual inherits certain susceptibilities and resistances peculiar to his species, his race and his family, and has his own personal combination of these. Presumably they depend upon his tissue chemistry, his superficial and cellular defence mechanisms, and possibly non-secific humoral agents. As he matures and ages, all of these factors vary through hormonal and other influences, so that immunity changes with age quite apart from the contributions mentioned in the next two paragraphs.

Acquired immunity

(1) **Naturally acquired**. Naturally occurring clinical or subclinical infection commonly provokes responses on the part of the host which enhance his ability to resist the causative organism when he meets it again. These responses involve production of agents that react specifically with the organism or its products—either special proteins called **antibodies** or special sensitized cells (the effect of which is described as **cell-mediated immunity**) or both. Immunity due to the host's own responses is called **active**, whereas immunity conferred by maternal antibodies that enter the infant's circulation via the placenta or via colostrum or milk is called naturally acquired **passive** immunity because the infant itself has made no contribution to it.

(2) **Artificially induced**. Artificial stimulation of specific resistance to infection is called **immunization**, and is discussed more fully in Chapter 28. This also may be **active** or **passive**, in that the subject may be provoked to make his own antibodies or may be given some ready-made.

Genetic control of immunity

Resistance to infection, innate or acquired, is under genetic control. The nature and intensity of immune responses to antigenic stimulation are determined by sets of **immune response genes**, and some of these are closely linked to the genes that control the histocompatibility antigens involved in graft rejection. There is an association between possession of some histocompatibility antigens and liability to certain forms of infection—e.g. between possession of antigen HLA-B27 and development of reactive arthritis following salmonella, yersinia or campylobacter infections (p. 303). The mechanisms of such connections are far from clear, but possible explanations of their existence include (a) bacterial antigens that resemble host HLA antigens and are therefore exempt from host attack (p. 62); or (b) close linkage between the genes determining the HLA antigens and those controlling susceptibility to such infections.

Antigens

An **antigen** is a substance capable of stimulating an immune response in a host. It is recognized as foreign by small lymphocytes that are **immunocompetent** (i.e. capable of mounting immunological responses). The response is specifically directed against particular determinants on the antigen molecule. As the molecular weight of antigen molecules increases, so their **immunogenicity** (i.e. ability to induce immune responses) is also increased. Small molecules which become immunogenic only when coupled to carriers are called **haptens**; although in the uncoupled state they cannot provoke immune responses, they can react directly with appropriate **antibodies** (host proteins with specific affinity for antigens—p. 52). Indeed, most naturally occurring antigens can be viewed as consisting of carrier portions (usually protein) that render them immunogenic, and haptenic portions (usually non-protein) that determine the specificity of the immune responses.

All but the smallest micro-organisms consist of hundreds or even thousands of antigens, but only a few of these are important in the induction of immunity. Those of greatest importance, and therefore of greatest relevance to our understanding of resistance to infection, are located superficially on the organisms and so are readily accessible to the host's immune mechanisms. They include many determinants of pathogenicity such as those that we considered on p. 41—capsular and other surface polysaccharides; the M proteins of group A haemolytic streptococci; adhesins; and influenza virus neuraminidases and haemagglutinins. Antibodies that react with these substances interfere with their contributions to pathogenicity. Similarly, bacterial toxins provoke the production of specific antibodies (**antitoxins**) that neutralize their toxicity to the host. Virus-infected host cells develop new surface antigens—either virus components or modified host components—that are recognized as foreign and attract the attention of cytotoxic T lymphocytes (p. 54).

The Basis of Immunological Responses

The lymphoid system

All cells involved in immunological reactions—lymphocytes, granulocytes and macrophages—are derived from a common ancestor, the pluripotent stem cell found in the embryo yolk sac, in the foetal liver and finally in the bone marrow. Immunological responses are initiated by cells that morphologically are small lymphocytes. Some stem cells mature in the **bone marrow** to become **B lymphocytes**, whereas others migrate to the **thymus** and there mature into **T lymphocytes**. From these **primary lymphoid organs** the mature lymphocytes move out to the lymph nodes and spleen—**the secondary lymphoid organs**. Most of the mature lymphocytes are then in constant migration, from lymph nodes via the lymph to the blood and back into the lymph nodes through the walls of the postcapillary venules. B and T lymphocytes occupy different regions while in the secondary lymphoid organs; B lymphocytes are found in the red pulp of the spleen and in the primary and secondary follicles of the lymph nodes, whereas T lymphocytes are found in the peri-arteriolar sheaths of the spleen and in the paracortical areas of the lymph nodes. B lymphocytes are responsible for **humoral immunity**—i.e. that immunity which is determined by antibodies. T lymphocytes are of central importance in **cell-mediated immunity**; but they can also modulate the activities of B lymphocytes in ways described on p. 54.

In defence against infection a most important part is played by lymphoid tissues which are associated with mucosal surfaces and so exposed to continual bombardment with micro-organisms and their products. They are responsible for initiating responses to local infections, whereas foreign antigens or organisms that enter the blood are removed by phagocytic cells of the spleen, liver, lungs and blood and the appropriate responses are initiated largely in the spleen.

Specificity and diversity

Host immunity to infection is based on adaptive responses of the lymphoid system, which eliminate or restrain the infecting organisms. Responses to a particular organism have a specificity that depends on the molecular structure of its surface antigens. Organism-specificity is often not absolute, as cross-reactions can occur with antigens of similar configuration on the surfaces of other organisms and elsewhere. The existence of specificity of response can be explained in terms of the **clonal selection theory**. Each B lymphocyte or T lymphocyte has on its surface antigen-binding sites specific for a particular antigenic configuration; in the case of B lymphocytes these sites are in fact on immunoglobulins. The first step in an immune response is the processing of the antigen by macrophages, which then present it in recognizable form to the immunocompetent lymphocytes. Among these there is such diversity of antigen receptors that any of a wide range of foreign susbtances can find at least a few of them capable of binding it. The 'selected' lymphocytes react to contact with the antigen by dividing rapidly, initiating clones of new cells, each clone producing a response that has the same antigen-specificity as the binding sites on the initiating lymphocyte. Because an antigen is bound most avidly by lymphocytes with receptors that give the best 'fit', a small amount of a new antigen stimulates highly specific responses, but a larger amount is more than enough for the most relevant lymphocytes and the excess will be available to bind with and stimulate others that are less appropriate.

Although all lymphocytes stimulated by a given antigenic configuration initiate responses of roughly or more precisely the same antigen-specificity, the nature of those responses may show great diversity in other respects. If both B and T lymphocytes are involved, both antibody production and cell-mediated responses will result. Furthermore, each **individual** B lymphocyte that is stimulated initiates production of antibody of only one Ig class (see below); but stimulation of a **population** of B lymphocytes may well result in production of antibodies of various immunoglobulin classes and subclasses. Still further diversity of responses to a given antigen molecule results if different areas of its surface can act separately as antigenic determinants.

Immunological memory

Immunological responses have a built-in 'memory' that is the basis of long-term immunity and of active immunization (Chapter 28). When specific immunocompetent cells are recruited by an antigen (on its own or as part of an organism) from the recirculating pool, they cease to circulate; instead, they set up an immunological response locally, by dividing and differentiating to form populations of effector cells. In the case of the B lymphocyte series these are antibody-secreting **plasma cells**, and in the case of the T lymphocyte series they are **specifically sensitized T lymphocytes**. Such antigenic stimulation also results in formation of **memory cells**—long-lived specifically sensitized recirculating lymphocytes that are responsible for the faster and greater response to the antigen that occurs when it is encountered again.

Antibodies

Antibodies are a distinct family of proteins with specific activity against antigens (p. 50). They are secreted by plasma cells (see above).

Structure

Antibodies are globulins, and are alternatively known as **immunoglobulins** (Igs). Five major classes are distinguished on the basis of physicochemical, serological and biological properties and of amino-acid sequences; they are designated IgG, IgM, IgA, IgD and IgE. The molecular structure of IgG is illustrated diagrammatically in Fig. 6.1. Two heavy and two light polypeptide chains, linked by interchain disulphide bonds, form a Y-shaped molecule. Papain cleavage of the molecule divides it into two Fab (antigen-binding) fragments with identical antigen specificity and an Fc

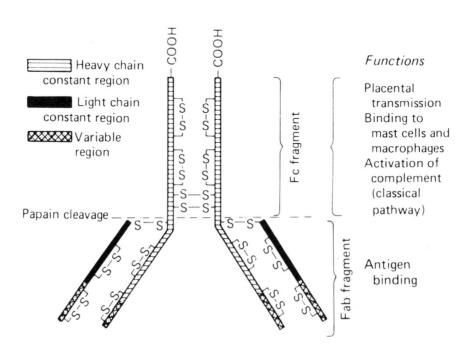

Figure 6.1 Schematic representation of the 4-chain structure of human IgG (molecular weight approximately 150 000).

(crystallizable) fragment which contains virtually all structures responsible for the other functions of the antibody. The amino-acid sequences of the N-terminal residues (the antigen-binding sites at the ends of the Fab fragments) vary considerably between immunoglobulin molecules, and determine their antigen-binding specificities; but the remaining parts of the chains are much more constant. IgA, IgD and IgE molecules in the blood are similar in structure to IgG, but **secretory IgA** found in mucus and other secretions consists of two such units bound to an additional component (p. 55). An IgM molecule consists of 5 Y-shaped units linked by their Fc regions, so that each such molecule has 10 antigen-binding sites; this structure makes it particularly efficient in binding complement (p. 57). Three-quarters of the total immunoglobulin of normal human serum is IgG, and this is the Ig class that passes most readily from blood into extravascular fluids. Such movement is greatly enhanced during inflammatory processes, when the permeability of blood-vessel walls is increased.

Functions of antibodies in protection against micro-organisms

At various points in this section reference is made to the complement system and its many components (designated C1, C2 etc.). These are explained in the section that begins on p. 56.

OPSONIZATION **Opsonins** are substances which combine with surface components of micro-organisms or other particles and increase their susceptibility to phagocytosis (p. 48). Particles with surfaces more hydrophilic than those of the phagocytes are resistant to phagocytosis; and many pathogenic bacteria have surface structures—usually proteins, glycoproteins or polysaccharides—which by virtue of their negative charges make the surfaces hydrophilic. They can be opsonized (by derivation the word means prepared for the table) by attachment of antibodies or complement factors that make their surfaces more hydrophobic. IgG and IgM can each be opsonic, and in each case the antibody molecules first become attached by their Fab portions to their specific antigens on the microbial surfaces. Then in the case of IgG the Fc portions of the molecules become attached to special Fc receptors on phagocyte surfaces. IgM on the other hand, after attachment to its antigen, does not bind directly to phagocytes but fixes and activates complement by the classical pathway (p. 57), generating C3b, which then binds to C3b receptors on the phagocyte surfaces. Binding of the (organism + antibody + complement) complex to the phagocyte surface triggers engulfment, degranulation and a burst of oxidative metabolism (p. 48). The importance of opsonization as a defence against capsulate pathogenic bacteria is illustrated by the high frequency and severity of pneumococcal infections in patients with antibody deficiencies (**hypogammaglobulinaemia**, pp. 316–7).

BACTERIOLYSIS In some circumstances attachment of antibodies to the surfaces of gram-negative bacteria is followed by activation of complement components C1–C9 and consequent production of 'holes' in the bacterial cell walls. Complement factors and host enzymes such as lysozyme can enter such a cell, causing damage to the cytoplasmic membrane, loss of its selective permeability and therefore death of the cell. Under the electron microscope, holes similar in appearance and method of production to those in gram-negative cell walls can be seen in the envelopes of viruses that possess such structures (p. 12). Gram-positive bacteria are resistant to complement-mediated lysis, even though complement is fixed on their surfaces; and capsulate gram-negative bacteria have decreased susceptibility in proportion to the thickness of their capsules. IgM, having more ability than IgG to fix complement, is more effective in bacteriolysis. The importance of complement-mediated bactericidal systems in protection is illustrated by the increased incidence of severe gonococcal or meningococcal infections in individuals whose serum is congenitally deficient in the late complement components C5–C8 and is therefore devoid of bacteriolytic activity.

TOXIN NEUTRALIZATION As exotoxins are important virulence factors for some bacterial species, so neutralization of such toxins by antibodies (**antitoxins**) is an important defence mechanism. IgG antibodies of appropriate specificity are highly effective for this purpose, but not IgM. Secretory IgA antitoxins are important in protection against toxins released by intestinal pathogens. Antitoxins act by steric inhibition of the reaction between toxins and target cells, and can do no good once the toxin is bound to its target. Consequently, therapeutic administration of antitoxin (p. 000) must be started as early as possible and becomes less effective the longer it is delayed; nor is it effective as soon as it is given intravenously, since it has to pass from the blood into the relevant tissues.

VIRUS NEUTRALIZATION The spread of virus infection can be reduced by the attachment of antibodies to free virus particles—secretory IgA when they are on mucous surfaces and mainly IgG when they are in blood or tissue fluids. Once inside host cells, viruses are protected from such neutralization until they emerge again. Neutralization is effected either by coating the virus particles with antibody or by aggregating them by antibody cross-linking, and so reducing the number of infectious particles. Neutralization is potentiated greatly by the participation of the early components of complement; unlike lysis, this effect does not require components C5–C9.

ANTIBODY-DEPENDENT CELL-MEDIATED CYTOTOXICITY (ADCC) 'Killer' cells of several types differ from 'natural killer' cells (p. 49) in that they are active only against IgG-coated micro-organisms or cells. Their killing action is extracellular—i.e. it does not depend on phagocytosis. They include monocytes, macrophages, cytotoxic T lymphocytes (p. 58) and also eosinophils, which are important in extracellular destruction of parasites too large for phagocytosis (p. 59).

Placental transmission of maternal antibodies

IgG is the only immunoglobulin class that crosses the human placenta. A neonate has in its blood antibodies that protect it only against organisms to which the mother is immune by virtue of making IgG antibodies. Thus neonates are poorly protected against *Escherichia coli* and other gram-negative bacilli that stimulate production mainly of IgM; and if removed to special care units or other environments that contain potential pathogens not previously encountered by their mothers, they are particularly liable to succumb to overwhelming infections. Placental transmission of antibodies in humans increases from about the 20th week of pregnancy to reach a maximum in about the 35th week. The length of time for which a neonate is protected against a particular infection depends, among other things, on the level of relevant IgG antibody in the mother's blood in the later stages of pregnancy. Antibody molecules undergo natural degradation, and not much maternal immunity is left by the time that the infant is 3 months old. The period between the loss of passive maternal protection and the establishment of active immunity (natural or artificial) is a time of special hazard for all infants. Congenital defects in the infant's own immunological systems often present themselves clinically as recurrent bouts of infection beginning at about 3 months old. This is also the optimum age for starting some forms of active immunization (see Chapter 28), as the presence of maternal antibodies before this time may interfere with the infant's own immune responses to antigens.

The role of T lymphocytes in antibody production

Activation of B lymphocytes by some antigens—e.g. influenza virus haemagglutinins—requires 'help' from a special subset of T lymphocytes called **T helper cells**. Production of IgG, IgA and IgE is particularly dependent in this way on the thymus. In contrast, other antigens—including bacterial capsular polysaccharides and the endotoxins of gram-negative bacteria—can stimulate B lymphocytes unaided. The antibody produced is then mainly IgM, and there is little or no memory following a thymus-independent response of this kind. Another subset of T lymphocytes, **T suppressor cells**, can moderate the intensity of immune responses, and these may be responsible for the immunosuppression (involving both humoral and cell-mediated immunity) seen in many chronic infections, including the acquired immune deficiency syndrome (AIDS, p. 128).

The mucosal antibody system

As we have seen (p. 51), lymphoid tissue associated with mucous membranes is of particular important in defence against microbial invasion. The lymphocytes of the intestinal epithelial layer are predominantly cytotoxic T lymphocytes (p. 58) or T suppressor cells, but in the lamina propria are found B lymphocytes and T helper cells. Aggregates of specialized lymphoid tissue in the intestine (Peyer's patches or gut-associated lymphoid tissue, GALT) and in the respiratory tract (bronchus-associated lymphoid tissue, BALT) are covered with 'microfold' epithelium adapted for taking 'antigenic samples' of the microbial flora of these two regions. B lymphocytes stimulated by antigen in these areas are transformed into large blast cells which migrate via the lymphatic system to the blood and thence to any of the secretory mucosal tissues in the gut, genito-urinary tract, respiratory tract, salivary and lachrymal glands and lactating breast. They thus provide the basis for **a common mucosal antibody system**. The antibodies are produced by plasma cells derived from the blast cells and now situated in the mucosal lamina propria. They are IgA molecules, and two such molecules coupled to a secretory piece provided by the epithelial cells constitute **secretory IgA**, which is highly resistant to attack by the many proteolytic enzymes of the gut. This is the dominant immunoglobulin in external body fluids, and the degree of surface immunity of an individual correlates with secretory IgA levels rather than with serum antibody levels. It is probable that secretory IgA protects by preventing adhesion of pathogens or binding of toxins to epithelial cells.

New-born babies for the first weeks produce little IgA but plentiful IgM. Some individuals have a permanent selective inability to make IgA. In both of these situations secretory IgM takes over the role of secretory IgA, and usually plays it very successfully. IgG penetrates to surface secretions only when the mucous membranes are inflamed; it may then provide a useful secondary defence system; but it may exacerbate the local inflammatory process by forming immune complexes (p. 61).

Secretory IgA and breast milk

Secretory IgA is the dominant immunoglobulin of human milk. It protects the infant's intestinal mucosa against a wide array of pathogens experienced by the mother, since the common mucosal antibody system described above ensures the presence of antibodies reflecting her intestinal and other exposure. This is probably the reason why breast-fed babies have fewer intestinal infections than those fed artificially; and why, particularly in countries with relatively poor standards of hygiene, weaning is often followed by gastro-intestinal infections.

Age and antibody responses

The human fetus begins to make antibody (IgM) at about 11 weeks of gestation and immunocompetent T cells are detectable at about 14 weeks. This is too late to deal with some infections that reach it through the placenta. For example, when a non-immune mother develops rubella during the first trimester of pregnancy, dissemination of the virus in the fetus is unchecked and serious damage may be done to the developing organs. Fetal IgM and maternal IgG appear in the fetal circulation after the damage is done, but the child is born with a congenital rubella infection (p. 322). Presence of IgM in the cord blood (necessarily fetal, as IgM cannot cross the placenta) is evidence of such infection.

From the time of birth the neonate begins to meet innumerable new antigenic stimuli and potentially pathogenic organisms. IgA production is negligible at first and does not reach adult levels till puberty; but the breast-fed infant enjoys protection given by maternal IgA (see above), as well as that from placentally transmitted IgG (p. 54). By degrees a pool of memory cells is built up that enables the infant to respond rapidly to subsequent infective challenge; but it seems that the ability to respond to some kinds of antigen takes months or years to develop. For example, bacterial polysaccharides from pneumococci, meningococci or capsulate

Haemophilus influenzae, encountered naturally or as vaccines, elicit little or no antibody response from children under the age of 18–24 months—a serious matter in relation to protection against *H. influenzae* type b, since this is a common cause of meningitis and other life-threatening infections in the first year or so of life.

Time scale of antibody responses

This subject is dealt with more fully on p. 362, in relation to artificial immunization; but for the purpose of understanding the host's immunological responses it is important to appreciate that production of an effective level of antibodies to an antigen not previously encountered by the host almost always takes at least 7–10 days.

Monoclonal antibodies

The diversity of clones of antibody-producing cells formed in response to any antigenic stimulus has been described on p. 51, and has presented many problems to research workers and to those who use antibody preparations in diagnostic tests (p. 175). Modern hybridization techniques have made it possible to produce, and maintain *in vitro*, clones of cells all derived from a single antibody-producing ancestor and therefore all producing the same antibodies. Applications of such monoclonal antibodies are mentioned at various points in this book, but will undoubtedly figure far more prominently in any subsequent editions, when more of the potential of this important technical advance has been realized.

Complement

Though it is often convenient, especially in the context of diagnostic tests involving complement fixation (p. 176), to refer to complement as though it were a single substance, it is in fact a complex system of some 20 different serum

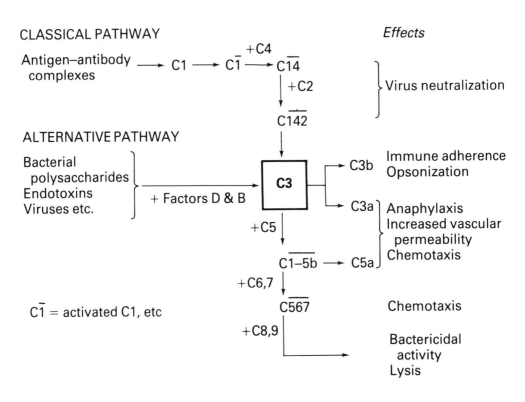

Figure 6.2 *Simplified scheme of complement activation, indicating those effects of complement components that relate to resistance to infection.*

proteins, most of them biochemically characterized, which are capable of interacting with each other, with antibodies, and with cell membranes. The individual proteins are present as functionally inactive molecules that are activated in sequence. The existence of complement has been known since the beginning of the century, but awareness of the important role of this complex biological amplification system in the body's antimicrobial defence is far more recent. An orderly sequence of interactions between the protein components of complement leads to the generation of many biologically active substances, and ultimately to the lysis of bacterial or other cells. The possible sequences are indicated in Fig. 6.2. The system can be activated in either of two ways:

(1) **The classical pathway**. IgM or some IgG antibodies, when combined with their specific antigens, can bind the C1 component, activating it ($\overline{C1}$ in the figure means activated C1, and so on) and so activating the whole sequence.

(2) **The alternative pathway**. A wide range of micro-organisms and their components, and of other substances, can by-pass the early stages of the sequence and activate C3 directly.

In either case the event of central importance in the system is the splitting of C3 into two biologically active components—C3a and C3b. The effects of complement activation which are of greatest importance in defence against infection are:

(a) **Immune adherence**. C3b in particular promotes adherence of micro-organisms or antigen-antibody complexes to phagocytes, as described under Opsonization on p. 53.
(b) **Biological action on host cells**. C3a and C5a both cause release of histamine from mast cells and C5a and $\overline{C567}$ cause chemotactic migration of polymorphonuclear cells to areas of antigen-antibody interaction, and thus promote both local inflammation and phagocytosis of invading micro-organisms.
(c) **Bacterial cell-wall lesions**. Activation of the complete system through to C9 leads to the formation of 'holes' in cell walls, as described under Bacteriolysis on p. 53. For bacteriolysis to occur, lysozyme is also required.

Complement also enhances virus neutralization by specific antibodies, and the antibody-mediated killing of some of the large parasites. In schistosomiasis, for example, alternative-pathway activation at the surfaces of larvae can attract eosinophils, which then kill the antibody-coated larvae (ADCC, p. 54).

Much of our information about the activities of complement has come from *in vitro* studies, but rare congenital deficiencies of individual complement components also throw much light on the role of the complement system in defence against infection. Opsonization, chemotaxis and lysis are all impaired in patients with total C3 deficiency, and they are liable to overwhelming infections with capsulate bacteria such as pneumococci. As we have seen on p. 53, patients with deficiencies of C5, C6 or C8, in whom bacteriolysis is much reduced but other antimicrobial functions are normal, are particularly susceptible to disseminated gonococcal or meningococcal infection.

Two other host mediator systems—the fibrinolytic system and the kinin system—are commonly activated along with the complement system, but homeostatic mechanisms usually limit the intensity of the reaction. However, massive activation sometimes occurs, with serious effects on the patient. Bacterial endotoxins can activate these three systems, with resulting tissue damage, shock and disseminated intravascular coagulation (DIC) and in many cases death of the patient. Such mechanisms probably underlie some of the most serious syndromes associated with microbial infections—acute haemorrhagic adrenal necrosis (the Waterhouse–Friderichsen syndrome, p 40), 'gram-negative shock' (p. 40) and the haemorrhages and shock associated with the terminal stages of infection with the haemorrhagic fever viruses (pp. 122, 124).

T Lymphocytes and Cell-Mediated Immunity

We have already considered on p. 54 two of the functional subsets into which T lymphocytes can be divided—T helpers and T suppressors, which respectively promote or inhibit the immune responses. Using monoclonal antibodies (p. 56) it is now possible to type the different functional subsets of T cells on the basis of their surface antigens (designated T1–T11). Thus T helper cells carry the T4 antigen (the receptor for HIV p. 128) and T suppressors have the T8 antigen. Other subsets are directly involved in cell-mediated immunity. Viruses, some bacteria—notably mycobacteria, brucellae and salmonellae—and various other micro-organisms can cause infections that are relatively resistant to the body's humoral defences (antibodies, complement, etc.) because the infecting organisms are within host cells, inaccessible to antibodies. Dealing with such intracellular infections is largely the job of T lymphocytes of appropriate antigen-specificity. They respond to the antigenic stimulus in a number of ways:

(1) On encountering their specific antigens some sensitized T lymphocytes secrete a range of soluble substances called 'lymphokines, which act as means of communication between different cells of the immune system. These include:
(a) macrophage chemotactic factor, which attracts macrophages to the site of infection;
(b) macrophage-migration-inhibiting factor (MIF), which prevents them from going away again;
(c) activators of macrophage cidal activity, against micro-organisms and infected or altered host cells in general, not just against the antigens that stimulated the T lymphocytes.

The 'angry' macrophages resulting from (c) themselves secrete other substances that increase the vigour and antimicrobial effectiveness of the inflammatory reaction.
These are:
(i) interleukin 1, which enhances the production of lymphokines by lymphocytes;
(ii) eosinophil and basophil (mast cell) chemotactic factors, which may be particularly important in helminth infections;
(iii) mitogenic factor, which non-specifically stimulates bystander T and B lymphocytes, thereby enlisting their help;
(iv) γ-interferon, which enhances the activity of T and B lymphocytes and of natural killer (NK) cells (p. 49) and also plays an important role in limiting the early spread of virus infections; and
(v) interleukin 2, which also enhances the killing capacity of NK cells.

(2) Such activities are often accompanied by T-cell-mediated Type IV hypersensitivity reactions to the infecting organisms (p. 62).

(3) Cytotoxic ('killer') T cells (p. 54) become detectable 2–4 days after infection with a virus not previously encountered by the host—i.e. several days before such a primary infection has produced any effective antibody response. Thus, while such cytotoxic cells can do nothing to prevent the initial infection, they are of major importance in preventing its further spread and eliminating it. They kill the infected host cells by direct contact, not through the mediation of lymphokines, though lymphokine-producing cells provide valuable help by promoting an inflammatory reaction, stimulating macrophage activity and producing γ-interferon.

When an infecting virus is not itself strongly cytopathic, the T cell response may be more important than the infection in determining the disease manifestations.

Local cell-mediated immunity is independent of its systemic counterpart, and local application of live vaccines is more effective than their parenteral administration in generating local cell-mediated immunity. However, immunity generated in this way is short-lived. Local administration of vaccines consisting of killed organisms or purified microbial products evokes little or no immune response.

Patients with defective T cell immunity are liable to severe generalized infections, particularly with herpes or pox viruses, mycobacteria or fungi (p. 315). It is dangerous to give them live vaccines.

Some Special Features of Immunity in Protozoan and Helminth Infections

The large size of most protozoa and of helminths means special difficulties for the host in killing and disposing of them. It also means that they have many surface antigens. This problem is compounded by variation in these antigens at different stages of the life cycles of the organisms and by the ability of some organisms—notably the trypanosomes—to produce a series of antigenic variants, that can evade the specific immunity induced by earlier forms and so cause waves of recurrent illness (cf. relapsing fever, p. 232, and influenza, p. 125). Another means of evasion, illustrated by the schistosomes, is to incorporate host components such as erythrocyte glycolipid into the parasite's outer layers, so that it is not recognized as 'foreign'.

Although infection with these complex organisms provokes formation of a myriad of different antibodies, few of them play any part in host defence— though some provide a basis for useful diagnostic tests. Massive antigenic challenge often leads to hypergammaglobulinaemia—notably in visceral leishmaniasis and in African trypanosomiasis. In malaria, IgG antibodies (helped by T lymphocytes) confer some protection by opsonizing the merozoite stage and reducing erythrocyte invasion, but they are ineffective against intra-erythrocytic forms. High IgE levels in the blood are common in helminth infections, as are increased numbers of eosinophils both in the blood and at infection sites. The IgE is one of the factors that stimulates the release of histamine and other pharmacologically active substances by mast cells (p. 58), with consequent local tissue inflammation, increased blood-vessel permeability and accumulation of eosinophils. The eosinophils, assisted by complement, can kill some antibody-coated helminths, such as schistosome larvae. Thus IgE may be useful in defence against helminths, but it can be harmful, as in the anaphylactic reaction (p. 61) that can follow rupture of a hydatid cyst and release of its contained antigens into a body cavity (p. 268).

Cell-mediated immunity is important in limiting the extent of infection in leishmaniasis, toxoplasmosis and schistosomiasis, and may be responsible for resistance to reinfection in American trypanosomiasis. Specifically sensitized and stimulated T lymphocytes have an important role in initiating and orchestrating the non-specific inflammatory reactions that deal with these larger parasites, notably the eosinophil and mast-cell responses mentioned in the previous paragraph. On the other hand, cell-mediated immune responses may be harmful (as they can be in infections due to other types of organism). For example, they are responsible for the fact that deposition of schistosome ova in the liver may be followed by granuloma formation and cirrhosis of the liver, leading to the formation of oesophageal varices and the possibility of fatal haemorrhage.

Herd Immunity

As we have seen, an individual's liability to become ill or to die as a result of exposure to a particular pathogen depends largely on his own immunity. His risk of being so exposed, however, depends largely on the level of immunity of those around him. This in turn depends on the community's experience of the pathogen and of artificial immunization. In a population unprotected by artificial immunization, each new generation is exposed to the endemic pathogens—with fatal results in some cases, but survivors are in general immune. Major epidemics can occur when the rate of exposure to the organism in question has fallen low enough (perhaps as a result of improvements in general hygiene) for there to be a large proportion of non-immune people in the population. Artificial immunization (discussed in Chapter 28) breaks this pattern, allowing maintenance of a high level of immunity in the community (herd immunity) without the risks of natural infection or the need for the organism to remain endemic. The proportion of the population who must be immune to avoid the risk of epidemics varies with the organism, but is always high. A constant danger is that, when a disease becomes uncommon, enthusiasm for immunization against it may flag, with resultant build-up of a susceptible population adequate to sustain epidemic spread. It is therefore a cardinal rule of preventive medicine that, when a potentially epidemic disease has been brought under control, herd immunity must be kept high by a comprehensive immunization programme until all risk of infection has been eliminated— i.e. until the disease has been eradicated, which so far has been achieved only in the case of smallpox.

Some Possible Results of Infection

A consideration of all possible outcomes of encounters between pathogens and host defences would be a review of the entire field of microbial diseases. The discussion would be further complicated by the need to take into consideration many different factors in the host's condition and circumstances which may affect the issue, notably those reviewed in Chapter 24 (The Compromised Host). However, it may be helpful to outline and illustrate a few of the possible outcomes here.

(1) **Elimination of the pathogen without any clinical lesion.** This is undoubtedly the commonest outcome of infection by a pathogen, but for obvious reasons it is virtually never observed. If the elimination is not too rapid, it may be possible to demonstrate retrospectively that infection has occurred, because shortly afterwards specific antibodies appear in the host's blood or specific tissue hypersensitivity develops (pp. 184–5).

(2) **Localization of the pathogen with production of a local lesion.** This is clearly illustrated by the common small staphylococcal pustule of the skin. There is no impairment of the patient's general health, tissue damage is confined to the immediate vicinity of the pathogen's point of entry, and as a rule the infection is soon eradicated.

(3) **Localization of the pathogen with production of distant lesions.** *Corynebacterium diphtheriae* is usually itself confined to the throat, but by means of its exotoxin it can produce distant lesions in the heart and nervous system. *Clostridium tetani*, growing as a rule in subcutaneous tissue or muscle, also does serious distant damage by means of a neurotoxin.

(4) **Extension of infection to surrounding tissues.** Because it produces hyaluronidase, *Streptococcus pyogenes* is particularly liable to spread rapidly through connective tissue surrounding a primary lesion, causing the diffuse inflammation known as cellulitis. Actinomycosis and caseating tuberculosis spread in quite a different way, advancing slowly through the tissues and causing severe destruction as they go.

(5) **General dissemination.** Entry of bacteria or viruses into the blood-stream (**bacteraemia, viraemia**) is common, but is often of little or no importance. They may arrive from local lesions via the lymphatic system (p. 47), but bacteria can enter the blood directly from body surfaces as the result of various events or procedures—e.g. dental extraction or even vigorous tooth-brushing or chewing; urinary tract catheterization or other instrumentation, particularly if the urine is infected; and difficult defaecation. Most such invasions are dealt with effectively by the reticulo-endothelial system; but virulent organisms or those that arrive in large numbers may overwhelm the defences and cause serious illness, sometimes rapidly fatal. The term **septicaemia** has had a confusing career, but is best taken to mean the serious clinical condition (including shock) associated with the presence of large numbers of pathogens in the blood (p. 47). Patients who survive the early stages of significant bacteraemias or viraemias are likely to develop multiple local lesions in various organs, the localization of these depending largely on the nature of the organism. Survival from bacteraemic illnesses is of course much commoner, and less likely to be complicated by local lesions, when it is possible to oppose the bacteria with lethal or inhibitory concentrations of drugs.

(6) **Chronic infection.** Failure of the host's defences to eliminate a pathogen soon after its arrival may result in persistent active disease. Often, however, there is a balance between the pathogen and the defences, and the infection may remain asymptomatic (latent, p. 31) for many years but turn into active disease again when the balance is shifted in favour of the pathogen. Failure of elimination may be due to defective defence mechanisms (Chapter 24), but in many cases it is the result of special properties of the pathogens. They may be able to cause immunosuppression themselves (e.g. lepromatous leprosy, p. 288), or to perform 'antigenic tricks' such as those described on p. 232, or even, in the case of the herpes simplex virus (p. 130), to

escape detection over long periods by being integrated into the host cell DNA. Some chronic infections are due to organisms that do relatively little harm to host tissues in the short term and are rather weak stimulators of immunity. Interesting variations in immune responses and consequent clinical manifestations result; examples are tuberculosis (p. 206), leprosy (p. 288), syphilis (p. 278) and leishmaniasis (p. 147).

(7) **Damaging immune responses**. These are described in the next section.

Hypersensitivity Reactions

Encounter with a foreign substance sometimes provokes immunological responses that lead to a state of **acquired hypersensitivity** to that substance. Such a state manifests itself in vigorous reactions that occur when the host again encounters the same substance, and such reactions cause damage to his tissues, varying in degree from minor local inflammation to severe illness or even death. Acquired hypersensitivity must be distinguished from **idiosyncrasy**, an abnormal sensitivity to a pharmacologically active compound (e.g. aspirin or morphine) which does not depend on previous experience of the compound and is often familial.

Gell and Coombs classified hypersensitivity rections into four main types, differing in their underlying mechanisms.

Type I (anaphylactic-type or immediate hypersensitivity) reactions

These reactions depend on special antibodies, known as reaginic or homocytotropic antibodies and belonging predominantly to class IgE; these become attached by their Fc regions to the host's mast cells. Cross-linking of two cell-bound IgE molecules by the sensitizing antigen (**allergen**) leads to release of histamine and other pharmacologically active substances from the mast cells. These substances are responsible for the clinical manifestations of this type of hypersensitivity, which range from mild eczema through hay fever, asthma and urticaria to acute anaphylactic shock (which is sometimes fatal). Anaphylaxis in man is rare, but can follow injection of foreign serum protein (e.g. antiserum from horses), of vaccines or of drugs (notably, in the microbiological context, of penicillins) in previously sensitized patients.

If a small amount of the sensitizing agent is injected intradermally into an already sensitized subject, a local **immediate-type** reaction consisting of a wheal with surrounding erythema develops within a few minutes. Such a procedure is used in identifying the agent or agents to which a particular patient is sensitized, and also illustrates the course of events which follows natural exposure to such agents. Type I reactions occurring in infected tissues can increase vascular permeability and so aid the protective mechanisms by enhancing diffusion of antibodies and other serum factors to the place where they are needed. In allergic lung disease, on the other hand, preceding Type I reactions are probably important in determining the site of deposition of immune complexes and the consequent local Type III reactions. Thus patients with allergic bronchopulmonary aspergillosis give a dual (Type I + Type III) response to skin or bronchial challenge with aspergillus antigen (p. 141).

Type II reactions

IgM or IgG antibodies, formed in response to infection, may react either with host cell surface antigens resembling their 'proper' targets or with micro-organisms or their products that have become attached to host cell membranes. Such a reaction may lead to complement-mediated cell damage or destruction. As an example, most patients with *Mycoplasma pneumoniae* infection develop 'cold agglutinins'—antibodies active only at low temperatures and with an affinity for the I antigen found on adult human red blood cells. These antibodies are useful in the laboratory diagnosis of such infections and are probably responsible for the mild haemolytic anaemia that may occur.

Type III (immune complex) reactions

Formation of antigen–antibody complexes is normally followed by their removal from the circulation by phagocytic cells of the mononuclear phagocyte (reticuloendothelial) system, and is then beneficial to the host.

However, the host may be damaged during this process. Many factors determine other possible sequelae: the nature of the provoking antigen(s); whether exposure is intermittent or continuous and how long it lasts (chronic virus, protozoan or helminth infections are frequently associated with immune-complex manifestations); and the quantity and quality of the antibodies produced (controlled by immune-response genes among other things). Immune complexes formed in the presence of slight antigen excess may circulate widely and be deposited in vascular basement membranes, with harmful effects. Renal glomeruli, the choroid plexus, synovial membranes, the uveal tract and the skin are particularly vulnerable because of their high blood flows and intricate capillary beds. Deposition of complexes can lead to fixation and activation of complement, with chemotactic accumulation of neutrophils that release vasoactive substances and powerful hydrolytic enzymes; local tissue damage results. Common clinical manifestations of such a process are fever, proteinuria, joint pains and skin rashes. The classical example is **serum sickness** which follows repeated injection of horse antiserum (e.g. for protection against tetanus) and the development by the patient of antibodies to horse serum proteins. Local challenge (e.g. by intradermal injection) with an antigen to which the patient is immune, either actively or passively (p. 50), may produce a local **Arthus reaction**, similar in mechanism to systemic Type III reactions. This is maximal after 4–8 hours and takes 12–24 hours to resolve. 'Farmer's lung' and other forms of allergic alveolitis resulting from repeated inhalation of fine organic dusts were formerly thought to be pure Type III reactions to antigens absorbed through the alveolar lining; but it is now clear that the minigranulomata of the lungs which are largely responsible for the typical radiological appearances of the chest in this disease are a form of Type IV reaction triggered by the presence of non-degradable complexes of fungal antigen and antibody. Circulating antigen–antibody complexes are formed in many protozoan and helminth infections and may be responsible for some disease manifestations—e.g. glomerulonephritis in quartan malaria. The measurement of circulating immune complex levels is helpful in the diagnosis of some diseases and in the monitoring of treatment—e.g. in infective endocarditis (p. 237).

Type IV (cell-mediated or delayed-type) reactions

The mechanism of these reactions is as described for cell-mediated immunity (p. 58). The hallmarks of a Type IV hypersensitivity reaction to a skin test (e.g. the tuberculin reaction (p. 184)) are that its development is delayed, taking 48–72 hours to reach the maximum response, and that there is palpable induration, due to lymphocyte and macrophage infiltration of the dermis. It is probable that hypersensitivity of this type arises in all acute and chronic infections, and could be demonstrated by skin tests using appropriate antigens.

While we have dissected out the different forms of hypersensitivity reaction for ease of explanation, it should be appreciated that the immune response of the body to a complex structure such as an infecting micro-organism is liable to include a mixture of some or all of them.

Immunological tolerance and auto-immunity

When an antigen binds to the surface of a lymphocyte, the usual result is stimulation of the cell and an immune response. Under certain conditions, however, the immunocompetent cell may be switched off or deleted, and **tolerance** of the antigen results. The fact that the body's immunological mechanisms do not normally take action against its own constituents (**self-antigens**) is due to suppression or deletion, during fetal life, of lymphocyte clones with receptors for self-antigens. **Auto-immunity** is a breakdown of self-tolerance.

Clone suppression operates mainly against T lymphocytes. B lymphocytes with self-antigen specificity circulate in small numbers in the blood of normal adults. They may be stimulated by certain micro-organisms (e.g. the EB virus of infectious mononucleosis) or by microbial products such endotoxin, since these are liable to stimulate B cells in general, not merely those with specific affinity for them. Thus patients recovering from infectious mononucleosis may have a wide spectrum of **auto-antibodies** in their blood. Micro-organisms with antigenic components closely resembling those on the surfaces of host cells, and viruses that can alter such host antigens so that they appear to be 'non-self', can stimulate T helper cells, which co-operate with the self-reactive B cells to produce auto-antibodies. Chronic massive tissue destruction, as in leprosy, releases intracellular antigens unfamiliar to the host's immunological system, and also enzymes that alter host cell surface antigens; such apparently non-self-antigens may provoke immunological reactions. The mere presence of auto-antibodies does not necessarily result in **auto-immune disease**. When tissue damage does occur, it is mediated through Type II, III or IV hypersensitivity reactions.

Fever

An increase in body temperature is a very common host response to infection. It may well be protective in some circumstances—e.g. by providing an environment too warm for optimal growth of the pathogen. The mechanisms that cause the febrile response are complex. An important part in its initiation in the early stages of an infective illness is played by interleukin 1 (p. 58). This polypeptide hormone-like mediator is produced by mononuclear phagocytes that have been activated by their ingestion of micro-organisms or their products; bacterial endotoxin (p. 40) is a particularly potent activator. Interleukin 1 travels in the blood to various target tissues in which it produces an assortment of effects. The most obvious of these is the induction of fever, which it achieves by stimulating increased prostaglandin synthesis in the thermoregulatory centre in the anterior hypothalamus—in other words, by resetting the body's 'thermostat' at a higher level. The highly effective antipyretic action of aspirin and non-steroidal anti-inflammatory agents depends not on any interference with the production or transport of interleukin 1 but on their potency as inhibitors of prostaglandin synthesis.

Suggestions for Further Reading

Roitt, I.M. 1980. *Essential Immunology*. 4th edn. Blackwell, Oxford.

Fudenberg, H.H. *et al.* 1980. *Basic and Clinical Immunology*. 3rd edn. Lange Medical Publications, Los Altos, California.

Lachmann, P.J., Smith, S.J. and Peters, D.K. 1982. *Clinical Aspects of Immunology*. 4th edn. Blackwell, Oxford.

PART II

Organisms that Infect Man

Bacteria

Taxonomy and Nomenclature

Bacterial taxonomy is the scientific approach to the separation of the vast array of different bacteria into manageable groups (taxa), the arrangement of related groups into hierarchical orders, the allocation of names to the groups and the development of methods for allocating new isolates to the appropriate named group. The three main aspects of taxonomy are **classification**, the grouping of organisms into taxa; **nomenclature**, the allocation of names to the taxa in the appropriate hierarchical series; and **identification** of fresh isolates in terms of the established classification.

Bacterial nomenclature can be a source of bewilderment and frustration to medical students, and even to some experienced clinicians. It is, however, a short-hand system of great value to those who learn to read it (and keep up with the changes necessitated by improvements in bacterial taxonomy). When a bacterium has been isolated from a clinical specimen and put through appropriate identification tests, the bacteriologist knows whether it is similar to previous isolates from the same or associated patients, and whether it is likely to be clinically important in the situation from which it was isolated. To communicate this information to his laboratory and clinical colleagues in a few words, he needs a name for the organism. The early device of naming a bacterium according to the disease that it causes ('the typhoid bacillus', 'the tubercle bacillus') was valid only for those few organisms that are always and exclusively associated with particular diseases. Naming organisms after their discoverers (Koch's bacillus, etc.) was even less useful. Clearly medical bacteriology needed a comprehensive system of bacterial classification and nomenclature—ideally one acceptable in all countries and shared by non-medical bacteriologists.

An obvious choice was the **Linnean** system of orders, families, genera and species, with appropriate Latin (i.e. international) names, used so successfully in other branches of biology. However, vigorous efforts over many years to fit bacteria into such a system have met with limited success, for a number of reasons. Unlike most other biologists, bacteriologists are dealing not with individual organisms but, at best, with what are euphemistically described as **pure cultures**—i.e. populations of individuals all derived from the same single organism, but no longer necessarily identical in genotype or phenotype (see pp. 18–21). The term **strain** is used for a group of pure cultures derived from a common source and thought to be the same—e.g. all apparently identical pure cultures derived from a single specimen, or from different specimens from the same patient, or even from a number of victims of a common-source outbreak of infection. A group of closely similar strains can be said to constitute a **species**, but defining species boundaries among bacteria is a peculiarly difficult problem; cross-fertility, a valuable criterion for this purpose among higher organisms, has no relevance to asexual creatures that are capable of interchange of genetic material between manifestly 'unrelated' individuals (see p. 20). Bacterial species have therefore to be defined according to other criteria, and there has been almost unlimited scope for disagreement as to what these should be. The larger the number of criteria applied, the greater the range of possible permutations becomes and the more apparent it is that the boundaries are in fact artificial. Grouping of 'related' species into **genera** also presents problems, and many bacterial taxonomists have given up trying to fit genera into families and families into orders.

Bacteriologists have tended to apply informed value judgements by selecting a small number of **important** criteria as the major basis for classification of strains into species and genera. An alternative is the **Adansonian** or **numerical** approach of applying to each bacterial strain the same large range of criteria, all regarded as of **equal importance**; the strain is given a score of + 1 if the character sought by a given test is present, and − 1 if it is absent. This approach has its merits, particularly as a means of sorting large groups of basically similar organisms into clusters of strains of much closer similarity, or as a basis for computer-matching of the properties of an unidentified strain with those of a large number of reference strains; but it has not yet produced an overall system of classification which is of value to the clinical bacteriologist.

Traditional methods of bacterial classification have relied on phenotypic characters as indicators of genotype. A more recent approach, useful in research but not applicable to routine diagnostic bacteriology, is direct analysis of DNA. Determination of the (guanine + cytosine) : (adenine + thymine) ratio in DNA (p. 19) provides useful information—not as a basis for positive classification, since similarity of base-pair ratios is no evidence of overall DNA similarity, but as a means of challenging inclusion in the same species or genus of two strains with markedly differing ratios. Similarities and differences in DNA composition can be demonstrated more precisely by studying homologous segments in DNA extracts. The strands of double-stranded DNA separate on heating and re-associate on cooling. If a mixture of DNAs from two bacterial strains is heated and cooled, the amount of cross-over in the re-association stage provides an indication of the degree of homology between the two DNAs; prior radio-labelling of one of the two DNAs allows this to be measured precisely. With classification soundly based on such research methods, appropriate simple tests for phenotypic characters can be selected for routine diagnostic work.

Despite all the problems, there is at present a useful level of international agreement about the classification into genera and species of most bacteria of medical importance, and about the standardization of Latin binomials for them. The International Committee on Systematic Bacteriology has made a major effort to tidy up outstanding problems, to put an end to arguments about historical priority of names, and to regulate future changes of names. Scope must of course be left for discovery of new organisms or of new information invalidating the classification of known organisms, but not for the individual bacteriologist to revise names or introduce new ones according to his personal fancy.

Meanwhile the medical student needs a working knowledge of the language. He must reconcile himself to the fact that a Linnean binomial, while it tells the genus (first name, capital first letter) and species (second name, no capital) to which the organism is assigned, may for historical reasons suggest something which is no longer to be believed—e.g., that *Haemophilus influenzae* is the cause of or related to influenza. (Names of humans can be equally inappropriate!) History has created many other nomenclatural problems. For example, the word bacillus, without an initial capital, means any rod-shaped bacterium, and at one time most of these were given the generic name *Bacillus*; now, however, the generic name is confined to aerobic spore-bearing rods (see p. 83), and other rod-shaped organisms (bacilli) are assigned to a large number of other genera. Many species have undergone several changes of name, and some still have alternative binomials in common use, as well as less formal names in many cases. Thus *Streptococcus pneumoniae* = *Diplococcus pneumoniae* = the pneumococcus. In this book we follow the general custom of using conventional abbreviations for generic names (e.g. *B.* for *Bacillus*, *Br.* for *Brucella*, *S.* for *Salmonella*, *Staph.* for *Staphylococcus*, *Str.* for *Streptococcus*) when the names are being frequently repeated or should have become familiar; and we sometimes employ widely used informal names instead of Linnaean names so that they also will become familiar to our readers. Alternative names are given in brackets following the headings of many sections dealing with individual species.

It is customary to print generic names in italics with a capital first letter when they are used in the singular (e.g. *Staphylococcus* or *Staph.*) but without italics and with a lower-case first letter when they are used as adjectives or in the plural as collective names for organisms belonging to the genus (e.g. staphylococcus or staphylococcal strains and staphylococci). Clearly the latter part of the convention cannot be applied to the generic name *Bacillus*.

Major Criteria for Classification and Identification

The bacterial characters that are used in classification and identification include a wide range of morphological features; staining reactions; growth characteristics; results in tests of metabolic (biochemical) function and end-product formation; and serological tests for specific antigenic markers.

Morphological features

The initial description of a bacterial strain is based upon microscopic examination of stained, and sometimes unstained, films. This provides knowledge of the **size, shape** and **arrangement** of cells, and may reveal other morphological features described below.

Size. Although all bacteria are small and measured in μm (p. 10), there may be quite marked differences in size between members of different genera. This is more evident with rod-shaped organisms (bacilli) than with spherical ones (cocci); the cells of the gram-positive rods of the genera *Bacillus* (aerobes) and *Clostridium* (anaerobes) are much larger than those of gram-negative rods (see Fig. 7.1, p. 70).

Shape. Bacteria are divided into two large groups according to their cell shape. Those with more or less spherical cells are **cocci** and those with rod-shaped cells are **bacilli**. Some bacilli form elongated or filamentous cells, which may show true branching. A further group are the spiral organisms or **spirochaetes**, which have a cork screw appearance. Distinction between cocci and bacilli is sometimes difficult if individual cells are studied, but is usually fairly easy if a large number of organisms from the same culture are examined together. Bacterial strains that are classified as cocci may show variation of individual shape from spherical to oval, but never to rod-shaped forms. Bacilli on the other hand may include many very short rods ('cocco-bacilli'), but indisputable rods are nearly always to be found, and filamentous forms may also be present. A culture that shows unusual diversity of size and shape is described as pleomorphic. When all organisms in a particular culture are of indeterminate cocco-bacillary shape, examination of the same strain grown on a different medium or incubated for a longer or shorter time will usually resolve the doubt.

Arrangement. The way in which particular bacteria divide determines the arrangement of their cells, and some species have characteristic patterns. Streptococci always divide in the same plane, thus producing chains of cocci, whereas staphylococci divide in random planes and appear as clusters. Some organisms usually appear as pairs (diplococci). Amongst bacilli, some remain attached after cell division and produce chains of bacilli (e.g. *B. anthracis*) whereas in other species the bacilli come to lie side by side or in bundles (e.g. corynebacteria).

Capsules. Some bacterial species (e.g. *Str. pneumoniae*, *Klebsiella pneumoniae*, *B. anthracis*) characteristically form capsules around their cell walls. In most cases these capsules consist of complex polysaccharides, but that of *B. anthracis* is predominantly a polypeptide. Capsules are not satisfactorily shown in preparations stained by ordinary methods, but can be demonstrated by 'negative staining', in which the background between the bacteria is filled with some opaque material such as indian ink; the capsules then show up as unstained zones around the bacterial cells, which can be made more visible by simple positive staining. Capsular development is determined by environmental conditions, and is usually best when the organism is growing in

Figure 7.1 Morphological features of some bacteria. (a) A cluster of staphylococci as seen in pus and tissues. (b) A chain of streptococci. (c) Pairs of capsulate pneumococci as seen in sputum or pus. (d) Pairs of kidney-shaped meningococci or gonococci as seen in pus (where, as a rule, most of them are inside pus cells). (e) Diphtheria bacilli showing 'Chinese character' arrangement. (f) Enterobacteria. (g) Lactobacilli in branching chains (false branching). (h) Members of the genus Bacillus, with spores narrower than the bacilli. (i) Clostridia with subterminal spores wider than the bacilli. (j) Cl. tetani with terminal drum-stick spores. (k) Yersiniae showing polar staining. (l) An actinomycete showing true branching and fragmentation. (m) Vincent's organisms (p. 191). (n) Leptospires. (o) Treponemes. (p) Vibrios. (q) A budding yeast for comparison of size (See also Fig. 13.2(a)).

living tissues. The protective value of capsules is discussed on p. 41 and their immunological significance on pp. 49, 50 and 53.

Flagella. These are long, thin thread-like appendages, about 0.02 μm in diameter, which project from the cells of certain bacteria. They have their origin in basal granules in the bacterial protoplast and pass through the cell wall. They are composed almost entirely of protein. The original meaning of *flagellum* is 'a whip', and it is to rotatory (propeller-like) movements of these appendages that flagellate bacteria owe their motility. Flagella cannot be demonstrated under the light microscope unless they are first considerably thickened by the deposition on their surfaces of special stains. They can be studied more satisfactorily by electron microscopy. Demonstration of motility due to flagella is mentioned on pp. 90 and 164.

Spores. These are round or oval structures formed by gram-positive bacilli of the genera *Bacillus* and *Clostridium*. Sporulation appears to be in general a reaction to conditions that are unfavourable for normal growth, in particular to deficiency of essential nutrients. The spore has a low water content, its metabolic activity is minimal, and it is surrounded by a thick protective coat which enables it to resist heat, desiccation and other harmful agenices far better than do the vegetative forms (see pp. 11 and 87). Bacterial spores, unlike those of fungi, are not reproductive; one vegetative cell usually produces one spore, which in turn germinates to form a single new vegetative cell. Certain trigger substances (e.g. L-alanine for some species) are needed to initiate the germination of spores. A spore may be narrower than the bacillus in which it originates, or it may distend it (see Fig. 7.1).Depending on the species, it may be at the end of the bacillus (terminal), near the end (subterminal) or in the middle (central). When mature it is freed from the bacillary cell, which then disintegrates.

Protoplasts, spheroplasts and L-forms. Some bacteria are particularly susceptible to the action of lysozymes (enzymes of human, animal or bacterial origin which attack mucopeptide—see p. 46) and by such action they are converted into **protoplasts**. These are complete cells apart from the loss of their cell walls, or at least of important wall components, and consequent loss of rigidity, shape and osmotic resistance. They survive only if kept in suitable hypertonic environments. They are metabolically active and can grow but cannot multiply. Similar structures without cell walls can be produced from suitable organisms by the action of an antibiotic such as penicillin, which prevents mucopeptide synthesis (see p. 389). But many bacteria, under the influence of penicillin or of any of a variety of other agents, *in vitro* or *in vivo*, produce either **spheroplasts** or **L-forms**. Spheroplasts have residual but damaged cell walls and assume bizarre shapes, but when transferred to suitable culture media free from the agent that caused them they may be able to produce orthodox colonies composed of normal individuals. L-forms, on the other hand, are not ordinary bacteria with damaged cell walls, but mutants that do not form cell walls and have been selected out by an environment unsuitable for those bacteria that do form them. As might be expected, these L-forms are delicate and fragile, but they differ from protoplasts in being able to multiply. When grown on solid media they form characteristic small 'fried egg' colonies, with central smooth portions deeply embedded in the medium and consisting mainly of minute round forms, and more superficial peripheral zones in which much larger irregular forms are found. It is possible that many kinds of bacteria, when confronted with unfavourable conditions in the body, may persist there as relatively undetectable L-forms and in that state be in some way better able to resist the host's defences.

Staining reactions

The refractility of bacterial cells differs little from that of water, and unstained cells can therefore be seen only with some difficulty in wet films examined with reduced illumination, or more clearly but without structural detail by phase-contrast microscopy (p. 164). The morphology of bacteria is generally studied in stained films examined with an oil-immersion objective (p. 164). The size and shape of bacterial cells can be seen with any of a wide range of simple stains but more essential information is obtained if **Gram's** method of differential staining is used, and this is the commonest staining procedure used in clinical microbiology. In this method, methyl violet or gentian violet is applied first to a fixed film, followed by iodine as a mordant. After this treatment, some organisms, known as **gram-positive**, resist decolourization by ethyl alcohol or acetone and remain violet or blue in colour. Others, known as **gram-negative**, readily give up the violet dye and are then stained red by a counterstain such as neutral red, dilute carbol fuchsin or safranin. This distinction is not absolute, in that faulty technique can give equivocal or wrong results and, even in the most skilled hands, some strains are difficult to classify. Very young or very old cultures may give anomalous results. The procedure is open to the criticism, in common with any similar staining method, that the stained objects seen down the microscope are distorted artefacts bearing little resemblance to the live bacteria. Nevertheless, Gram's technique provides a division of cocci and bacilli into two main groups; and this is of great practical value because the distinction revealed by the gram-staining reaction, empirical though it was when Gram first introduced it in 1884, depends on a major difference in cell-wall structure which determines various important properties of the bacteria (p. 10).

Other special staining procedures are useful in the characterization of some bacteria. In particular, the **Ziehl–Neelsen** method (p. 103) is fundamental to the demonstration of acid- and alcohol-fast bacilli of the genus *Mycobacterium*, notably *Myco. tuberculosis*. A modified Ziehl–Neelsen stain with less vigorous decolourization is used to stain bacterial spores. Another type of differential stain, **Albert's** or **Neisser's** staining method, is used to demonstrate the metachromatic (volutin) granules characteristic of *Corynebacterium diphtheriae* (p. 84).

Growth characteristics

Pathogenic and commensal bacteria have a wide range of growth requirements and of typical appearances in cultures; their diversity in these respects is useful in their characterization. Nutritional requirements (p. 16) will determine the ability of isolates to grow on different culture media—some grow well on simple **nutrient agar** (p. 166), whereas others require enriched media such as **blood agar** (p. 166) or **'chocolate' agar** (p. 81). The addition of selective agents for certain pathogens is based upon known resistance characteristics of the species being sought—e.g. the use of bile salts in **MacConkey's agar** (p. 167) to select for intestinal bacteria and of potassium tellurite to select for *C. diphtheriae*.

The most important growth characteristic is whether the organism requires aerobic or anaerobic conditions for growth—i.e. whether it is aerobic, facultative or strictly anaerobic (p. 15). The effect of incubation temperatures on growth may also give useful clues as to the organism's identity (p. 16).

Once a bacterial strain has been grown in culture under optimal conditions, the appearance of its colonies and their effect on the growth medium are useful basic characters. In interpreting these we must make allowances for factors such as composition of the medium, duration and conditions of incubation and genotypic or phenotypic variations in the organism; a 'pure' culture may show colonial variants on one plate, give different appearances on different media, and undergo progressive changes of colonial morphology on repeated subculture. However, for a given species grown under defined conditions on a specified medium the size, shape, surface appearance, elevation, colour, opacity and consistency of a colony are often characteristic and, taken with other simple microscopic and growth characters, may give a good indication of the identity of an isolate.

Changes in the culture medium around the colonies may also be significant. Clear, colourless zones on blood agar are due to lysis of the red blood cells (**haemolysis**). Also, many media contain ingredients to detect particular biochemical activities—e.g. MacConkey's medium contains (as well as bile salts) lactose and neutral red, so that colonies of lactose-fermenting bacteria become red due to acid production.

Tests of metabolic (biochemical) functions

When a strain has been isolated in pure culture

and allocated to a general group on the basis of microscopic and growth characteristics, tests of biochemical activity are often used to confirm its specific identity. These include tests for the utilization of substrates, the production of acid from carbohydrates ('fermentation' p. 15), the production of specific metabolites (e.g. H_2S, indole) or the breakdown of organic macromolecules (e.g. gelatin, casein). Traditionally, such tests have required overnight or longer incubation so that the bacteria can multiply in the presence of the substrates and produce enough enzymes to cause a detectable change; in the interest of more rapid reporting some can now be carried out by adding heavy bacterial inocula to small volumes of substrate solution which allows preformed enzymes to give a detectable result within a few hours.

End-product analysis

Many of the above tests depend upon the detection of metabolic end-products, such as acid or indole, from special substrates, but detailed analysis of the products of normal metabolism may also be used to characterize some groups. The end-products of the energy metabolism of aerobes are only water and CO_2, but those of anaerobes include various volatile short-chain fatty acids and alcohols. These can be detected by gas–liquid chromatography (GLC) which gives product patterns characteristic of groups, and sometimes of individual species.

Serological tests

Antigens characteristic of genera, species or subgroups (serogroups) within a species are carried on bacterial surfaces, and some are released into their environment (e.g. exotoxins). Tests for detection of such antigens are widely used in identifying pathogenic bacteria—e.g. tests for agglutination of suspended bacteria by antisera directed against surface antigens (p. 176); tests for the ability of specific antiserum to precipitate toxin liberated into culture medium (Elek's test for *C. diphtheriae*, p. 84) or to neutralize it (Nagler's test for *Cl. perfringens*, p. 89).

Systematic Classification of Bacteria

On the basis of three fundamental characters—reaction in Gram's stain (positive or negative), cell morphology (coccus or bacillus) and oxygen requirement (aerobic/facultative or anaerobic)—the majority of bacteria parasitic for man can be allocated to eight main groups. A minority of species of parasitic bacteria belong to another four groups—the acid-fast bacteria, branching bacteria, spirochaetes and mycoplasmas. The 11 groups described in the rest of this chapter are listed in Table 7.1 Genera from the first nine groups of major importance in medicine are listed in Table 7.2.

Gram-Positive Cocci

The patterns of arrangement of cocci, as seen under the microscope, are a valuable guide to generic distinctions in this group (p. 69). The grape-like clusters from which the genus *Staphylococcus* gets its name are best seen in pus and other body fluids (Fig. 7.1, p. 70); films made from cultures may show small clusters and short chains, but as a rule most of the cocci are

TABLE 7.1 Major groups of bacteria parasitic for man

Group	Cell shape	Other major features
1	Cocci	gram-positive, aerobic
2	Cocci	gram-positive, anaerobic
3	Cocci	gram-negative, aerobic
4	Bacilli	gram-positive, aerobic
5	Bacilli	gram-positive, anaerobic
6	Bacilli	gram-negative, aerobic
7	Bacilli	gram-negative, anaerobic
8	Bacilli	acid-fast
9	Bacilli	branching
10	Spirochaetes	
11	Mycoplasmas	

distributed at random. Sets of four cocci in squares and larger geometrically arranged packets are formed by members of the usually non-pathogenic genus *Micrococcus*. Chain formation is characteristic of the genus *Streptococcus* (Fig. 7.1, p. 70).

The Genus *Staphylococcus*

Staphylococci make a large contribution to man's normal commensal flora and also account for a high proportion of suppurative lesions. Most strains isolated from such lesions produce golden-yellow colonies on common culture media and so they have long been classified as *Staph. aureus*. However, pigment production is a variable property, somewhat at the mercy of growth conditions, and some strains that form white colonies are highly pathogenic. The capacity to produce **coagulase** (see below) is more closely correlated with pathogenicity and the name *Staph. aureus* is currently applied to all strains which do so (coagulase-positive). Strains of *Staph. aureus* are common as commensals, as well as in lesions, but should always be regarded as potential pathogens. On the other hand most staphylococci found on the skin and in the upper respiratory tract are coagulase-negative. These almost invariably form white colonies and are at most only low-grade pathogens. They are now classified as *Staph. epidermidis*, though the name *Staph. albus* (originally including all white-colonied staphylococci) is still in common use as an alternative.

Staphylococcus aureus

OCCURRENCE AND PATHOGENICTY *Staph. aureus* is found on the anterior nasal mucosa of 40–50% of healthy adults, in the throats of many of them, in the faeces of about 20% and on the skin of 5–10%. As well as being carried in these situations it may also multiply profusely there, notably in the nose and on the perineal skin. New-born babies are rapidly colonized by this species, and 90% or more of these born in hospital carry it in the nose and around the umbilicus within two weeks of birth. Droplet spread and shedding of skin scales result in widespread distribution of this species in the human environment, notably in air, dust, clothing and bedding.

TABLE 7.2 Some medically important genera of cocci and bacilli

Gram-positive	Gram-negative			
Cocci	Cocci			
Staphylococcus	Neisseria			
Streptococcus				
Bacilli	Bacilli			
Corynebacterium	(a)	Enterobacteria	(c)	Vibrio
Lactobacillus		Escherichia		Campylobacter
Bacillus		Klebsiella	(d)	Parvobacteria
Clostridium		Salmonella		Haemophilus
(Mycobacterium)*		Shigella		Bordetella
		Proteus		Brucella
		Yersinia		Legionella
	(b)	Pseudomonas	(e)	Bacteroides

* Some mycobacteria, including *Myco. tuberculosis* are stained only faintly or not at all by Gram's method.

Staph. aureus is the commonest cause of pyogenic infections of man. It is also responsible for a form of food-poisoning due to the eating of food in which staphylococci have multiplied and produced a toxin (p. 257).

MICROSCOPY Staphylococci are gram-positive spherical organisms, 0.7–1 μm in diameter, arranged in clusters (p. 69).

CULTURE Staph. aureus grows well on common media under aerobic conditions, but less well anaerobically. Its optimal growth temperature is around 37°C, but it will grow within the range of 10–44°C. On nutrient or blood agar, colonies are 2–4 mm in diameter after 18 hours incubation at 37°C, and are smooth, shiny, opaque, yellow-to-white domes resembling small drops of gloss paint. The ability of staphylococci to grow in concentrations of sodium chloride which are inhibitory to other genera (e.g. 7.5% in nutrient agar or 10% in nutrient broth) is useful for their isolation from specimens likely to contain large numbers of other bacteria.

COAGULASE AND DEOXYRIBONUCLEASE PRODUCTION The thrombin-like enzyme coagulase may play a part in the survival of Staph. aureus in the host and enhance its pathogenicity (p. 17). Production of coagulase and that of another enzyme deoxyribonuclease (DNAase) are closely associated in staphylococci, so tests for either can be used as the basis for recognition of Staph. aureus in the routine laboratory.

In the **tube coagulase test** diluted plasma and a broth culture of the organism are incubated together. Formation of a clot indicates that the organism is coagulase-positive. The **slide 'coagulase'** test is quicker and simpler to carry out, and usually gives the same result as the tube test, though it does not depend on the same mechanism; part of a colony is emulsified in a drop of water on a clean slide and a loopful of plasma is added. The test is positive if visible clumping of the bacteria occurs, provided that spontaneous clumping (auto-agglutination) does not occur in a comparable suspension without plasma. **DNAase production** is easily demonstrated by growing the staphylococcus on a nutrient agar made opaque by adding DNA; the enzyme causes clearing around the growth.

TOXIN PRODUCTION Staph. aureus produces various exotoxins and exo-enzymes (aggressins). Effects demonstrable include lysis of red cells, leukocidal action (killing of granulocytes and macrophages by lysis of their cell membranes) and vasoconstriction with resultant tissue necrosis. These are probably components of the pathogenicity of Staph. aureus to man. Human diseases in which Staph. aureus toxins are undoubtedly involved include: food-poisoning (p. 257), scalded skin syndrome (p. 306) and toxic shock syndrome.

PHAGE TYPING Bacteriophages are virus parasites of bacteria (see pp. 13, 23). Many Staph. aureus strains carry phages which, when spotted on to plate culture of other staphylococcal strains, cause lysis of some of them. This phenomenon makes it possible to divide the species Staph. aureus into phage types—i.e. groups of strains having the same or closely similar phage susceptibilities and resistances. For this purpose a basic set of 2 dozen or so phage cultures is used. These are numbered according to international agreement, and are divisible into 4 groups (I–IV), according to their antigenic composition. Any one strain of Staph. aureus is likely to be susceptible to the lytic action of several phages, which often belong to a single antigenic group. The phage type of the strain is then designated by listing the numbers of the phages which can lyse it under specified conditions—e.g. type 52/42B/42C/44A/80/81, notorious for causing outbreaks of hospital sepsis, and commonly given the less precise but more manageable designation 'type 80/81'. A minority of strains are not typable because they are resistant to all of the phages used, but new phages are found from time to time which, when added to the standard set, increase the number of strains that can be typed.

IMMUNOLOGY Various antibodies are formed in response to staphylococcal infection, but their protective value is uncertain and their measurement seldom gives much diagnostic help. Staph. aureus strains that have **protein A** as a surface component are resistant to phagocytosis, because protein A binds to the Fc region of antibody molecules, which would otherwise attach to the phagocytic cells (p. 48). These strains can be coated with antibody molecules specific for other bacteria and can then be used as diagnostic reagents for their detection; mixing the coated staphylococci and the appropriate other bacteria results in **co-agglutination**. This method is used, for example, in the grouping of β-haemolytic streptococci (p. 78).

ANTIBACTERIAL TREATMENT Since antibiotics became available, *Staph. aureus* has shown exceptional ability to produce resistant variants (p. 382). However, most strains are still sensitive to cloxacillin and related drugs (p. 390).

Other staphylococci

Staph. epidermidis resembles *Staph. aureus* in most respects except that it is coagulase-negative, usually forms white colonies, and is generally non-pathogenic. It is found in large numbers all over human skin and on many mucous surfaces. It may play some part in the pathogenesis of acne and in other minor skin lesions, and has occasionally been incriminated as a cause of bacterial endocarditis. It is an important and common cause of bacteraemia associated with intravenous catheters, and of infections around cerebrospinal fluid shunts, artificial heart valves and other prostheses (p. 319). *Staph. epidermidis* strains are also commonly responsible for urinary tract infections in elderly males, particularly after bladder instrumentation or prostatectomy. A common form of acute cystitis in young women is due to coagulase-negative staphylococci that are recognizable by their resistance to the antibiotic novobiocin and are now called *Staph. saprophyticus* (p. 274).

The Genus Streptococcus

Members of this genus are widely distributed in nature, largely as parasites of man and animals. They make a large contribution to the normal bacterial flora of the human respiratory, alimentary and female genital tracts. *Str. pyogenes* and *Str. pneumoniae* are important pathogenic species, though even these may be carried as harmless commensals. Some other commensal streptococci are pathogenic in certain circumstances.

Streptococci are spherical or oval gram-positive cocci, about 1 μm in diameter, non-motile, non-sporing, sometimes capsulate, and characterized by their tendency to form chains (p. 69 and Fig. 7.1, p.70). Chain formation is best seen in pathological materials or in fluid cultures, less well in films made from cultures on solid media. Chain length also depends upon the species involved; some form chains containing scores of cocci, whereas, at the other extreme, *Str. pneumoniae* characteristically appears on diplococci (Fig. 7.1, p. 70).

Streptococci are facultative organisms (p. 15). Those that grow well in **aerobic** conditions include the principal pathogens of the genus, and can be classified (apart from the **enterococci**, see below) according to their **haemolytic** activities, as follows:

(1) **β-haemolytic** streptococci produce, around a colony (e.g. of *Str. pyogenes*) on a blood agar plate, a clear colourless zone in which the red cells have been lysed and the haemoglobin decolourized. The degree of haemolysis by the same strain may vary with blood from different mammalian species and is often more pronounced after anaerobic incubation.

(2) **α-haemolytic** streptococci produce around the colonies a zone in which the red cells are partly destroyed (so that the medium becomes somewhat less opaque) and haemoglobin is converted to a green pigment. This is the reaction given by streptococci of the viridans group and by *Str. pneumoniae*

(3) **Non-haemolytic** streptococci do not produce either form of haemolysis. A few cause other changes, such as partial destruction of the red cells, with no green pigmentation or a brownish discolouration of the medium but many have no visible effect on the medium. Most are commensals.

β-haemolytic streptococci are divided into Lancefield groups A–H and K–V by serological identification of the carbohydrate antigen. The species name *Str. pyogenes* is given to Lancefield group A streptococci, the predominant human pathogens of the genus (see below). Group B (p. 78) and group D (which contains many of the enterococci—p. 80) are also of medical importance, and strains of other groups (notably C, F, G, and R) are occasionally pathogenic to man.

Streptococcus pyogenes (β-haemolytic streptococci, group A)

OCCURRENCE AND PATHOGENICITY Healthy human beings may carry in their throats, or less commonly in their noses, *Str. pyogenes* strains which are potential pathogens and may cause various diseases. The commonest picture is acute sore throat, which may spread to the middle ears, mastoids and even meninges. Scarlet fever is due to a strain that produces erythrogenic toxins (see below) which, in the absence of protective antibodies, cause the characteristic skin rash. *Str. pyogenes* also causes infections of the skin (p. 305) and soft tissues (p. 386) and of the uterus following childbirth (puerperal sepsis). Spread into the lymphatic system (lymphangitis and lymphadenitis) and into the blood stream (septicaemia) are common and characteristic complications of untreated *Str. pyogenes* infections. The non-suppurative complications rheumatic fever (p. 191) and acute glomerulonephritis (p. 191) are hypersensitivity reactions that develop 2–3 weeks after the initial infection.

MICROSCOPY Chain-formation is usually well marked. The cocci are spherical and conform to the general description given for the genus.

CULTURE Growth is poor on nutrient agar but better on media containing serum or blood, and is better in aerobic than in anaerobic conditions. The optimal temperature for growth is 37°C, and none occurs below 18°C. Even under optimal conditions on blood agar, colonies are usually only 1 mm or so in diameter after 24 hours; they may be smooth and shiny, but are commonly dry and irregular in contour and outline, especially when they belong to virulent strains. They are greyish-white and opaque or semi-transparent. β-haemolysis is best seen after anaerobic incubation, and the zones may then be up to 5 mm in diameter (see below). In fluid cultures there is usually a granular deposit and a relatively clear supernate, in contrast to the more uniform turbidity of broth cultures of staphylococci and of most other streptococci.

TOXIN AND ENZYME PRODUCTION *Str. pyogenes* produces many exotoxins and aggressins.

(1) *Streptolysin O* is a haemolysin which is oxygen-sensitive and is active only in the reduced state. It is responsible for the larger zones of β-haemolysis after anaerobic incubation. It is powerfully toxic to animals (and presumably to man), acting mainly on the heart.

(2) *Streptolysin S* is also a haemolysin. It is not inactivated by oxygen and is not demonstrably antigenic. It is responsible for β-haemolysis on aerobic blood agar plates.

(3) *Streptokinase* activates plasminogen to the proteolytic enzyme plasmin, which then breaks down fibrin. This process facilitates the spread of the streptococci through the fibrin barrier laid down as part of the host's defence mechanism.

(4) *Hyaluronidase* also facilititates the spread of *Str. pyogenes* by breaking down the hyaluronic acid of the connective tissue cement substance.

(5) *Deoxyribonucleases A, B, C and D* depolymerize DNA and deoxyribonucleoprotein. Since the latter material is largely responsible for the viscosity of purulent exudates, these enzymes make streptococcal pus less viscous. They have been used therapeutically to liquefy pus and hasten the cleaning up of wounds and abscess cavities. Streptokinase solutions have been similarly used.

(6) *Erythrogenic toxins* cause the characteristic rash of scarlet fever (see above). Their production depends on a lysogenic relationship (p. 23) between *Str. pyogenes* and a bacteriophage.

ANTIGENIC STRUCTURE Lancefield grouping depends on the cell-wall C polysaccharides possessed by most β-haemolytic and some other streptococci. C polysaccharide can be removed from the organisms and obtained in solution by acid, formamide or enzymic extraction, and can then be identified by precipitation tests (p. 175), using antisera for the various Lancefield groups.

Many diagnostic laboratories now use simpler and more rapid methods—latex agglutination (p. 176) or co-agglutination (p. 76); the latter has the advantage that it uses streptococcal suspensions from solid media and dispenses with the need for extraction. Group A (*Str. pyogenes*) strains can be subdivided into over 60 Griffith types by study of their surface proteins, notably the M proteins which are important contributors to the virulence of the organisms. Such fine subdivision is sometimes useful in studies of the spread of infection, and certain types (notably type 12) are particularly associated with infections that lead to acute glomerulonephritis (p. 194).

IMMUNOLOGY Immunity after natural infection is highly type-specific, and immunization is impracticable because of the large number of types involved. Infection with a strain that produces erythrogenic toxin is followed by the development of immunity to such a toxin. Useful diagnostic information can be obtained by measuring levels of antibodies to extracellular products of *Str. pyogenes*—notably antistreptolysin O (ASO) and antideoxyribonuclease B (antiDNAase B). The ASO titre invariably rises soon after *Str. pyogenes* throat infection (but can also do so after infection with group C or G streptococci, since these can produce streptolysin O); the antiDNAase B titre rises more slowly and is a more reliable indicator of recent skin infection. These tests are particularly valuable in confirming or refuting the diagnosis of rheumatic fever or acute glomerulonephritis.

ANTIBACTERIAL TREATMENT *Str. pyogenes* strains are always sensitive to penicillin, which is the agent of choice for the treatment of infections due to them. They are usually sensitive to most other antibiotics except the aminoglycoside group, and to the sulphonamides. Tetracycline-resistant strains are now fairly common in many areas.

Group B streptococci

OCCURRENCE AND PATHOGENICITY These organisms are carried in the female genital tract and in the lower intestinal tract, with carriage frequences of the order of 5–30%. They can be important pathogens during the neonatal period when two disease patterns can be distinguished; early onset disease (<5 days), with a fulminating septicaemia, and late onset disease (>7 days), which typically results in meningitis (p. 324).

CULTURE Most group B streptococci are β-haemolytic. Their isolation from the genital and alimentary tracts is made easier by using a selective medium that contains an aminoglycoside to suppress many other bacterial species. They produce a characteristic orange pigment when grown on serum-starch agar.

IMMUNOLOGY Many infants are protected against group B streptococcal infection by transplacental transmission of maternal antibodies (p. 54), but the mothers whose infants are affected have no such antibodies. These can be given passively or stimulated by immunization with a group B polysaccharide vaccine, and maternal immunization is therefore a possible means of preventing group B streptococcal neonatal infections.

The viridans group of streptococci (Str. viridans)

The bacterial flora of the human upper respiratory tract and mouth normally includes large numbers of α-haemolytic streptococci. Some of these may be *Str. pneumoniae* (see below). The rest were formerly lumped together under the name *Str. viridans*, but are now divided into a number of more precisely defined species. However, they are still collectively referred to as **the viridans group** or **viridans streptococci**—though some of the new species incorporate non-haemolytic as well as α-haemolytic strains. Some are important in dental caries—notably *Str. mutans*, which turns sucrose into a sticky layer of dextran on surfaces of teeth (p. 191). Apart from this the viridans streptococci are rarely pathogenic to man in their normal habitat, but they are liable to gain access to the blood stream, particularly during dental filling or extraction. They are nearly always eliminated from the circulation without causing any trouble, but in patients with rheumatic endocarditis or congenital heart lesions they may invade the fibrinous vegetations attached to the valves or deformed structures, and they then cause subacute bacterial endocarditis (p. 237). The species most likely to do this are *Str. sanguis*, *Str. mutans*, *Str. bovis* and *Str. mitis*.

The microscopic appearances of members of this group are in general similar to those described for *Str. pyogenes*. So are their cultural characteristics, except that haemolysis, if present, is of the α type and usually less extensive than the zones of β-haemolysis around *Str. pyogenes*.

Streptococcus pneumoniae (Diplococcus pneumoniae, the pneumococcus)

OCCURRENCE AND PATHOGENICITY This species is both a normal commensal and a common pathogen of the human respiratory tract, and it is therefore difficult at times to assess the significance of its isolation from sputum or other respiratory tract specimens. It is the commonest cause of pneumonia and shares with *Haemophilus influenzae* responsibility for acute suppurative exacerbations of chronic bronchitis. Other pneumococcal infections include pleurisy and pericarditis (usually associated with pneumonia), meningitis, otitis media, paranasal sinusitis and (in girls or young women) peritonitis.

MICROSCOPY In tissues, pus or sputum pneumococci are typically arranged in pairs (diplococci). Each coccus is somewhat elongated, and pointed at one end but rounded at the other (lanceolate); the two members of a pair point away from each other. They are surrounded by a polysaccharide capsule (Fig. 7.1, p. 70). However, in artificial culture short chains are common and capsules tend to be less evident, so that distinction from other streptococci is not easy on the basis of microscopic morphology.

CULTURE In its nutritional and environmental requirements *Str. pneumoniae* resembles the streptococci already described except that its growth is more definitely enhanced by the addition of 5–10% CO_2 to the incubated atmosphere. It is α-haemolytic, like the viridans group, but in general the colonies themselves are larger and more disk-shaped. Typically they have raised edges and concentric ridges on their surfaces which have earned for them the name of 'draughtsmen'. Some strains form moister, more mucoid colonies, whereas others have rough, granular surfaces.

OPTOCHIN SENSITIVITY Pneumococci are most easily distinguished from viridans streptococci in the routine laboratory by their inability to grow on blood agar in the vicinity of a paper disc containing optochin (ethyl hydrocuprein hydrochloride).

ANTIGENIC STRUCTURE Pneumococcal strains form polysaccharide capsules of over 80 immunologically distinct kinds; and on this basis 46 capsular types are recognized, some of them divided into sub-types. Using type-specific antisera, strains can be identified by agglutination or capsule-swelling tests of the organisms themselves or by precipitation of capsular polysaccharide from solution. Surveys in various countries have shown that 12–14 of the 46 types are between them responsible for the great majority of serious pneumococcal disease, and that the relative importance of individual types varies somewhat according to the disease—e.g. type 1 causes pneumonia more often than meningitis, whereas the reverse is true (in some series, at least) of type 18. Infections due to type 3 tend to have a particularly high mortality rate.

IMMUNOLOGY Pneumococcal capsular polysaccharide prevents phagocytosis (p. 41), but anticapsular opsonins (p. 53) overcome this protection and provide the host with type-specific immunity. The appearance of type-specific antibodies 7–10 days after the onset of pneumococcal pneumonia was responsible for the 'resolution by crisis' which was the outcome of many such infections in pre-antibiotic days. Type-specific passive immunization, the only effective antibacterial treatment for pneumococcal infections in the 1930s, was abandoned when sulphonamides and penicillin became available, as were attempts to produce a vaccine for active immunization. The continued high mortality of serious pneumococcal infections, despite the *in vitro* sensitivity of pneumococci to penicillin and other antibacterials, prompted renewed interest in active immunization during the 1960s; and this was reinforced by the emergence of antibacterial resistance among pneumococci (see below). A 14-type vaccine is now generally available. Protection is particularly desirable for patients with hypogammaglobulinaemia (p. 316) and those who have had splenectomies or whose splenic function is impaired (e.g. by sickle cell anaemia), since such patients are particularly susceptible to fulminating pneumococcal infections.

ANTIBACTERIAL TREATMENT Pneumococci in Britain are almost invariably sensitive to penicillin, and usually to the sulphonamides and to most antibiotics except the aminoglycosides, though resistance to tetracyclines is now quite common. Strains relatively resistant to penicillin were first reported in 1967 from Australia, and consituted 12% of all pneumococcal strains isolated in a survey in New Guinea, but have fortunately remained rare elsewhere. Strains with a higher degree of penicillin resistance, many of them also resistant to most of the obvious alternative drugs, have been reported from South Africa since 1977, and some have been encountered in Europe.

Streptococci dependent on carbon dioxide

From many specimens received in routine diagnostic laboratories two blood–agar plate cultures are set up; one is incubated in air and the other in an anaerobic atmosphere containing 5–10% of CO_2. Many of the streptococcal commensals of the vaginal and other mucous surfaces grow only on the second plate—not because they are anaerobes but because they need CO_2 in excess of the amount found in air. Though usually harmless, they can cause puerperal and various other forms of sepsis. *Str. milleri* is important as a cause of large abscesses in the brain, thorax or abdomen, and of some cases of bacterial endocarditis. Some strains of this species are β-haemolytic, and it includes Lancefield group F streptococci, some strains from other groups and some that are not groupable. Penicillin is usually effective against all of these streptococci.

The enterococcus group (faecal streptococci)

Streptococci of this large and ill-defined group are normal intestinal organisms. They differ from other streptococci in being able to grow in the presence of moderate concentrations of bile salts, e.g. on MacConkey's medium, on which they form characteristic small colonies of deep magenta colour. They tend to be oval cocci, and to form short chains. On blood agar their colonies are usually somewhat larger than those of *Str. pyogenes*. Some strains are β-haemolytic, some are α-haemolytic, some are non-haemolytic, and some produce a brown discolouration of the medium. They mostly survive heating at 60°C for 30 minutes, treatment which is lethal to most other streptococci. Many of them belong to Lancefield group D; such strains are not necessarily β-haemolytic. The specific name *Str. faecalis*, at one time applied to all group D enterococci, now strictly belongs to one of several species into which such organisms are divided.

Enterococci may cause urinary tract infections and may invade wounds and ulcerative skin lesions. They may also cause subacute bacterial endocarditis after gynaecological or genito-urinary instrumentation or surgery, and this gives special importance to the fact that they are resistant to penicillin and often to other antibiotics effective against other streptococci (p. 78).

Anaerobic cocci

There is no satisfactory classification of the gram-positive anaerobic cocci. Two genera (*Peptococcus* and *Peptostreptococcus*) are recognized, but the group is best considered as a whole. They are commensals of mucous surfaces, notably in the vagina and the large intestine, and are sometimes associated with other anaerobic or facultative bacteria in deep abscesses or in the spreading gangrene around surgical wounds known as Meleney's or synergic gangrene (p. 289). They can also cause puerperal sepsis and septicaemia, or chronic endometritis. Pus from lesions in which they are involved has an unpleasant smell because they are markedly proteolytic and produce much H_2S. Most strains are sensitive to penicillin, and all to metronidazole.

Gram-negative anaerobic cocci are assigned to the genus *Veillonella*. They are commensals of the mouth and large intestine, but have no known pathogenicity.

Gram-Negative Cocci

The Genus Neisseria

Members of this genus are non-motile and mostly non-capsulate aerobes. They include two important pathogens (*N. meningitidis*, the meningococcus and *N. gonorrhoeae*, the gonococcus) and a number of species that are common commensals of the human respiratory tract. In stained films made from cultures of any of the species, the cocci are usually arranged in pairs, each coccus being somewhat flattened or concave on the side facing its partner. In films of pathological material, such as cerebrospinal fluid or pus, the pathogenic species are characteristically found as pairs of kidney-shaped cocci (Fig. 7.1, p. 70), many of which are inside pus cells. The pathogens are more exacting than the commensals in their nutritional, atmospheric and temperature requirements.

Neisseria meningitidis (the meningococcus)

OCCURRENCE AND PATHOGENICITY This organism is an obligate human parasite. It is an important cause of meningitis and of septicaemia (p. 210). The meningococcus may be carried in the upper respiratory tracts of healthy people and during epidemics more than half of those at risk may become carriers especially if they are living and sleeping in overcrowded conditions.

MICROSCOPY (See above) In stained films made from purulent cerebrospinal fluid the meningococcus can as a rule be rapidly recognized and distinguished from other likely causes of meningitis by its characteristic shape, arrangement and predominantly intracellular situation.

CULTURE *N. meningitidis* will grow on some of the richer varieties of nutrient agar, but does better on blood agar, and better still on medium in which the blood has been heated and which is called **chocolate agar** because of its brown colour. The species is aerobic; its growth is often enhanced by the presence of about 5% of CO_2 in its atmosphere (p. 167). The optimal growth temperature is about 37°C, but some growth usually occurs anywhere between 25 and 42°C. The colonies are rather small (around 2 mm in diameter after 24 hours), smooth, greyish, semi-transparent and devoid of striking positive features. Cultures usually die within a few days.

BIOCHEMICAL REACTIONS In common with most other neisseriae, *N. meningitidis* produces an **oxidase** which can be detected by pouring a 1% solution of tetramethyl-*p*-phenylenediamine over the culture plate on which it is growing. Neisserial colonies become pink and then purple within a few minutes. This helps their detection in a mixed culture, but in order to survive they must be subcultured as soon as the colour change becomes apparent. A single colony can be tested by transferring part of it to a strip of filter paper that has been impregnated with the indicator; the colour then develops on the paper.

Carbohydrate oxidation by *N. meningitidis* and other neisseriae can be tested by growing them on a specially enriched nutrient agar to which have been added a sugar and an indicator. *N. meningitidis* produces acid from glucose and from maltose but not from sucrose, *N. gonorrhoeae* (see below) from glucose but not from the other two, and most of the commensal neisseria from all three sugars or from none of them.

ANTIGENIC STRUCTURE Seven or more serogroups are currently recognized; most epidemics of meningococcal meningitis are due to group A strains but most sporadic cases in Britain to those of group B.

IMMUNOLOGY Natural active immunity follows asymptomatic carriage of meningococci, and effective polysaccharide vaccines are available for groups A and C; but infants, the group most needing protection, fail to respond adequately to the vaccines (p. 55).

ANTIBACTERIAL TREATMENT Penicillin, to which all strains are sensitive, is the treatment of choice for meningococcal disease, but is not effective in prophylaxis. Sulphonamides were highly effective in treatment before bacterial resistance became a problem, and are still of great value as cheap and acceptable prophylaxis in major epidemics due to sulphonamide-sensitive strains. The only other effective prophylactic drug is rifampicin.

Neisseria gonorrhoeae (the gonococcus)

OCCURRENCE AND PATHOGENICITY This is also an obligate human parasite. It causes the sexually transmitted disease gonorrhoea (p. 279) and neonatal ophthalmia (p. 295).

CARRIERS The healthy carrier state probably does not exist, but women with chronic infections may have no symptoms.

MICROSCOPY The gonococcus is indistinguishable from the meningococcus by ordinary microscopy, but can be identified by the fluorescent-antibody technique (p. 165). Characteristic intracellular gram-negative diplococci are to be seen in urethral or cervical pus in the acute stages of gonorrhoea.

CULTURE The gonococcus behaves like the meningococcus in culture except that it will not grow on nutrient agar, has a more stringent requirement for CO_2 and has a narrower temperature range; its colonies are usually slower to appear and smaller.

BIOCHEMICAL REACTIONS See under *N. meningitidis*.

IMMUNOLOGY Immunity does not follow natural infection and cannot be produced artificially. The gonococcal complement-fixation test (GCFT) previously used, as an 'aid' to the diagnosis of chronic infections, is unreliable and can be misleading.

ANTIBACTERIAL TREATMENT Emergence of sulphonamide resistance was much more rapid and widespread among gonococci than among meningococci. The penicillins have remained effective far longer, but over many years it has been apparent in various parts of the world that the prevailing gonococcal strains have become less sensitive to penicillins, so that progessively higher doses have been necessary; more recently frankly penicillin-resistant strains, producing β-lactamase as a consequence of acquiring transmissible plasmids (p. 20), appeared in several parts of the world, particularly in Africa and Asia, and have become established in many countries, especially in 'international' cities such as London, Amsterdam, New York etc. Spectinomycin (p. 394) is commonly used as an alternative treatment when such strains are around.

Neonatal ophthalmia need never occur, as it can be prevented by the administration of silver nitrate or penicillin drops to the eyes of babies are soon as they are born (pp. 295–6).

Commensal neisseriae

Other species of the genus (e.g. *N. sicca, N. pharyngis*) make a large contribution to the normal commensal flora of the human upper respiratory tract. They are virtually never pathogenic. They are distinguishable from the pathogenic species by their growth on simpler media and at lower temperatures and by their different sugar reactions.

Branhamella (formerly Neisseria) catarrhalis

OCCURRENCE AND PATHOGENICITY Long recognised to be one of the commonest of nasopharyngeal commensals, *Bran. catarrhalis* occasionally causes acute otitis media and low-grade purulent infections of the lower respiratory tract in compromised patients (p. 204).

MICROSCOPY Branhamellae are indistinguishable from neisseriae in gram-stained films. In films of respiratory secretions from patients with branhamella infections, the diplococci are both intra- and extracellular.

CULTURE *Bran. catarrhalis* grows well on blood agar or 'chocolate' agar, producing grey-white glistening colonies that can be pushed across the agar surface.

BIOCHEMICAL REACTIONS Unlike neisseriae, *Bran. catarrhalis* does not ferment any common sugars. It does produce DNAase.

ANTIBACTERIAL TREATMENT Many strains produce β-lactamase and are thus resistant to penicillin and ampicillin. They are generally sensitive to cephalosporins and erythromycin.

Gram-Positive Bacilli

The genera of medical interest have the following characteristics:

Corynebacterium. Non-sporing; bacilli mostly in palisades or in 'Chinese characters' (see below); aerobic. (*Listeria* and *Erysipelothrix* are similar.)
Propionibacterium. Non-sporing; similar morphology to corynebacteria; anaerobic.
Lactobacillus. Non-sporing; bacilli commonly in chains; mostly micro-aerophilic or anaerobic.
Bifidobacterium. Non-sporing; branched or Y-shaped ('bifid') bacilli; anaerobic.
Bacillus. Sporing, with spores usually not exceeding the bacilli in diameter; aerobic.
Clostridium. Sporing, with spores usually wider than the bacilli; anaerobic.

Their differences in microscopic morphology are illustrated in Fig. 7.1 (p. 70).

The Genus Corynebacterium

The genus owes its name to the club-shaped swellings often seen at the ends of the bacilli, especially in old cultures. It includes one important human pathogen, the diphtheria bacillus. Other species cause suppurative diseases of animals. Various non-pathogenic corynebacteria, known as **diphtheroid** or **coryneform bacilli**, are normal commensals of human skin, upper respiratory tract, external ears and conjunctivae. Some of these have been given specific names—e.g. *C. hofmanni* and *C. xerosis*. Their properties will be mentioned only incidentally as part of our discussion of *C. diphtheriae*, but they are important to the clinical bacteriologist because they are present in a high proportion of the specimens that he receives.

Corynebacterium diphtheriae (the diphtheria bacillus)

OCCURRENCE AND PATHOGENICITY Diphtheria is now a rare disease in Britain, where widespread immunization against it was introduced in the early 1940s, and in other countries with comparable immunization programmes. However, it remains a common disease in other parts of the world; and rapid bacteriological confirmation of suspected cases is important in countries where it is rare. *C. diphtheriae* is an obligate parasite, with man as its only natural host. Typically, diphtheria is an infection of the pharynx (p. 192); more remote effects of the disease are due to an exotoxin which enters the patient's circulation and damages the heart, nervous system, liver, kidneys and adrenals. The toxin is a protein with two active sites—one part of the molecule attaches to specific receptors in the target cell membrane and the other, active part then inactivates elongation factor (transferase II), stopping polypeptide and protein synthesis and ultimately leading to cell death. One toxin molecule will inevitably kill any cell to which it has attached. Occasionally the primary infection is in the skin, a wound or the vagina; systemic disturbance is usually not severe in these cases. Three biotypes of *C. diphtheriae* (*gravis*, *intermedius* and *mitis*) can be differentiated in the laboratory and in very general terms are associated respectively with severe, intermediate and mild illnesses. Non-toxigenic (i.e. non-pathogenic) strains of all three varieties occur (see below); they are commonest in the *mitis* variety.

C. diphtheriae is not naturally pathogenic to animals, but guinea-pigs and rabbits are highly susceptible to injections of toxigenic strains or their toxins, and many other species are less markedly so.

CARRIERS These play an important part in the spread of the disease. Some are convalescent, but many have had only subclinical infections. While the throat and nose are the common sites of carriage, the organism is sometimes carried in an ear or elsewhere.

MICROSCOPY Corynebacteria divide like snapping sticks in which the bark fails to break on one side. In most diphtheroid bacilli the resulting arrangement tends to resemble a stake fence or palisade—rows of bacilli of rather irregular lengths lying side by side. The rods of C. diphtheriae are slightly curved, and therefore form less tidy bundles, conventionally likened to Chinese characters (Fig. 7.1, p. 70). An average diphtheria bacillus is of the order of 3 μm by 0.5 μm, and has terminal **volutin granules** which are clearly visible as metachromatic dots at the end of the bacilli in films stained by special procedures such as Albert's or Neisser's. Some of the diphtheroid bacilli also form such granules, but rarely in the characteristic bipolar distribution seen in C. diphtheriae. The appearances of C. diphtheriae in smears from pharyngeal or other lesions are not clearly distinguishable from those of commensal corynebacteria, and cannot be relied upon for diagnosis.

CULTURE C. diphtheriae grows in simple media, but better on those which contain serum or blood. On blood agar its inconspicuous small grey colonies are easily overlooked, or mistaken for those of some of the diphtheroid bacilli. Many *mitis* and some *gravis* strains are more noticeable because they are haemolytic.

For the isolation of the diphtheria bacillus from clinical material, two special media are commonly used. **Loeffler's serum** is an inspissated mixture of ox or horse serum and glucose broth. A wide variety of bacteria will grow on it, and colonial differentiation is very poor. However, the microscopic morphology and staining reactions of C. diphtheriae after incubation for 18 hours or less on this medium are so characteristic that it can be recognized even in films containing many other organisms, giving a presumptive diagnosis of diphtheria. Confirmation comes from the use of a selective **tellurite medium**—a blood or chocolate agar containing 0.04% of potassium tellurite, which suppresses the growth of most bacteria. It also slows even that of corynebacteria, so that they may take up to 48 hours to produce recognizable colonies. By naked-eye examination of these it is possible to distinguish C. diphtheriae from diphtheroid bacilli and from the few other species which are not inhibited by the tellurite. The three varieties of C. diphtheriae form colonies of somewhat different sizes and shapes, but all are predominantly dark slate-grey in colour, as a result of metabolism of the tellurite, whereas diphtheroid colonies are usually black, light grey or brown.

BIOCHEMICAL REACTIONS Grown in a suitable medium C. diphtheriae ferments glucose and maltose but rarely sucrose, whereas most diphtheroid bacilli ferment all three or none of these. Only *gravis* varieties of C. diphtheriae ferment starch. C. ulcerans, an uncommon cause of diphtheria-like throat infection, gives the reactions of C. diphtheriae var. *gravis* in these tests but can be differentiated by other biochemical tests.

TOXIN PRODUCTION This depends on a lysogenic relationship (see p. 23) between C. diphtheriae and a bacteriophage; strains that lack the phage are non-toxigenic. The presence of diphtheria toxin in a broth culture can be demonstrated by showing that it produces typical lesions when injected into guinea-pigs, and that these can be prevented by prior administration of diphtheria antitoxin to the animals. An *in vitro* method of demonstrating toxin production is **Elek's** double-diffusion procedure (p. 175). A strip of filter paper soaked in diphtheria antitoxin is incorporated in a plate of serum agar, and known and suspected toxigenic C. diphtheriae cultures are streaked in single lines across the plate at right angles to the strip. After incubation for 48 hours, during which time antitoxin diffuses out from the paper strip and toxin diffuses out from the growing toxigenic cultures, fine white lines of toxin–antitoxin precipitate can be seen in the medium, radiating out from the points at which toxigenic cultures cross the strip. Lines due to antigens other than diphtheria toxin may occur; but a true positive reaction is shown by the fact that the line from the suspected culture, on meeting that from the known toxigenic control strain, does not cross it but fuses with it to form an arch between the two cultures (the **reaction of identity**—p. 175).

IMMUNOLOGY Immunity follows natural infection, often without any clinical illness, and can be detected by studying the patient's response to intradermal injection of a small dose of toxin (the Schick test, p. 183). Antitoxin can also be measured in serum. Active immunization (p. 366) has played a large part in reducing the incidence of diphtheria. Passive immunization (i.e. the administration of serum containing antitoxin) is the only effective treatment for an established infection, and must be given at once to any patient suspected of having diphtheria, without waiting for laboratory confirmation of the diagnosis.

ANTIBACTERIAL TREATMENT *In vitro* the diphtheria bacillus is sensitive to most antibiotics, but these are of limited value in treatment because they cannot deal with toxin already in the patient's body. However, penicillin and other antibiotics can be used to prevent the spread of infection and to stop cases from becoming carriers, and they are occasionally effective in dealing with established carriers. Prophylactic erythromycin is recommended for susceptible close contacts.

The Genus Propionibacterium

Anaerobic bacilli, morphologically 'diphtheroid', with propionic acid as the major product of their metabolism, are common commensals of man and animals. They are found in the intestine and on the skin, where *P. acnes* in particular is an important member of man's microflora. This organism breaks down lipid components of sebum, and the acid products of this process are inhibitory to many potentially pathogenic bacteria; but in excess that may be partly responsible for the inflammatory element of acne (p. 306). Such bacilli are virtually never directly pathogenic, but may colonize artificial heart valves and other prostheses, with serious consequences.

Listeria monocytogenes and Erysipelothrix rhusiopathiae

These two species, which have many properties in common with one another and with the corynebacteria, are both animal pathogens that occasionally cause disease of man. Both are sensitive to most of the common antibiotics.

Listeria monocytogenes

This aerobic non-sporing gram-positive bacillus differs from the corynebacteria in being flagellate and feebly motile (p. 90), in being agglutinable by known anti-listerial sera and in causing monocytosis in rabbits. Unlike most parasites of man, it grows well at temperatures down to 4°C; and it is only at temperatures of around or below 20°C that it is motile (in an unusual tumbling style). In man it causes meningitis or septicaemia in neonates (p. 325) or in those who are elderly, debilitated, suffering from malignant disease or on immunosuppressive drugs. Aysmptomatic vaginal carriage by the mother is the probable source of most neonatal infections. It is said to cause mild fever in some pregnant women, sometimes followed by abortion or still-birth. Most listeria infections respond to treatment with a combination of ampicillin and gentamicin.

Erysipelothrix rhusiopathiae

This organism resembles *Listeria monocytogenes* but is non-motile and is usually micro-aerophilic. It is widely distributed in nature, causing swine erysipelas and other animal diseases, and it is present on the skin and scales of many kinds of fish. Erysipeloid (p. 306), the human condition for which it is responsible, occurs mainly among those who handle meat, poultry or fish. Penicillin is the drug of choice for treating these infections.

The Genus Lactobacillus

This genus of long gram-positive bacilli, commonly occurring in chains (Fig. 7.1, p. 70), deserves to be ranked as important in medical bacteriology because its members make a substantial contribution to man's normal flora. They are found in the mouth, the stomach, the intestine and the vagina (where they perform an important protective function—see pp. 16, 29), and are particularly abundant in the faeces of milk-fed babies. They are mostly anaerobic or micro-aerophilic, some of them will multiply only in an acid environment and they have unusual nutritional requirements. Consequently the average clinical bacteriologist does not grow many of them, and he has little incentive to try to do so, since they are generally regarded as non-pathogens.

The Genus Bifidobacterium

These anaerobic bacilli have much in common with the lactobacilli, but differ in that their cells are characteristically branched or Y-shaped and that they produce more acetic and less lactic acid. They form a major component of human faeces, where their numbers equal those of *Bacteroides* species (p. 30) at about 10^{11} organisms per gram of wet faeces. They are particularly numerous in the faeces of breast-fed infants where they help to maintain the acetic acid/acetate buffer system and so the acidic pH. They have virtually never been incriminated as pathogens.

Gardnerella vaginalis

This small bacillus that generally appears to be a mixture of gram-positive and -negative cells is associated with anaerobic vaginosis (p. 281). In the vaginal discharge it is present in large numbers, many *Gardnerella* cells adhering to epithelial cells ('clue' cells). It can also be found in women without vaginosis.

The Genus Bacillus

The sporing aerobic bacilli include one highly pathogenic species, *B. anthracis*, and a large assortment of other species that are almost always saprophytic.

Bacillus anthracis (the anthrax bacillus)

OCCURRENCE AND PATHOGENICITY Mainly a pathogen of herbivorous animals, in which it produces illnesses varying from a fulminating haemorrhagic septicaemia to a chronic fever with pustules, *B. anthracis* is an occasional cause of human disease. This may take the form either of cutaneous anthrax (p. 307) or of pulmonary anthrax ('wool-sorter's disease', p. 202). Whereas man does not acquire anthrax readily, cattle and sheep are highly susceptible to natural infection. It also occurs in other herbivorous mammals, including goats, horses and camels. All of these are infected by ingesting spores from contaminated pasture.

MICROSCOPY Anthrax bacilli are large—4–8 μm long and 1 μm or more wide. Their spores are near the centres of the bacilli, and do not distend them (Fig. 7.1, p. 70); they can be seen as non-staining areas in gram-stained films and can be demonstrated even more clearly by special spore stains. The rods are non-flagellate and, in culture, non-capsulate. They are square-ended and are often arranged in chains.

In smears of vesicle fluid or other pathological material the presence of chains or large gram-positive rods is strong evidence of anthrax, but not conclusive, since some of the other members of the genus can present similar appearances. In the body of its host *B. anthracis* does not form spores, but usually does have a capsule. Whereas most other bacterial capsules are composed of polysaccharides, that of *B. anthracis* contains a polypeptide. It gives a characteristic purple staining reaction with polychrome methylene blue (McFadyean's reaction).

CULTURE *B. anthracis* grows well aerobically and anaerobically on most common media, but forms spores only in aerobic conditions. Its opaque, greyish-white, rough-surfaced colonies resemble tangles of fine hairs, often with loose curls protruding at the edges ('Medusa-head' colonies).

ANIMAL INOCULATION The identity of the cultured organism is confirmed by production of the characteristic disease in inoculated mice or guinea-pigs.

IMMUNOLOGY Pasteur's conclusive and dramatic demonstration, at Pouilly-le-Fort in 1881, of the possibility of immunizing animals against anthrax is one of the milestones of bacteriology. Pasteur used a vaccine of *B. anthracis* attenuated by prolonged growth at 42°C or higher. It has not been possible to produce in this way vaccines that are consistently effective and safe enough to be given to man, but dockers, factory workers and others whose work exposes them to special anthrax hazards can now be safely immunized with a non-living protein antigen precipitated from bacterium-free filtrates of broth cultures of *B. anthracis*.

ANTIBACTERIAL TREATMENT Passive immunization, using serum of immunized animals, has been superseded by antibiotic treatment. The organism is usually sensitive to penicillin and to several other antibiotics.

Disposal of animal carcasses and other control measures are discussed in Chapter 26.

Other members of the genus Bacillus (sometimes called anthracoid bacilli or aerobic spore-bearers)

This is a large and varied group. Its members all form spores and grow better in air than in anaerobic conditions. Many strains are only weakly gram-positive, often with a characteristic mottled blue-and-red appearance, and a few are frankly gram-negative. Species differ in the shapes and sizes of their bacilli. The spores are usually narrower than the bacilli, but in a few species are wider and cause distension. Colonial appearances are diverse, and in some cases bizarre. In general, organisms of this group are harmless saprophytes; but they are responsible for very rare cases of meningitis, endocarditis, pneumonia or septicaemia in debilitated patients, and *B. subtilis* occasionally causes severe eye lesions (iridocyclitis and panophthalmitis). *B. cereus* is an important cause of food-poisoning (p. 258). Members of this genus can be troublesome contaminants of bacterial culture media, blood intended for transfusion, other fluids that should be sterile, specimens from patients, and foods. They can grow in varied conditions, and their spores survive adverse conditions that kill most other bacteria. The bacteriologist makes use of these features in testing the efficiency of sterilization methods. Heat or irradiation or other physical treatment that can kill organisms, including spores, of this group—usually represented by *B. subtilis*, the 'hay bacillus', or by *B. stearothermophilus*—is also sufficient to kill all pathogens, including spore-bearers of the next genus to be discussed.

The Genus Clostridium

This genus consists of anaerobic gram-positive bacilli which form spores that in most cases distend their bodies (Fig. 7.1 (i and j), p. 70). Such organisms are widely distributed in nature as soil saprophytes and as intestinal commensals of mammals. They include the causative organisms of some very serious human diseases—botulism, tetanus, gas-gangrene and several intestinal infections.

Clostridium botulinum

OCCURRENCE AND PATHOGENICITY This species is to be found in soil in many parts of the world. Botulism is due not to bacterial infection but to eating food that contains the very powerful neurotoxin of *Cl. botulinum* (p. 258). This toxin causes paralysis by blocking the release of acetylcholine from motor neurones (neuromuscular blockade). A different and usually much less severe form of botulism occurs in infants (p. 327), and botulism following *C. botulinum* infection of wounds has been described.

There are at least five different types of *Cl. botulinum*, producing slightly different toxins. Types A, B and E cause botulism in man, whereas types C and D cause similar natural diseases of animals. Large outbreaks occur in birds living on and around estuaries and lakes. Laboratory animals of many sorts are susceptible to the effects of these toxins given by injection.

MICROSCOPY The individual bacilli, like those of most clostridia, are intermediate in size between those of the genera *Corynebacterium* and *Bacillus*. They are flagellate and have oval spores which are subterminal in position—i.e. near one end of the rod.

CULTURE This species is strictly anaerobic. It grows best at temperatures below 37°C and has simple nutritional requirements, as would be expected of a saprophyte. Like many other members of the genus, it forms diffuse, greyish, semi-transparent colonies of irregular shape.

BIOCHEMICAL REACTIONS The different clostridial species can be distinguished by their patterns of carbohydrate fermentation (cf. the enterobacteria, p. 91) and by their ability to liquefy coagulated serum and to digest meat. They also have characteristic patterns of volatile fatty acid production, as detected by GLC (p. 73).

IMMUNOLOGY Active immunization is not indicated for such a rare disease, and serological diagnosis is inappropriate for one of such short duration. Antiserum, produced in animals, can be given to counteract the toxin, but its efficacy is low even in those patients who receive it between eating suspected food and developing symptoms. Unless the type of the *Cl. botulinum* strain is known, a polyvalent serum that contains antibodies to the toxins of all types should be used.

ANTIBACTERIAL TREATMENT Since the usual adult form of botulism is not an infection, there is no indication for using antibiotics.

Clostridium tetani (the tetanus bacillus)

OCCURRENCE AND PATHOGENICITY Tetanus spores are particularly common in soil containing animal manure. When introduced by trauma into the tissues of man, the horse or various other animals, they may cause tetanus, but as a rule only if certain conditions are fulfilled. These include interference with the blood supply to at least a small amount of tissue, and multiplication of other organisms so that all oxygen is used up; the presence of soil increases the risk of tetanus, probably because calcium salts encourage germination of the spores. However, tetanus may follow puncture wounds or abrasions so trivial that they go unnoticed, and it is hard to see how the necessary conditions are then fulfilled (p. 290).

TOXIN Pathogenicity of *Cl. tetani* is determined by the production of a neurotoxin. Days, weeks or even months after infection, local multiplication of the organism occurs, with liberation of toxin. This does no harm locally, but travels along peripheral nerves to the central nervous system; here it blocks certain synapses, notably those that modify or suppress the effects of limb reflex arcs, and so causes the characteristic spasms in response to even mild stimuli. Toxin also spreads by the bloodstream to nerves elsewhere in the body and travels up these. In the laboratory this powerful exotoxin (neurotoxin), second in potency only to that of *Cl. botulinum*, can be identified by the characteristic spasticity and convulsions that follow injection of broth-culture filtrate into mice and guinea-pigs, and by the neutralization of these effects by specific antitoxin. Most mammals are susceptible to the toxin.

MICROSCOPY In films from cultures or from pathological material *Cl. tetani* has a typical 'drum-stick' appearance, due to the presence of large, round, terminal spores (Fig. 7.1 (j), p. 70). However, this appearance can be mimicked by other clostridia, such as the non-toxigenic *Cl. tetanomorphum*, and by some of the aerobic spore-bearers. *Cl. tetani* is usually flagellate and motile (p. 90).

CULTURE A strict anaerobe with an optimal growth temperature of about 37°C and growing well on routine media, *Cl. tetani* forms flat, translucent spreading colonies with fine finger-like projections. Its swarming growth and the heat resistance of its spores are the basis of procedures for isolating it from the mixture of organisms often present in material from lesions. The material, or a suspension or mixed broth culture made from it, is heated to 65°C for 30 minutes to destroy all non-sporing organisms, and is then used to inoculate one side of a plate or the bottom of a slope of blood agar. When the growth begins to swarm across the plate or up the slope the advancing edge is subcultured to another plate or slope. This is repeated if necessary.

BIOCHEMICAL REACTIONS *Cl. tetani* is unique among medically important clostridia in that it usually fails to ferment any carbohydrates.

IMMUNOLOGY Active immunization with toxoid, when established and maintained, prevents tetanus and avoids the risks of antiserum treatment (p. 367).

ANTIBACTERIAL TREATMENT Penicillin and some other antibiotics inhibit *Cl. tetani* in the laboratory, but they have little relevance to the established disease, except in preventing other bacterial infections. However, there is a place for their prophylactic use, along with careful surgical elimination of suitable sites for clostridial germination, in some types of high-risk injury.

Clostridium perfringens (Cl. welchi)

OCCURRENCE AND PATHOGENICITY This species is common in the intestines of animals and of man and in soil. It may gain entry into dirty accidental wounds from contaminated soil but its entry into surgical wounds is usually from the patient's own intestines, via the skin. It is unlikely to cause trouble if there is little or no damaged devitalized tissue in the wound, or if such tissue is promptly and thoroughly excised, but in the presence of damaged ischaemic tissue gas-gangrene (clostridial myositis) may develop. This is due to the multiplication of *Cl. perfringens* or other clostridia (or less commonly of anaerobic streptococci), which produce exotoxins that digest muscle and subcutaneous tissues. This leads to liberation of gas and noxious metabolites, and further reduces the blood supply to the tissue compartments by increasing their internal pressure. Clostridial cellulitis is a less serious infection, which causes much gas-formation and crepitation of the tissues and so is often confused with gas-gangrene but does not destroy muscle. Further details of these infections are given on p. 291.

Numerous *Cl. perfringens* exotoxins have been identified, and strains producing different combinations of these are classified into six types, A–F. Those that cause human gas-gangrene belong to type A. Types B–E are all associated with enterotoxaemia in sheep or cattle; types C and F occasionally cause a similar disease in man, enteritis necroticans; and type C also causes pig-bel (p. 252). Type A strains are atypical in that they survive prolonged boiling (whereas ordinary strains are killed in about 5 minutes) and are responsible for many outbreaks of food-poisoning (p. 257).

MICROSCOPY The bacilli are stouter than those of most clostridia, measuring about 4–8 μm by 1 μm. Capsules may be present in films made from pathological materials but are not formed in culture. Most strains do not form spores in culture, unless fermentable carbohydrate is absent and the medium is alkaline; but small oval subterminal spores are formed freely in natural conditions. The species is not flagellate.

CULTURE. *Cl. perfringens* is a strict anaerobe with an optimal growth temperature of about 37°C. Its colonies on blood agar are usually large round moderately opaque discs surrounded by large zones of haemolysis. Selective media containing an aminoglycoside antibiotic (usually neomycin) are often used for primary culture.

BIOCHEMICAL REACTIONS See under *Cl. botulinum*

ALPHA-TOXIN Of the many exotoxins formed by this species, the α-toxin is common to all types and is believed to be largely responsible for the toxaemia of human gas-gangrene. It is a relatively heat-stable phospholipase C (p. 41), and is the basis of the **Nagler reaction**, by which *Cl. perfringens* can be recognized even in mixed culture (p. 90). Material suspected of containing this organism is streaked across a plate of a transparent medium containing human serum or egg yolk, a small amount of *Cl. perfringens* α-antitoxin having first been spread over half of the plate. After overnight incubation, if *Cl. perfringens* was present in the inoculum, a zone of increased opacity will have been produced by the phospholipase in the medium around and beneath the growth on the untreated side of the plate; but on the other side, although the growth is equally good, there is no increased opacity because the phospholipase has been neutralized by the antitoxin. Some other organisms produce phospholipase, but only *Cl. perfringens* α-toxin and the closely related toxin of *Cl. bifermentans* are inhibited by *Cl. perfringens* antitoxin.

IMMUNOLOGY There is no indication for active immunization. Passive immunization, both prophylactic and therapeutic, with polyvalent sera against the toxins of all of the common gas-gangene organisms, may be used but is of limited value.

ANTIBACTERIAL TREATMENT Penicillin is the antibiotic of choice for treatment of gas-gangrene; metronidazole is also effective.

Other gas-gangrene bacilli

Cl. novyi (Cl. oedematiens), Cl. septicum, and, less frequently, *Cl. bifermentans, Cl. histolyticum* and *Cl. fallax* are all capable of causing gas-gangrene similar to that caused by *Cl. perfringens*. In view of the nature of the infection and the similar habitats of the various species, it is not surprising that the same lesion often yields two or more of these species. They are distinguishable by the shapes and sizes of their bacilli, the shapes and situations of their spores, their cultural and biochemical properties and the toxins that they form.

Clostridium difficile*

Disturbance of the ecological balance of the intestinal flora by antibiotics—notably lincomycin or clindamycin—or by other factors sometimes permits proliferation of *Cl. difficile*. This species, commonly found in the faeces of healthy neonates but usually hard to detect in those of healthy adults, produces an exotoxin and causes pseudomembranous colitis (PMC—p. 259), which may be severe and even fatal if not treated promptly and appropriately. Oral vancomycin is the antibiotic treatment of choice for PMC, with metronidazole as a less well-tried alternative.

* This is a Latin word, pronounced as four syllables, not three as in French.

Gram-Negative Bacilli: (a) Enterobacteria

Many species of aerobic, facultatively anaerobic gram-negative bacilli that ferment (in the strict sense—p. 15) glucose and other carbohydrates and are oxidase-negative (p. 81) are commonly present in the human intestine, usually as commensals but some of them as actual or potential pathogens. Such organisms are also widely distributed in nature. They are collectively known by the informal name **enterobacteria** or the official family name Enterobacteriaceae. The name 'coliform' bacilli is used by some to indicate the same range of organisms.

To save repetition, we shall deal with the common properties of the enterobacteria and the chief procedures used for differentiating between them before we discuss the individual genera *Escherichia, Klebsiella, Salmonella, Shigella,* *Proteus* and *Yersinia*, with brief references to others.

MICROSCOPY In gram-stained films all enterobacteria are indistinguishable medium-sized bacilli, about 0.5 μm by 1–3 μm (Fig. 7.1, p. 70), with some (at times many) filamentous forms. Some of the group—notably the kelbsiellae and many *Esch. coli* strains—have capsules. Possession of flagella can be an important criterion for differentiation—e.g. between the genera *Salmonella* (flagellate) and *Shigella* (non-flagellate), or *Klebsiella* (non-flagellate) and *Enterobacter* (flagellate); or between the species *Yersinia pestis* (non-flagellate) and other yersiniae (flagellate). Direct demonstration of flagella by light microscopy is difficult (p. 71) but **motility**, the effect of flagella, is relatively easily detected by direct microscopy of a drop of unstained fluid culture under a microscope, using reduced lighting (p. 164). An alternative non-microscopic method depends on the ability of motile organisms to make their way through a semi-solid medium (0.2% agar).

CULTURE Enterobacteria grow well over a wide range of temperatures on or in simple media such as nutrient agar or peptone water (1% commercial peptone with 0.5% NaCl) under aerobic conditions, but somewhat less well anaerobically. After aerobic incubation in blood agar, colonies are usually rather large (3–4 mm in diameter after 18 hours) and greyish-white, smooth and moderately opaque; however, dry rough-surfaced colonies, large mucoid colonies and other variants occur. Haemolysis may occur round the colonies, but is not a useful differentiating feature in this group. Having modest nutritional requirements and being tolerant of bile (understandably, in view of their intestinal habitat), enterobacteria grow on MacConkey's agar (p. 167). This medium, frequently used to separate them from a mixture of other bacteria, also provides a useful distinction between the red colonies of lactose-fermenters (the usual behaviour of *Esch. coli* and klebsiellae) and the straw-coloured colonies of non-lactose-fermenters (salmonellae, most shigellae, proteus and yersiniae).

BIOCHEMICAL REACTIONS Lactose is only one (though the most useful) of a wide range of carbohydrates that can be employed in fermentation tests for the identification of enterobacteria. If such tests are carried out in liquid media, small inverted tubes or other means can be used to detect generation of gas bubbles during fermentation and so to increase the information derived. For example, fermentation of glucose and failure to ferment lactose and sucrose are properties common to all salmonellae; but *S. typhi* almost alone in this genus fails to produce visible amounts of gas during glucose fermentation. Fermentation tests using 6 or 8 sugars or alcohols, supported by some other biochemical tests (see below) permit identification of most enterobacteria to species level.

Other biochemical tests commonly used are those for the production of H_2S and indole when growing in peptone water; for splitting of urea with liberation of ammonia (the urease test); for multiplication in a medium in which citrate is the only carbon source; for decarboxylation of various amino-acids; for formation of acetylmethylcarbinol when growing in glucose-phosphate broth (the Voges–Proskauer test); and for lowering of the pH in glucose broth to 4.5 or less (the methyl red test).

ANTIGENIC STRUCTURE Of the many enterobacterial antigens that have been studied, most can be classified according to their location as H (flagellar), O (somatic) or K (capsular or envelope). (The letters H and O are taken from the German terms *Hauch* and *ohne Hauche*, because flagellate proteus strains form spreading films across culture plates, whereas non-flagellate forms are 'without film'.) The relevance of antigenic analysis to our understanding of the various enterobacterial genera is indicated at appropriate places below.

The Genus Escherichia (See also above)

Esch. coli is the only species currently assigned to this genus.

OCCURRENCE AND PATHOGENICITY *Esch. coli* is a normal inhabitant of the intestine of man and animals, but is not always harmless there. In Britain and other temperate countries strains of a limited but growing number of O and K serotypes have been shown to be responsible for gastro-enteritis in infants and young children, though by no means all strains of those types are potential pathogens (p. 252). The pathogenic mechanism of these 'infantile enteropathogenic' strains is still debatable, but strains of other serotypes ('enterotoxigenic') cause gastroenteritis in adults as well as children, mainly in tropical countries, by producing enterotoxins similar to that of *Vibrio cholerae* (p. 97); and still other strains ('entero-invasive') resemble the shigellae in their ability to invade and destroy intestinal epithelial cells. Adult infections are rare in Britain, but account for many cases of 'traveller's diarrhoea'. The other main contribution of *Esch. coli* to human ill-health is made in the urinary tract; it is responsible for 80% or more of all infections of this tract that occur outside hospital, and a smaller proportion of those acquired in hospital (p. 274). *Esch. coli* can also cause biliary tract infections; and it is often associated with *Bacteroides* species (p. 102–3) in intra-abdominal abscesses and infected wounds, particularly those of the abdominal wall. From such sites it may reach the bloodstream, and may cause septicaemia. It is a common cause of meningitis in the first few weeks of life.

ANTIBACTERIAL TREATMENT Enterobacteria, including *Esch. coli*, are resistant to penicillin, cloxacillin, erythromycin, clindamycin and fucidin. Many *Esch. coli* strains are sensitive to ampicillin and some other 'broad-spectrum' penicillins and to the cephalosporins; but strains which produce enzymes (β-lactamases—p. 382) that destroy some of these drugs are increasingly common, especially in hospitals because these provide a selective environment for their propagation. Similarly, many strains are sensitive to sulphonamides, trimethoprim, tetracyclines, aminoglycosides, chloramphenicol and various other drugs; but resistance to one or more is common, especially in hospital strains. When septicaemia or some other serious infection has to be treated urgently, without waiting for results of sensitivity tests on the causative *Esch. coli* strain, a drug to which it is unlikely to be resistant should be used—e.g. gentamicin or one of the newer cephalosporins. Nitrofurantoin or nalidixic acid are sometimes effective in urinary tract infections (p. 000).

The Genus Klebsiella (See also pp. 90–91)

Klebsiellae are non-motile, which distinguishes them from the less important genus *Enterobacter*. They have polysaccharide capsules by means of which, like pneumococci, they can be divided into a large number of serotypes. The more pathogenic members of the genus are included in the first three types. Type 3 includes the organism, originally known as Friedlander's bacillus, which in some countries (though not in Britain) is a quite common cause of pneumonia (p. 202); it can also cause meningitis, otitis and sinusitis. Some use the species name *K. pneumoniae* for this organism, and the names *K. edwardsi* and *K. atlantae* for organisms of somewhat similar pathogenicity that belong to types 1 and 2; whereas others include all of these in *K. pneumoniae*. Most other klebsiellae are classified as *K. aerogenes*, a species that includes strains of many serotypes. These are widely distributed in nature, are commonly found in the human intestine, and occasionally cause urinary tract, wound or other infections. Heavy growths of klebsiellae (or other enterobacteria) are commonly found in purulent sputum from patients whose respiratory tract ecological patterns have been disrupted by treatment with antibacterial drugs to which these gram-negative bacilli are resistant (as klebsiellae commonly are, for example, to ampicillin and the older cephalosporins). They are seldom pathogenic in such situations, and further antibiotic treatment is seldom advisable; stopping all antibiotics usually results in a fairly rapid restoration of the normal bacterial population.

The Genus Salmonella (See also pp. 90–91)

Virtually all members of this genus are motile, though non-flagellate variants occur. They are all intestinal pathogens. *S. typhi*, *S. paratyphi* A, *S. paratyphi* B and *S. paratyphi* C cause the septicaemic illnesses collectively known as **enteric fever** (p. 231). Other members of the genus occasionally produce a similar picture, but far more commonly they cause the more localized condition known as gastro-enteritis or food-poisoning (p. 256).

The subdivision of this genus is unusual in that it is based mainly upon antigenic analysis. Most strains possess two or more somatic (O) antigens and one or more flagellar (H) antigens. On the basis of O antigens alone the species can be divided into ten major and some minor groups, and even then the members of one group are not identical in their O antigenic structure but simply have one or more antigens in common. Each group can then be subdivided according to the H antigens of its members. In this way over 1800 salmonella types have been distinguished and many have been allotted bimomials which suggest that they are species—e.g. *S. typhimurium*, *S. enteritidis*. However, antigen analysis is a much more delicate means of subdivision than those applied to other genera, and the majority of these different salmonellae, which cannot be distinguished by their microscopic or colonial appearances or, in most cases, by their biochemical reactions, are more reasonably described as **serotypes** rather than as separate species.

Analysis of the H antigenic structure of salmonellae is further complicated by the fact that many types exist in either of two phases, differing in their H antigens. Types which differ when in phase 1 may have identical phase-2 H antigens. Thus *S. typhimurium* has the same phase-2 antigens (designated 1 and 2) as *S. derby*, *S. heidelberg* and a number of other types that are also identical with it in their O antigen composition. When in phase 2, therefore, these types are indistinguishable. However, an apparently pure phase-2 culture nearly always contains a very small minority of phase-1 forms (and vice versa), and there are ways of extricating a pure culture of the minority phase. In phase 1, *S. typhimurium* has antigen i, *S. derby* has f and g; *S. heidelberg* has r; and so on.

This fine differentiation of salmonellae, which in some cases can be carried still further by detecting variations of phage type or of biotype within a single serotype, increases the precision and certainty with which the sources and methods of spread of infection can be traced. The finding of a salmonella carrier may or may not be relevant to a particular outbreak; the finding of a carrier of a strain identical with that which caused the outbreak is far more likely to be relevant; and that likelihood increases in step with the accuracy with which the identity of the two strains is established.

Salmonella typhi (the typhoid bacillus)

OCCURRENCE AND PATHOGENICITY Like all salmonellae, this species is entirely parasitic. It differs from many of the others in that man is its only natural host, and that even in the laboratory it is of low virulence for mice and other animals. Although epidemics are usually spread via water supplies or food, the source of the organism is always a human patient or carrier.

In its early stages enteric fever (whether due to S. typhi or to one of the other salmonellae—p. 231) is predominantly a septicaemia (p. 60) rather than an alimentary disorder.

CARRIERS A small proportion of those who recover from typhoid continue to harbour the bacilli in their gall bladders or kidneys, and many excrete them intermittently for many years.

CULTURE Salmonellae, like all enterobacteria, grow well on simple media and are therefore easy to isolate from such specimens as blood and urine. Isolation from faeces is more problematical, because of the many other bacteria present; even MacConkey's medium (on which, being non-lactose-fermenters, salmonellae form colourless colonies) is of little use for this purpose, as the salmonellae are often heavily outnumbered by other enterobacteria that grow equally well on this medium. Various special **selective media** have been devised, such as **deoxycholate citrate agar** (DCA), which suppress the growth of most bacteria other than salmonellae and shigellae. When the number of salmonellae in the specimen is very small, even such a medium as this may not permit their detection, and the chances of isolating them are then considerably enhanced by using a **selective enrichment medium**, such as **selenite F broth**. In this, the lag phase (p. 21) of salmonellae is considerably shorter than that of other bacteria. From being a very small minority they may therefore become, within a few hours, a considerable proportion of the population, so that they are easily detected when the broth is subcultured.

During the first week of typhoid the organism can usually be grown from the blood. Faecal culture is more often positive in the second and third weeks. The bacillus may also be found in the urine after the second week.

PHAGE TYPING Some dozens of S. typhi types have been identified by means of Vi phages, so called because they are effective only against strains that still possess the Vi antigen (see below). The typing procedure is in most respects similar to that for staphylococci (pp. 23 and 75).

ANTIGENIC STRUCTURE S. typhi has only a single phase-1 H antigen, which it shares with a few other salmonellae but not with any that have the same O antigens; and it has no phase-2 antigen. Freshly isolated strains have a surface antigen, designated Vi (for virulence). Some strains of S. paratyphi B and a few other salmonellae have the same Vi antigen.

IMMUNOLOGY There is no conclusive evidence that useful immunity follows natural infection. Active immunization, using suspensions of killed bacilli (p. 366), causes pronounced antibody responses, but whether it also gives a useful degree of protection has been far more difficult to establish, and depends on the method of preparation of the antigen (p. 366). H, Vi and O antibodies can be measured separately in a patient's serum (the **Widal test**, p. 180). They may give useful diagnostic information from the second week of illness onwards.

ANTIBACTERIAL TREATMENT Chloramphenicol has been the mainstay of antibacterial treatment of typhoid; it controls the acute illness and greatly reduces the mortality. Unfortunately it fails to eradicate the organism, so that relapse and carriage are common after its use; and in recent years its usefulness has been much reduced in many parts of the world by the emergence of chloramphenicol-resistant strains of S. typhi.

Alternative drugs for treatment are amoxycillin (which is better than chloramphenicol at eradicating carriage) or cotrimoxazole.

The paratyphoid bacilli

S. paratyphi A, B and C, which also cause enteric fever but usually a milder form than that due to *S. typhi*, differ from that species and from one another in their biochemical properties and antigenic structure and in their geographical distribution. *S. paratyphi* B is the only one to occur at all commonly in Britain. Like *S. typhi* it can be divided into many phage types.

Other salmonellae

Most of the other salmonellae are primarily animal pathogens that occasionally attack man, usually following transmission in food, causing vomiting and diarrhoea (gastro-enteritis or food-poisoning—p. 256).

S. typhimurium, which has many other animal hosts besides the mice from which it takes its name, is the commonest cause of human salmonellosis in Britain. Many other types make their contributions, and their relative frequencies vary from year to year. Treatment of salmonella enteritis with antibiotics is seldom beneficial (unless there is accompanying bacteraemia) and, probably because of its disruptive effect on the normal intestinal flora, it is liable to prolong the period of salmonella carriage.

The Genus Shigella (See also pp. 90–91)

OCCURRENCE AND PATHOGENICITY The dysentery bacilli are obligate parasites of man or occasionally of chimpanzees or monkeys. They cause illnesses which vary in severity according to the species involved—in the general order *Sh. dysenteriae*, *Sh. flexneri*, *Sh. boydi* and *Sh. sonnei*. Their pathogenicity is thought to depend on destructive invasion of the intestinal mucosa, but the fiercer onslaught of *Sh. dysenteriae* includes production of a cholera-like enterotoxin. Systemic invasion is rare. *Sh. sonnei*, the mildest of the four, is the commonest in Britain and causes many epidemics in institutions. Young children are particularly susceptible.

CARRIERS Intestinal carriage is usually of short duration but may persist for years. In most cases excretion is intermittent, but that of *Sh. dysenteriae* may be continuous.

CULTURE Growth on most media is similar to that of other enterobacteria. Isolation of shigellae from mixed cultures is helped by selective media such as DCA (p. 93), on which they form colourless or, in the case of *Sh. sonnei*, pale pink ('late lactose-fermenting') colonies.

TYPING Shigellae have no H antigens, but they can be divided into serogroups and serotypes based on their O antigens. They can also be typed according to their production of **colicines**—antibiotics of a class produced by many enterobacteria, effective only against limited ranges of other enterobacteria, and therefore classifiable by determining the action of each one on a standard set of test strains.

IMMUNOLOGY Immunity following natural infection is type-specific and transitory. There are no generally accepted procedures for either active or passive immunization.

ANTIBACTERIAL TREATMENT Many antibacterial drugs are active against shigellae *in vitro*, but their clinical value is less certain. Widespread use of almost any of them is liable to be followed by rapid emergence of strains resistant not only to the drug used but often to various others as well (see p. 382—transferable multiple antibiotic resistance). Since many cases, especially of *Sh. sonnei* infection, recover spontaneously and completely within a few days, antibacterial treatment should be used only for seriously ill or very frail patients or persistent carriers.

The Proteus Group (See also pp. 90–91).

The members of this group are commonly found in faeces, in soil and in other moist situations. They resemble *Esch. coli* in their ability to cause urinary tract infections, and in their involvement in wound infections, especially in debilitated patients. Members of the genus *Proteus* swarm over the surfaces of many solid culture media, burying the colonies of other organisms, to the annoyance of the bacteriologist who is trying to obtain those other organisms in pure culture; he has to resort to special inhibitory media or procedures. The genus consists of two species, *Pr. mirabilis* and *Pr. vulgaris*, of which the commonest, in spite of their names, is *Pr. mirabilis*. Other members of the group have been allocated to the genera *Morganella* and *Providencia*. Many isolates are resistant to various commonly used antibiotics, and are therefore liable to persist in wounds and other lesions after treatment has disposed of staphylococci or other primary pathogens.

The Genus Yersinia (See also pp. 90–91)

This genus contains three species of medical interest: *Y. pestis* and *Y. pseudotuberculosis* (both previously classified in the genus *Pasteurella*) and *Y. enterocolitica*.

Yersinia pestis

OCCURRENCE AND PATHOGENICITY The causative organism of plague (the Black Death of the Middle Ages, still endemic in many countries) is primarily a flea-borne pathogen of rats and other rodents, among which it causes highly lethal epidemics. Transfer to man occurs when an infected rat-flea (*Xenopsylla cheopis*) abandons the dead body of its rodent host and then bites its new human host. The human host rapidly becomes ill with bubonic plague (p. 244), which may lead to pneumonic plague (p. 202). (**Bubo** is the name given to the characteristic swelling of regional lymph nodes.)

MICROSCOPY In tissues and other pathological materials, such as bubo aspirate and sputum, *Y. pestis* is a short oval capsulate non-motile gram-negative bacillus. With methylene blue and various other dyes it shows bipolar staining—i.e. deeper staining at the ends of the rods than in their centres (Fig. 7.1, p. 70). In culture the capsule is lost, bipolar staining is less obvious and the bacilli are often longer and pleomorphic.

CULTURE This species is unusual among human pathogens in that it grows best at about 27°C (p. 16). In other respects its laboratory behaviour is similar to that of other enterobacteria.

Yersinia pseudotuberculosis and Yersinia enterocolitica

Each of these organisms has been isolated from many species of domesticated and wild animals. In man, *Y. pseudotuberculosis* can cause a fulminating typhoid-like septicaemia, which is fortunately rare; *Y. enterocolitica* can cause enterocolitis (p. 252); and either of them can cause mesenteric adenitis (p. 245). The organisms (both resembling *Y. pestis* in most laboratory properties) can be isolated from blood, faeces, lymph nodes or other appropriate material, and rising levels of specific antibodies can be detected in serum. Isolation of *Y. enterocolitica* from faeces is facilitated by 'cold enrichment'—i.e. keeping the specimen for three weeks at 4°C (a temperature that suits yersiniae better than most other bacteria present) and culturing it from time to time during that period.

Other Enterobacteria

Organisms of genera such as *Enterobacter*, *Citrobacter* and *Serratia* were formerly of interest to the medical bacteriologist only because they are liable to be found in the same situations as potential pathogens and had to be distinguished from them. However, such organisms have with increasing frequency been incriminated as pathogens, mainly in seriously debilitated patients in whom they may even cause septicaemia. They are commonly resistant to many antibiotics.

Gram-Negative Bacilli: (b) The Genus Pseudomonas

Members of this genus resemble the enterobacteria in microscopic appearance (apart from having polar rather than lateral flagella) but differ from that group in being strict aerobes, in oxidizing carbohydrates instead of fermenting them, and in being oxidase-positive (p. 81). Many produce fluorescent pigments. They are widely distributed, and include plant pathogens and many saprophytic species. With the exception of *Ps. aeruginosa* they are rarely pathogenic to man.

Pseudomonas aeruginosa (*Ps. pyocyanea*)

OCCURRENCE AND PATHOGENICITY This species, commonly present in the human intestine, can grow in almost any moist situation over a wide temperature range, needing only oxygen and a modest supply of nutrients. Consequently it can mutiply in such preparations as eye-drops, ointments, lotions or even weak disinfectant solutions, and so be applied in large numbers to patients' surfaces or wounds. In such circumstances it can cause serious local or even systemic infections. Similarly, it can multiply in the warm moist conditions inside baby incubators or ventilating machines and so be inhaled in large doses and colonize the patients' respiratory tracts. Here it usually does little harm, but it sometimes causes a necrotizing pneumonia, which may be fatal. It is resistant to many antibiotics and is liable to take over when the primary causes of infection in wounds, burns etc. have been eliminated by treatment. It also colonizes the damaged airways of patients with bronchiectasis or cystic fibrosis who have received antibiotics for treatment of more orthodox respiratory tract pathogens; again, it is usually harmless but may cause pneumonia. A curious feature of such invasion of the lower respiratory tract is that any strain of this species which has established itself there is liable to undergo a change that causes it to form mucoid colonies on plate cultures. In the urinary tract, pseudomonas infection is a common and often intractable complication of neurological or other conditions that interfere with bladder emptying (p. 273). *Ps. aeruginosa* is frequently present in inflamed external auditory canals, and may be pathogenic there, particularly in swimmers and divers. Outbreaks of pseudomonas skin and ear infection sometimes occur among users of swimming pools and heated whirlpool baths with inadequate chlorination (p. 356).

CULTURE This species grows well on simple media under aerobic conditions. Typically its colonies are rough and irregular and produce green pigment (p. 18), which diffuses into the medium. (This pigment accounts for the characteristic blue-green colour of pus due to this organism, and colonies and pus share also a distinctive musty smell).

TYPING The species may be subdivided by **pyocine typing** (analogous to colicine typing— p. 94), by serotyping or by bacteriophage typing.

IMMUNOLOGY *Ps. aeruginosa*, an opportunist pathogen (p. 27), can cause severe infections in immunocompromised (notably neutropenic) patients. Immunization with polyvalent vaccine (i.e. one containing many *Ps. aeruginosa* serotypes) may give some protection to patients with extensive burns. Neutropenic patients with *Ps. aeruginosa* infections may require granulocyte transfusion (p. 314).

ANTIBACTERIAL TREATMENT Widespread resistance is the rule in this species. The penicillins ticarcillin and azlocillin, the aminoglycosides gentamicin and tobramycin and the newer cephalosporins ceftazidime and cefsulodin are the most useful drugs currently available, but resistance to them is not uncommon. Hence it is all the more important to minimize the frequency of serious pseudomonas infections by avoiding high-risk situations and by taking care not to transmit pseudomonas strains from one patient to others.

Gram-Negative Bacilli: (c) Vibrios, Campylobacters and Mobiluncus

Vibrios are comma-shaped bacilli (Fig. 7.1, p. 70) that are aerobic, facultatively anaerobic gram-negative bacilli and ferment carbohydrates. They have polar flagella and are oxidase-positive. They are widely distributed in nature. The species of medical importance are

V. cholerae and *V. parahaemolyticus*. Micro-aerophilic curved rods are now in separate genera, *Campylobacter* and *Mobiluncus*.

Vibrio cholerae

This species includes the organism responsible for cholera and a large number of biochemically identical or closely similar vibrios that are found in fresh or brackish (but not salt) surface waters in many parts of the world, including Britain. The true cholera vibrio has its own distinctive O antigen, designated O1. The other *V. cholerae* strains are agglutinable by sera to their own O antigens, and are known as 'non-cholera vibrios' (NCVs), though some of them cause diarrhoea when ingested by man and on rare occasions this is severe enough to resemble cholera. True cholera follows ingestion of an effective dose (p. 33) of cholera vibrios, usually in water or food. They multiply in the intestine and produce an enterotoxin which stimulates adenylcyclase activity in the intestinal epithelium and so provokes a sustained outpouring of water and electrolytes, manifested clinically as a profuse watery diarrhoea (p. 254).

Up to 1961 cholera was predominantly due to the 'classical' cholera vibrio, which produces severe disease in a large proportion of those infected but does not establish a stable carrier relationship with its host—i.e. it is excreted only during incubation and the period of illness and for perhaps a few weeks after that. The **eltor** biotype of *V. cholerae*, named after the Sinai village of El Tor where it was first identified in 1906, became established as a cause of a milder form of cholera in and around Indonesia during and after the second world war, and then from 1961 onwards it spread westward across Asia into Africa and Europe. Cholera was unknown in Britain from 1909 to 1970; but in the latter year a holiday-maker returned from North Africa infected with the eltor vibrio, and another 25 such cases had been recognized by 1980. The eltor vibrio produces severe disease less frequently than the classical strain. It spread so rapidly and so widely because many of those infected are only mildly ill or symptomless but are nevertheless vigorous excreters of the organism.

In the laboratory, cholera vibrios can sometimes be recognized by microscopy of the fluid stools, as slightly curved rods with characteristic darting motility; but diagnosis in this way is unreliable except in cases of severe and clinically obvious cholera. Isolation of the vibrios is made easier by their ability to grow in alkaline peptone water (pH 8.0 or more) and to form yellow colonies on thiosulphate–citrate–bile salts–sucrose (TCBS) agar. The identity of classical and eltor biotypes is confirmed by agglutination with the same specific O1 antiserum.

Natural immunity to *V. cholerae* depends on secretory IgA, which prevents both the adherence of vibrios to the intestinal epithelium and the binding of enterotoxin to cell receptors. Immunization with killed *V. cholerae* has been widely practised but is of little value.

Vibrio parahaemolyticus

This salt-water vibrio is a cause (in Japan a common cause) of acute gastro-enteritis that comes on some 10–20 hours after eating raw or inadequately cooked sea-food and lasts for a day or two. In Britain the organism has been isolated from locally bred oysters, but imported sea-foods such as prawns are more likely sources. Its isolation from foods, vomitus or faeces is helped by its ability to grow in media containing 3% NaCl and to form green colonies on TCBS agar (as used for *V. cholerae*).

Campylobacter jejuni and Campylobacter coli

The clinical importance of these two micro-aerophilic species has become apparent since the publication in 1977 of a simple method by which diagnostic laboratories can isolate them from faecal specimens. Unlike most other human intestinal bacteria, including other *Campylobacter* species, they grow at 43°C (i.e. they are thermophiles—p. 16), and on plates containing selective antibiotics and incubated at this temperature in an atmosphere containing 10% CO_2 and only 5% O_2 they form distinctive colonies. They are oxidase-positive gram-negative bacilli that resemble vibrios. In Britain they cause gastro-enteritis at least as often as salmonellae (p. 251). A few patients have developed septicaemia. The thermophilic *Campylobacter* species are carried in the intestinal tracts of many animals, especially birds and cattle. Many isolated cases or small outbreaks of infection have been associated with eating contaminated poultry or other meat and large water-borne and milk-borne outbreaks have been described. Pet dogs or cats may suffer from and transmit the disease. Human carriers are rare in Britain, but much commoner in some warmer countries. It is not yet clear how widely *Camp. jejuni* and *Camp. coli* differ in clinical significance and epidemiological behaviour. Measurement of antibodies after the infection may be helpful for retrospective diagnosis or to identify the factor that precipitated a reactive arthritis (p. 302). A third species, *Camp. pyloridis*, has been implicated in the pathogenesis of gastritis (p. 251).

Mobiluncus species

These micro-aerophilic motile curved rods, described in the early 1980s, are present in large numbers in the vaginal condition known as anaerobic (or bacterial) vaginosis (p. 281). Two species are distinguished by cell morphology—*Mob. curtisi* (short curved rods; gram-variable) and *Mob. mulieris* (long curved rods; gram-negative). Their role in the pathogenesis of vaginosis has not been established.

Gram-Negative Bacilli: (d) Parvobacteria

Whereas the enterobacteria are a group of essentially similar genera, the name 'parvobacteria' has no sound taxonomic basis; we retain it merely as a convenient collective term for a rather heterogeneous group of aerobic gram-negative bacilli that differ in a number of respects from those described so far. They are smaller, usually 0.3–0.4 μm in width, and are often cocco-bacillary. They are not intestinal parasites. They are generally exacting in their nutritional requirements, forming much smaller colonies on blood agar than do the enterobacteria, and with a few exceptions they do not grow on MacConkey's medium. Included among them are a number of important pathogens.

The Genus Haemophilus

Members of this genus are characterized by, and classified according to, their need to be supplied with X and V factors (haemin and nicotinamide-adenine dinucleotide, p. 16). Both factors are absent from some forms of nutrient agar, and strains can therefore be tested for their requirements by growing them on such a medium in the presence of filter-paper discs soaked in one or both factors. On blood agar, which contains X factor and a little V factor, enhanced growth of V-factor requiring organisms occurs around colonies or streaks of *Staph. aureus* and various other species which liberate V factor into the medium—a phenomenon known as **satellitism**. Heating of blood agar to produce chocolate agar destroys a substance in the red cells which is inhibitory to V factor, and the amount of this factor present is then sufficient to support a good growth of all *Haemophilus* species.

Haemophilus influenzae (Pfeiffer's bacillus, the influenza bacillus)

OCCURRENCE AND PATHOGENICITY The species occurs purely as a parasite of man, and is very commonly carried in healthy nasopharynges. Most of these 'respiratory' strains are not capsulate, but a small percentage of healthy people carry capsulate strains, divisible into six serotypes (a–f). Type b strains are of great clinical importance, being responsible for many cases (in some countries, notably the USA, for the great majority of cases) of bacterial meningitis in young children (p. 212), and for various other serious acute infections, mainly in the same age group. These include two conditions virtually always due to *H. influenzae* type b: epiglottitis (p. 197), and a characteristic cellulitis, usually on the face and neck or on a limb. Other infections that may be due to *H. influenzae* type b include otitis media, suppurative arthritis and lobar or segmental pneumonia. Capsulate strains of the other five types are much less often incriminated as pathogens. Non-capsulate strains are the principal cause of bronchial suppuration, either persistent or during acute exacerbations, in patients with chronic bronchitis or bronchiectasis (pp. 200 and 201).

The name of the species is derived from the claim of its discoverer, Pfeiffer (in 1892), that it was the cause of influenza. This suggestion was not finally discredited until the influenza virus was discovered (1933).

CARRIERS When a child has meningitis or any of the other acute illnesses caused by *H. influenzae* type b, there are commonly several nasopharyngeal carriers of this organism among his family or other close contacts; secondary cases of illness occur during the next few days or months in a small minority of such situations.

MICROSCOPY *H. influenzae* is a small non-flagellate non-sporing and usually non-capsulate gram-negative bacillus. Films from young cultures grown under favourable conditions usually consist of cocco-bacilli; pleomorphism with filament formation is commoner in cultures on less adequate media.

CULTURE Factors X and V are both required for growth. Colonies on blood agar after incubation for 24 hours vary according to the quality of the medium from minute pin-points to translucent domes approximately 1 mm in diameter. In mixed cultures, satellite clusters of colonies around V-factor-producing colonies of other species are characteristic, and routine addition of a staphylococcal streak to blood agar cultures of respiratory tract specimens can be used as an aid to rapid recognition of the species. Chocolate agar is also a useful primary medium, since colonies formed on it by *H. influenzae* are considerably larger than those formed on blood agar. The colonies of capsulate strains grown on a transparent medium (e.g. **Levinthal's agar**) exhibit iridescence when examined by strong obliquely transmitted light.

ANTIGENIC STRUCTURE The six capsulate types have distinct polysaccharide capsular antigens. They can be recognized by capsule-swelling tests, by agglutination or co-agglutination (p. 76) of the bacilli, or by precipitin tests using aqueous extracts of the bacilli with type-specific rabbit antisera.

IMMUNOLOGY Antibodies to somatic and capsular antigens are common in the blood of adults, and it is thought that acquired immunity is responsible for the restriction of serious *H. influenzae* type b infections very largely to young children. Because of the significant mortality from fulminating type b infections and the high rate of permanent disabilities following haemophilus meningitis, development of a vaccine for active immunization has been vigorously pursued in the USA; type b polysaccharide, an excellent antigen in other respects, fails to stimulate antibody formation in children under 18 months old, but a conjugate with diphtheria toxoid has given promising early results.

ANTIBACTERIAL TREATMENT Chloramphenicol is the agent of choice for treating haemophilus meningitis and other life-threatening *H. influenzae* type b infections; its action against this species is bactericidal and chloramphenicol-resistant haemophili are as yet rare. Ampicillin had proved somewhat less reliable even before the emergence of penicillinase-producing *H. influenzae* type b strains against which it is disastrously ineffective. 'Third generation' cephalosporins such as cefotaxime and latamoxef may be useful, especially if strains resistant to both chloramphenicol and ampicillin are encountered. As yet no antibiotic has proved fully effective for treating carriers of *H. influenzae* type b or protecting children at risk of infection with this serotype. Long-term prevention or control of haemophilus infections in chronic bronchial disease is a difficult problem; amoxycillin, the tetracyclines and cotrimoxazole are the main weapons for this purpose.

Other haemophilus species

H. aegyptius (the Koch–Weeks bacillus), an organism which was first incriminated as a cause of outbreaks of acute conjunctivitis in Egypt in 1883, resembles *H. influenzae* very closely.

H. para-influenzae differs from *H. influenzae* in not requiring X factor. It is a common mouth commensal and (like many other species, especially in this site) it causes occasional cases of bacterial endocarditis. There is some evidence that haemolytic variants of this species (*H. parahaemolyticus*) may sometimes cause acute pharyngitis.

H. ducreyi causes the veneral disease chancroid (p. 280) and is part of the mixed pathogenic flora of other genital ulcers. It requires X but not V factor, but it also has other special nutritional needs.

The Genus Bordetella

Bordetella pertussis (the whooping cough bacillus)

OCCURRENCE AND PATHOGENICITY Purely a human parasite, this organism causes whooping cough (p. 201) in children, and sometimes affects adults.

MICROSCOPY The bacilli resemble cocco-bacillary cultures of *H. influenzae*.

CULTURE Primary isolation of this species requires special media—charcoal blood agar or the time-honoured Bordet–Gengou medium (containing glycerol, potato extract and 50% horse blood), usually with penicillin added to suppress other more sensitive species. Aerobic incubation at 37°C needs to be continued for 2–3 days before the typical 'bisected pearl' or 'aluminium paint' colonies of *Bord. pertussis* appear. Once grown in this way, the organism can be transferred to serum agar or blood agar.

Specimens are collected by pernasal swabbing of the nasopharynx (p. 201), sometimes supplemented by holding a culture plate in front of the mouth of the coughing patient ('cough plate').

IMMUNOLOGY Natural infection is followed by lasting immunity. Active immunization is discussed on p. 365.

ANTIBACTERIAL TREATMENT Although *Bord. pertussis* is sensitive *in vitro* to numerous antibiotics, these drugs have little effect on the disease once symptoms have developed—presumably because after that it is too late to prevent epithelial damage. Erythromycin given during the incubation period may be of value; and antibiotics have an important role in dealing with secondary invasion of the lower respiratory tract by other bacteria, which may result in bronchopneumonia.

The Genus Brucella

OCCURRENCE AND PATHOGENICITY The three closely related members of this genus which cause human brucellosis (undulant fever—p. 232)—*Br. melitensis*, *Br. abortus* and *Br. suis*—are primarily pathogens of goats, cattle and pigs respectively. Brucellosis in man, of which *Br. abortus* is virtually the sole cause in Britain, is a systemic infection; the organism is distributed throughout the body, including the blood stream, and multiple small granulomatous nodules and micro-abscesses are found in affected tissues. The bacilli are mainly intracellular. Human infections with *Br. melitensis* occur chiefly in the Mediterranean area (Malta fever), Africa and parts of the Far East and America. *Br. suis* is found in the USA and in Denmark. Ingestion of unpasteurized cows' or goats' milk or of milk products may disseminate infection widely, but in Britain today infection is mostly restricted to those in close contact with farm animals, notably veterinary surgeons, abattoir workers and farm workers and their families.

Contagious abortion of cattle due to *Br. abortus* is one of the very few microbial diseases in which the localization of the pathogens can be explained. Unusually high concentrations of erythritol are present in certain layers of the bovine placenta and in the fetal fluids, and are believed to account for the vigorous multiplication of *Br. abortus* in precisely these sites which is a feature of contagious abortion.

MICROSCOPY The bacilli are small and usually short. They are non-flagellate, non-sporing, non-capsulate (except possibly in some fresh isolates) and gram-negative.

CULTURE *Br. abortus* requires 5–10% CO_2 in its atmosphere for primary isolation (p. 167). Otherwise all three species are aerobes, with an optimal growth temperature of 37°C. They are not particularly exacting in their nutritional requirements, but grow slowly, producing translucent and undistinguished colonies. Various special media for their primary isolation, including selective media, are available. Isolation from patients is usually attempted by culture of large amounts of blood (up to 40 ml) added to a suitable broth; CO_2 must be provided for growth of *Br. abortus*, and cultures should be incubated for at least three weeks. However, repeated attempts to isolate *Br. abortus* are often unsuccessful; the other two species are easier to isolate from blood. Culture of aspirated bone marrow is sometimes successful when blood culture has failed. The organism may also be isolated by inoculating the buffy coat from centrifuged blood into a guinea-pig. Similar culture procedures and guinea-pig inoculation are used in isolating brucellae from milk and other animal materials.

SPECIES DIFFERENTIATION Recognition of species, and of biotypes within species, is based on biochemical tests (mostly different from those used for other genera), on susceptibility to a bacteriophage specific for *Br. abortus* and to the inhibitory action of low concentrations of dyes in culture media, and on agglutinability by antisera specific for this genus. Using these criteria *Br. abortus* can be divided into 9 biotypes and *Br. melitensis* and *Br. suis* into 3 each.

IMMUNOLOGY Some immunity follows natural infections. An attenuated *Br. abortus* strain has been widely used in Russia for immunization of humans exposed to high occupational risks of brucella infection. Measurement of antibody levels in the blood of patients is useful for diagnostic purposes (p. 181).

ANTIBACTERIAL TREATMENT Probably because of the intracellular situation of the organisms, antibacterial treatment must be continued for several weeks to have a good chance of success, and the course may have to be repeated. The most commonly used drugs are the tetracyclines (sometimes with streptomycin, on empirical grounds) or cotrimoxazole. If these fail, a course of gentamicin may be beneficial.

PREVENTION Transmission of brucellae in milk can be prevented by efficient pasteurization (p. 358). Eradication of the disease has been, or is near to being, achieved in some countries by vaccination of young animals and slaughter of animals with serological or other evidence of brucella infection. It is costly, but far less so than persistence of infection among cattle; this causes abortions, still-births, infertility and reduced milk yield, and inevitably results in transmission to some humans, in whom it produces long periods of obscure ill health and impaired working capacity.

The Genus Pasteurella

Among the organisms (small gram-negative bacilli, mostly oxidase-positive and unable to grow on MacConkey's agar) that remain in this genus after the removal of those now classified as *Yersinia* are some that colonize the mouths of a wide range of animals hosts and may cause haemorrhagic septicaemia and other infections in them. All of these bacilli are now included in a single species *P. multocida*. Such organisms cause cellulitis and abscesses in man following animal (especially dog and cat) bites. The wounds may fail to heal until pasteurellae are eliminated by suitable antibiotic treatment—usually penicillin. This and other *Pasteurella* species may occur in the sputum of patients with chronic bronchial disease, but their significance there is uncertain.

Francisella tularensis (*Pasteurella tularensis*, *Brucella tularensis*)

This small gram-negative bacillus is an insect-borne pathogen of rabbits and other rodents in many countries, notably in the western USA and in Russia and Siberia. In such hosts it causes a plague-like illness known as tularaemia. Occasionally humans who handle infected animals are themselves infected, through skin abrasions, and develop a brucellosis-like illness which responds to tetracycline therapy. In the laboratory the organism resembles a pasteurella in some of its properties and a brucella in others.

Legionella pneumophila

In 1976 there were 183 cases of pneumonia or other respiratory illness, some of them severe and 29 fatal, among over 3500 delegates to an American Legion convention in Philadelphia. A previously unrecognized small bacillus was shown to be the cause of this 'legionnaire's disease' or legionellosis, and is now known to have caused numerous other outbreaks and sporadic cases of illness in many countries, both before and since 1976. In 1985, 38 patients died in an outbreak in Stafford, England. The bacillus is called *Legionella pneumophila*. It is gram-negative, though it hardly stains at all unless the method is suitably modified; and it was difficult to grow until special media were devised. There is an obvious paradox about an organism that is now known to be widely distributed in water—e.g. in water-tanks, in shower-heads and in the cooling towers of air-conditioning systems—and yet is difficult to grow in the laboratory; the explanation may lie in its dependence on other organisms, such as amoebae, found in its natural environments. The clinical features of the infection are described on p. 203. The diagnosis can be made by using immunofluorescence to demonstrate the presence of the organism in sputum or bronchial secretions, or in biopsy or autopsy material. The organism can be grown from such specimens, either directly on a selective medium or following passage through a guinea-pig. Detection of specific antibodies in the patient's blood is currently the simplest way of establishing the diagnosis, but they often take several weeks to appear there. Erythromycin has been the most successful antibiotic for treatment of legionellosis. *L. pneumophila* also causes Pontiac fever (p. 203).

Gram-Negative Bacilli: (e) The Bacteroides Group

The genera *Bacteroides*, *Fusobacterium* and *Leptotrichia* consist of anaerobic bacilli, most non-motile, that differ from clostridia (p. 87) in being gram-negative and in not forming spores.

OCCURRENCE AND PATHOGENICITY These organisms are obligate animal parasites, forming part of the intestinal flora of all animals from termites to primates. In man, *Bact. fragilis* and related species (the fragilis group) are the predominant gram-negative bacteria in faeces, where they are as numerous as bifidobacteria (p. 86) and outnumber *Esch. coli* by by 1000 to 1; *Bact. bivius* and *Bact. melaninogenicus* (members of the melaninogenicus/oralis group) are normal inhabitants of the adult vagina; and *Bact. melaninogenicus*, *Bact. intermedius* and *Bact. oralis* (melaninogenicus/oralis group), fusobacteria and *Leptotrichia buccalis* are found in the mouth, colonizing the gingival crevice and forming part of dental plaque (p. 191). These gram-negative anaerobes are frequently involved in necrotic or gangrenous lesions with a distribution related to their normal habitats. *Bact. fragilis* is the one most commonly found in abdominal infections (p. 349). *Bacteroides* of the melaninogenicus/oralis and asaccharolytic groups are frequently associated with infections of the male and female genital tracts (p. 282). Oral species, notably *Bact. oralis*, *Bact. melaninogenicus*, *Bact. gingivalis* (a pigment-producing member of the asaccharolytic group) and fusobacteria, take part in infections of the head or neck, such as cerebral, dental or soft tissue abscesses and in the condition known as ulcerative gingivitis (p. 191) in the mouth or as Vincent's angina in the throat (p. 196). The asaccharolytic species *Bact. asaccharolyticus* and *Bact. ureolyticus* (formerly known as *Bact. corrodens*) are commonly involved in necrotic or gangrenous lesions of the perineum, genitalia, lower abdomen or lower limb, especially in diabetics. Gram-negative anaerobes are seldom isolated in pure culture from septic lesions; they are usually accompanied by aerobic or facultative organisms, notably *Esch. coli* (p. 349), and in many cases by other anaerobes. Various forms of synergy between these different organisms have been postulated, but the anaerobes are probably responsible for most of the tissue damage. Discharges from lesions in which they are involved characteristically have a foul, putrid smell. In the absence of adequate antibacterial treatment gram-negative anaerobes are liable to spread from the original lesions via the blood to give serious generalized infections.

MICROSCOPY Bacteroides are in general small round-ended bacilli, sometimes uniformly cocco-bacillary but often pleomorphic; they are as a rule clearly gram-negative. Bacilli of the other two genera are commonly long and tapered at both ends ('fusiform'), and may be difficult to see in gram-stained films.

CULTURE In general gram-negative anaerobes are rapidly killed by exposure to oxygen or by desiccation. They survive much better in pus than on swabs (p. 163); if there is no pus and swabs must be used, they should be sent in a transport medium (p. 163). Most strains from human sources grow well on blood agar if incubated anaerobically (p. 167); the presence of 5–10% CO_2 enhances the growth of most strains. The *Bact. fragilis* group produce small nondescript colonies overnight as a rule, but other species may take several days to do so. *Bact. melaninogenicus* and *Bact. asaccharolyticus* colonies on blood agar turn brown or black on prolonged incubation through accumulation of altered haemoglobin.

IDENTIFICATION Three groups of *Bacteroides*—fragilis, melaninogenicus/oralis and asaccharolytic—and the genera *Fusobacterium* and *Leptotrichia* can be distinguished by GLC analysis of acid end-products of metabolism (p. 73) and by tests for sugar fermentation, indole production and resistance to bile and antibiotics.

ANTIBACTERIAL TREATMENT Removal of pus and necrotic tissue is an essential part of treatment of anaerobe infection; appropriate antibacterial treatment is important, but often of little avail on its own. Gram-negative anaerobes are resistant to many commonly used antibiotics, notably the aminoglycosides—an important point when gentamicin is used for treatment of septicaemia associated with intestinal lesions or surgery. The most commonly pathogenic species, *Bact. fragilis*, is also penicillin-resistant. Metronidazole, to which virtually all of these organisms are sensitive, is currently the treatment of choice, and its prophylactic use has greatly reduced the frequency of infective complications following intestinal or female genital tract surgery (p. 376).

Acid-Fast Bacilli

The Genus Mycobacterium

The tubercle bacilli (*Myco. tuberculosis* and *Myco. bovis*), the leprosy bacillus (*Myco. leprae*) and the other members of this genus are distinguished by their acid-fast staining—i.e. their resistance, after being stained with hot carbol fuchsin, to decolourization with acid (see below). This property is at least partly dependent upon their high content of certain lipids, notably mycolic acid, but this is not the whole explanation. Mycobacteria are gram-positive, but some of them are difficult to stain at all by Gram's method or by most other common procedures.

Mycobacterium tuberculosis and Myco. bovis (the tubercle bacilli)

OCCURRENCE AND PATHOGENCITIY Although these two species have different primary hosts (man and cattle), they have much in common and it is convenient to retain the name 'tubercle bacilli' as a means of referring to them collectively. Both are pathogenic to man, but they differ in their pathogenicity to other species (p. 38). Man acquires infections with *Myco. tuberculosis* from his fellow human beings, usually by inhalation, and the resulting disease commonly involves the lungs (p. 206). The risks of airborne infection are considerably increased by the ability of *Myco. tuberculosis* to survive for months outside the host—e.g. in dust and in books. Human infection with *Myco. bovis* usually results from drinking milk. It has been almost eliminated from Britain and many other countries by pasteurization of milk and control of tuberculosis in cattle. Because it enters the human body by the alimentary tract, *Myco. bovis* characteristically causes cervical and mesenteric adenitis rather than pulmonary lesions. Either species may travel via the blood to attack the meninges, bones, joints, skin and almost any other part of the body.

MICROSCOPY Mycobacteria do not form flagella, capsules or spores. Bacilli of this genus are best seen in films stained by the **Ziehl–Neelsen** (ZN) method. In this they are stained with hot carbol fuchsin for 5 minutes and then decolourized. Most mycobacteria other than *Myco. leprae* (p. 105) are resistant to decolourization by 20% sulphuric acid and are therefore called **acid-fast**, but the tubercle bacilli are **acid–alcohol-fast bacilli** (AAFB)—i.e. they cannot be decolourized by acid or by 95% ethyl alcohol. Films are counter-stained with methylene blue or malachite green, which stain micro-organisms that have not retained carbol fuchsin and also host cells and other structures, providing a suitable background against which to see the red mycobacteria. In pathological materials such as sputum and pus, tubercle bacilli are fine slightly curved bacilli, measuring about 3 μm by 0.3 μm, and often appear beaded. As a rule they are scanty in such preparations, so that a prolonged search may be necessary before they are found. Staining with the fluorescent dye auramine (p. 165) is better than the ZN method for their rapid detection, though less good for demonstrating their morphology. Films from cultures usually show shorter, straight bacilli arranged parallel to one another in 'cords' or 'ropes'.

The finding of tubercle bacilli in pathological material signifies that the patient has an active tuberculous infection. If they are found in sputum, they also indicate that his lesion is 'open'—i.e. is discharging into his respiratory tract—and that he is a danger to those around him. Because they are of such great significance, and also because they may be present only in very small numbers in a specimen, smears must be searched carefully for them. Smears from a concentrate of the specimen (see below) may give positive results when no bacilli are to be seen in those from the untreated specimen. It is very important to appreciate that **acid–alcohol–fast bacilli found in smears of pathological material are not necessarily tubercle bacilli**. Their identity must always be confirmed by culture and appropriate further tests.

CULTURE Tubercle bacilli are strict aerobes with a rather narrow range of growth temperature around 37°C, are exacting in their nutritional requirements and will not grow on ordinary media. The widely used **Lowenstein–Jensen medium** is made from eggs, glycerol, asparagine, potato starch and mineral salts; it also contains malachite green which inhibits other organisms and colours the medium so that the slightly yellow dry wrinkled colonies of *Myco. tuberculosis* are more easily detected. Solid and fluid culture media of simpler composition are also used. Even on optimal media growth is very slow, and colonies take ten days at least, more commonly several weeks, to become visible.

Most pathological materials to be examined for the presence of tubercle bacilli contain them in small numbers, irregularly distributed throughout the specimen and mixed with larger numbers of faster-growing organisms. Various **concentration methods** are used to overcome these problems. They homogenize and liquefy such materials as sputum, so that the tubercle bacilli can be concentrated into a small volume by centrifugation; and they also kill virtually all of the other bacteria present. For this they rely upon the high resistance of tubercle bacilli to various forms of chemical treatment, but there is rather a narrow margin between the minimum treatment that will achieve the desired ends and that which will also kill all of the tubercle bacilli.

Myco. tuberculosis can be distinguished from *Myco. bovis* in culture by differences in its colonial appearance, by its enhanced growth in glycerol-containing media and by its ability to synthesize niacin.

GUINEA-PIG INOCULATION This procedure was in general use for many years as the most sensitive way of detecting the presence of tubercle bacilli in clinical specimens, and also for distinguishing between tubercle bacilli, which cause typical tuberculosis in guinea-pigs, and other mycobacteria, which fail to do so. Improvements in *in vitro* techniques have been such that guinea-pigs are now rarely used except for specimens thought likely to contain tubercle bacilli in very small numbers (e.g. CSF).

IMMUNOLOGY Immunity to tuberculosis is cell-mediated, and depends largely on interaction between sensitized T lymphocytes and macrophages (p. 58). Patients present a spectrum of clinical, bacteriological and immunological features ranging between states of low and high resistance. At one extreme, in miliary tuberculosis, cell-mediated immunity and tuberculin hypersensitivity are absent, and mycobacteria are numerous. At the other extreme, when there is a healed localized tuberculous lesion, cell-mediated immunity and tuberculin hypersensitivity are strong, and bacteria are few or absent. Patients may swing between these extremes—e.g. waning of immunity from age or disease may be associated with reactivation of disease, whereas successful drug treatment is correlated with increased cell-mediated immunity and elimination of the mycobacteria. Active immunization, using an attenuated bovine strain (BCG), is discussed on p. 365, and the tuberculin test on pp. 184–5.

ANTIBACTERIAL TREATMENT The prognosis of tuberculous infections, especially of progressive primary infections, was revolutionized by the introduction of streptomycin in the 1940s; but using it alone commonly resulted in the development of resistance to it by the tubercle bacilli, and the treatment then became ineffective. Present-day drug treatment of tuberculosis with combinations of antibiotics is discussed on pp. 397–8.

Mycobacterium leprae (the leprosy bacillus, Hansen's bacillus)

OCCURRENCE AND PATHOGENICTY This organism was first described by Hansen, as early as 1874. It is an obligate parasite of man. It causes leprosy (p. 287), a disease now largely restricted to tropical countries, where the total number of patients is of the order of 10 millions and is probably still rising. The disease has been difficult to investigate because it has only recently become possible to grow *Myco. leprae* in non-living culture media, and until 1960 there was no known means of infecting animals.

MICROSCOPY *Myco. leprae* resembles *Myco. tuberculosis* in its morphology, but it is not alcohol-fast and is less strongly acid-fast—5% sulphuric acid is therefore substituted for 20% in the decolourization of ZN films aimed to detect this organism. It is gram-positive.

For diagnostic purposes, smears are made from scrapings of the nasal mucosa of patients with lepromatous disease, and from subcutaneous material obtained by making small skin incisions into actual lesions and at a number of standard sites such as the ear-lobes and forehead. Staining and microscopy of such smears is useful not only for establishing the diagnosis but also for assessing the effectiveness of treatment, since a satisfactory response is indicated by granularity and fragmentation of the bacilli.

ANIMAL INOCULATION AND CULTURE During the past 20 years it has become possible to propagate *Myco. leprae* in the laboratory in various ways: (1) in the foot-pads of mice, where it produces granulomatous lesions, followed by 'borderline' lesions (p. 288) in other parts of the body after many months; (2) in thymectomized irradiated mice, which develop 'lepromatous' disease (p. 288); (3); in armadillos and hedgehogs, which have appropriate body temperatures (see below); (4) in human nerve-tissue culture; and (5) in artificial media. In any circumstances the organism grows very slowly, with an optimal temperature of 30°C.

IMMUNIZATION In Uganda in 1960–64 BCG vaccination of children was found to give a degree of protection against leprosy comparable with that which it gives against tuberculosis. Subsequent reports from other countries have been less encouraging. Other aspects of the immunology of leprosy are discussed on p. 288.

ANTIBACTERIAL TREATMENT Dapsone (diaminodiphenylsulphone), continued for 2–4 years, has been the mainstay of antibacterial treatment for leprosy. However, dapsone resistance has become an increasingly serious problem, and has necessitated multiple-drug treatment as in the case of tuberculosis (pp. 397–8). Rifampicin and clofazimine are currently added to the dapsone regimen. Rifampicin has been shown to be bactericidal to *Myco. leprae* and so to eradicate the infection in mice, and it may be able to achieve total cure of human lepromatous leprosy.

Other mycobacteria

Some cases of tuberculosis-like illness in man (particularly in immuno-compromised patients) are due to organisms that have been given the unsatisfactory collective name **atypical mycobacteria**. Their chief clinical importance is that primary (i.e. pre-treatment) resistance to some of the standard antituberculous drugs (pp. 397–8) is common among them. Unlike tubercle bacilli, which have a narrow range of growth temperatures around 37°C, some of these other mycobacteria can grow at 25°C, some at 42 or even 45°C and some throughout the range 25–45°C. Some produce orange-pigmented colonies, but only when growing in light (**photochromogens**) and some do so in light or in darkness (**scotochromogens**). Some are fast-growing by mycobacterial standards, forming visible colonies in a few days. With the possible exception of *Myco. ulcerans*, the atypical mycobacteria are not transmitted from man to man but are derived from environmental sources—including in some cases birds, fishes or amphibia.

The photochromogenic *Myco. kansasi* is the mycobacterium that most often causes lung lesions resembling those of true tuberculosis; but it can also cause disease of lymph nodes or elsewhere. Conversely, the non-chromogens cumbrously called the *scrofulaceum/avium/intracellulare* group (because they are less easily divisible into 3 species than was once thought) are now responsible for most cases of mycobacterial cervical lymphadenitis in children in Britain, but can also attack the lungs. *Myco. ulcerans*, which grows very slowly with little or no pigment and only within the range 31–34°C, causes skin ulcers in Africa, Australia and elsewhere, and may prove to be rather closely related to *Myco. leprae*. *Myco. marinum*, a photochromogen with a preference for low temperatures, is a pathogen of fish and amphibia but also causes 'swimming bath granuloma' in man. Fast-growing mycobacteria are widely distributed in man's environment and are mostly non-pathogenic to him; but *Myco. chelonei* can cause abscesses (e.g. at injection sites) or even disseminated infections, usually in patients with impaired defence mechanisms.

Branching Bacteria

Some bacillary species form branched chains, but the branching occurs at the meeting-points of bacilli, not within individual bacilli. True branching, in which the branches are parts of the same cell, is shown by the filamentous fungi and by certain bacteria known as **higher bacteria**. (Fig. 7.1, p. 70). These include the genera *Streptomyces*, *Actinomyces* and *Nocardia*. The first genus is important to us because its members include many of the important antibiotic-producing organisms (see Chapter 29). The genera *Actinomyces* and *Nocardia*, sometimes referred to collectively as the actinomycetes, include a number of pathogenic species. They are non-motile, non-capsulate, non-sporing gram-positive filamentous organisms. The filaments, which are much narrower than those of filamentous fungi, tend to fragment into conventional bacilli when grown in artificial cultures.

The Genus Actinomyces

Members of this genus are micro-aerophilic or virtually anaerobic, though their metabolic pathways are not those of strict anaerobes. They are obligate parasites and, unlike some *Nocardia* species, are not acid-fast. *A. israeli* causes human actinomycosis, and the closely related *A. bovis* causes 'lumpy jaw' in cattle. The non-pathogenic *A. naeslundi* can cause confusion, since it is found in human mouths and resembles *A. israeli* in many ways, but it is recognizable by its ability to grow aerobically.

Actinomyces israeli

OCCURRENCE AND PATHOGENICITY The normal habitat of this species is the human mouth. It can be found around the teeth, gum margins and tonsils of many healthy individuals as well as in the lesions and discharges of actinomycosis— an acute, sub-acute or chronic granulomatous infection (p. 286)—but has not been isolated from sources outside the body. A small gram-negative bacillus dignified with the disproportionately long name of *Actinobacillus* (or *Haemophilus*) *actinomycetemcomitans* is sometimes found in large numbers in closed actinomycotic abscesses, together with *A. israeli*, but its significance is unknown.

The cervico-facial region is involved in over 50% of cases of actinomycosis, the abdomen in 20%, the thorax in 15% and other parts of the body in the remaining few cases. Colonization of intra-uterine contraceptive devices by actinomycetes is being recognized with increasing frequency.

MICROSCOPY On close naked-eye examination of actinomycotic pus, small yellow bodies—'sulphur granules'—can often be seen. These are colonies of A. israeli, and if one of them is crushed between two microscope slides and stained, it can be seen to consist of a tangled mass of gram-positive branching filaments. This appearance differs from that of a film made from a culture only in that short 'V' or 'Y' forms (a few of which are shown in Fig. 7.1, p. 70) are common in the latter. In gram-stained sections of actinomycotic tissue, colonies resembling the sulphur granules are seen, surrounded by radiating 'clubs' of gram-negative lipoid material produced by the host's tissues, probably as a form of protection.

CULTURE A. israeli cannot grow in air. Some strains are micro-aerophilic and others require anaerobic conditions. The addition of 5–10% CO_2 to the atmosphere often stimulates growth. The optimal growth temperature is 37°C, and growth does not occur at temperatures much below this. Raised, irregular, opaque colonies become visible after 3 or 4 days' incubation on blood agar or serum agar. They adhere firmly to the medium. Good growth also occurs in cooked meat medium and the other fluid media for anaerobic growth mentioned on p. 167, or in a glucose agar shake culture. This last is set up by inoculating the organism into a tube of melted medium at 50°C and dispersing it throughout the medium by shaking. When it has set, the medium is incubated for several days at 37°C. Colonies of A. israeli develop best about 10–15 mm below the surface, and also in the depths of the medium, but not near the surface, where the oxygen tension is too high.

For isolation of this species from pus, sulphur granules should be used as the inoculum. The pus is shaken up with water, and the granules are allowed to sediment. The diluted pus is then removed and the granules are repeatedly washed by shaking them up with more water, in order to rid them of accompanying bacteria, before they are used to inoculate media.

ANTIBACTERIAL TREATMENT The prognosis of abdominal and thoracic actinomycosis, formerly often fatal conditions, was radically altered by the advent of antibiotics, to many of which A. israeli is sensitive. Penicillin is the agent of choice, but prolonged high dosage may be required, since in chronic cases extensive deposition of fibrous tissue may restrict access of antibiotic to the lesion; ordinary dosage may give inadequate tissue levels and so allow the organism to become antibiotic resistant. Good results have been obtained with the tetracyclines, erythromycin or clindamycin.

The Genus Nocardia

These organisms differ from those of the genus Actinomyces in that all of them are aerobes, many are acid-fast and the majority are soil saprophytes. A few species are pathogenic to man, causing chronic granulomatous lesions.

Nocardia asteroides

This causes a rare pulmonary infection which may be mistaken for tuberculosis, particularly as acid-fast bacillary fragments may be found in the sputum. Systemic spread with abscess formation in liver, kidney, brain and other tissues sometimes occurs. Noc. asteroides is resistant to many antibiotics but treatment with drug combinations, often including sulphonamide, trimethoprim and amikacin, for 6–8 weeks may be effective.

Nocardia madurae

This is one of the causative agents of the chronic granulomatous, sinus-forming disease of the human foot known as Madura foot or mycetoma (p. 287). The form due to Noc. madurae is characterized by the presence of white or yellowish granules in the pus, and responds to treatment with sulphonamides or with dapsone.

Spirochaetes

The bacteria included in this group differ markedly in structure from any of the others that we have described. They consist of spiral filaments, in many cases too slender to be seen by ordinary microscopy of stained preparations (see below). They have no flagella, but are motile —often vigorously so—by means of whip-like flexion movements of their bodies or by screw-like rotation around their long axes. Partial digestion and electron-micrography have revealed the presence of one or more fine axial fibrils intertwined with the coils of their bodies. It seems likely that these are contractile and are responsible both for maintaining the spiral shapes of the organisms and for the movements that result in locomotion.

The spirochaetes of medical importance belong to the genera *Treponema*, *Leptospira* and *Borrelia*.

The Genus Treponema

Treponema pallidum

OCCURRENCE AND PATHOGENICITY Apart from experimental infections of apes, monkeys and rabbits, this organism is purely a pathogenic parasite of man. It is transmitted almost exclusively by sexual intercourse or by intra-uterine infection, and causes syphilis (p. 278). *T. pallidum* is present in profusion in the exudate from an early chancre (ulcerating primary lesion) and also in the red macular skin lesions, the moist peri-oral and ano-genital papules (condylomata) and the 'snail-track' mouth and throat ulcers of the secondary stage. It is less profuse, but often demonstrable, in the granulomatous lesions of many organs (gummata) which characterize the tertiary stage and in the arterial and nervous system lesion of late syphilis. Congenital syphilis involves many tissues and may kill the fetus.

MICROSCOPY The organism is a delicate thread, only about 0.15 μm thick, which is wound into a neat spiral 5–15 μm long and 1–5 μm wide with a 'wavelength' of about 1 μm (Fig. 7.1, p. 70). Being difficult to stain and of low refractility, *T. pallidum* is best seen by dark-ground microscopy (p. 164). It can be distinguished from the non-pathogenic treponemata (see below) and from other spirochaetes by its delicate structure and by its leisurely motility, which involves flexion and rotation. In dried smears of sections its presence can be demonstrated by the silver-impregnation methods of Fontana and Levaditi, but these obscure fine structural detail. Fluorescent-antibody staining is a better way of demonstrating the morphology of *T. pallidum* and also establishes its identity (p. 165).

CULTURE It is doubtful whether *T. pallidum* has ever been grown in laboratory cultures. Spirochaetes have been isolated from lesions and maintained in culture, but these were almost certainly contaminating saprophytes. Nelson devised a medium in which *T. pallidum* can be kept alive for several days, but it does not multiply. It can also survive for several days in refrigerated blood; and might therefore be transmitted by blood transfusion. It can be propagated in the laboratory by intratesticular inoculation of a rabbit and transfer to a new rabbit every three weeks or so.

IMMUNOLOGY The natural history of syphilis, including long periods in which the organism remains latent in the tissues, indicates that some form of partial immunity does develop. Furthermore, reinfection of a person whose tissues already contain live *T. pallidum* does not result in a fresh primary lesion. However, immunity seldom, if ever, progresses to the stage of spontaneous eradication of the organism, and resistance to reinfection disappears following adequate bactericidal treatment of the first infection. Immunization is not practicable. The serological diagnosis of syphilis is discussed on pp. 181–2.

ANTIBACTERIAL TREATMENT Chemotherapy with arsenical compounds, introduced by Ehrlich in 1910, was effective but tedious and not without dangers. Antibiotic treatment is more rapidly effective and safer. Penicillin is the drug of choice because *T. pallidum* is always sensitive to it and because the possibility of giving a single injection of a long-acting penicillin preparation has overcome the great problem of default from treatment at venereal diseases clinics. Other antibiotics can be used for patients who are hypersensitive to penicillin.

Other treponemata

The skin diseases yaws (framboesia) and pinta,

occurring in dark-skinned races in hot countries, are caused by spirochaetes respectively named *T. pertenue* and *T. carateum*. Like the causative organism of bejel (a highly infectious skin disease occurring in Arabia), they are indistinguishable from *T. pallidum* in their morphology, in the serological reactions that they evoke and in their response to treatment (p. 279). Since they are also impossible to grow in culture, it is not clear whether they are simply variants of *T. pallidum*. In addition to their pathological effects, they give rise to problems of serological diagnosis and can cause difficulties for would-be immigrants into countries whose laws exclude those with positive serological tests for syphilis.

Non-pathogenic treponemata are found in the mouth and around the genitalia, but usually differ in morphology from *T. pallidum* and stain more easily by ordinary methods.

The Genus Leptospira

OCCURRENCE AND PATHOGENICITY Like the salmonellae, these organisms have been divided serologically into a large number of types. These are now all classified within the single species *L. interrogans*, which is subdivided into two complexes. The **biflexa complex** includes saprophytic strains, which are numerous and widely distributed, particularly in water (even in domestic supplies); and the **interrogans complex**, consisting of some 130 serotypes arranged in 16 serogroups, includes most of the pathogenic and parasitic strains. These occur throughout the world, but many individual serotypes are geographically restricted. Many are pathogens of animals—e.g. rats (serotype *icterohaemorrhagiae* and many others), mice (*grippotyphosa, hebdomadis*), dogs (*canicola*), cattle (*hebdomadis* serovar *hardjo*) and pigs (*pomona*)—but can survive for long periods in neutral or alkaline (though not in acid) water. They live and multiply in the urinary tracts of convalescent or unaffected carrier members of their various host species. Excreted in the urine, they enter new hosts through skin abrasions or mucous membranes. The clinical picture produced in man depends to some extent upon the serotype responsible. Thus *icterohaemorrhagiae* infection typically results in classical Weil's disease (p. 267), and *canicola* infection typically produces a lymphocytic meningitis (p. 218). However, many leptospiral infections produce nondescript pyrexial illnesses. In Britain, leptospirosis used to affect principally fish workers, sewage workers and coal miners, who were exposed to water contaminated with rat urine and developed Weil's disease. More cases occur now in those who work with cattle. They develop a 'flu-like' illness, rarely with jaundice or meningitis, due to serovar *hardjo* infection.

MICROSCOPY The tightly coiled fine spirals, similar in length to those of *T. pallidum*, are often bent into hooks at one or both ends (Fig. 7.1, p. 70). The organisms are vigorously motile, spinning so rapidly around their long axes that the hooks often have the appearance of closed loops.

CULTURE Leptospires grow well just below the surfaces of various fairly simple fluid or semi-solid media. The optimal temperature for growth is about 30°C. Blood culture, by adding a few drops of patient's blood to a few millilitres of one of the fluid media, is often successful in isolating the organism during the first week of the illness. It can profitably be supplemented by intraperitoneal injection of blood into a young guinea-pig or hamster. If leptospires are present they can be recovered in pure culture a few days later by cardiac puncture of the animal, which also goes on to develop the characteristic and fatal haemorrhagic disease. Leptospires can sometimes be recovered from human urine in the second or later weeks of the illness by animal inoculation, but rarely by culture.

IMMUNOLOGY Immunity to many or all serotypes follows recovery from natural infection. Artificial immunization has been used successfully in Japan. Complement-fixation tests and various other serological procedures are used in diagnosis.

ANTIBACTERIAL TREATMENT Treatment with penicillin (in high doses) or one of the tetracyclines is sometimes helpful if used early in the disease but is valueless later on.

The Genus Borrelia

These large motile spirochaetes with only a few loose irregular waves are distinctly gram-negative.

Borrelia vincenti and *Leptotrichia buccalis* (Vincent's organisms—see p. 102 and Fig. 7.1, p. 70), are numerous in smears from the lesions of acute gingivitis and Vincent's angina (ulcerative conditions of the lips, mouth or throat—pp. 191 and 196), but are also found in small numbers in normal mouths and throats. Both are strict anaerobes.

A similar combination of organisms is also frequently found in material from lung abscesses, and in superficial ulcerating and necrotic lesions, especially those related to mucocutaneous junctions and sometimes known as noma.

Borr. recurrentis and *Borr. duttoni* cause relapsing fever in Europe and in West Africa respectively. Closely related organisms have been described as causing the same disease in other parts of the world. They are transmitted from rodents to man or from man to man by lice or ticks. They can be recognized in the peripheral blood as long coarse spiral threads, and can be grown in blood-containing media under anaerobic conditions.

Borr. burgdorferi causes Lyme disease, an immune mediated inflammatory disorder characterised by a distinctive skin lesion—erythema chronicum migrans—and recurrent attacks of arthritis. Other sequelae include carditis and neurological abnormalities. The infection is transmitted to man by the bite of an *Ixodes* tick from field mice, sheep and deer. Tetracycline or penicillin are recommended for treatment.

Mycoplasmas

These are very small bacteria that do not form cell walls (p. 10). They closely resemble L-forms of more orthodox bacteria (p. 71), and it has been repeatedly suggested, but never confirmed, that they are stable derivatives of these. All mycoplasmas are exacting in their nutritional requirements, and the parasitic species (see below) can be grown only on very rich media that must contain sterols, usually supplied by incorporating 20% or so of blood or serum in the medium. Even on such media they grow very slowly, taking weeks rather than days to produce their typical minute 'fried-egg' colonies.

Saprophytic mycoplasmas are found in soil and sewage and elsewhere. Parasitic species can be isolated from plants and from moist mucosal sites in animals. Many human beings carry them in their respiratory and genito-urinary tracts. The only species undoubtedly pathogenic to man is *Mycoplasma pneumoniae*. This was first isolated (and described as a virus) in 1944 from patients with primary atypical pneumonia—a condition so called largely because it differed from 'typical' (pneumococcal) pneumonia in its failure to respond to penicillin treatment. *M. pneumoniae* infection of humans is common in many parts of the world (p. 203). Laboratory confirmation of diagnosis is usually by detecting rising levels of antibodies. 'Cold agglutinins' may also be detected, often before the specific antibodies (pp. 61 and 203). The organism can be cultured in some cases from pharyngeal swabs, sputum or blood. Since mycoplasmas have no cell walls, they are resistant to β-lactams (penicillins and cephalosporins), but treatment with tetracyclines or erythromycin is usually effective.

Other mycoplasmas frequently isolated from man include *M. hominis* and strains formerly called T strains (because they form particularly tiny colonies) and now known as *Ureaplasma urealyticum* (a name based on their characteristic urea-splitting activity). Both species frequent the human genital tract of either sex, are transmissible by sexual intercourse, and are frequently found in association with genital tract infections (p. 280). However, it remains uncertain whether they play a part in producing or maintaining these infections.

Suggestions for Further Reading

Gillies, R.R. (reviser). 1984. *Gillies and Dodds Bacteriology Illustrated*. Churchill Livingstone, Edinburgh.
See also books mentioned in the Preface.

Rickettsiae, Coxiella and Chlamydiae

The general features of these very small, obligately intracellular micro-organisms are described on pp. 11–12.

Rickettsiae

The rickettsiae which cause typhus and related diseases are named after Dr H.T. Ricketts, who died of typhus while investigating its cause and transmission. They are primarily intestinal parasites of blood-sucking arthropods such as ticks, mites, rat-fleas and lice, to which they are not usually harmful and indeed may in some cases be necessary for survival. They are pathogenic to man, though except in the case of *R. prowazeki* (the cause of epidemic typhus) he is only an accidental host for them. Man becomes infected by direct inoculation into bites, by contamination of bites or scratches with arthropod faeces or by inhalation of dried arthropod faeces. The hallmarks of rickettsial infection are high fever, severe headache, a skin rash usually appearing after 7–10 days and hepatosplenomegaly. The symptoms are due to the proliferation of rickettsiae in the endothelial lining of small blood vessels, producing an inflammatory vasculitis that often affects skin, heart, brain and lungs. In severe untreated infections, toxaemia with disseminated intravascular coagulation (p. 40) may be present. In any patient with evidence of a tick bite (*tache noire*) who presents with fever followed 4–5 days later by a maculopapular rash, it is important to establish whether there is a history of travel to a rickettsial endemic area within 14 days of the onset of illness. Tetracycline and chloramphenicol are the antibiotics of choice but they do not eradicate the organisms.

The rickettsial diseases can be classified into two groups according to their epidemiology, clinical features and serological responses (Table 8.1, p. 112).

(1) Typhus Group

Epidemic typhus (classical, famine or European typhus)

This is purely a human disease, due to *R. prowazeki* and transmitted by the human body louse. After being ingested in human blood by the louse and multiplying in its intestine, the organism appears in large numbers in its faeces and enters the body of the next human victim by contamination either of a louse bite or of the scratches which he inflicts on himself in response to the bites. A severe febrile illness follows, characterized by cerebral disturbances and a rash which commences in the axillary folds and then spreads centrifugally. The death rate in the absence of antibiotic treatment is high, especially in older patients.

Louse infestation is a product of overcrowding and poor hygiene, and typhus epidemics are consequently associated with war, drought and famine. Spread can be prevented by 'delousing' threatened populations, their clothing and their bedding by spraying them with an insecticide. A vaccine prepared from a formalin-killed yolk-sac culture of *R. prowazeki* is available but it is recommended only for high-risk groups such as scientists or workers visiting endemic areas and

TABLE 8.1 Summary of the rickettsial diseases

Diseases	Organism	Reservoir	Vectors	Eschar	Weil–Felix reaction		
					OX 19	OX 2	OX K
Typhus group							
Epidemic typhus	R. prowazeki	Men	Lice	No	+++	(+)	–
Endemic typhus	R. typhi	Rats	Fleas, lice	No	+++	+	–
Scrub typhus	R. tsutsugamushi	Rodents	Mites	Yes	–	–	+++
Spotted fever group							
Rocky Mountain spotted fever	R. rickettsi	Rodents, dogs	Ticks	No	++	++	–
Boutonneuse fever	R. conori	Rodents, dogs	Ticks	Yes	++	++	–
Rickettsial pox	R. akari	Mice	Mites	Yes	–	–	–

doctors and nurses working in those areas.

Brill's disease is a recrudescence of typhus in a mild atypical form in a patient in whose tissues *R. prowazeki* has remained dormant, sometimes for many years, following a previous typical attack. The rapid IgG antibody response clearly distinguishes this disease from primary epidemic typhus, in which a predominantly IgM response is found.

Endemic typhus (murine typhus)

This is primarily a disease of rats, due to *R. typhi* (*mooseri*) and transmitted by the rat-flea and the rat-louse. Sporadic cases of human infection occur throughout the world in places where the rat population is high.

It is predominantly a disease of urban populations with a peak incidence in late summer and autumn. The disease is much less severe than epidemic typhus and the rash when present affects the chest and abdomen and remains central in distribution throughout.

Scrub typhus (tsutsugamushi fever)

This occurs in Japan, Malaysia and the Pacific area. The causative organism, *R. tsutsugamushi*, is transmitted to man by larval mites. Adult mites become infected by biting field mice, rats and other rodents, and then pass on the infection to succeeding generations of mites. Mite larvae are common on scrub in low-lying damp areas; human beings who walk there are liable to be bitten unless they wear protective clothing and use insect repellants. Civilians in rural endemic areas and military personnel active in such areas are often affected. The resultant disease, in which generalized lymphadenopathy and lymphocytosis are common features, resembles epidemic typhus with the addition of a local black-scabbed ulcer or eschar at the point of entry of the organism.

(2) The Spotted Fevers

These are tick-borne diseases of man, horses, dogs and rodents. The ticks themselves are the main reservoirs for the causative organisms, passing them on from generation to generation via their eggs. All of the human diseases are characterized by fever and by rashes which appear first on the wrists and ankles and then move centripetally; but they differ in geographical location, rickettsial species and severity of illness.

Rocky Mountain spotted fever is the name given to the disease caused by *R. rickettsi*, found in North and South America. The vectors are dog ticks in the south-eastern part of the USA but wood ticks in the west, particularly in rural wooded areas. The disease, which can be severe, has a seasonal peak incidence in spring and summer.

Boutonneuse fever is the name given to the type of tick typhus found in the Mediterranean area and in Africa. It is caused by *R. conori*, transmitted by the dog tick. It is a relatively mild illness.

Rickettsialpox is a mite-borne disease caused by *R. akari* and transmitted to man from rodents. There is again an eschar at the point of entry, but the disease is mild. The rash in this disease, which occurs in Russia, Korea, the USA and South Africa, is vesicular and resembles chickenpox (p. 131).

Diagnosis of rickettsial infection

The diagnosis of rickettsial infection depends on clinical presentation allied to a history of exposure or travel to an endemic area. Serological evidence of infection is not detectable until the second week of the illness.

The **Weil–Felix test**, which depended on the agglutination of special strains of *Proteus* (OX-19, OX-2 and OX-K) by serum from patients with rickettsial infection, was for many years the mainstay of diagnosis. It has now been superseded in specialized laboratories by more sensitive and specific complement-fixation tests with soluble antigens derived from the typhus and spotted fever groups of rickettsiae, and by indirect immunofluorescence tests with purified rickettsial suspensions.

Coxiella burneti

An influenza-like febrile illness, with variable manifestations that usually include patchy pneumonic consolidation, was first recognized as an entity in 1935 in Queensland, Australia. It was named **Q fever**, not because of its place of origin but because of the original query about its aetiology. When the causative agent was identified, it was found to be a rickettsia-like organism now known as *Coxiella burneti*. It occurs throughout the world, and affects various birds and wild animals as well as the domesticated goats, sheep and cattle from which man usually acquires his infection. It is carried between animals largely by ticks, but human infection is usually by inhalation of dust or droplets contaminated from the excreta or other products of infected animals, by ingestion of infected milk or (particularly in abattoirs) by transconjunctival entry. The organism's spread is therefore greatly assisted by its resistance to desiccation, and by its ability to survive heat treatment only slightly less than that generally recommended for the pasteurization of milk. Veterinary workers, slaughterhouse men and farm workers are among those most at risk, and there have been instances of accidental infection of laboratory workers. Epidemics have occurred among soldiers operating in areas where Q fever abortion was common in flocks of sheep. There is serological evidence that subclinical infection is common in rural communities.

Infection frequently results in an influenza-like illness with high fever, headache, myalgia and atypical pneumonia, but there is a spectrum of disease from mild subclinical illness to death from fulminant pneumonia or meningo-encephalitis. Hepatitis is an increasingly often recognized component of acute infection. Chronic Q fever may follow clinical or subclinical infection; its commonest manifestations are osteomyelitis and endocarditis, the latter usually involving abnormal or prosthetic heart valves (p. 235–6).

Complement-fixation tests of the patient's serum have supplanted dangerous isolation techniques in the diagnosis of Q fever. The tests employ the two different antigenic preparations of *Cox. burneti*—phase 1 and phase 2. In acute Q fever there is a four-fold rise in antibodies only to phase 2 within two weeks, whereas in chronic infection antibodies to both phases appear and persist.

Tetracyclines are the most useful antibiotics for treatment of Q fever, but their effects are unreliable, especially in endocarditis. Addition of either lincomycin or cotrimoxazole is said to enhance their efficacy.

Chlamydiae

Two *Chlamydia* species are recognized. Strains belonging to *Chl. psittaci* are primarily bird pathogens but sometimes infect man or animals. *Chl. trachomatis* includes the **lymphogranuloma venereum** (LGV) and **trachoma and inclusion conjunctivitis** (TRIC) agents, and also the strains involved in various genital tract infections. The spectrum of clinically distinct diseases produced by infection with the different serotypes of *Chl.*

trachomatis is summarised in Table 8.2. The members of this species are natural parasites of man only. Tetracyclines or erythromycin are the most appropriate antibiotics for their treatment.

Chlamydia psittaci

Psittacosis in the strict sense is an infection of psittacine birds (parrots etc.) or a zoonotic infection of man with chlamydiae derived from such birds. Similar infections of many other birds, including pigeons, ducks, turkeys and gulls, are collectively known as **ornithosis**; when transmitted to man they tend to produce illnesses milder than psittacosis. Human infection usually results from inhalation of dust containing dried droppings from infected birds (which may be apparently healthy or only mildly ill). Less commonly it follows a bite from such a bird, or is acquired from a laboratory culture. Psittacosis is a recognised occupational hazard for people working in the poultry business. Droplet transmission between human beings is possible. Human infection may be subclinical, or may cause symptoms that vary from a mild influenza-like illness to a severe and sometimes fatal pneumonia. Respiratory tract carriage of the organism may persist long after recovery from the disease. Since the re-introduction in 1976 of controls on the importation of psittacine birds, the number of human cases of *Chl. psittaci* infection in Britain attributable to exposure to such birds has fallen. However, the overall number of reported human cases of ornithosis has risen.

Other strains of *Chl. psittaci* can cause disease in animals, notably endemic abortion and arthritis in sheep and cattle. A given strain is likely to cause the same disease when it crosses from one species to another and people whose occupation brings them into close contact with infected animals may acquire the infection; *Chl. psittaci* acquired from sheep during the lambing season has caused abortions in women.

Chlamydia trachomatis

With the exception of the strains causing lymphogranuloma venereum, which have a predilection for lymph nodes, *Chl. trachomatis* strains grow only in the columnar epithelial cells found in the conjunctiva, the cervix, the urethra, the respiratory tract, the gastro-intestinal tract and the rectal mucosa. This is reflected in the spectrum of diseases that they cause.

Lymphogranuloma venereum (LGV) is caused by *Chl. trachomatis* serotypes L 1–3, which are more invasive than other serotypes and cause

TABLE 8.2 Chlamydial species, serotypes and diseases

Species	Serotypes	Diseases	
Chl. psittaci	Many	Psittacosis, ornithosis	
Chl. trachomatis	L1, L2, L3	Lymphogranuloma venereum	
	A, B, C	Trachoma	
	D–K	Inclusion conjunctivitis	
		Non-gonococcal urethritis	
		Post-gonococcal urethritis	
		Epididymitis	in adults
		Proctitis	
		Cervicitis	
		Salpingitis	
		Perihepatitis	
		Inclusion conjunctivitis	
		Pneumonia	in neonates
		Otitis media	

systemic disease instead of being restricted to mucous membrane surfaces. LGV is solely a human disease, transmitted as a rule by sexual intercourse, though non-venereal infection can occur, e.g. through the conjunctiva. The disease occurs throughout the world, but its highest incidence is in tropical and subtropical countries. In its common form it produces painless ulcerated genital lesions with enlarged unilateral regional lymph nodes (**buboes**—the name also used for similar lesions in plague, p. 95) that suppurate and discharge pus through multiple sinus tracts. Generalised dissemination follows, with fever, diffuse aches and sometimes conjunctivitis, arthritis or encephalitis. Chronic infection can lead to proctitis (particularly in women and homosexual men), anal and genital strictures and elephantiasis. Diagnosis is based on physical examination supported by a serological response. Contact tracing and early treatment and control of those infected are important in limiting spread of the disease.

Trachoma is a form of conjunctivitis in which formation of fibrous tissue in the conjunctiva and cornea commonly leads to lid deformities and to blindness, often some 15–20 years after the initial infection in childhood. It is the world's commonest cause of blindness, affecting particularly rural communities with poor hygiene and poor socio-economic conditions, such as those living in the arid regions of the Middle East, Africa and South East Asia. It is transmitted mainly by direct and close contact, e.g. from mother to baby, or by transfer of infected discharge from eye to eye by contaminated hands, clothing or towels. Flies attracted to the discharge are also important vectors.

Inclusion conjunctivitis is a similar but milder condition of worldwide distribution, caused by *Chl. trachomatis* serotypes D–K, which differ from those causing trachoma in that their primary habitat is the human genital tract rather than in the eye. Adult infection (**chlamydial ophthalmia**) consists of a chronic conjunctivitis which is often unilateral. Systemic treatment for the patient and sexual contacts is essential, tetracycline being the drug of choice. **Neonatal conjunctivitis (inclusion blenorrhoea)** occurs when chlamydiae from the mother's genital tract contaminate an infant's eyes during delivery; 40% of infants born to women with chlamydial infection of the cervix develop conjunctivitis 3–13 days after delivery. If neonatal conjunctivitis of unconfirmed aetiology is treated empirically with chloramphenicol drops, chlamydial infection will not be cured but only suppressed, and will flare up later when treatment is stopped. Since *Chl. trachomatis* can also cause neonatal pneumonia and otitis media, systemic erythromycin is the recommended treatment when infection with this organism is a possibility.

Neonatal pneumonia follows colonization of an infant's nasopharynx during passage through the birth canal. It has a gradual onset, with eosinophilia, diffuse lung involvement and tachypnoea, 2–12 weeks after birth. Confirmation of the diagnosis can be obtained by detecting the presence of *Ch. trachomatis* in the respiratory secretions and by demonstrating the development of a specific IgM antibody response. Both the mother, whose infection is often unrecognized, and the infant should be treated with erythromycin.

Female genital tract infection with *Chl. trachomatis* is the commonest form of sexually transmitted disease of that tract. It affects the cervix, the urethra and Bartholin's ducts, particularly in the sexually active age group. Women may be asymptomatic carriers, discovered only as a result of being investigated as partners or contacts of men with non-gonococcal urethritis (NGU—see below), or they may present with cervicitis and discharge or with dysuria and frequency without significant bacteriuria—the 'urethral syndrome'; 70% of sexual partners of men with NGU have *Chl. trachomatis* cervical infection. Ascending infection can cause pelvic inflammatory disease, particularly salpingitis, with resultant infertility.

Curtis-FitzHugh syndrome is a *Chl. trachomatis* infection, usually affecting young sexually active women, which consists of a combination of acute severe upper abdominal pain, perihepatitis and genital tract infection.

Male genital tract infection with *Chl. trachomatis* is initiated by invasion of the columnar epithelium of the distal urethra following sexual intercourse. It can then spread to the proximal urethra, causing urethritis, or via the vas deferens to the epididymis. This species is the most important cause of **non-gonococcal urethritis** (NGU, p. 280), an extremely and increasingly common sexually transmitted disease. *Chl. trachomatis* can be isolated from the urethra of 30–60% of men with NGU; they present with dysuria and urethral discharge after an incubation period usually of 1–3 weeks. Pus cells are seen in urethral smears or in urine. *Chl. trachomatis* is also responsible for about 70% of cases of **post-gonococcal urethritis** (PGU, p. 280), the urethritis that persists when patients with gonorrhoea have received antibiotic treatment sufficient to eliminate the gonococcal component of their double infection. In all genital tract infections with *Chl. trachomatis* it is important to treat both sexual partners, to prevent reinfection.

Sexually acquired reactive arthritis (SARA) is one form of the category of reactive arthritis more fully considered on p. 302. It is a sterile synovitis of one or more joints following primary *Chl. trachomatis* infection in the genito-urinary tract. SARA occurs in 1–2% of patients, most of them male, following *Chl. trachomatis* infection; 35% of the patients also have conjunctivitis and other components of Reiter's syndrome (p. 303); and 80% of them have the histocompatibility antigen HLA-B27.

Laboratory diagnosis of chlamydial infection

Laboratory confirmation of the diagnosis of chlamydial infection can be provided by direct demonstration of the organism in clinical material, or by its isolation in cell cultures, or by serological evidence. For isolation of the organism, carefully collected swabs from the conjunctiva, cervix or other relevant sites should be sent to the laboratory in a special transport medium containing antibiotics to suppress contaminating bacteria and fungi. Intracellular chlamydial inclusions, in smears of material from patients or in monolayer cultures, can be detected by Giemsa or iodine staining or by direct or indirect immunofluorescence tests with high-titre specific chlamydial antisera. Chlamydial antigens can be detected rapidly (in less than four hours) in material from patients by using highly specific (preferably monoclonal— p. 56) antibodies in enzyme immuno-assay.

The complement-fixation test based on a heat-stable lipoprotein–carbohydrate antigen common to all chalmydiae is rather insensitive, but nevertheless adequate for the diagnosis of systemic infections—psittacosis and lymphogranuloma venereum. It offers no differentiation between *Chl. psittaci* and *Chl. trachomatis* infections. However, the more sensitive micro-immunofluorescence test using type-specific antigens detects antibodies produced against individual serotypes, and so can differentiate between infections due to different species, strains and serotypes. It can detect antibodies in secretions as well as in blood, but the high background rate of antibody positivity in populations at risk (60% among patients attending genito-urinary clinics in the UK) limits its value in the diagnosis of genital infection. The finding of IgM antibody to chlamydiae in an infant's serum is useful as confirmation of a diagnosis of neonatal chlamydial pneumonia.

9
Viruses

As we have seen in Chapter 2 (pp. 12–13), nucleic acid composition (DNA or RNA) and virion structure are the chief criteria for primary division of viruses into major groups. Recognition of smaller groups and subgroups relies to a varying extent on similarities of habitat, pathogenicity, mode of transmission, antigenic composition and laboratory behaviour. This diversity of criteria is reflected in the names that have been compounded for the groups and subgroups—e.g. picornaviruses = small RNA viruses; enteroviruses = intestinal viruses; polioviruses = viruses of poliomyelitis. In some cases such names are retained although they are no longer true descriptions of all viruses currently grouped under them, and they should therefore be regarded as convenient labels rather than as statements about the members of the groups (cf. bacterial nomenclature, p. 68). Fine subdivisions into types, designated by letters or numbers, depend on demonstration of antigenic diversity among viruses that are similar in their other properties, including in many cases possession of common group antigens.

This chapter deals with the medically important viruses listed in Table 9.1. The virus-isolation and other diagnostic procedures mentioned at many points in this chapter are described in Chapter 13 (pp. 168–171). Antivirus drugs are reviewed at the end of the present chapter.

Picornaviruses (Small RNA Viruses)

These are small icosahedral RNA viruses without envelopes. They include two groups that are of considerable importance in medicine—the enteroviruses and the rhinoviruses—and the virus of foot-and-mouth disease of cattle, which attacks man on rare occasions.

Enteroviruses (Intestinal Viruses)

These viruses are primarily inhabitants of the human intestine, though they are also commonly found in the upper respiratory tract. They occur throughout the world, their frequency in temperate climates being higher during the warmer part of the year. They are excreted in faeces, in which they can sometimes survive for many days even in the presence of disinfectants. Not much is known about their transmission, but flies can certainly play a part. Infection occurs via the mouth, and the viruses establish themselves in the lymphoid tissue of the upper respiratory and alimentary tracts before travelling via the blood stream to other parts of the body, such as the central nervous system where many of them can produce lesions. Symptomless carriage of enteroviruses is common, and therefore isolation of such an organism, unsupported by a rise in appropriate antibody levels, is not evidence of its involvement in the patient's illness. Enteroviruses can be divided into polioviruses (3 types) and coxsackieviruses and echoviruses (about 30 types each).

Polioviruses

The epidemiology and clinical features of **poliomyelitis** are discussed on p. 221. It is

TABLE 9.1 Classification of the more important viruses affecting man

Nucleic acid	Capsid symmetry	Envelope	Group	Approx. size (nm)*	Members of the group	Principal diseases caused
RNA	Cubic	−	Picornaviruses	24–35	Enteroviruses: polioviruses coxsackieviruses echoviruses → Rhinoviruses	Poliomyelitis Aseptic meningitis Aseptic meningitis Hepatitis A Common cold
		−	Reoviruses	60–80	Reoviruses Rotaviruses	Gastro-enteritis
		+	Togaviruses	40–70	Alphaviruses Flaviviruses →	Encephalitis Yellow fever, dengue Rubella (German measles)
		+	Retroviruses	100	Human T cell lymphotropic virus	Acquired immune deficiency syndrome
	Helical	+	Orthomyxoviruses	80–120	→	Influenza
		+	Paramyxoviruses	100–120	Parainfluenza viruses → → Respiratory syncytial virus	Croup, other respiratory infections Mumps Measles Bronchiolitis
		+	Rhabdoviruses	80×180	→	Rabies
		+	Coronaviruses	80–130	Coronaviruses	Common cold
		+	Arenaviruses	110	→ →	Lymphocytic choriomeningitis Lassa fever
		+	Bunyaviruses	90–100	Sandfly fever virus Hantavirus	Haemorrhagic fever with renal syndrome
	Unclassified				Marburg virus Ebola virus	Viral haemorrhagic fevers
DNA	Cubic	−	Parvoviruses	20		Erythema infectiosum (fifth disease), aplastic crisis
		−	Papovaviruses	50	Papilloma virus	Warts,
		−	Adenoviruses	50		Respiratory infections, conjunctivitis
		+	Herpesviruses	120	Herpes simplex virus Varicella-zoster virus Cytomegalovirus Epstein-Barr virus	Herpes simplex, genital herpes Chickenpox, shingles Cytomegalic inclusion disease Infectious mononucleosis
	Complex	−	Poxviruses	230×300	Variola Vaccinia →	Smallpox Cowpox Orf
	Unclassified			42	→	Hepatitis B

* Most viruses are roughly spherical, except rhabdoviruses (bullet-shaped) and poxviruses (brick-shaped).
→ = virus(es) named after the disease listed in the right-hand column – e.g. hepatitis A virus, rubella virus.

caused by three antigenically distinct types of poliovirus, of which type 1 is responsible for the majority of epidemics. Infection is often subclinical. In a minority of cases an **aseptic meningitis** or true **paralytic poliomyelitis** develops.

VIRUS ISOLATION Polioviruses can be isolated by growing them in human or monkey tissue cultures. They are recoverable from throat swabs or washings during the first few days of illness and from the faeces for some weeks longer, but rarely from cerebrospinal fluid. They are to be found in affected parts of the central nervous system in fatal cases, and in the faeces of symptomless carriers.

DIAGNOSTIC SEROLOGY Neutralizing and complement-fixing antibodies appear in the blood following infection. As a rule they are specific enough in their action to indicate the type of the infecting poliovirus, though there may also be rises of antibodies to other types previously encountered.

IMMUNITY Natural infection results in lasting immunity, but only against the type involved. Active immunization is discussed on pp. 368–9.

Coxsackieviruses (prototype isolated in Coxsackie, USA)

These are distinguished from other enteroviruses by the fact that they are pathogenic to newborn mice. They are divided into two groups, A and B, according to the lesions which they cause in these animals. They are further divided into about thirty types according to their antigenic composition. Group B contains fewer types than group A but they are more often incriminated as pathogens. Diseases caused by coxsackieviruses include **aseptic meningitis** (group A and B, p. 219); **herpangina**, an acute pharyngitis with vesicle formation (group A, p. 190); **hand, foot and mouth disease**, characterized by vesicles in all the three sites indicated (group A, p. 190—not the same as foot and mouth disease, p. 8); **endemic myalgia** or **Bornholm disease**, a febrile illness associated with severe pain in the chest muscles and elsewhere (group B); and **myocarditis** of newborn babies and adults (group B, p. 239). Also they are among the miscellaneous viruses that cause **colds** (p. 192). Individual types are associated with particular forms of disease. Diagnostic procedures are similar to those used for poliomyelitis, except that inoculation of newborn mice is used as well as tissue culture, and that virus may be recovered from the cerebrospinal fluid in cases of meningitis.

Echoviruses (= *E*nteric *C*ytopathic *H*uman *O*rphan viruses)

The presence of these viruses in human faeces was first recognized because of their cytopathic effect in tissue cultures; since they did not appear to 'belong' to any disease they were described as orphans. However, it has since then been established that at least some of the thirty or so types can cause **aseptic meningitis, febrile illnesses with or without rashes, diarrhoea** or **mild upper respiratory tract infections**. Isolation procedures are as for polioviruses. Virus may also be recovered from the cerebrospinal fluid in cases of meningitis. Diagnostic serology is seldom practicable until the virus has been isolated, because of the number of viruses in the group, the lack of any common antigen and the impossibility of predicting which type is involved in a particular illness.

Hepatitis A virus

This is a small (25–28 nm) RNA virus with the morphological features of a picornavirus (p. 117). It has not yet been grown in culture, but has been transmitted from man to marmosets and chimpanzees. It differs widely from the hepatitis B virus, which is a DNA virus of unusual morphology (p. 135). The diseases hepatitis A and B are discussed together in Chapter 19 (pp. 264–6).

Rhinoviruses and the Common Cold

The all too familiar symptom-complex known as **coryza** or the **common cold** was for many years an insoluble problem to virologists. Most colds remained unexplained until 1960, when the rhinoviruses were discovered as a consequence of growing tissue cultures in a slightly acid culture medium at 33°C (a better approximation to human nose temperature than 37°C). Still more rhinoviruses were discovered when organ culture was introduced (p. 170), and over 100 types have now been distinguished. They conform to the description of picornaviruses, and are similar in size and structure to the enteroviruses. Other viruses have also been identified as causing colds, notably coxsackieviruses, echoviruses, some of the myxoviruses (influenza, parainfluenza and respiratory syncytial viruses), some adenoviruses and coronaviruses. Indeed it is now possible to isolate causative agents from most patients with colds, and quite often to isolate several from one patient. The next obviously desirable step—production of vaccines that will prevent colds—seems to be a long way off. This is because of the embarrassingly large number of different cold-producing viruses (with the further complication that mycoplasmas are also involved—p. 110). Immunization with a vaccine prepared from one strain may protect the recipient against infection with that strain, but it leaves him still liable to the attacks of some hundreds of other viruses that can give him colds. Immunity resulting from natural infection is equally strain-specific.

Reoviruses (= Respiratory Enteric Orphan viruses)

The original members of this group of relatively large icosahedral RNA viruses without envelopes are found in the human respiratory or intestinal tract, often in association with mild inflammatory diseases, but have never been shown to be responsible for these conditions.

Rotaviruses and Virus Gastro-enteritis

Rotaviruses closely resemble reoviruses. They get their name from the characteristic electron-microscopic appearance of the virion, which resembles a wheel with an inner hub-like core and spikes radiating to the outer double-shelled capsid. These viruses, which can cause infection in all age groups, are the commonest cause of gastro-enteritis in children, with a peak incidence between 6 months and 3 years (p. 252). Infection is commonest during the winter. Outbreaks of infection are particularly common in special-care baby units and among elderly people in hospitals and other residential institutions; and rotaviruses are also among the recognized causes of traveller's diarrhoea (p. 249). Excretion of rotaviruses in faeces is heavy during the acute illness and may continue after clinical recovery, and such excreted viruses retain their activity for months if not dried up. Transmission is faecal–oral—direct from person to person or via the hands of attendants or via food handled by an excreter. The same rotaviruses can also cause diarrhoea in newborn animals (e.g. piglets and calves) which may then be sources of infection for man. Rotaviruses are resistant to gastric acidity, and infect mature epithelial cells lining the small bowel, replicating there and impairing the cells' absorptive and transfer functions; this is one contribution to the diarrhoea. The damaged cells slough, shedding numerous virus particles into the gut lumen. The destruction of brush-border epithelial cells reduces lactase production, with a resultant failure to split and absorb saccharides. Disaccharide build-up in the lumen draws fluid from the tissues and further interferes with absorption, leading to diarrhoea, fluid loss and dehydration.

LABORATORY DIAGNOSIS Human rotaviruses can be grown only with difficulty in cell culture but the diagnosis can be rapidly confirmed by latex agglutination or enzyme-linked immuno-assay (ELISA) of faeces suspensions; these methods have largely replaced electron microscopy as screening tests for infection.

IMMUNITY The role of immune mechanisms in resistance to rotavirus infection is not fully elucidated. Immunity probably depends on IgA antibodies produced in the gut wall. Virtually all babies have maternal anti-rotavirus antibodies at birth, but the decline of this protection by the age of six months presumably contributes to the increased susceptibility of children to rotavirus infection during the next few years. Breast feeding confers some protection if anti-rotavirus antibodies are present in the milk. By age six years almost all children have antibodies to the common rotavirus serotypes. The presence of serotype-specific antibody does not necessarily confer resistance to infection, but the severity of symptoms is less if it is present, and this is the probable reason why adults and older children in general have a milder form of illness.

Other Viruses Causing Gastro-enteritis (p. 253)

Caliciviruses are larger than enteroviruses and are thought to be implicated in gastro-enteritis, having been observed in the faeces of children with diarrhoea. **Norwalk virus**, which is believed to be similar to calicivirus, causes outbreaks of 'winter vomiting disease' particularly in children but also in adults; despite the name many of those affected have diarrhoea as well as vomiting. **Astroviruses** are small viruses which may be implicated in outbreaks of diarrhoea and vomiting. Infections caused by these and other **small round structured viruses** (SRSV) have been reported after drinking or swimming in contaminated water, and after eating contaminated food such as oysters or cockles (p. 359) or food prepared by someone who was excreting the virus. As well as these viruses, certain serotypes of adenovirus and coronaviruses can be detected by electron microscopy in the faeces of patients with gastro-enteritis and occasionally in those of healthy people.

Arboviruses

This name was originally given to a group of viruses which had in common that they were arthropod-borne. Its validity is now questionable, because of the recognition of other viruses that are morphologically similar but epidemiologically different. Thus of the groups discussed here under this heading, the togaviruses and the sandfly fever virus (part of the bunyavirus group) are arthropod-transmitted, but not the hantaviruses of the bunyavirus group, nor the rubella virus (which is morphologically a togavirus). The true arboviruses are transmitted from one vertebrate host to another by blood-sucking mosquitoes or by ticks. One virus species may have more than one major host—e.g. birds as well as horses are commonly infected with the equine encephalitis viruses. Only for a few species—e.g. dengue—is man the sole known host. He plays an important part in the cycle of transmission of some others—e.g. yellow fever—but with the majority human infection is incidental, the disease being maintained chiefly in other hosts. One virus species may have more than one cycle of transmission (e.g. in yellow fever). The arthropod vectors are blood-suckers, and their infection depends on the occurrence of viraemia in the vertebrate hosts. The viruses multiply within the bodies of the vectors but cause no disease in them. In man they are responsible for a spectrum of disease ranging from subclinical infections or self-limiting non-specific influenza-like illnesses (fever, malaise, headache, nausea, vomiting and myalgia, sometimes accompanied by a maculopapular rash and polyarthritis) to severe and frequently fatal encephalitis or haemorrhagic fever.

Togaviruses

This group consists of some 300 icosahedral RNA viruses mostly transmitted by arthropods. They are unstable outside the bodies of their hosts. Those that are pathogenic to man are included in the alphavirus and flavivirus subgroups.

Alphaviruses

These include the organisms of **eastern equine, western equine**, and **Venezuelan encephalitis** and others which cause **dengue-like illnesses** (see below). They are mosquito-borne.

Flaviviruses

Some of these are mosquito-borne—e.g. **St. Louis, Japanese B** and **Murray Valley encephalitis; dengue** and the somewhat similar **West Nile fever**; and **yellow fever**. Ticks are the vectors of **Russian spring–summer** and **Central European encephalitis, louping ill** and some haemorrhagic febrile illnesses which occur in Russia and in India.

DENGUE This disease of tropical and subtropical areas is characterised by fever, severe and widespread muscle and joint pains (hence its other name of 'breakbone fever'), lymphadenopathy and rashes. It is rarely fatal except in outbreaks of **dengue haemorrhagic fever**; in this severe form of the disease, seen mostly in children in SE Asia, there is diffuse intravascular coagulation and shock, possibly triggered by an immune complex reaction with activation of the complement, coagulation and kinin systems (p. 57). Man is the only known vertebrate host for the virus, which is transmitted by *Aedes aegypti* (the vector of urban yellow fever) and related mosquitoes.

YELLOW FEVER This is a febrile illness in which an incubation period of 3–6 days is followed by jaundice and proteinuria (due to hepatic and renal necrosis) and by haemorrhages in various sites. It occurs mainly in equatorial Africa and South America, and caused the death of many early explorers of such regions. Monkeys of many species are susceptible to it, and the **jungle yellow fever** cycle, maintained by *Haemagogus* mosquitoes in South America and by various *Aedes* species other than *A. aegypti* in Africa, does not include man unless he penetrates into the jungle and allows himself to be bitten. **Urban yellow fever** has man as its host and *A. aegypti* as its vector. Fortunately this mosquito species, which breeds mainly in small accumulations of water around human habitations, is relatively easy to control, and the urban cycle can therefore be broken. However, the virtually uncontrollable jungle cycle remains as a menace to human beings who enter the jungle, and to nearby communities which allow their *A. aegypti* population to recover.

Vaccination with the egg-adapted 17D strain of yellow fever virus is safe and is effective for 10 years or more following a single injection. Natural infection is followed by lifelong immunity. In endemic areas almost the whole population may be immune as a result of subclinical infection.

LOUPING ILL The only arthropod-borne togavirus infection known to occur in Britain is primarily a disease of sheep, causing cerebellar damage and the characteristic ataxic movements from which its name is derived. It is transmitted by a tick, *Ixodes ricinus*, which sometimes bites man, who may then develop a mild encephalitis.

Laboratory diagnosis of togavirus infections

ISOLATION Nearly all of these viruses can be detected and isolated by intracerebral inoculation of suckling mice, in which they cause encephalitis. Most of them also grow readily in the yolk-sacs, or on the chorio-allantoic membranes of fertile hens' eggs or in tissue culture. Virus isolation may be possible from the blood of patients in the very early days of the illness, especially in yellow fever, but otherwise it may be impossible unless the patient dies and tissue from the brain or other affected organs can be used.

DIAGNOSTIC SEROLOGY Haemagglutination-inhibition, complement-fixation and neutralization tests are used to detect antibodies in patients' sera. These tests may indicate only the subgroup to which the infecting organism belongs, since there is much cross-reaction between different members of the group.

Bunyaviruses

This group of RNA viruses includes one which is transmitted to man by sandflies, and the hantaviruses, which man acquires from infected rodents.

Sandfly fever virus

Sandfly fever is an acute illness with high fever, muscle aches and pains behind the eyes. Recovery is invariable, rapid and complete. The disease occurs around the Mediterranean sea and in parts of Africa, India, Russia and China. The vector is the sandfly *Phlebotomus papatasi* which, because of its small size, can pass through screens and mosquito nets. Diagnostic tests are similar to those for togaviruses.

Hantaviruses

Members of this bunyavirus subgroup are responsible for causing **haemorrhagic fever with renal syndrome** (HFRS). A severe form of the disease occurs in China, Korea and Japan, and a milder, often subclinical form, occurs in Europe—particularly Scandinavia and Russia. The incubation period is 11–25 days. The mild form is characterized by sudden onset of fever, headache, myalgia and lumbar and abdominal pain followed by signs of renal involvement—oliguria, proteinuria and, in severe cases, renal failure, haemorrhagic complications and shock. The reservoirs of infection are wild and laboratory rodents that excrete virus via the lungs and in saliva and urine. Man-to-man transmission has not been described.

Rubella Virus

The virus of **rubella** (**German measles**) is morphologically a togavirus, but differs from those discussed earlier in being purely a parasite of man, with no arthropod or other vector. Rubella infection in childhood may be too mild to attract any attention, or it may be recognizable by the presence of posterior auricular and occipital lymphadenopathy and a maculopapular rash. The rash starts on the face and spreads over the trunk and extremities. The incubation period is 14–21 days. The virus can be recovered from the nasopharynx. Rubella antibodies appear in the patient's serum as the rash fades. Infection is spread by airborne droplets; patients are infectious from one week before to four days after the rash appears.

In adults there may be more constitutional disturbance, and joint pains are a common feature in women, but the illness is still mild, brief and followed by a high level of lasting immunity. However, when a non-immune woman becomes infected during early pregnancy and the infection is transmitted to the fetus, the story is far more serious; it is to be found in Chapter 25 (p. 322).

SEROLOGICAL DIAGNOSIS By no means every illness with clinical features suggesting rubella is due to rubella virus; and conversely, rubella infection in a pregnant woman can be subclinical so far as she is concerned but still damage the fetus. The two main reasons for measuring rubella antibodies are (a) to see whether a woman already has an antibody level that indicates that she is immune; and (b) to see whether a rubella-like illness results in a rising titre of rubella antibodies. If a woman who is not known to be immune is exposed to the risk of rubella infection, or develops an illness that might be rubella, during the early months of pregnancy, her blood should be tested for antibodies before there is time for them to rise as a result of infection—i.e. the blood sample should be collected within 14 days of exposure or within the first day or two of the illness. Such a test is likely to show that she is immune. If it does not, a further sample should be taken 16 days or more after exposure or 7–10 days after the onset of symptoms, and this may show an antibody rise, indicating that she has been infected and that the fetus is at risk. To establish that she has not been infected, it is necessary to obtain negative results from samples taken over the next few weeks. If no sample is taken early enough to allow demonstration of a rising titre, the interpretation of anitbody levels found in a late sample may be difficult, but may be made easier by determining whether the antibodies are IgG or IgM (p. 179). The purpose of all these antibody tests is to determine whether there is a risk of fetal infection and consequent malformation, and much anxiety on this score can be avoided by **checking the rubella antibody level as part of the routine early care of a pregnant woman**, rather than waiting until there is a known risk of infection. Indeed, there is no need to wait until the woman is pregnant; there is a strong case for routine screening of women who might become pregnant, especially schoolteachers and others particularly at risk of exposure to rubella infection. Immunization is discussed on p. 369.

Arenaviruses

These are medium-sized enveloped RNA viruses characterized by granularity of appearance in electron micrographs.

Lymphocytic Choriomeningitis Virus

A virus that causes endemic infection in wild mice and is excreted in their urine and faeces is occasionally transmitted to man, by inhalation of contaminated dust or by eating contaminated food. The resultant human illness may be influenza-like, but characteristically it is an aseptic meningitis with a high lymphocyte count (up to $1000/mm^3$) in the cerebrospinal fluid. The virus may be isolated from the patient's cerebrospinal fluid or blood by tissue culture or by intracerebral inoculation into mice. Antibodies can be detected in patients' sera by complement-fixation tests.

Lassa Fever Virus

A similar virus causes the much more serious

human illness first described in Lassa, Nigeria, in 1969, and subsequently seen in outbreaks and as sporadic cases in various parts of West Africa. Infection with this virus is endemic in certain rodent species in West Africa, and human infection is probably due to direct contact with body fluids, secretions or tissues of such animals. The illness is one of the **viral haemorrhagic fevers**; it begins with non-specific symptoms—fever, malaise, muscle pains, headache and sore throat—but after a few days suddenly becomes more severe, with high fever, prostration, diarrhoea and vomiting, chest and abdominal pains, and a distinctive inflammation and ulceration of the throat. Leukopenia and proteinuria are common. About a quarter of the patients die. The others may remain febrile for a week or two. Serum from those who have recovered from the disease is the only form of specific therapy. The infection is transmissible to a patient's close attendants, either as it is transmitted from rodents or in respiratory droplets, so special isolation of patients is essential (see below). The virus can be isolated in tissue culture, and can be recognized in the patients' tissues (the usual means of autopsy diagnosis) by electron or immunofluorescence microscopy. Special precautions should be taken (see below) in carrying out such investigations.

Marburg and Ebola Viruses

These as yet unclassified RNA viruses are morphologically similar (resembling elongated rabies viruses, p. 127), but antigenically distinct. Like Lassa fever virus they have caused outbreaks of **viral haemorrhagic fever**, characterized by fever, headache, muscle pain, diarrhoea and vomiting, maculopapular rashes and evidence of liver, kidney and central nervous system involvement, and with high mortality rates. The first such outbreak occurred in 1967, and affected laboratory workers in Marburg, West Germany, followed by other workers in Frankfurt and Belgrade; all of those involved had been dealing with tissues and tissue cultures from the same batch of African green monkeys from Uganda. Another outbreak due to the Marburg virus, affecting only three people, occurred in South Africa in 1975. Ebola virus disease outbreaks occurred in 1976 in Sudan and Zaire, and again in 1979 in the Sudan. The reservoir for this organism is unknown. Tranmission of these viruses from patients to medical and nursing staff can occur if special isolation facilities are not employed. The viruses can be grown in tissue culture, and can be recognized by electron or immunofluorescence microscopy. They form inclusion bodies resembling Negri bodies. Antibodies can be detected in patients' sera by complement-fixation or immunofluorescence tests.

Imported cases of suspected viral haemorrhagic fever

Viral haemorrhagic fevers due to the Lassa fever, Marburg or Ebola viruses, hantaviruses or other viruses are endemic in Africa, Asia and elsewhere, but from time to time a patient with what might be one of them arrives by air in Britain or some other country in which such illnesses are not familiar. Their non-specific influenza-like presentation and their varied incubation periods makes early diagnosis difficult, and it is therefore important that they should be suspected, particularly in medical or nursing staff who may have been in contact with appropriate patients but also in anyone who has recently been in an endemic area—especially if there is any history of relevant animal contacts or arthropod bites. Any suspected patient should immediately be admitted to a designated hospital with special isolation facilities, and virological investigations should likewise be carried out only in a designated laboratory specially equipped for handling dangerous pathogens.

Myxoviruses (Ortho– and Para–)

These two groups of enveloped RNA viruses with helical symmetry derive the shared part of their names from their affinity for mucus. By means of mushroom-shaped projections consisting of the enzyme neuraminidase they are able to penetrate surface mucus, and then by means of spikes projecting from their envelopes (known as haemagglutinins because they cause agglutination of red blood cells) they attach to mucoprotein receptors in host-cell surfaces.

Orthomyxoviruses

These are the **influenza** viruses. Human influenza is an acute febrile illness of world-wide distribution, occurring as sporadic cases or in epidemics or pandemics. Transmission is mainly by droplets, and the incubation period is only a day or two because the respiratory epithelium is both portal of entry and final target. Influenza is a systemic illness; coryza, sore throat and cough are accompanied by fever, myalgia, lethargy and malaise. It is typically of short duration (though recovery of full health may be slow, particularly in the elderly). Though usually it is not a serious illness, myocarditis, pericarditis and severe pneumonia due to the virus itself can occur; and in some outbreaks there is a high incidence of secondary bacterial infection of the damaged bronchial epithelium and lungs, notably by *Staph. aureus*, which can be rapidly fatal (p. 204). Rare central nervous system complications include the Guillain–Barré syndrome, in which post-infection demyelination causes paralysis that usually resolves after some weeks or months; and Reye's syndrome—encephalopathy and liver failure with high mortality associated with virus infection (particularly influenza B) in children. Salicylate ingestion during the infection has been implicated in the pathogenesis of Reye's syndrome.

Except during epidemics it is often difficult or impossible to make a clinical diagnosis of infection by an influenza virus rather than by one of the other viruses that attack the respiratory tract, and many sporadic cases or small outbreaks of 'flu' are not in fact influenza.

Influenza viruses are relatively stable at room and refrigerator temperatures and can survive for some weeks in dust.

ANTIGENIC STRUCTURE Viruses that cause influenza in man are divided according to their ribonucleoprotein (S) antigens into three types—A, B and C. Strains belonging to type A can be recovered from various animals and birds as well as from man, and these may well be important sources of new strains causing human outbreaks. Most epidemic strains belong to type A, and the antigenic structure of this type is particularly complex and unstable. Nucleic acid mutation may result in progressive minor antigenic changes ('antigenic drift'), but much bigger changes ('antigenic shifts') also occur as a result of the unusual structure of the virus. Its RNA consists of a 'team' of 8 pieces, each coding for a peptide, which need one another for production of viable virions but are to some extent independent in that each can be replaced by a corresponding member of another team. This can occur when two influenza A viruses meet in one host, and results in changes in the haemagglutinin (H) and neuraminidase (N) proteins—the antigens against which host immunity is directed. So far 16 H and 10 N subtypes are known, though only a few of them have been found in strains isolated from man and by no means all of the 160 possible HN combinations are known to exist. Individual strains are designated according to a complicated code—e.g. A/Hong Kong/1/68 (H3N2) was the first influenza A strain isolated in Hong Kong in 1968 and was of H subtype 3 and N subtype 2. Subtyping allows study of the movements of particular strains in epidemics and pandemics. After pandemics particular antigen combinations may disappear from circulation among humans (though it may be that they are maintained in animal or bird reservoirs) for long periods—e.g. H1N1 reappeared in the late 1970s after being absent for 30 years, and the pandemic of 1957 was due to a strain with a combination of antigens to which antibodies were found in those who had lived through the 1889 pandemic. Such periods of retirement are probably reflections of herd immunity (p. 59). Immunity following natural infection with a type A virus is highly strain-specific and therefore of limited value. Antigenic shifts and up-to-date knowledge of prevailing antigenic types are of great importance to the planning of immunization programmes for influenza A (p. 370). Type B strains undergo some antigenic variation, but type C strains are stable.

VIRUS ISOLATION The viruses can be found in the throat during the first few days of illness. Influenza viruses can be rapidly demonstrated in nasopharyngeal secretions by direct immunofluorescence. They can be grown in the amniotic cavities of fertile hens' eggs, where they are detected by their formation of haemagglutinins, or they can be grown in monolayers of human or monkey cells, in which they cause cytopathic changes and endue the cells with the property of haemadsorption.

SEROLOGICAL DIAGNOSIS Haemagglutination–inhibition tests are valuable for precise identification of virus isolates, but they are too highly strain-specific to be suitable for routine diagnostic examination of patients' sera. For this purpose it is better to employ complement-fixation tests, using type-specific antigens each of which will detect antibodies formed in response to infection with any virus strain belonging to its type (A, B or C). Such an antigen is formed when any virus of the appropriate type is grown in hens' eggs, and being soluble it can be separated from the virus particles. Since influenza antibodies are to be found in the blood of healthy people, it is particularly important in this disease to look for a **rising titre** (pp. 179, 180), though an unusually high titre in a single specimen may be diagnostic of recent infection.

Paramyxoviruses

Parainfluenza viruses

These include the various myxoviruses formerly called Sendai, croup-associated or haemadsorption viruses. They cause minor respiratory ailments, and can be distinguished from viruses of sundry other groups found in similar conditions by the haemadsorption phenomenon which they cause in tissue cultures (p. 170). They are divisible into four antigenic types.

Mumps virus

Mumps occurs mainly in childhood, when it produces acute and painful inflammatory swelling of one or more salivary glands, the parotids being most commonly involved. About 20% of post-pubertal males who have this disease develop orchitis. Central nervous system involvement is not uncommon, and may occur in the absence of parotitis; it may take the form of an aseptic meningitis or of meningo-encephalitis, but is rarely serious. Less common complications include pancreatitis and a self-limiting polyarthritis. Transmission of mumps virus is thought to be mainly by airborne droplets; patients are infectious from 7 days before symptoms appear to a week after. There is no certain explanation for the long incubation period—commonly 18–21 days or more—but it seems probable that generalized dissemination, along the lines indicated on p. 42, precedes localized disease. One third of those infected have subclinical infections and are subsequently immune. In most cases life-long immunity follows natural infection.

In most of its properties the mumps virus has a general resemblance to the influenza virus. It differs in that it rapidly becomes inactive at room temperature, and has a distinct and stable antigen structure. It can be isolated during the first few days of illness, from the mouth or the salivary ducts or in appropriate cases from the cerebrospinal fluid. It grows in the amniotic cavities of hens' eggs, or in monkey kidney tissue culture; in the latter it induces the formation of syncytia (giant cells). Complement-fixing antibodies to the S (soluble) antigen of the mumps virus appear in the patient's blood during the acute phase of the illness, but are short-lived, whereas antibodies to the V (virus) antigen appear during convalescence and are more persistent. Active immunization with a live attenuated strain of mumps virus gives good protection and is recommended in the USA and other countries for children over one year old (often given in a combined live virus vaccine with measles and rubella) and for susceptible adults, but it is not part of the recommended schedule in Britain.

Measles virus

In most parts of the world measles is a common, usually mild disease of children; epidemics occur every 2–3 years as the number of susceptible individuals in the community increases. However, isolated communities in which the disease is not endemic are liable to severe outbreaks affecting all age groups and with mortality rates as high as 25%. Infection is spread by droplets and is commonest in the late winter and spring. Children with measles are infectious 1–2 days before the onset of symptoms until 4 days after the rash has appeared. The portal of entry for measles infection is the respiratory tract. The viruses reach the local lymph nodes and spread to other lymphoid tissues, notably the spleen. Six days after infection the virus enters the blood and seeds all the epithelial surfaces of the body. At 9–10 days the infected epithelial cells in the respiratory tract and conjunctiva break down, shedding large numbers of virus particles, and causing symptoms of acute respiratory illness—cough, running nose and red conjunctivitis. The oropharyngeal mucosa becomes necrotic with the formation of small ulcers (Koplik's spots) by day 11. Finally the characteristic maculopapular rash appears about 14 days after infection and spreads over the whole body at the same time as the fever disappears and circulating antibodies appear. The rash is largely a manifestation of cell-mediated immunity and is characteristically absent in patients who have impaired T cell immunity. The necrotic areas in the nasopharynx and lungs are susceptible to secondary bacterial infection; hence bronchopneumonia and otitis media are common complications. In malnourished children or immunodeficient patients an unusual form of measles characterized by giant-cell pneumonia may occur (p. 204). Encephalitis (p. 221) is an infrequent complication (1 per 1000 cases) with a high mortality and a high incidence of permanent sequelae, such as epileptic seizures, in survivors. **Subacute sclerosing panencephalitis** (SSPE, p. 221) is a rare fatal disease of children or adolescents who have had measles some years earlier; it is apparently due to reactivation of measles virus latent in the brain, with resultant slowly progressive demyelination.

Measles virus can be isolated from the nasopharynx in the early phase of the illness, and rising levels of complement-fixing and haemagglutination-inhibiting antibodies may be found in patients' sera, but a single finding of a high level means little because of the likelihood of past infection. These problems are of little practical importance, since laboratory confirmation of the diagnosis is seldom required.

Lasting immunity follows natural infection. Passive immunization with human immunoglobulin is indicated when young or delicate infants have been exposed to infection. Active immunization is discussed on p. 369.

Respiratory syncytial (RS) virus

Because it survives for only a short time outside the host and cannot tolerate freezing, this common pathogen of the human respiratory tract escaped detection until 1957. It is now known to be widely disseminated in Britain, the USA and many other countries during winter months, presumably by means of droplets, and to be a major cause of respiratory tract infections. It causes minor respiratory ailments in adults; but young children, particularly those less than one year old, may have a severe illness, with laryngotracheitis, croup, bronchiolitis and pneumonia, which may be fatal (p. 199). This virus readily spreads amongst children during the winter months, and cross-infection in paediatric wards can be a considerable problem.

The virus can be detected in nasopharyngeal exudate by immunofluorescence and grows in human or monkey tissue cultures. Complement-fixing and neutralizing antibodies are formed in response to infection even in very young patients.

Rhabdoviruses

Rabies virus

The causative agent of rabies is a helical enveloped RNA virus which differs from the myxoviruses in that its virions are bullet-shaped (typically) or rod-shaped or filamentous. The clinical presentation, management and prevention of rabies are discussed on pp. 222–3 and 334.

Coronaviruses

These pleomorphic enveloped RNA viruses take their name from their crown-like ring of petal-shaped projections. Many distinct types within this group attack the human respiratory tract, usually producing illnesses indistinguishable from common colds produced by rhinoviruses. Their incrimination in outbreaks of acute diarrhoea and vomiting, often associated with fever and pharyngitis, has added coronaviruses to the list of viruses that may cause gastroenteritis.

Retroviruses

RNA viruses of this unusual group have been suspected of causing some rare human leukaemias and lymphomas; but one has achieved notoriety as the presumptive cause of **acquired immune deficiency syndrome** (AIDS), a disease first recognized in 1980 among male homosexuals on the west coast of the USA. It was characterized by a break-down of cell-mediated (T cell) immunity and consequent development of opportunist infections, which were eventually fatal. In 1983 French workers isolated the virus from the lymph node of an AIDS patient and called it lymphadenopathy associated virus (LAV); almost simultaneously workers in the USA isolated it from leukocytes of AIDS patients and called it human T cell lymphotropic virus no. III (HTLV-III), now re-named human immunodeficiency virus (HIV). It has been isolated from patients with the various clinical presentations of AIDS (p. 283) and from clinically healthy people in high-risk groups; and serological studies in the USA have shown antibodies to it in the blood of over 90% of AIDS patients and their sexual partners and of nearly 40% of healthy homosexuals, compared with less than 1% of healthy blood donors. Sero-epidemiological studies have shown that infection with this or a closely related virus is endemic in parts of central Africa, where it is spread by heterosexual intercourse, causing only relatively mild immune deficiency; it seems likely that the virus spread from central Africa to the USA (possibly via Haiti) and to Europe.

HIV infection has spread rapidly, mainly among promiscuous homosexuals, since 1980 and has become a major health problem in the USA and other countries. It can also be spread by heterosexual intercourse and vertically from mother to baby, and among drug addicts who used contaminated syringes or needles (cf. hepatitis B, p. 265). Another important and worrying mode of spread has been to haemophiliacs via factor-VIII concentrate prepared for their treatment from plasma pools to which infected homosexual (and possibly other) donors had contributed. Only a proportion of people infected become ill. Present evidence suggests that this occurs in 5–10% of exposed homosexuals. The incubation period, assessed in patients infected by blood products given for therapy, has ranged from 9 months to 6 years (mean 28 months). The clinical manifestations of AIDS are considered in Chapter 20 (p. 283).

HIV has a specific tropism for T helper (T4) lymphocytes (p. 58), the T4 surface marker acting as a virus-receptor. The many roles of T helper cells in the immune mechanisms are reflected in the complex immunological deficiencies found in AIDS. Humoral immunity is depressed as well as cellular, but paradoxically patients have hypergammaglobulinaemia. The reduction in the numbers of T helper cells is largely responsible for the overall lymphopenia and for reversal of the normal T4:T8 ratio.

LABORATORY DIAGNOSIS Serological tests (ELISA, RIA and immunofluorescence) for HIV are now used to screen all blood donors for evidence of HIV infection. A positive serological test for HIV indicates only that the patient has been exposed to the virus and does not indicate whether that patient will develop AIDS or transmit the infection.

Parvoviruses

These are small spherical DNA viruses. Some are defective and can cause infection only when they are in mixed infections with other viruses such as adenoviruses; but others are now recognised as causing distinct disease.

Aplastic crises

Parvoviruses are responsible for the aplastic crises that can occur in sickle cell disease, hereditary anaemias and spherocytosis. In these conditions the marrow is already under stress, trying to compensate for the shortened red-cell survival. Parvovirus infection causes an acute cessation of erythropoiesis, and since these patients have very limited marrow red-cell reserves there is a precipitate onset of anaemia, manifest clinically as acute distress and collapse. Healthy individuals infected with parvovirus do not experience such a dramatic outcome, probably because they can compensate by drawing on their marrow reserves. Such crises characteristically occur only once in childhood (after which the patient is immune) and are often preceded by general symptoms of a virus illness.

Fifth disease

A common parvovirus infection which usually occurs in childhood is known as **fifth disease** (or **erythema infectiosum**). After a prodromal illness with headache, sore throat and malaise a vivid red facial rash develops, giving a 'slapped cheek' appearance. This is followed 1–2 days later by a maculopapular rash on the trunk and outer aspect of the limbs, which fades leaving a widespread reticular pattern. In children the rashes of fifth disease and rubella are often clinically indistinguishable. The virus particles can be demonstrated in the blood during the second week after infection. The presence of IgM antibodies to parvovirus in serum indicates recent infection; 60–70% of adults have IgG antibodies to parvovirus, indicating past infection. Studies of outbreaks of fifth disease, which are common in schools and closed communities, have shown that adult infection is frequently subclinical. In other cases the first manifestation may be the development of arthritis; like the arthritis associated with rubella this is usually mild and self-limiting and affects mainly young women.

Papovaviruses

Papovaviruses (**PA**pilloma, **PO**lyoma and monkey **VA**cuolating viruses) are small icosahedral DNA viruses without envelopes. A virus of this morphology is demonstrable by electron microscopy in **common human warts**, and since such warts are transmissible to other humans by means of cell-free filtrates they are presumed to be due to the virus. It has not been certainly grown in tissue culture or transmitted to other species. Morphologically similar viruses have been demonstrated by electron microscopy in (but not yet grown from) brain tissues of patients with **progressive multifocal leuco-encephalopathy**, a rare form of encephalitis occurring in the terminal stages of neoplastic diseases; and in the urine of immunosuppressed kidney-transplant patients, who also show an increased incidence of warts. The other members of this group are all animal viruses. They are the causative agents of benign and malignant tumours in their various host species, which suggests that similar DNA viruses may be involved in the aetiology of human malignant diseases (c.f. pp. 131 and 133). Such speculation is not limited to this group of viruses, as some RNA viruses also cause malignant tumours in animals—notably the first two viruses ever shown to do so, the fowl leukaemia virus and the rous sarcoma virus.

Adenoviruses

These also are icosahedral DNA viruses without envelopes. Those isolated from humans share a common complement-fixing antigen but can be separated into at least 39 types by haemagglutination–inhibition and neutralization tests. Some (types 3, 4, 7, 14 and 21) cause **acute febrile illnesses** with **inflammation of various parts of the respiratory tract**, of which **sore throat** is a frequent component. Epidemics of such infection may occur, and have been particularly noted among large communities of new recruits to the US forces. **Conjunctivitis** is often associated with the respiratory tract infections ('pharyngo-conjunctival fever') in children, and outbreaks have been associated with swimming in inadequately chlorinated swimming pools. Types 8, 19 and 37 have been incriminated in many countries as causes of **epidemic kerato-conjunctivitis**, notably among shipyard workers and others exposed to abnormal risks of minor corneal abrasions. Outbreaks of this condition have also resulted from cross-infection between patients who were being investigated and treated at the same ophthalmological clinics. There is evidence that in children enlargement of Peyer's patches as a result of adenovirus infection may lead to **intussusception**. Other serotypes are responsible for **acute haemorrhagic cystitis**. Certain adenovirus serotypes, which are anomalous in that they do not replicate in tissue culture, cause a form of virus gastro-enteritis, particularly in children, which has a long incubation period of 8–10 days; large numbers of adenovirus particles can be demonstrated in the faeces of such patients by electron microscopy.

LABORATORY DIAGNOSIS Adenoviruses can be detected by direct immunofluorescence in respiratory secretions or in conjunctival cells. Electron microscopy is used to demonstrate their presence in faeces and, with the exception of these intestinal strains, adenoviruses can be recovered from appropriate sites by growing them in human-tissue cultures. Both group- and type-specific antibodies appear in the blood following infection; complement-fixation tests for the former are used to provide evidence of adenovirus infection and neutralization tests for the latter to indicate the type of virus involved.

Herpesviruses

These are icosahedral DNA viruses with envelopes. They multiply inside the nuclei of host cells, producing intranuclear inclusion bodies. The group includes the herpes simplex viruses (and their simian equivalent, the monkey B virus, an occasional cause of fatal central nervous system infection in humans who work with monkeys or monkey tissues); the varicella–zoster (chicken-pox and shingles) virus; the cytomegalovirus; and the Epstein–Barr virus of infectious mononucleosis.

Herpes Simplex Viruses

These are probably man's commonest virus parasites. Two types of herpes simplex virus can be distinguished by serological methods and by some of their cultural features—notably the pocks that they produce on chorio-allantoic membranes. Type 1 (HSV-1) used to be responsible for virtually all oral and central nervous system infections, whereas type 2 (HSV-2) was found mainly in the genital tract. With the increase in orogenital sex this division is no longer clear-cut. Exposure to HSV-1 occurs early in life. Children acquire the infection from asymptomatic excreters or by contact with contaminated items such as cutlery, toys or communal towels. (The virus can survive at room temperature for 24–48 hours.) The prevalence of HSV-1 is greatest in families of low socio-economic status with low standards of hygiene. Antibodies to HSV-2 are seldom demonstrable before adolescence, which supports the view that this type is sexually transmitted. Most primary HSV-2 infections (c. 75%) are subclinical but the patient may nevertheless excrete the virus and be infectious for about 2 weeks.

Primary HSV–1 infection commonly occurs early in childhood, often with few specific manifestations other than mild fever and malaise, but it can cause **aphthous stomatitis** (p. 190). Primary infection in eczematous children may cause **eczema herpeticum**, in which large numbers of vesicles appear on the eczematous skin. **Herpetic whitlow** is a painful and often destructive finger infection, resulting from direct inoculation of herpes virus into the skin of the finger—sometimes the finger of a doctor, dentist or nurse attending to a patient with herpes. Other possible results of primary infection include mild **aseptic meningitis** (p. 219) and a more severe **encephalitis** (p. 221)—a disease with a high mortality and high incidence of neurological sequelae in survivors.

Genital herpes (p. 282) is one of the commonest sexually transmitted diseases and the least tractable. Recurrent genital herpes is common in pregnant women, and if lesions are present at term infection may be transmitted to the baby during vaginal delivery, causing severe keratitis, meningo-encephalitis or generalized infection with high mortality (see Chapter 25).

Despite antibody formation by the host during the primary infection, the virus is not eliminated and establishes itself in a latent state in the sensory ganglia, particularly the trigeminal

(HSV-1) and lumbosacral (HSV-2) ganglia. From time to time this host–parasite relationship is disturbed and a crop of vesicles may appear on the skin near the lip margin, near the nose, on the genitalia or elsewhere. These commonly recur many times, possibly as a result of widely different stimuli—e.g. exposure to sunlight, fever, menstruation or emotional stress—and nearly always occupying the same site in a given patient. This is by far the commonest disorder caused by this virus, and is known as **herpes simplex** (or as **herpes labialis** when it is at its commonest site, as **herpes genitalis** when on the genitalia, or as **herpes febrilis** on the frequent occasions when the stimulus is a febrile illness). The skin lesions are preceded by itching or burning pain in the affected area. Recurrent lesions are infectious for about 5 days. Herpetic lesions, either primary or recurrent, may involve the cornea (**dendritic ulcers**) or conjunctiva (**conjunctivitis**) or both (**kerato-conjunctivitis**). Treatment of such lesions and of other herpetic infections is discussed on p. 135.

LABORATORY DIAGNOSIS Herpes simplex virus may be detected in material from lesions by electron microscopy or by direct immuno-fluorescence. It can be grown on hens' egg chorio-allantoic membranes or in tissue culture, in which it forms inclusion bodies and is cytopathogenic. Detection of rising levels of neutralizing and complement-fixing antibodies may be of diagnostic value in primary infections, but such antibodies are common in the blood of healthy adults.

HSV-2 and carcinoma

Women with carcinoma or precancerous changes of the cervix uteri have antibodies to HSV-2 in their blood more commonly than do women from appropriate control groups. It is therefore possible that the virus, though seldom isolated from such lesions, is responsible for inducing neoplastic changes. Alternatively, it may be transmitted in company with a carcinogenic agent.

Varicella–Zoster (VZ) Virus

Chickenpox (**varicella**) is a common and usually mild illness that occurs mainly in childhood and is characterized by a vesicular rash, usually widespread. It is highly infectious, with an incubation period of 13–17 days. **Shingles** (**zoster**) is a much less common condition, almost confined to adults; its main features are fever, malaise and severe pain in the distribution of one or more nerve roots, with clusters of vesicles in the same distribution. It is not transmissible as shingles, but close contacts—particularly children—may develop chickenpox.

VZ infection is acquired by inhaling airborne droplets from patients who have developed chickenpox within the past 10 days, or from vesicle material from patients with chickenpox or shingles. The virus enters the body via the respiratory tract mucosa, and after the incubation period and a viraemic phase it localizes in the skin and mucous membranes, where degeneration of epithelial cells and accumulation of tissue fluid produce the characteristic skin lesions. These start as pinhead macules, turn rapidly into papules and then become highly irritant vesicles that ultimately form crusts. The lesions are unlike those of smallpox (p. 134) in that they appear in crops on successive days, so that new ones develop when others are well advanced. The trunk is usually affected before the face and limbs. Lesions may become secondarily infected with staphylococci or streptococci. Encephalitis and pneumonia are rare complications in childhood, but primary infection in adults may lead to a severe form of chickenpox with pneumonia. Immunodeficient children or patients on immunosuppressive drugs are liable to develop disseminated VZ infection with a haemorrhagic rash, and this condition is often fatal. An ordinary attack of chickenpox commonly confers life-long immunity against that disease, but the virus may remain latent in sensory nerve cells for years after the attack and then, reactivated by lowering of immunity (e.g. by irradiation or a neoplasm), it may produce an attack of shingles; the pain is due to posterior horn cell infection.

Despite the clinical resemblance between chickenpox and the poxvirus diseases, the VZ virus is morphologically not a poxvirus but a herpesvirus. Furthermore, characteristic giant cells with intranuclear inclusions (Tzank cells) are seen on microscopic examination of material from the skin lesions that VZ virus produces, and are also found in herpes simplex lesions, but not in those of poxvirus infections. The VZ virus can be differentiated from both poxviruses and herpes simplex virus by electron microscopy and by its failure to produce lesions on the chorio-allantoic membranes of eggs. It can be grown in human-embryo tissue culture, where it produces cytopathic effects and intranuclear inclusions. Immunofluorescence and immunodiffusion can be used to detect virus antigen. Complement-fixing antibodies for VZ virus antigen can be detected in patients' blood following either chickenpox or shingles, but levels are higher after the latter—a finding which fits in with the concept that shingles is a disease produced by the chickenpox virus in subjects who are already partly immune. Drugs that can be used in treatment of VZ infections are discussed on p. 135. Human VZ-specific immunoglobulin can be used to prevent or treat severe VZ infection in immunodeficient patients.

Cytomegalovirus (CMV)

This virus is indistinguishable morphologically from herpes simplex and varicella–zoster viruses. The names cytomegalovirus and cytomegalic inclusion disease are derived from the characteristic changes found in infected cells, whether in the body or in tissue culture—namely, marked cellular enlargement and the formation of intranuclear inclusions.

Primary infection, which occurs mainly during adolescence and adult life, may be asymptomatic and detectable only by an antibody rise, or may cause a mild fever or occasionally hepatitis, pneumonitis or heterophil-antibody-negative infectious mononucleosis (p. 193). The virus may also be associated with localized or more generalized disease in debilitated elderly patients or in those with impaired immunological mechanisms, particularly those on immunosuppressive or cytotoxic drugs. After symptomatic or asymptomatic primary infection, CMV remains latent in the body, often for many years. CMV infection may be transmitted by kissing, sexual contact or blood or cell transfusion, and in transplanted tissues; the evidence of widespread infection provided by serological surveys may indicate that other means of transmission (e.g. by droplets) are also common. Many patients infected by transfusion are asymptomatic, but others develop fevers, splenomegaly and lymphocytosis (with abnormal blood lymphocytes, as in infectious mononucleosis, p. 133) 3–6 weeks later. Healthy children may excrete virus in the urine and saliva, and so infect other children and adults. Homosexual men too have a high rate of virus excretion in the urine and also in semen, infection being transmitted by sexual contact.

Cytomegalic inclusion disease is common in transplant recipients whose cell-mediated immunity is depressed by immunosuppressive drugs, and in patients with leukaemia receiving similar therapy. The infection may be primary or may result from reactivation of latent virus. Symptoms of fever, leukopenia, hepatitis and arthralgia appear 1–2 months after a transplant operation and have to be distinguished from the symptoms which accompany graft rejection. Interstitial pneumonia caused by a combination of CMV and *Pneumocystis carini* is a serious and often fatal complication following bone-marrow transplantation or in patients with AIDS (p. 283).

A pregnant woman with either a primary or a recurrent CMV infection may transmit it transplacentally, but more often the baby is infected from her genital tract during delivery or postnatally via milk. Most infected babies (including virtually all of those infected during or after birth) are unaffected, but 10–15% of intra-uterine infections are symptomatic, with manifestations including damage to the nervous system (which may result in mental retardation or deafness), liver and spleen enlargement, rashes and thrombocytopenia. The virus can be isolated from the mother at the time of delivery, and cytomegalic inclusions can be identified in the placenta. CMV infection is now commoner than rubella as a cause of fetal damage, since intra-uterine rubella is largely preventable.

LABORATORY DIAGNOSIS CMV infection can be diagnosed in the laboratory by finding typical infected cells, with 'owl's eye' inclusions, in the patient's saliva or urine or in portions of infected tissue. The virus can be grown in human-embryo tissue culture, and antibodies can be detected by complement-fixation tests or immunofluorescence. The presence of IgM antibodies to CMV in cord serum is diagnostic of intra-uterine infection. It is important to remember that the mere presence of this virus in the human body does not mean that it is causing disease.

No effective drug is available for use against this virus. A vaccine for active immunization to prevent maternal infection is being evaluated.

Epstein–Barr (EB) Virus

In 1962 Burkitt drew attention to the occurrence in children in certain parts of Africa of an assortment of malignant tumours that had many features in common. Their epidemiology was such as to suggest that they are transmitted by insects, and this idea led to a vigorous pursuit of a possible virus agent, both in material from the African tumours and in that from similar conditions recognized in many other parts of the world. As in other attempts to find causative agents for tumours and leukaemias, the results were confusing because too many viruses were found. Presumably some of them at least were innocent 'passengers' in the tissues examined. The most convincing contender for the role of the Burkitt's lymphoma virus is the herpesvirus detected in cell-cultures from lymphoma tissue by Epstein and Barr in 1964. The same virus has been shown to be associated with nasopharyngeal carcinomas in Chinese patients, and its role in the aetiology of these malignant conditions is still a matter for debate and vigorous research. Meanwhile a quite different field of research was opened up when a technician developed **infectious mononucleosis** after working with the EB virus, and was observed to produce antibodies to that virus during her illness. This disease, alternatively known as **glandular fever**, is described in Chapters 14 (p. 193) and 17 (p. 245). Its main features are low fever, pharyngitis, lymphadenopathy and the presence of abnormal mononuclear cells in the peripheral blood. It is not caused exclusively by EB virus; some cases are due to other viruses (notably CMV—p. 132) or to *Toxoplasma gondi*. Heterophil antibodies, so called because they agglutinate cells from other species (in this case, sheep or horse red cells) are found in the blood of most patients with infectious mononucleosis, and give a strong indication of the aetiology, since EB virus is responsible for the illness in nearly all of those in which they are present (and also for a few in which they are not). These antibodies are detected by the Paul–Bunnell test (p. 193) or a variant of it. This test gives an answer earlier in the illness than do specific tests for EB-virus antibodies.

EB virus resembles other herpesviruses not only in its morphology but in its epidemiology. Natural infection is common in childhood, and indeed almost universal in many communities, but is frequently asymptomatic. The salivary glands are the primary site of infection and once infection has occurred the carrier state is established. In primary infection EB virus produces a lytic infection of B lymphocytes, with shedding of new virus particles into the saliva. A substantial minority of those infected have persistent buccal infection with virus shedding, and are therefore sources of infection to others, particularly by kissing. Persistence of virus can also be demonstrated in some of the circulating B lymphocytes of these people. Infection may also be transmitted by transfusion of blood cells or by bone-marrow transplantation.

The abnormal mononuclear cells in the blood which give the infectious mononucleosis that name are T lymphocytes, proliferating in response to virus-induced 'foreign' antigens on the surfaces of infected B lymphocytes. This proliferation accounts for various features, in particular the enlarged lymph nodes and spleen. A small group of patients do not recover from primary infection and symptoms persist for months or even years. Activation or reactivation of latent EB virus infection may occur during pregnancy but rarely produces disease. In those who are immunodeficient or on immunosuppressive drug treatment, EB virus infection can result in unchecked B lymphocyte proliferation, which may be fatal.

Poxviruses

By virus standards the poxviruses are large, and

their enveloped particles, which appear brick-shaped on electron microscopy, have a relatively complex structure that is not yet fully elucidated. Members of this group of DNA viruses of greatest interest to us are those of **smallpox, cowpox** and **vaccinia**. **Contagious pustular dermatitis** ('orf', p. 308) is a poxvirus infection occasionally transmitted to man from sheep or goats. **Molluscum contagiosum** (p. 308) is a transmissible human disease which produces multiple warty skin nodules in many parts of the body.

Smallpox (Variola), Cowpox and Vaccinia Viruses

Smallpox, which is said to have killed 60 million people in the eighteenth century, was still endemic in many countries and liable to appear anywhere in the world until a few years ago. However, a vigorous World Health Organization eradication programme has now eliminated the disease, and the world was officially declared free of it in 1980 (p. 332). The next paragraph is therefore largely of historic interest only.

Smallpox was a febrile illness, with macular skin lesions that became papular, then vesicular, then pustular, and finally crusted after about 10 days. In severe cases they were confluent and even haemorrhagic, and there was then much systemic disturbance and many patients died. The lesions differed from those of chickenpox in being more peripherally distributed and in tending to be synchronized rather than appearing in crops.

Cowpox is a disease of cattle. Its lesions resemble those of smallpox, but the infection causes little or no systemic disturbance, either to cattle or to humans who acquire it from them—and are thereby immunized against smallpox, as Jenner discovered (p. 7).

Vaccinia is the condition resulting from infection with the virus used for immunization against smallpox. The origins of this virus are uncertain. It may well be that some of the strains used were derived from smallpox viruses and some from cowpox viruses, both types having been so modified by animal passage that they became identical. Although the vast majority of people vaccinated intradermally with live vaccinia virus produced only single local lesions (similar to those of smallpox, as described above), a few developed either a systemic illness with widespread skin lesions (generalized vaccinia) or vaccinial encephalitis, and some of these died. Vaccinia virus should therefore not be used any more, unless smallpox unexpectedly reappears. The possibility of its doing so as an act of microbiological warfare has sadly to be borne in mind, and it may therefore be considered necessary to maintain stocks of vaccinia virus.*

Poxviruses can be detected in fluid or crusts from skin lesions by electron microscopy, and although they cannot be distinguished from one another in this way, it does permit their rapid distinction from chickenpox virus, which despite its name is not a poxvirus but a herpesvirus (p. 131). Poxvirus antigens can be detected in material from lesions by double diffusion (p. 175). Methods for distinguishing between the viruses of this group include growing them on hens' eggs chorio-allantoic membranes (where they have different ceiling temperatures for growth and produce different lesions) and DNA analysis using bacterial restriction endonucleases.

Hepatitis Viruses

A number of viruses already mentioned can cause liver disease, as indicated in Chapter 19 (p. 267). There are also at least five other viruses that are specifically associated with hepatitis—hepatitis A and B viruses, delta virus and at least two others currently classified as non-A non-B hepatitis viruses. Clinical features of virus hepatitis are discussed in Chapter 19 (pp. 264–6).

Hepatitis A virus is an RNA virus of the enterovirus group (p. 119) which has not been cultivated in the laboratory but can be detected in the stools of patients with a clinical diagnosis of infective hepatitis. It is spread by the faecal–oral route and causes an acute, self-limiting disease that does not lead to chronic hepatitis or chronic carriage. Infection with hepatitis A virus is common and many of those who become infected remain asymptomatic; serological surveys have shown that a large proportion of the general population have antibodies against

---

*Recent demonstration of very long term persistence of smallpox virus in corpses suggests another possible source of recurrence of the disease.

hepatitis A virus, indicating previous infection, but most have not been jaundiced.

Hepatitis B virus is a DNA virus which causes the disease known as serum hepatitis. The virus has never been grown in culture or transmitted to animals but it can be seen by electron microscopy of the blood of patients with hepatitis of the long incubation type (as well as others for whom there is no clue as to the incubation period) and of some other people who are carrying it without symptoms. Three characteristic types of particle are recognized— 42 nm double-shelled spheres (Dane particles) believed to be the virions proper, 22 nm spheres and 22 nm-diameter filaments. The latter two types consist merely of capsid material (HBsAg). Presence of any of the particle types is associated with detectable HBsAg; and HBsAg detection, using RIA or ELISA or a form of passive haemagglutination, is more practicable than electron microscopy as a means of checking blood samples for the presence of virus. The 42 nm particles contain a core antigen, HBcAg, but this cannot be detected in patients's blood samples; it is found after disruption of the particles in the laboratory. A third antigen, HBeAg, can be found in blood; its presence there correlates with presence of 42 nm particles, with a high level of infectivity and with a propensity to chronic infection. Subtypes of HBsAg have been defined, and are useful in epidemiological studies.

Delta virus is believed to be a defective RNA virus. It is always associated with hepatitis B virus infection and it appears that delta virus can replicate only with the aid of hepatitis B virus in a mixed infection.

Non-A non-B viruses cause at least two clinically distinct forms of hepatitis—one resembling hepatitis A and the other resembling hepatitis B, but with negative serological tests for these and other viruses known to cause hepatitis. These viruses have not been isolated nor detected serologically.

Chemotherapy of Virus Infections

Most antimicrobial drugs owe their therapeutic usefulness to the fact that they interfere with the metabolic processes of the pathogens but not to any serious extent with those of the host's cells. Clearly there is not much scope for such discrimination in virus infections, though the action of interferons (p. 49) shows that it is possible. Much work has gone into the search for antivirus agents, but currently available compounds are limited by their poor solubility and varying degrees of bone-marrow toxicity. **Idoxuridine** (IDU), one of a series of thymidine analogues that interfere with DNA synthesis, gives good results when applied in aqueous solution to herpetic corneal ulcers. Dissolved in dimethyl sulphoxide, which allows it to penetrate skin, it has been used successfully in treatment of herpesvirus skin lesions and even of herpetic whitlows, vaccinia and orf, but it is no longer the first choice for treatment of herpes infection and is too toxic for parenteral use. **Vidarabine** is a purine nucleotide active against certain DNA viruses. It is used as an ointment for treating herpetic eye lesions and has been given intravenously with benefit to immunosuppressed patients suffering from severe varicella–zoster infections. **Acyclovir**, the least toxic of the new antivirus compounds, is a guanosine analogue active against DNA viruses such as herpesviruses. It is harmless to uninfected host cells, but inside infected cells it is a substrate for virus-encoded thymidine kinase and becomes incorporated into the virus DNA. However, some naturally occurring virus mutants are acyclovir-resistant because their DNA synthesis does not involve thymidine kinase. Acyclovir is effective only against replicating virus and does not eliminate latent virus infection. As an ophthalmic ointment it is probably the drug of choice for the treatment of ocular herpes and zoster infections. Topical creams or oral acyclovir are indicated for the treatment of herpetic infection of the skin and mucous membranes. In primary genital herpes the duration of pain, healing time and period of virus shedding can be shortened by topical, oral or in severe cases intravenous administration of acyclovir, but it does not affect the recurrence rate. In recurrent infection virus shedding is reduced but the other benefits are less marked than in primary infection. Prompt intravenous administration of acyclovir can bring about clinical improvement in herpes encephalitis and other severe herpetic infections, and so should be started as soon as a presumptive diagnosis has been made. Such intravenous treatment can also be effective as prophylaxis against severe herpetic infection in compromised hosts, but since the latent virus is not damaged it may become active when the treatment is stopped. **Amantadine** is useful only against influenza A viruses. As a prophylactic, given orally after exposure to the virus, it is as effective as previous immunization. It also may shorten the duration of symptoms when given for an established infection, though it has little effect on influenzal pneumonia.

Suggestions for Further Reading

Williams, G. 1960. *Virus Hunters*. Hutchinson, London—For the history of virological research.

Timbury, M.C. 1978. *Notes on Medical Virology*. 6th edn. Churchill Livingston, Edinburgh.

Fungi

The main properties of fungi are outlined on pp. 13–14 and 24. Study of these organisms is called **mycology** and diseases caused by them are known as **mycoses**. Those affecting man are conveniently discussed in two groups—superficial mycoses and deep mycoses. The drugs used to treat fungus infections are reviewed at the end of the chapter (p. 142).

Superficial Mycoses

These are infections confined to the body surfaces, principally the skin, the hair and the nails, and do not involve living tissue.

Pityriasis Versicolor

This common and benign form of skin infection produces non-inflamed brown scaling patches, particularly on the body and limbs of young adults. It is caused by *Malassezia furfur*, a dimorphic fungus which in its yeast form is part of the normal skin flora. The incidence of the disease is increased in warm climates and in patients on corticosteroid drugs. The diagnosis can be confirmed by using a **Wood's light** (ultraviolet), which causes yellow fluorescence of affected areas, and by direct microscopy of KOH-treated skin scales, which reveals the fungus as a characteristic mixture of round thick-walled budding cells and short irregular lengths of mycelium. Pityriasis versicolor usually responds well to treatment with a sulphur ointment or with a topical antifungal agent.

Superficial Candidiasis

This is an infection of mucous membranes or skin with yeast-like fungi of the genus *Candida*—of which *C. albicans* is by far the most important in clinical practice. *Candida* species are not part of the normal skin flora but are commonly present in the mouth (14%), gut (24%) and vagina (10%) of healthy subjects. During pregnancy and in women using oral contraceptives the carriage rate rises, possibly because of hormonally induced increases in the amount of glycogen in the vaginal epithelium. These resident yeasts may from time to time become pathogenic, usually in consequence of malnutrition, general debility, diabetes, antibiotic suppression of the bacterial flora which normally inhibit proliferation of these and other fungi, use of oral contraceptives, steroid therapy, immunosuppression or other predisposing factors. They can then cause a wide variety of disorders, including the following:

(1) **Acute oral candidiasis (thrush)**—a superficial infection of the mucous membrane of the mouth, characterized by white adherent patches on the buccal mucosa or tongue (p. 192). Candida may also be responsible for the inflammation associated with dentures or dental prostheses.

(2) **Candida oesophagitis**, which occurs in immunosuppressed patients and causes retrosternal pain exacerbated by swallowing. The white plaques and erosions in the oesophagus are seen as small filling defects on barium swallow.

(3) **Vulvovaginitis (vaginal thrush)**, with white curd-like discharge and much irritation and inflammation (p. 282). This is particularly common during pregnancy and in sexually

active women on oral contraceptives. Their partners often have erythematous rashes and white penile pustules (balanitis—p. 282).

(4) **Dermatitis**, usually affecting warm moist skin folds—e.g. in the groins and the submammary regions. The skin becomes red, exudative and irritant. In infants *C. albicans* frequently colonizes areas affected by napkin dermatitis and causes further inflammation.

(5) **Chronic paronychia** of housewives, barmaids, fruit packers and others whose hands are constantly in water. Candida infection of the nail-fold causes redness, swelling and loss of attachment of the cuticle to the nail plate which is often grossly distorted.

(6) **Candida peritonitis**—an increasingly common complication of peritoneal dialysis and of gastro-intestinal surgery, notably after large-bowel perforation.

(7) **Candida enteritis**—almost always a consequence of suppression of the normal bowel flora by broad-spectrum antibiotics.

(8) **Chronic mucocutaneous candidiasis**—a rare condition, consisting of persistent and disfiguring infection of skin, mucous membrane and nails. There may be underlying defective cell-mediated immunity and endocrine hypofunction.

The main laboratory features of *Candida* species are outlined on p. 171.

Treatment

The first step is to remove the predisposing cause if possible. Local appliction of a polyene or imidazole antifungal agent is then usually effective in skin and mucous membrane candidal infection. For thrush, lozenges or suspension are recommended, and for vaginal infection pessaries or creams, with treatment of sexual partners if there is recurrent infection. In chronic mucocutaneous candidiasis initial treatment is often with oral ketoconazole to induce a remission, followed by intermittent maintenance with therapy as required.

Ringworm (The Dermatomycoses)

Fungi of the genera *Trichophyton*, *Epidermophyton* and *Microsporum*, collectively known as the **dermatophytes**, cause superficial infections of the keratinized layers of the skin, hair and nails (**ringworm** or **tinea**) but never attack deeper tissues (p. 309). There is considerable overlap in the clinical syndromes which they produce, but *Epidermophyton floccosum* (the only species in its genus) does not attack hair and *Microsporum* species do not attack nails. Infection is commonly from man to man. For example, **tinea pedis** (athlete's foot), usually due to *T. rubrum* or *T. mentagrophytes*, is frequently transmitted via wet floors around swimming pools and bathrooms; and **tinea capitis** (scalp ringworm) due to *M. audouini* or *T. tonsurans*, can be transmitted by direct contact or via infected combs and brushes. Other species are acquired from animals—e.g. *M. canis*, which affects dogs, cats and other animals, may be transmitted to humans, particularly children, in close contact with them; and *T. verrucosum*, a common cause of cattle ringworm, is also transmissible to man. Human infections with such animal pathogens tend to provoke vigorous local reactions and to be self-limiting. A highly inflammatory pustular lesion can occur in the scalp or beard area, particularly in infection with an animal pathogen, and is called a **kerion**. Scalp ringworm is mainly a disease of children. Hairs infected with *M. canis*, *M. audouini* and *T. schoenleini* show green fluorescence under a Wood's light. Nail-plate infection is usually due to *T. rubrum* or *T. mentagrophytes*, and produces discoloured, thickened and distorted nails. (For laboratory diagnosis of ringworm see p. 171.)

Treatment

For skin lesions this may be topical or systemic, but systemic therapy is essential for hair or nail infection. In some dermatophyte infections, notably athlete's foot, secondary bacterial infection may have to be dealt with before the fungal infection can be controlled. Ointments of benzoic, salicyclic or undecylinic acid or combinations of these, for many years the best treatments for skin ringworm, still have a place in topical treatment of localized skin infections, as have the imidazoles. The antibiotic griseofulvin, which is valueless in treatment of bacterial or candida infections, cures most forms of ringworm. Taken by mouth for at least 3 weeks in the case of skin lesions, 4–8 months for fingernails and 8–12 months for toe-nails, it is incorporated in newly formed cells of the skin and its appendages and makes them unsuitable media for the growth of most dermatophytes. These cells then gradually replace the existing keratinized epithelial layers, hair and nails, causing the dermatophyte infection to die out. Nail infection is the form of ringworm that most commonly fails to respond to such treatment. Surgical removal of the nails may then be necessary.

Deep Mycoses

These fall into two fairly distinct groups, according to whether the main factor in their production is the virulence of the fungus or the dimished resistance of the host.

(1) Infections due to Virulent Fungi and Commonly Occurring in Previously Healthy People

Blastomycosis, paracoccidioidomycosis, histoplasmosis and **coccidioidomycosis** are rarely seen in Britain, and each has its endemic areas. The dimorphic fungi responsible are found mainly in soil, and man is usually infected by inhaling spores. The yeast phases then invade his tissues, producing spectra of disease patterns similar to and often mistaken for those of tuberculosis. Lesions are usually deep-seated, and demonstration of skin hypersensitivity to a specific antigen may be the simplest means of confirming the suspected diagnosis—except that in endemic areas such hypersensitivity is common as a sequel of asymptomatic infection.

Blastomycosis, caused by *Blastomyces dermatitidis*, was formerly believed to be restricted to eastern parts of the USA and Canada but is increasingly being found in many parts of Africa. Chronic pulmonary lesions are common, but the disease may spread to involve the skin, bones, viscera and meninges. Amphotericin B is the most effective form of treatment, with hydroxystilbamidine as a less toxic and often successful alternative.

Paracoccidioidomycosis, which is restricted to South America with its highest reported incidence in Brazil, is caused by *Paracoccidioides brasiliensis*, of which the natural habitat is not known. Infection may be via the respiratory tract, or by direct implantation of the organism into the oral or nasal mucous membranes with production of local ulceration. Systemic spread mainly involves lymphatic tissues. Oral ketoconazole therapy for 6 months has proved effective in inducing remissions and has supplanted combined amphotericin B and sulphonamide as the treatment of choice.

Histoplasmosis is caused by *Histoplasma capsulatum*, which is found in North and South America in soil contaminated by bird or bat excreta. In endemic areas many people have symptomless infection with consequent skin hypersensitivity to an extract of the organism called **histoplasmin**. Miliary lung lesions, similar to those of tuberculosis but healing spontaneously, may be found in such people. In a few subjects the infection progresses to involve lymph nodes and other organs, causing a febrile and often fatal illness, for which amphotericin B is the treatment of choice.

Coccidioidomycosis is caused by *Coccidioides immitis*. It is largely restricted to certain areas in the south-west of the USA. Almost all inhabitants of those areas are infected, most of them without symptoms but with the development of **coccidioidin** hypersensitivity. A very few develop a generalized tuberculosis-like illness with a bad prognosis, for which the treatment is either amphotericin B or an imidazole.

(2) Infections Caused by Opportunist Fungi, Mainly in Compromised Patients

These infections are becoming increasingly common throughout the world. The main reasons for this increase are the widespread use of immunosuppressive drugs, the advances in cardiac valve surgery, the increasing use of long-line intravenous cannulae particularly for parenteral nutrition, and the growing number of patients with haematological diseases (e.g. leukaemia, lymphoma) whose life expectancy has greatly increased as a result of advances in medicine. The outcome of such an infection is largely determined by the degree of the patient's immunological deficit or other underlying disorder. This group of mycoses includes **systemic candidiasis, aspergillosis, cryptococcosis** and **mucormycosis**.

Systemic candidiasis is the commonest of the opportunist deep mycoses. It occurs in compromised patients of all ages, and is an increasing problem in neonatal practice. *C. albicans* is the species most frequently isolated but others (notably *C. parapsilosis* and *C. tropicalis*) may be involved, particularly in endocarditis. Isolation of *Candida* from the blood of a 'high-risk' patient, especially in the presence of intermittent or continuous fever, has serious implications and calls for a search for the source (e.g. by culture of sputum and urine) and for removal of possible infected intravenous or other catheters, because **candida septicaemia**, involving abscess formation in the kidneys and brain and elsewhere, has a bad prognosis. **Candida endocarditis** may occur as part of such a septicaemia, or as a complication of heart-valve replacement, or in drug addicts who introduce the organisms into their blood while giving themselves injections.

TREATMENT OF SYSTEMIC CANDIDIASIS In general, unless the underlying immune defect is corrected (e.g. by reducing the dose of immunosuppressive drugs or by removing a predisposing factor such as an infected catheter) no amount of specific antifungal therapy will produce successful resolution. Infection of a heart valve requires both replacement of the valve at the earliest opportunity and effective drug therapy. For this purpose amphotericin B can be used alone or in combination with the anti-yeast agent flucytosine. By giving the two it is possible to lower the amphotericin dosage and hence minimize its nephrotoxicity. *Candida* strains vary in their sensitivity to flucytosine, and this should be checked before instituting therapy. The imidazoles are still being evaluated as agents for treating systemic candidiasis. Oral anti-fungal drugs are given to some 'high risk' patients (notably those with neutropenia) to reduce their risk of developing systemic candidiasis caused by organisms from their own intestinal tracts.

Aspergillosis is the collective name for diseases caused by members of the genus *Aspergillus*. These common saprophytic moulds, such as grow on stored foods, are easily recognized under the microscope by their conidiophores—swollen hyphal ends from which radiate large numbers of sterigmata (short lengths of narrower hyphae) ending in short chains of spores. As aspergilli are widely distributed in the environment and multiply at room temperature, their presence in clinical specimens needs cautious interpretation, particularly if the specimens are not freshly collected. However, species of this genus are from time to time clearly responsible for human disease. **Otomycosis** and **allergic aspergillosis** do not belong under the heading of opportunist deep mycoses. The former is a chronic superficial infection of the external auditory meatus, which becomes filled with fungal mycelium; imidazoles are useful in its treatment. Allergy to aspergilli develops in some people who repeatedly inhale them—e.g. in dust in farm buildings, or from air-conditioning systems; the consequences are discussed on p. 61. **Invasive aspergillosis** (usually due to *A. fumigatus*) occurs in some immunocompromised patients, most often in those with neutropenia, and consists of invasion of lung parenchyma and other tissues. It is a serious and often fatal infection, which in some cases can be eliminated by withdrawing immunosuppressive drugs and giving amphotericin B intravenously in high dosage. **Aspergilloma** is another opportunist aspergillus infection of lung, also usually due to *A. fumigatus*, but there is typically no invasion of lung parenchyma; the lesion is a spherical mass or giant colony of aspergillus that forms in a lung cavity left by an abscess, an old tuberculous infection or an infarct. Radiography shows the ball moving freely in the cavity so that there is always a crescent of air above it in the picture, regardless of the position of the patient. Detection of serum antibodies (multiple precipitin lines) gives further confirmation of the diagnosis. Treatment is by surgical removal of the affected portion of the lung.

Cryptococcosis is a systemic infection which can occur in either normal or immuno-compromised people. It is caused by the true yeast *Cryptococcus neoformans*, which is found in pigeon droppings throughout the world. Human infection begins as a local cutaneous or pulmonary lesion but usually becomes generalized, its commonest feature being a form of subacute or chronic meningitis which is liable to be mistaken for a brain abscess or tumour (p. 220). Immunocompromised patients, especially those with Hodgkin's disease or AIDS (p. 283), are particularly liable to develop this disease. The yeasts can be seen by microscopic examination of the CSF and their characteristic large capsules are well demonstrated if indian ink is first added to the fluid to provide a dark background. They grow well on Sabouraud's glucose agar, forming creamy colonies and differing from other saprophytic yeasts in that they grow rapidly at 37°C. The latex agglutination test for capsular antigen in CSF and blood is extemely helpful in diagnosis and in monitoring treatment. Cryptococcosis is best treated with both amphotericin B and flucytosine.

Mucormycosis is a rare opportunist infection with with sporulating saprophytic moulds such as *Absidia*, *Rhizopus* or *Mucor* which are found in soil or in decaying vegetable matter. The infections are characterized by blood-vessel invasion and tissue necrosis. Dissemination may occur, with fatal outcome. Amphotericin B in full dosage should be commenced as soon as possible.

Respiratory Allergies due to Fungi

Many fungi produce numerous spores less than 5 μm in diameter, which can penetrate to the lung alveoli (p. 47). As we have seen, many of the deep mycoses are initiated in this way. Three distinct forms of respiratory allergy are recognized. (For a reminder of the nature and types of hypersensitivity reactions see pp. 61–62.)

Allergic asthma

Some atopic individuals form IgE antibodies to moulds such as *Aspergillus*, *Alternaria*, *Cladosporium* or *Penicillium* species, and then develop Type I hypersensitivity reactions, with extrinsic asthma and rhinitis, when they again inhale the offending spores.

Allergic bronchopulmonary aspergillosis

This condition presents as a complication of asthma, with acute dyspnoea, cough and fever, blood and pulmonary eosinophilia and fleeting pulmonary infiltrates shown by radiography. In addition to elevated serum aspergillus-specific IgE levels, precipitating antibodies are formed; patients therefore give a dual skin-test reaction to aspergillus antigen, the immediate Type I reaction being followed after 6–8 hours by a Type III immune-complex-based vasculitic reaction. Corticosteroid treatment is used to produce a remission.

Extrinsic allergic alveolitis (hypersensitivity pneumonitis)

Inhalation of fungal spores by previously sensitized individuals can induce episodes of fever, cough, chest tightness and dyspnoea, occurring some 4–6 hours after exposure. Repeated exposure over a long period may lead to progressive pulmonary fibrosis and incapacitating limitation of respiratory function. This is the pattern in the prototype disease of this group, **farmer's lung**, which is caused by inhalation of spores of thermophilic actinomycetes such as *Micropolyspora faeni* or *Thermoactinomyces vulgaris* that have grown in mouldy straw, hay or grain (p. 16). Crops can become mouldy during storage, particularly if they have been harvested in damp conditions. The temperature inside a compressed bale of hay can reach 45–60°C, favouring growth and sporulation of thermophiles. Drying of hay, straw or grain before storage and spraying them with propionic acid to prevent moulding helps to minimize the problem even in the dampest harvest season.

The blood of a patient with extrinsic alveolitis contains precipitating antibodies against the offending fungus. Eosinophilia is not a feature of this disease. Pulmonary function testing shows a pure restrictive ventilatory defect, and chest radiography shows diffuse nodularity of the lungs. It was formerly believed that extrinsic allergic alveolitis was due purely to a Type III immune-complex reaction in the lung; most patients with the acute illness have elevated levels of circulating immune complexes. However, it is now clear that the minigranulomata seen on lung biopsy and largely responsible for the typical chest radiographic appearances are a form of Type IV reaction, triggered by the presence of non-degradable complexes of fungal antigen and antibody. Corticosteroids are effective in treatment, but in addition efforts should be made to prevent further exposure of the patient by changing his occupation or in the case of a farm worker by equipping him with an air-stream helmet that filters out the offending spores.

Many other examples of extrinsic allergic alveolitis (p. 62) have been described, differing in the nature of the antigens involved—e.g. malt workers' lung (*Aspergillus clavatus*). Outbreaks of extrinsic allergic alveolitis sometimes occur in premises with poorly maintained air-conditioning or humidification systems; for example, thermophilic actinomycetes may contaminate the water in a humidifier reservoir, multiply in the warm environment and be disseminated throughout the building as an aqueous aerosol of spores.

Antifungal Drugs

As we have stressed, many mycoses—notably superficial candidiasis and our secondary category of systemic mycoses—owe their existence to underlying diseases or immunological deficits. If these can be ameliorated and sources of infection such as infected catheters removed, antifungal drugs have far more chance of success.

Griseofulvin, a metabolite of *Penicillium* species, inhibits the formation of cellular microtubules in susceptible fungi. It is given orally, absorbed into the blood and then preferentially concentrated in the stratum corneum of skin—hence its efficacy in treatment of ringworm (p. 138).

Amphotericin B is a polyene that acts on the fungal cell membrane, causing increased permeability and disruption of the cytoplasm. It is nephrotoxic to man, causing decreased renal clearance, tubular damage and hypokalaemia. It has a broad spectrum of antifungal activity, which does not include dermatophytes. It is poorly absorbed after oral administration, and has to be given intravenously for the treatment of deep mycoses, of which it is currently the mainstay. It is also used in topical preparations for treatment of superficial candidiasis, as is **nystatin**, another member of the polyene group.

Flucytosine is a synthetic analogue of pyrimidine which, when converted to its active form 5-fluorouracil, is incorporated into RNA and inhibits protein synthesis. The *in vitro* spectrum of activity is largely confined to yeasts. Serious side effects such as thrombocytopenia and neutropenia are related to blood levels, so these should be monitored. Fungal resistance to the drug may develop during a course of therapy, particularly if it is protracted.

The **imidazoles**—e.g. **clotrimazole** (topical), **miconazole** (topical and intravenous), **ketoconazole** (oral)—are synthetic compounds that appear to inhibit ergosterol biosynthesis. As a group they are effective in both superficial and systemic fungal infections. Hepatitis has occurred during treatment with ketoconazole, and liver function should therefore be monitored regularly during prolonged courses.

Suggestions for Further Reading

Emmons, C.W. *et al.* 1977. *Medical Mycology*. 3rd edn. Lea & Febiger, Philadelphia.

Also Section V of *Bacteriology Illustrated*, see p. 110.

Protozoa

The protozoa, including all of the parasitic species, are unicellular; those non-parasitic forms that appear to be multicellular are in fact colonies of single cells. Parasitic protozoa are a small proportion of the whole subkingdom protozoa, but representatives of one group or another infect most vertebrate and invertebrate animals. Parasitic protozoa (unlike most of the parasitic worms) can multiply either sexually or asexually (or both) in the host. In particular their capacity for very rapid asexual reproduction accounts in a large part for their ability to produce clinically overwhelming infections.

Protozoa of medical importance are placed in three phyla—*Sarcomastigophora, Apicomplexa* and *Ciliophora*. A simplified version of a classification including the medically important species is shown in Fig. 11.1 (p. 144). The drugs used to treat protozoan infections are reviewed at the end of this chapter (p. 152).

Amoebiasis

Entamoeba histolytica is the causative organism of amoebiasis in man (p. 253) and is world-wide in distribution, occurring wherever sanitation and hygiene are poor, although it is most common in moist, warm areas. The amoebae (**trophozoites**) normally live in the lumen of the large intestine, especially the transverse and descending colon, where they ingest bacteria and food particles and reproduce by simple asexual division. As the faecal material passes down the intestine, the trophozoites eject any undigested food particles, and each individual rounds up and secretes a protective cyst-wall around itself. **Cysts** passed in the faeces are about 12 μm in diameter and can resist adverse environmental conditions for two or three weeks, or for up to four weeks in water. During this period the parasite nucleus divides to form four nuclei, and only mature cysts containing four nuclei are capable of infecting a new patient. When the mature cyst is ingested, the parasite excysts in the small intestine, divides into four daughter amoebae (trophozoites) and starts to feed, grow and multiply. As a rule only the cysts are found in formed faeces, but both cysts and trophozoites may be present in loose stools and only trophozoites in diarrhoeic stools.

Entamoeba coli is a non-pathogenic amoeba with a distribution similar to that of *E. histolytica*. Its cysts contain eight nuclei when mature. Infection with *E. coli* does not require treatment, but indicates faecal contamination of food or drink and the need for higher standards of public health.

Naegleria and Acanthamoeba Infection

Naegleria fowleri and *Acanthamoeba* species are free-living amoebae that occasionally cause meningo-encephalitis (p. 220). Although a rare condition in man, *Naegleria* infection has been described from widespread parts of the world. The epidemiological picture is consistent. Young people are mostly affected, and there is generally a history of swimming in fresh or brackish water within 7–10 days before the onset of symptoms. Infection is acquired via the nasal mucosa and the amoebae penetrate the cribriform plate to infect the central nervous system. The main pathological changes are in the frontal lobes close to the cribriform plate and olfactory bulbs; the temporal and cerebellar areas may also be affected. There is an acute meningo-encephalitis with neutrophils and monocytes in the subarachnoid spaces and with large numbers of rounded amoebae, 10–11 μm in diameter, in the CSF and in the grey matter at the edges of areas of haemorrhage and advancing necrosis. In tissue cultures, naegleriae have been seen to phagocytose the cells and this may occur in the natural infection. *Acanthamoeba* is an even rarer cause of similar infections.

Giardiasis

Giardia intestinalis (formerly *lamblia*) is a flagellate protozoan parasite of the intestinal tract with a world-wide distribution and is transmitted by the faecal–oral route, often via drinking water. It is the most commonly reported parasite in Britain. The **trophozoite** attaches itself to the wall of the small intestine by means of a sucking disc, and a layer of *Giardia* may cover the intestinal villi. They multiply asexually by binary fission. In the intestinal lumen, the trophozoites develop into oval **cysts** measuring about 10–12 μm which are passed in the faeces.

Cryptosporidiosis

Cryptosporidium was principally known as a cause of diarrhoea in animals, especially cattle, and only in recent years has it been recognized as causing human illness. It was first recognized (in 1976) as a cause of diarrhoea in immuno-suppressed patients and was subsequently shown to be a cause of sporadic diarrhoeal disease in previously healthy people, especially children. Infection is found most commonly in rural areas. It may be acquired by contact with animals, or by drinking contaminated unpast-eurized milk, or from contaminated food or water, or by person to person spread. The organism

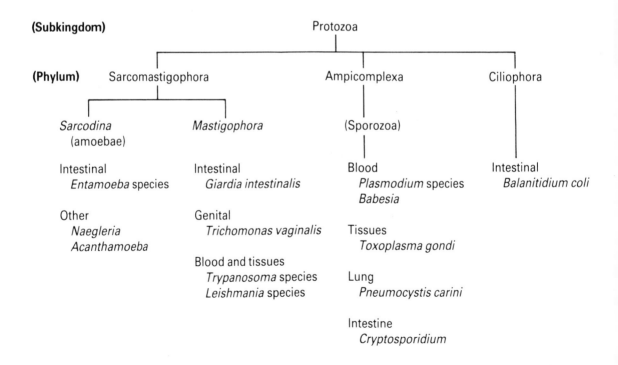

Figure 11.1 Simplified classification of protozoan parasites of man.

attaches to the intestinal mucosa and then penetrates the epithelial cells, leading to impaired digestion, malabsorption and profuse watery diarrhoea. Infection in immunocompetent patients causes a 'flu-like' illness with watery diarrhoea which usually resolves spontaneously in 7 days. By contrast, infection in immunocompromised patients, particularly those with AIDS, results in severe protracted diarrhoea with dehydration and prostration.

The diagnosis is made by seeing the acid-fast cysts of *Cryptosporidium* in films of faeces stained by a modified Ziehl–Neelsen method (decolourization with 0.5% H_2SO_4); alternatively they retain the fluorescent stain auramine when films are stained by the method used for tubercle bacilli (p. 165). There is no effective treatment for cryptosporidiosis.

Trichomoniasis

Trichomonas vaginalis is a parasite of the urogenital tract. It is common throughout the world, and in the developed world is perhaps man's commonest protozoan pathogen. It is transmitted by sexual intercourse and causes vaginitis in women and non-gonococcal urethritis in men (see Chapter 20). There are reported to be 3–4 million cases annually in the USA and about half this number in Britain.

The parasite is oval or pear-shaped, with an undulating membrane along one side and four anterior flagella. It measures about 15 μm long by about 7 μm wide. No cyst stage is known, and the parasites presumably multiply by simple binary fission. Transmission is direct during sexual contact, and no non-human hosts are known. Infection of the vagina is usually accompanied by a purulent frothy discharge and by changes in the surface epithelium and submucosa of the vagina and cervix. Parasites are usually found on the mucosal surface and in fluid exudate but do not invade. The severity of infection in women ranges from asymptomatic carriage to severe vaginitis. In contrast, the infection in men is generally mild or asymptomatic and often self-limiting. Successful treatment (p. 281) depends on tracing and treating **all** sexual contacts of a case.

Trypanosomiasis

Human trypanosomiasis occurs in two geographical areas—in Africa where the biologically similar species *Trypanosoma gambiense* (*T. brucei gambiense*) and *T. rhodesiense* (*T. brucei rhodesiense*) cause two clinically distinct forms of sleeping sickness; and in parts of Central and South America, where *T. cruzi* causes Chagas' disease, which differs widely from sleeping sickness. All three causative species are flagellate protozoa transmitted by blood-sucking insects. In sleeping sickness the parasites are found in man in the **trypomastigote** (flagellate) form only, and are always extracellular; they occur first in lymph glands, then in the blood and then in the CSF in the late stages of infection. *T. cruzi* is found in man both as trypomastigotes in lymph glands and circulating blood, and as **amastigotes** (non-flagellate, similar to *Leishmania*, p. 147) in intracellular pseudocysts in the heart and other involuntary muscles.

Sleeping sickness (p. 246)

T. gambiense and *T. rhodesiense* are morphologically indistinguishable. *T. gambiense* causes the more chronic form of sleeping sickness in which death occurs in the untreated patient 12–18 months after infection. It is found in the western half of Africa extending northwards from Angola to the latitude of The Gambia. It is transmitted by riverine tsetse flies of the *Glossina palpalis* and *G. tachinoides* groups. *T. rhodesiense* causes an acute form of sleeping sickness in which death occurs in 6–12 months. It is found in central and eastern Africa, where it is transmitted by the savanna tsetse fly *G. morsitans*. The course of the infection in man is described in Chapters 16 (p. 224) and 17 (p. 246). It is similar with either species, although it generally proceeds more quickly with *T. rhodesiense*. Gambian sleeping sickness tends to be endemic, whereas the Rhodesian form tends to be epidemic; but the clinical, geographical and epidemiological distinctions between the two forms are not clear-cut. Cases of the chronic form of trypanosomiasis occur from time to time in areas where the acute form is more typical and vice versa, and there are large areas of apparently suitable land in which no trypanosomiasis occurs. *T. gambiense* infects only man but *T. rhodesiense* infects cattle and antelopes, which provide a reservoir of infection. Trypanosomiasis in animals, mostly due to other *Trypanosoma* species, causes severe economic losses, rendering some areas unsuitable for farming. In areas where *T. gambiense* occurs, less than 1% of the human population is infected but there are foci of more frequent transmission, and there may have been some overall increase in rates of infection in recent years. Because *T. rhodesiense* infection tends to be more acute in onset, asymptomatic carriers do not play the major role in maintenance of transmission that they do with *T. gambiense*.

Laboratory diagnosis of sleeping sickness is based on the demonstration of trypomastigotes in thick and thin blood films, in aspirates from enlarged lymph nodes or from the chancre, or in CSF. If parasites are scanty, concentration methods such as triple centrifugation of blood may be necessary. Approximately 10 ml of venous blood containing anticoagulant are centrifuged in a conical centrifuge tube at 1000 rpm for 10 minutes. The supernatant plasma and the white cell layer are removed from the packed cells and transferred to a clean conical tube and recentrifuged at 1000 rpm for 10 minutes. The supernate is again removed and centrifuged once more at 2500 rpm for 10 minutes. The sediment from this third centrifugation should contain any trypanosomes present and can be examined either as a wet film or as a stained thin film. In epidemiological work it may be necessary to distinguish between *T. rhodesiense* and *T. gambiense*. While this cannot be done on parasite morphology and is not always reliable on clinical grounds, inoculation of blood containing *T. rhodesiense* into laboratory rats almost always produces an infection whereas blood containing *T. gambiense* does not.

Treatment of clinical cases in the early stages of the disease is generally successful, but the chances of a complete return to normality decrease as the nervous system becomes increasingly involved.

Chagas' disease (p. 246)

South American trypanosomiasis is due to *T. cruzi*, transmitted principally by blood-sucking reduviid bugs. These are particularly common in semi-urban and rural areas of Central and South America, various species living in the cracks in the mud walls and in thatched roofs of houses, in sheds and outbuildings used to house domestic animals, in bush and palm trees around rural houses, etc. Several species of bug are capable of transmitting the infection; all require blood meals, either from man or from various domestic or semi-domestic animals, in order to develop through the various larval stages to the adult insect. Some species feed more readily on man and so are more important vectors than others. **Trypomastigotes** are taken up by the bug during a blood meal, which is usually taken at night when the patient is asleep, and multiply and develop in the insect intestine. Infective trypomastigotes are eventually passed in the faeces of the insect. Bugs defaecate during a blood meal, and if the trypomastigotes are rubbed into the bite or an abrasion, or on to a mucous membrane, infection results. The conjunctiva is a common site of infection.

Parasites multiply initially at the site of infection, producing a focus of intense inflammatory reaction with polymorphs and lymphocytes, and with interstitial oedema and focal lymphangitis that together cause the swelling known as a **chagoma**. The parasites then spread through the lymphatic system to the regional lymph nodes. They are found in the blood in the trypomastigote stage early in the infection (by about the tenth day) but become less numerous as they enter mesenchymal cells, particularly in cardiac muscle and the reticulo-endothelial system and neuroglia. Inside these cells the parasites change to the **amastigote** form and multiply by asexual division to form intracellular pseudocysts containing many hundreds of amastigotes; the host cells are eventually destroyed, releasing amastigotes to invade other host cells. Heart muscle is frequently damaged (p. 239) with consequent loss of muscle bulk, and the involuntary muscles of the intestine may also be damaged. It has been postulated that in chronic Chagas' disease an auto-immune process affects muscular junctions of the digestive system autonomic nerves (p. 45).

Improving houses by replastering walls and replacing thatched roofs with corrugated metal eliminates the breeding sites of the bugs, and regular (e.g. twice yearly) insecticide treatment of houses, furniture, outbuildings, etc. helps to reduce the number of insects. However, such steps towards eradication of the disease are not entirely successful because insect species capable of transmitting it live in the bush surrounding houses in rural areas. In addition to infection in man, various domestic and peridomestic animals (e.g. dogs, cats, chickens, pigs, armadilloes, etc.) can also act as hosts for the parasite. Control of transmission is therefore difficult.

T. cruzi may also be transmitted via blood transfusion or by the communal use of hypodermic syringes and needles. The addition of 25 ml of a 1 in 4000 solution of gentian violet or of a 1 in 5000 solution of crystal violet to 5000 ml of blood 24 hours before use kills the parasites and has been advocated as a routine procedure in endemic areas.

Leishmaniasis

In the past it has been difficult to distinguish on other than clinical grounds between the various species of *Leishmania* which cause human disease. Recent studies have suggested biological groupings of the organism which do not always correlate closely with groupings based on clinical aspects. What follows is a simplified account of leishmaniasis based on the clinical aspects, and it therefore does not relate closely to current concepts of the taxonomy of the organisms.

In one clinical form or another, leishmaniasis occurs in China, India, Afghanistan, southern Russia, Arabia, the Middle East, the Mediterranean coasts of Europe and Africa, Africa immediately south of the Sahara, Ethiopia and Kenya, central America, north-western and northern South America and Brazil. The infection is transmitted to man by the bite of infected female sandflies (*Phlebotomus* species in the Old World, *Lutzomyia* in the New World). When the sandfly takes a blood meal from an infected host, macrophages containing the parasites are taken up. The stages contained in the macrophages (the **amastigotes**) are freed from the macrophages in the insect's gut, undergo at least one asexual division and transform into morphologically distinct **promastigotes**. These multiply asexually several times before spreading to the mouth parts of the insect. Small numbers of promastigotes are injected when the fly next feeds. The promastigotes invade cells of the host's reticulo-endothelial system, transform to amastigotes, and multiply asexually within the host cells. Amastigotes escape from the host cells and invade further reticulo-endothelial cells. At no stage of the parasite's life cycle is there any sexual reproduction.

There are two distinct forms of leishmaniasis—visceral and cutaneous.

Visceral leishmaniasis (kala azar) (p. 270)

This is caused by *Leishmania donovani*. Children up to 10 years old are most commonly affected, although infection also occurs in older people; males are more frequently affected than females. The incubation period ranges from 10 days to over one year. Amastigotes are present in reticulo-endothelial cells in all parts of the body but particularly in the spleen, liver and (to a lesser extent) the bone marrow. Common symptoms include malaise, weight loss, hepatomegaly and splenomegaly (more marked than in malaria). Infection can be acute and fatal, especially in children. Visceral leishmaniasis occurs in Mediterranean coastal areas, SW Asia, India, China, East Africa and Latin America.

Cutaneous leishmaniasis (p. 310)

This occurs in two distinct forms in the Old World and in another two distinct forms in the New World.

Old World: (1) *Leishmania tropica* (urban oriental sore). In urban areas on the Mediterranean coast, and in Arabia, northern India and Afghanistan, *L. tropica* has an incubation period of 2–8 months and causes a chronic infection lasting a year or more. It is characterized by a dry cutaneous lesion, usually on the face, which ulcerates after several months and eventually heals, leaving disfiguring scars.

(2) *Leishmania major* (rural oriental sore). In rural areas of North Africa, the Middle East, southern Russia, Sudan and West Africa, *L. major* has an incubation period of less than 4 months and causes acute infection lasting 3–6 months. It is characterized by a moist lesion, usually on the limbs, which ulcerates early. The lesions are slow to heal and may become multiple, and they leave large disfiguring scars.

New World: (1) *Leishmania mexicana* (chiclero's ulcer). In Central America (particularly often in Belize) and northern South America, *L. mexicana* causes a single painless lesion that generally heals within a few months, but chronic lesions may develop on the ear, progressively destroying tissue.

(2) *Leishmania braziliensis* (espundia). In Brazil, Surinam, Venezuela and other parts of South America *L. braziliensis* produces, in the primary cutaneous stage, single or multiple ulcers which seldom heal spontaneously. Up to 80% of untreated primary lesions are followed by the appearance of mucocutaneous lesions of the palate and nose, which are progressive and intensely disfiguring.

Sources and acquisition of leishmaniasis

The reservoir hosts for *Leishmania* species are as follows:

For *L. donovani*: Man only in India, southern China and central Kenya; dogs in other areas, occasionally other small carnivores.

For *L. tropica*: Predominantly man, but it has been isolated from dogs and rats.

For *L. major*: Desert rodents.

For *L. mexicana*: Forest rodents and other mammals.

For *L. braziliensis*: Sloths, ant-eaters and various other forest animals.

Man is liable to become infected when he is bitten by sandflies in an area where transmission is occurring among animal hosts—e.g. he may acquire *L. major* when working on an irrigation programme on the edge of a desert, or *L. mexicana* or *L. braziliensis* when living in a forest clearing or gathering a forest product (e.g. chicle—hence the name chiclero's ulcer).

Diagnosis

Infection is confirmed by the demonstration of amastigotes in macrophages in tissue scrapings taken from the edge of a cutaneous lesion, or in biopsies from the spleen, liver or bone marrow (visceral leishmaniasis). Immunological methods (a fluorescent antibody test and a skin test in which promastigotes are injected intradermally and the area of induration measured 48–72 hours later) are useful for diagnosis but not for determining cure.

Malaria (pp. 241–3).

Four *Plasmodium* species cause malaria in man—*P. falciparum*, *P. vivax*, *P. ovale* and *P. malariae*; they differ in details of life cycle and in clinical effects. All are transmitted by bites of female *Anopheles* mosquitoes. The parasite has its sexual reproductive cycle in the mosquito's digestive system and then reproduces asexually in its body cavity, forming **sporozoites** that migrate to the salivary glands and are inoculated with the saliva into the blood of the human host when the mosquito feeds. The time taken for the parasite to develop to the infective stage in the mosquito depends partly on the species of *Plasmodium* but mostly on the environmental temperature. Once infected, mosquitoes probably remain so throughout their lives, which may vary from a few days to weeks or even months.

Sporozoites injected into man by the mosquito disappear from the circulation in about 30 minutes and penetrate the parenchymal cells of the liver, where a cycle of asexual reproduction takes place. The parasite divides repeatedly within the host cell to form a liver **schizont** containing up to 40 000 **merozoites**, depending on the species of *Plasmodium*. The merozoites are released from the liver schizont and invade red blood cells in which another asexual reproductive cycle occurs. The merozoite develops through a **ring stage** (or **trophozoite**) to form a schizont which divides into 8–24 merozoites, depending on the species. When this erythrocytic schizont is mature, the infected red cell bursts, releasing the merozoites which invade further red cells, thus repeating the **erythrocytic cycle**. Instead of developing into schizonts some merozites develop into immature sexual stages (**gametocytes**) which are the only form that can transmit infection to the mosquito. There are differences between *Plasmodium* species in the morphology of the various developmental stages and in their time sequence. Some of these differences are important in precise diagnosis of the infecting species (which is necessary to ensure that the best treatment can be selected) and others are of epidemiological importance.

The most important species is *P. falciparum*, which causes **malignant tertian** or **subtertian** malaria, characterized by cycles of fever occurring every 36–48 hours. *P. falciparum* is the dominant species throughout most of the tropics and is responsible for 80–90% of malaria infections in these areas. The liver stage of this species in man lasts for 8–10 days, and there is thought to be only one population of parasites developing synchronously at any one time. The fully developed liver schizont contains about 40 000 merozoites, considerably more than in the other species. The cycle of infection in the erythrocytes is more rapid (36–48 hours) in *P. falciparum* than in the other species, and the mature erythrocytic schizont contains an average of 24 merozoites, more than in the other species. The increase in numbers of parasites in this species is, therefore, much more rapid than in the others, and this, together with the ability of the parasite to attack red blood cells of all ages, gives rise to a rapid deterioration in the clinical condition in the non-immune patient; hence the term 'malignant tertian' for falciparum malaria. Parasitaemia is often heavy; 20–40% of erythrocytes may be infected, compared with perhaps 5% in infections with the other species, and the sudden destruction of this proportion of red cells has serious consequences for the patient (p. 241). Although erythrocyte schizonts occur in *P. falciparum*, the great majority of these forms, particularly when parasitaemia is relatively low, are in the blood vessels of the deeper tissues and are not seen in peripheral blood. Schizonts are normally seen only in blood films from moribund patients.

P. vivax causes **benign tertian** malaria, in which the cycle of fever recurs every 48 hours. It is the second most important species and occurs over a much wider geographical area than *P. falciparum*, from the tropics and subtropics to some temperate areas of the world; it has caused endemic infections in parts of southern Britain in the fairly recent past, and in the 1920s was transmitted as far north as the Arctic Circle. The erythrocytic stages are morphologically distinct from those of *P. falciparum*, and the liver and erythrocyte schizonts contain fewer merozoites. Older erythrocytes are preferentially infected, in contrast to *P. falciparum* infection, and the percentage of cells infected may be much lower than in that condition. Vivax malaria is therefore less severe than falciparum infection in the non-immune patient; hence the term 'benign tertian malaria'. Relapses of parasitaemia and fever are frequently seen in *P. vivax* infection; until recently these were thought to be due to a secondary cycle of invasion of the parenchymal cells by merozoites from the primary liver cycle, so that infection was maintained there for months or years after most of the merozoites produced by that primary cycle had entered the circulation. Recent evidence suggests that the relapses may be due to the presence of two or more populations of parasites in the liver cells, one population developing as the typical liver schizonts and the other (termed **hypnozoites**) persisting for some time as small, non-developing uninucleate parasites. The latter may initiate cycles of development weeks, months or years after the initial infection, giving rise to the so-called relapses. Whatever the explanation, it is well established that in vivax malaria, treatment of the erythrocyte stages alone is not adequate, and drugs effective against both erythrocyte and hepatic stages are required. This contrasts with falciparum malaria, in which drugs effective against the erythrocyte stages alone provide successful treatment.

P. ovale also produces a 48-hour fever cycle. It is generally less common than *P. vivax* in areas where both occur, although in West Africa it is more common than *P. vivax*, and it also occurs, though more rarely, in parts of South America and Asia. It is thought that two populations of parasite may occur in the liver in *P. ovale* infection, as has been suggested for *P. vivax*. Relapses occur, as in vivax malaria, and treatment of erythrocyte stages alone is again not adequate; therefore specific identification of the parasite is necessary.

P. malariae produces a fever cycle of 72 hours—**quartan malaria**. It is generally found in tropical and subtropical areas but has a much lower prevalence than either *P. falciparum* or *P. vivax* which occur in roughly the same areas. Liver and erythrocytic schizonts both produce fewer merozoites than in the other species, so parasitaemia is generally low—of the order of one infected cell per 100–200 red cells.

While the four species differ widely in the degrees of parasitaemia that they produce, it should be borne in mind that factors other than the *Plasmodium* species affect the severity of clinical malaria. These include the size of the inoculum of sporozoites injected by the mosquito, and the immune status of the host. Parasitaemia may develop particularly rapidly and to a high level in *P. falciparum* infection, but even low levels (e.g. less than 1%) often cause clinical malaria. High levels of parasitaemia are life-threatening and require immediate treatment.

The epidemiology of malaria is complex, involving among other variables a considerable range of mosquitoes; about 60 species of *Anopheles* are known to be capable of transmitting malaria, in various parts of the world. Their efficiency as vectors varies, and each has specific ecological preferences. The epidemiology of clinical malaria is also affected by human factors such as general well-being and, particularly, the degree of immunity developed by patients. Immunity in malaria is only relative; it is short-lived in the absence of repeated reinfection, and is generally strain-specific. Children exposed to repeated infections in hyperendemic areas either succumb to the infections early in life (as many as 10% may die before they are 5 years old) or gradually (e.g. by the age of 5 years) develop a degree of immunity which, while not preventing infection, limits the severity of its effects to a subfatal but not subclinical level. As long as exposure to infection remains frequent, immunity remains relatively high and severe clinical infection does not occur. If, however, regular and frequent exposure ceases, as happens after a control campaign or when people move to a non-malarious area, the level of immunity falls fairly quickly during a period of about 1 year and any subsequent exposure results in a relatively severe, possibly fatal, infection unless adequate prophylaxis (p. 242) is employed. There is thus an argument for maintaining endemic exposure if prolonged protection cannot be assured, although in the absence of treatment there is likely to be a relatively high mortality in children under five years who have not acquired adequate immunity. Because immunity wanes so quickly, breakdown in a long-standing control programme often results in an outbreak of severe malaria in the community. Similarly, many people who have left a malarial area for a period of two or three years or more (e.g. students) contract repeated bouts of clinical malaria when they return home until their immunity is restored to its previous level. Immunity is typically a feature of the indigenous population of an area: expatriates, immigrants or visitors do not generally develop significant immunity.

Certain genetic abnormalities affecting the red blood cells alter susceptibility to malaria. Deficiency of glucose-6-phosphate dehydrogenase (homozygous genotype) is associated with enhanced resistance of red cells to parasite invasion in comparison with normal red cells. Increased resistance is also seen clinically in girls who are heterozygous for the G-6-PD deficiency gene. Various haemoglobinopathies are also associated with considerably reduced levels of mortality due to falciparum malaria, and it has been suggested that people with such defects have an evolutionary advantage in areas of high endemicity of falciparum malaria. The most important of these haemoglobinopathies is sickle cell trait, which is common in West African populations. As with G-6-PD deficiency, both homozygotes with sickle cell disease and heterozygotes with the sickle cell trait are less susceptible to malaria.

Malaria can also be transmitted mechanically without the intervention of a mosquito—for instance by blood transfusion or by communal use of hypodermic needles and syringes. In these cases erythrocytic infection develops without the preceding liver cycle.

Toxoplasmosis

Toxoplasma gondi is an intracellular protozoan parasite with a world-wide distribution. The asexual phase of the organism (**trophozoite**) is able to develop in the tissues of a wide variety of vertebrate hosts (including man) but the definitive host is the domestic cat, in which the sexual cycle occurs in the intestine after a prolonged latent period.

Human infection rates may be as high as 90% or more in some populations. Infection is most often acquired by ingesting trophozoites in undercooked meat, though it may follow ingestion of **oocysts** resulting from the sexual cycle in the intestine of a cat and excreted in its faeces. After ingestion, the parasites are distributed to many organs and tissues via the blood stream, and invade nucleated cells in all parts of the body. They multiply within the host cells, disrupting and finally destroying them. Focal areas of necrosis occur in many organs, particularly the muscles, brain and eye, and contain infective cysts. Infection in man is usually subclinical but may produce a glandular-fever-like syndrome (p. 193) or choroidoretinitis (p. 224). Transplacental infection may occur during an acute but undiagnosed infection in the mother, and may result in serious disease in the fetus (p. 321).

Babesiosis

Many species of *Babesia* parasitize the erythrocytes of a wide variety of wild and domestic animals including cattle, horses, sheep, deer and rodents. They are ingested by hard ticks (*Ixodidae*) during blood meals, and a cycle of multiplication occurs in the tick. Parasites are found throughout the body of the tick, including the ovaries, and are passed through the egg to the developing larvae (transovarial transmission). In the mammalian host, the parasite is found only in the erythrocytes, in which it multiplies by asexual division and looks like a small malarial trophozoite. When the infected cell bursts, the released parasites invade other erythrocytes and their multiplication continues. Red-cell destruction is responsible for most of the clinical effects. Babesiosis is a serious disease of livestock in many tropical and subtropical areas, but human infections have been reported only from Europe and North America (p. 243).

Pneumocystis Infection

Pneumocystis carini is a protozoan generally classed with the sporozoa, and is one of several organisms that may cause an interstitial plasma-cell pneumonia in immunosuppressed patients and in infants (p. 205). It is transmitted by the airborne route and causes inapparent pulmonary infection in a large proportion of the population. The infection is not eliminated but remains latent, suppressed by host immunity. Reactivation, often taking the form of a severe pneumonia, occurs when host immunity is suppressed.

Balantidiosis

Balantidium coli is a large ciliate protozoan parasite occurring throughout the world in the intestines of pigs, cattle and horses—most commonly in pigs, which are usually only mildly affected by it but are the main source of human infection. Multiplying asexually (rarely sexually) in the lumen of an animal's intestine, the parasites are passed in the faeces as cysts which remain viable for several weeks in moist conditions. Transmission is by the faecal–oral route. Human infection is uncommon, but has been reported from Russia, northern Europe, North and South America and East Asia. The parasites are present in the lumen of the human large intestine and may invade the intestinal mucosa, causing necrosis and ulceration. Infection may be chronic and asymptomatic, or it may cause acute dysentery with blood and mucus in the faeces. The diagnosis, in most cases suggested by history of close association with pigs, is confirmed by finding the typical large trophozoites in the patient's faeces.

Antiprotozoan Drugs

The nitro-imidazole drug **metronidazole** was first introduced (in 1959) into medical use for the treatment of protozoan infections. Only subsequently, in the mid-1970s, was its value in the treatment of anaerobic bacteria discovered (p. 388). It is the drug of choice for the treatment of amoebiasis, trichomoniasis and giardiasis. However, it is relatively ineffective in chronic intestinal amoebiasis, for which **diloxanide furoate** is the drug of choice.

Emetine hydrochloride is also an effective amoebicide which can be used for treatment of intestinal amoebiasis, but it is much more toxic to the patient than metronidazole (a very safe drug). It is used, usually in combination with metronidazole, for the preliminary treatment of amoebic liver abscess before surgical drainage (p. 269).

Sodium stibogluconate is the drug usually used to treat kala-azar. **Pentavalent antimony compounds** are alternative drugs for treatment of all forms of leishmaniasis.

Pentamidine is effective in the treatment of toxoplasmosis and *Pneumocystis* infection and in the early stages of trypanosomiasis (before the CNS is involved). **Trimethoprim**, alone or combined with a sulphonamide as cotrimoxazole, is a much less toxic alternative for treating toxoplasmosis or *Pneumocystis* infection; it is used prophylactically in immunosuppressed patients to prevent *Pneumocystis* pneumonia (p. 205). Chemotherapy of Chagas' disease is unsatisfactory. **Nifurtimox** and **benzonidazole** are the most useful drugs in acute infections; an 80% cure rate may be achieved with courses of treatment lasting between 30 and 120 days. Steroids must be avoided because they make the infection much worse.

Suramin and **melarsoprol** are used in the treatment of sleeping sickness; melarsoprol is the only drug effective once the CNS is involved.

Drugs are largely ineffective in treating amoebic meningo-encephalitis but **amphotericin B** has shown some evidence of efficacy in *Naegleria* infections.

Anti-malarial drugs are reviewed in Chapter 17 (p. 242–3).

12
Helminths (Worms)

The helminth parasites of man belong to three zoologically distinct groups—trematodes (flukes), cestodes (tapeworms), and nematodes (roundworms), subdivided as in Fig. 12.1 (p. 154). For other general features see pp. 14 and 24. Treatment of helminth infections is considered in the appropriate clinical chapters.

Trematodes (Flukes)

All trematode parasites of man are transmitted by fresh-water snails. There are four groups that infect man—intestinal flukes, liver flukes, lung flukes and schistosomes.

Intestinal and Liver Flukes

The stages of *Fasciolopsis* and *Fasciola* infective to man encyst on vegetation. *Fasciolopsis* infection occurs in East Asia in pigs and man. The adult flukes are attached to the wall of the host's small **intestine**. Eggs are passed in the faeces and hatch in fresh water to produce larvae which infect certain fresh-water snails and in them develop through additional larval stages. Finally **cercariae** are produced, and these escape from the snail and encyst on aquatic vegetation. Infection is by ingestion of the encysted **metacercariae** (in human infection usually on water caltrop or water chestnut). The metacercariae excyst in the intestine and develop into adults during the following 3–4 months. The adults are relatively large flukes (up to 7.5 cm long), and the gut wall becomes eroded where they attach themselves.

Fasciola hepatica is usually a parasite of the bile ducts of sheep and cattle, but it occasionally infects man. It occurs throughout the world. The usual vehicle of human infection is wild watercress on which the metacercariae are encysted. These excyst in the small intestine, and the miniature flukes penetrate the gut wall and enter the peritoneal cavity and then the **liver** parenchyma, migrating from there to the bile ducts. Eggs are passed in the faeces. The usual sources of infection are cattle or sheep whose faeces contaminate the water supply to the cress beds; *Lymnaea* snails in the cress beds act as intermediate hosts. Localized outbreaks of human infection occur occasionally, and such an outbreak can usually be traced to a common source.

Opisthorchis sinensis infection is common in East Asia. The adult parasites are relatively small flukes living in the **bile ducts** of man and of various wild and domestic fish-eating mammals. The life cycle differs slightly from those of *Fasciolopsis* and *Fasciola*: the eggs (which are small) are passed in the faeces and the intermediate host is a fresh-water snail, but then the cercariae encyst on or under the scales of fresh-water fish. Man is infected by eating these fish raw or improperly cooked. The pathological effects of infection are confined to the bile ducts; changes are slight in light infections but can be more severe, including biliary obstruction in heavy or long-standing infections.

Heterophyes and *Metagonimus* are two very small flukes with life cycles similar to that of *Opisthorchis* except that as adults they attach themselves to the mucosa of the **intestine** and sometimes burrow into the wall. They are common parasites of man in the eastern Mediterranean and East Asia; infection is

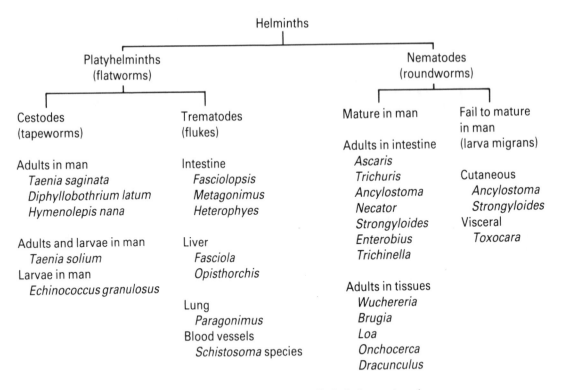

Figure 12.1 Simplified classification of helminth parasites of man.

acquired by eating raw or improperly cooked fish.

Lung Flukes

Paragonimus species are trematodes that cause cyst-like lesions in the lungs of people living in parts of the Far East, West and Central Africa and central and northern South America. Humans are infected with *Paragonimus* far less often than with nematodes such as *Ascaris* and hookworm (which can also cause lung disease—p. 209), but in some regions paragonimiasis is frequent enough to be an important disease. Whereas nematode respiratory disease is due to worm larvae, it is the adult stage of *Paragonimus* that occurs in the lungs. This is a fleshy worm, resembling a large coffee bean in shape, and colour, and measuring about 0.8–1.5 cm long by 0.4–0.8 cm wide and 0.3–0.5 cm thick. Human infection is acquired by eating larvae encysted in fresh-water crabs or crayfish. In many endemic areas (e.g. East Asia) crabs and crayfish are preserved for human consumption with pickling agents which do not kill the encysted larvae. When the pickled flesh is eaten, the larvae excyst in the small intestine, penetrate the gut wall and travel via the peritoneal cavity, the diaphragm and the thoracic cavity to the lungs. A fibrous cyst develops around each parasite as it grows in the lung. The adult produces eggs, and these escape from the cyst into an adjacent bronchiole. Eggs pass up the respiratory tree and are either swallowed and excreted in faeces or coughed up in sputum. The eggs are golden brown in colour and measure about $90 \times 60 \, \mu m$. Groups of eggs appear in sputum as flecks resembling rusting iron filings. On reaching water, the eggs hatch and release **miracidia**, which enter a suitable species of snail and develop into **cercariae**. These are released into the water and penetrate into the flesh of the fresh-water crabs and crayfish mentioned above as the sources of human infection.

Schistosomes

The schistosomes are by far the most important of the fluke parasites of man—it is estimated that 200–300 million people are infected in Africa, South America and East Asia, with perhaps 1.5 million deaths per year being attributed to the direct or indirect results of schistosomiasis.

Three species (*Schistosoma mansoni*, *S. japonicum* and *S. haematobium*) infect man, and their life cycles are somewhat different from those of the other flukes. The adult parasites are small and thread-like; they live within the abdominal veins of man (and also of some wild and domestic mammals in the case of *S. japonicum*); and the sexes are separate. Eggs are laid by the female worm in the capillaries draining the intestine (*S. mansoni* and *S. japonicum*) or bladder (*S. haematobium*). Most pass through the wall of the intestine or bladder and are excreted in the faeces or urine. A few are carried in the blood-stream to the liver (*S. mansoni* and *S. japonicum*) or lungs (*S. haematobium*) and these eggs, together with a proportion that remain trapped in the wall of the intestine or bladder, provoke a series of pathological changes ending with the formation of a fibrous granuloma around each egg (p. 276). The escape of eggs from the gut or bladder wall is often accompanied by a small amount of haemorrhage; when the infection is heavy, blood loss may be fairly severe and easily visible as flecks on the faeces or as haematuria. This blood loss and pathological changes in the gut, bladder, liver and lungs are responsible for the clinical manifestations of infection (pp. 261, 270, 276 and 209).

Those eggs that are passed in the excreta hatch in fresh water, and the resultant larvae infect certain fresh-water snails in which there is a cycle of asexual reproduction. The cercariae which are eventually produced, unlike those of other flukes, do not encyst but are capable of penetrating unbroken human skin immersed in contaminated water; this usually occurs during domestic, recreational, agricultural or fishing activities. Although some mammals can be infected with *S. mansoni* and *S. haematobium*, they are probably not important reservoirs of infection. However, a wide range of wild, semi-domestic and domestic animals are important reservoirs of infection with *S. japonicum* in East Asia.

Cestodes (Tapeworms)

Taenia

Two species of *Taenia* infect man—*T. solium* and *T. saginata*; their larvae are found in pigs and cattle respectively. Human infection results from eating raw or improperly cooked pork or beef in which there are infective larvae (**cysticerci**). The cysticercus excysts in the intestine and the **scolex** (p. 14) attaches itself to the intestinal wall. The **strobila** (p. 14) consists of a chain of **proglottides** and the terminal proglottides, which are full of eggs, become detached from the strobila. Although most eggs in the host's faeces are contained within the gravid proglottides, some escape and are passed free. Eggs are small and resistant; if ingested as appropriate by either a pig (*T. solium*) or a bovine (*T. saginata*), the embryo hatches from the egg in the gut, migrates through the gut wall and eventually transforms into a cysticercus, complete with a small scolex inverted into the cyst, in muscle. There it can remain dormant for several months or years, until it is ingested in meat that has not been properly cooked.

Adult *Taenia* parasites are of little consequence in man (p. 261); their diversion of nutrients has little effect on the nutrition of their hosts. However, eggs of *T. solium* can develop into cysticerci in man, causing **cysticercosis**. The cysticerci may be found in most tissues, including muscles, liver and, most dangerously, the brain (p. 226). Heavy infection can cause severe diseases. Human cysticercosis can arise from two sources: either the accidental ingestion of eggs passed in human faeces (e.g. as contaminants of green vegetables), in which case the infection is likely to be light; or by reverse peristalsis, in which the eggs or proglottides are passed up the gut by upward peristaltic waves, many eggs hatch in the stomach or upper small intestine, and a widespread cysticercosis may result. Cysticerci cannot develop further in man. They die, and become calcified within a year or two.

Echinococcus

Human infection with the larval stages of *E. granulosus* is also accidental. The normal cycle of infection is between sheep (or goats, cattle or horses) and dogs (or other carnivores). The adult parasites live in the gut of dogs. They are small tapeworms, each comprising a scolex and usually three segments. Their eggs are released from the terminal gravid segments and are passed in the faeces. They give rise to larval stages in the viscera of the intermediate hosts (sheep etc.) when the eggs are ingested. The larva grows steadily until it eventually becomes a substantial **hydatid cyst** (5–20 cm in diameter). The cyst is filled with fluid, and is lined with germinal epithelium that produces groups of young scolices inside it. If these scolices are eaten by a carnivore they grow into adult worms in its gut.

Human hydatid infection (p. 271), in which man becomes the intermediate host, is found mostly in people closely involved with sheep rearing (e.g. shepherds and dog handlers) and to a lesser extent in those working with goats, camels and horses. Infection in the dogs is maintained by scavenging carcasses of infected sheep etc.

Hymenolepis

H. nana is the commonest of the tapeworm parasites of man, particularly in Asia; it is also the smallest (1.5–4 cm long) and is the only one which does not require an intermediate host. The scolex has four suckers and a rostellum of hooks, and the strobila consists of about 200 short wide proglottides. Each gravid proglottis contains 100–200 eggs in its uterus. Most eggs are released from the proglottides in the ileum, by which stage they are capable of infecting either the intestinal mucosa of the same host (auto-infection) or another host via the faecal–oral route. In either case, the eggs hatch in the small intestine and the released larvae penetrate the mucosa, where they develop into immature adult tapeworms in about 2 weeks. Eggs are produced after a further 2 weeks. The eggs have relatively thin shells, and can survive for only about 2 weeks after being passed in the faeces.

Pathological changes in man depend on the intensity of the infection, on the immune status of the host, and on any concomitant disease. Infection is common and intense in malnourished, immunodeficient children in many parts of the world; it may start as early as the first year of life but becomes clinically more important after about 5 years.

Diphyllobothrium

This broad fish tapeworm of man, *D. latum*, is transmitted through fresh-water fish. Human infection occurs in the more northern temperate areas of the world and is particularly common in Scandinavia and in northern parts of North America. The adult worm is several metres long and has a characteristic flattened muscular scolex which attaches to the wall of the small intestine by burrowing into the villous crypts. Large numbers of operculate eggs are produced in the gravid proglottides; some escape and are passed free in the faeces, others remain in the proglottides. Unlike those of other tapeworms of man, the eggs hatch in fresh water and infect a fresh-water copepod. This is then eaten by a fresh-water fish in which the larva develops to the stage infective for man, who becomes infected by ingesting raw or improperly cooked or preserved fish. The larva attaches to the intestinal mucosa and the adult worm develops in the intestine.

Nematodes (Roundworms)

Intestinal Nematodes (See also pp. 260–1)

Ascaris and Trichuris

Adult *Ascaris lumbricoides* and *Trichuris trichiura* live in the human intestine, the former principally in the lumen of the small intestine and the latter with its anterior embedded in the wall of the large intestine. Although infection is most frequent in moist tropical areas, it can occur wherever poor sanitation provides opportunities for faecal–oral transmission. Eggs are excreted in the faeces, and their thick walls enable them to survive for many months or years in cold and dry conditions. Temperatures above 60°C are necessary to destroy *Ascaris* eggs. (When human faeces are used as fertilizer, they should first be composted, but although the temperature may exceed 60°C in the centre of a well-made heap, it is unlikely to do so in the outer layers or in poorly made heaps.) The larvae develop inside the eggs, and when these have been ingested they hatch out in the host's intestine.

Ascaris larvae penetrate the gut wall and are carried via the blood-stream to the lungs. There they break into the alveolar spaces and mature as they migrate up the respiratory tree, before being swallowed and reaching the intestine for the second time. Maturity is complete about 10 weeks after infection and the adults live for 1–2 years.

Trichuris larvae do not undertake a similar migration. Instead, they penetrate the wall of the colon, where they develop into miniature adults. They then re-enter the large intestine, where the anterior becomes embedded in the gut wall but the posterior remains in the lumen. Egg-laying commences about 12 weeks after infection and adults live for several years.

Ancylostoma, Necator and Strongyloides

In hookworm (*Ancylostoma duodenale*, *Necator americanus*) and *Strongyloides stercoralis* infections of man the adult worms live in the intestine, and eggs (hookworm) or larvae (*Strongyloides*) are passed in the faeces. In warm moist conditions eggs of hookworms mature and hatch in 24–28 hours, and the free-living larvae of both hookworms and *Strongyloides* develop quickly in soil through several larval stages to the infective stage. Infective larvae can penetrate unbroken skin when any part of the body (e.g. an unprotected foot, buttock or hand) comes into contact with ground in which they are present. They penetrate into blood vessels and are carried to the lungs, where they break into the alveolar spaces. The larvae grow and mature, pass up the respiratory tree and are swallowed, finally developing into adults in the intestine. Adult hookworms suck blood from the wall of the small intestine, whereas adult *Strongyloides* become embedded in the mucosa of the intestine (particularly the small intestine) where they occasionally produce serious damage. Hookworm infection is virtually limited to the tropics and is a rural disease. It is particularly common in plantations of cocoa, banana, sugar or coffee, where the crops provide privacy for defaecation, where dense shade, high humidity and good fertile soil provide ideal conditions for larval developments and where agricultural work is often carried out by hand. *Strongyloides* has a similar distribution to that of hookworms but it is less common. As *Strongyloides* larvae migrate through the tissue they cause inflammation and irritation—**systemic larva migrans** (p. 260). *Strongyloides* infection may remain quiescent for many years and become reactivated during immunosuppressive chemotherapy, causing severe, even fatal, disease.

Hookworm infection is readily diagnosed by microscopy of freshly passed faeces for the presence of the characteristic eggs; those of the two species are indistinguishable. Hatching of the eggs occurs in about 24 hours if faeces are kept at average day temperatures in the tropics, and it is then possible to distinguish *Ancylostoma* larvae from those of *Necator* by examining their mouth parts. Rhabdatidiform larvae of *Strongyloides* are readily detected in the faeces by low-power microscopy.

Enterobius

The threadworm or pinworm (*Enterobius vermicularis*) is also transmitted by the faecal–oral route but its epidemiology is different from that of *Ascaris* or *Trichuris*. *Enterobius* is probably the commonest helminth parasite in temperate areas with relatively high standards of hygiene, but it is also common in the tropics. It is typically associated with crowded living conditions: infection is most frequent in young children, though simultaneous infection in several members of a family is common, as is widespread infection in schools and other institutions with dormitory accommodation. The adult parasites are small pointed worms (hence pinworms) and live in the lumen of the large intestine. Adult females emerge from the anus when the patient is at rest (chiefly at night), and several thousand eggs are laid on the skin, after which the female dies. The eggs develop rapidly and are very light, being readily distributed in dust (e.g. during bed-making). The adult worms irritate the peri-anal region, and patients frequently get eggs under their finger-nails as they scratch the irritation. These eggs may then be transferred to the patient's own mouth, thus increasing the heaviness of his infection, or to the hands or mouths of others, either directly or via furniture, door knobs, etc. Contaminated clothes or bedding can also be fomites. Eggs can survive for several weeks in moist warm conditions, but are rapidly killed by sunlight. Eggs hatch in the small intestine and adults mature in the colon, where they live for 6–8 weeks.

Trichinella

Tr. spiralis infection is rare in man, the life cycle being mainly in omnivorous, cannibalistic or carnivorous rodents, or in domestic or wild pigs, polar bears or hyenas. Human infection usually follows eating of raw or inadequately cooked meat from domestic or wild pigs. In any host the adult nematodes live in the intestine for about 6–8 weeks, during which time the females produce large numbers of larvae (not eggs). These are not passed in the faeces, but penetrate the host's intestinal wall and are carried in blood to the tissues, where they encyst between cells in voluntary muscle. Here they gradually provoke the host tissues to form oval outer cysts; these gradually calcify, and the larvae eventually die, several years after invading the muscle. If such encysted larvae are ingested while still alive, they excyst in the intestine, become mature adults and start the cycle again. Thus each parasitized individual acts as definitive host for the adult and as intermediate host for the larval stages.

Filarial Nematodes

These include members of the genera *Wuchereria, Brugia, Onchocerca* and *Loa*, and the less closely related guinea worm (*Dracunculus*).

Bancroftian and Brugian filariasis (p. 246), caused by the nematode parasites *Wuchereria bancrofti* and *Brugia malayi*, are very widely distributed throughout the tropics and subtropics. *Wuchereria* occurs in much of Central Africa, coastal areas of the Middle East, the Indian subcontinent, SE Asia, the Pacific Islands, the West Indies and central and northern South America. *Brugia* occurs in East Asia, particularly Malaysia, Indonesia and the Philippines. Infection is transmitted by *Culex*, *Aedes* or *Anopheles* mosquitoes; the immature nematode larvae escape from the mouth-parts of the feeding insect and are deposited on human skin before entering the body through the puncture wound. The larvae gain access to the peripheral lymphatics and migrate to the regional lymph nodes and the large lymph vessels, where they mature into adults. Immature larval stages (microfilariae) are normally found in the blood and are taken up by and infect the feeding mosquito. The developing adults obstruct the lymph flow in the vessels in which they live, and this eventually leads to oedema of the affected area (usually a limb) and finally to fibrosis of the swollen tissues (elephantiasis).

Onchocerca volvulus and *Loa loa* are commonest in West Africa, and *Onchocerca* also occurs in Central America. Both are transmitted by blood-sucking insects, *Onchocerca* by *Simulium* flies and *Loa* by *Chrysops*. *Onchocerca* causes 'river blindness'. The blackflies (*Simulium*) that transmit the infection in Africa breed in fast-flowing well-oxygenated rivers, and become infected when taking a blood meal from an infected person. Larvae develop in the flies and infect man when a fly takes another blood meal after they are mature. They penetrate human skin via the insect bite, and undergo development as they migrate around the body for the next year. Eventually adult worms of the two sexes meet and mate, usually in the subcutaneous tissues, and the female then produces large numbers of immature larvae. The larvae pass into the subcutaneous tissues, where they provoke various changes, partially of an allergic type. The adult male worms, smaller than the females, die shortly after fertilizing them. Adult females are thread-like and measure about 40 cm long. Sometimes they live in the deeper tissues, lying between fascial layers; in other cases they are more superficial, becoming enclosed in fibrous capsules or nodules in the subcutaneous tissues. In Africa the usual location of these nodules is over the joints or trunk, whereas in South America they are more frequently found on the head. The adult worms do not cause the host any major problems but the larvae may cause river blindness and other diseases (p. 292).

Loa rarely causes severe disease in man. The

adult worms wander through the tissues and may sometimes be seen crossing the conjunctiva, hence the name 'eyeworm'; but they do not cause blindness.

Guinea worm (*Dracunculus medinensis*) is distantly related to the filarial parasites. Infection with it is found chiefly in West Africa, the Middle East and the Indian subcontinent. It is transmitted by small fresh-water crustacea called *Cyclops*. Infection is acquired when infected *Cyclops* are swallowed in drinking water taken from pools or shallow wells. The larval worm is released from the *Cyclops* in the human stomach, and begins a period of migration around the body. At the end of this period the female worm has grown to about 100 cm in length; the male is much smaller and dies shortly after fertilizing the female. A year after infection the female comes to lie just below the surface of the skin, usually of the foot, ankle or leg, and a blister forms in the skin over its anterior tip. This blister is irritating and painful and bursts if it comes into contact with cold water (e.g. when standing in a shallow pool to collect water for domestic purposes). The uterus of the worm prolapses through the base of the blister and many thousands of larvae are expelled into the water, where they are eaten by freshwater organisms (including *Cyclops*). Human infection occurs in rural communities that depend on natural water supplies of seasonal nature, and is usually contracted when supplies are low, immediately before the start of the rainy season. The latent period is nearly a year, so that the infection (which is painful and incapacitating) becomes apparent at the peak times of planting and harvesting. It reduces work capacity when agricultural activity is at is height, and so can seriously affect the domestic economy of those infected. Infection rates are very high in some communities (20–60%), with multiple infections common, and so the economy of whole villages can suffer.

Nematodes from Other Animals

Occasionally eggs or larvae of nematodes normally parasitic in other animals may infect man. Although they are unable to complete their development in man, larval stages can cause pathological changes. Larvae of dog hookworms (*Ancylostoma braziliensis*) and of *Strongyloides* normally found in non-human primates can penetrate human skin and migrate for several months through the subcutaneous tissues, stimulating a severe localized allergic reaction as they migrate (**creeping eruption**). The larvae die after several months. Eggs of *Toxocara* (a worm related to *Ascaris* and common in dogs and cats) can hatch in the human gut; the larvae migrate around the body like *Ascaris* larvae but cannot mature. Larvae migrating in the viscera cause allergic symptoms known as **visceral larva migrans** (p. 293). They are a rare cause of blindness in children (p. 297).

Suggestions for Further Reading

Muller, R. 1974. *Worms and Disease*. Heinemann, London.

Crewe, W. and Haddock, D.R.W. 1985. *Parasites and Human Disease*. Arnold, London.

Laboratory Investigation of Infection (and Skin Testing)

Using and Understanding The Laboratory

When a patient is suspected of having a microbial infection, the primary aims of investigations in the microbiology laboratory are to establish whether an infection is in fact present (e.g. by demonstrating pus cells in urine or other body fluids); to identify the causative organism as fast as possible and with as much precision as is likely to be helpful; and to find out which antimicrobial drugs are likely to be effective against it. Such information may allow more accurate diagnosis, better treatment, clearer prognosis and the recognition of steps that should be taken to prevent spread of the infection to others—e.g. isolation of the patient, immunization of contacts. The discovery that the infection is due to a virus or other organism for which there is no effective drug treatment may save the patient from being given irrelevant antibiotics with their attendant hazards (p. 373).

Proof that a patient is carrying, or has recently carried, a potential pathogen does not necessarily amount to proof of its responsibility for his illness. Due allowance must always be made for its known rate of carriage by healthy people and for the likelihood of its producing the type of illness from which the patient is suffering. Even a rising level of antibodies, specific for the organism in question, in the patient's blood is evidence only that it has been present, not that it has been acting as a pathogen—though the case against it is stronger if the pattern of antibody rise suggests that it arrived at the right time to initiate the illness.

Quality of input

The amount of help that a microbiology laboratory can give to a clinician depends largely on what it receives. As has been said of computers: 'garbage in, garbage out'. The laboratory needs:

(1) Careful and well-informed selection of the right specimens. If in doubt about this, it is wise to ask the laboratory staff; this often saves time and money, and avoids the tragedy of discovering too late what should have been done.

(2) Proper collection of such specimens. **(Whenever possible, this should be done before antimicrobial treatment is given.)** Even the best of specimens may be useless without a label showing what it is and the patient's name.

(3) Rapid transmission of the specimens to the laboratory.

(4) Completion of an appropriate laboratory request form, giving the name, whereabouts, age, sex and (when relevant) occupation of the patient; a legible indication of the name of the doctor concerned (often not the same as a signature); and clear concise indications of the diagnostic problem, relevant treatment already given (particularly antimicrobial drugs) and the reason for sending the specimen.

The importance of these points is illustrated by the following all-too-common situations:

(1) A vaginal swab is often the only specimen sent to the laboratory from a woman with suspected gonorrhoea. Because of the acidity of the adult vagina (p. 16), gonococci do not thrive there and are better sought in the cervix uteri or in urethral discharge.

(2) In the absence of supervision or clear instructions, a patient who has difficulty in producing sputum will often spit out saliva instead. This is, of course, useless as a source of information about his lower respiratory tract. Indeed, if it is examined and reported upon as though it were sputum, the clinician may be totally misled.

(3) Urine is a good culture medium, particularly for the common urinary tract pathogens. If it is left to stand at room temperature for some hours after collection, its bacterial content may increase considerably, and subsequent culture may give a highly inaccurate picture of what was in the specimen when it left the patient, since a lightly contaminated urine can come to resemble a heavily infected urine.

(4) The more delicate pathogenic species—e.g. the gonococcus and many anaerobes—are liable to die before reaching the laboratory if the specimen is allowed to become dry or is excessively delayed in transit. Their failure to grow in cultures may then give a misleading impression.

(5) The presence of a large number of *Esch. coli* or related organisms in the upper respiratory tract can be interpreted in various ways. In infants it is a common finding, probably a result of regurgitation, and is of no importance. It is also common in patients of all ages who have received antibiotic treatment sufficient to derange their normal flora; in such circumstances the most that is usually called for is modification or cessation of the antibiotic treatment. But in the absence of either of these explanations such a finding requires further investigation, and the enterobacteria may require specific treatment. The bacteriologist who has not been told anything about the patient's age or treatment is in no position to give intelligent co-operation.

Routine approach to specimens and reporting

Having received a satisfactory specimen in good condition and with adequate accompanying information, the laboratory staff have to decide what to do with it. Routine procedures in busy laboratories are planned to extract the maximum of useful information from each specimen and yet keep the amount of work within bounds. When the specimen comes from a part of the body that is normally sterile, any organism found is abnormal—though it may have entered the specimen as a contaminant during or after collection. However, it is clearly impossible to apply to each such specimen a routine calculated to detect any known micro-organism; the best that can be done is to look for those which are not excessively rare in such situations and are likely to be relevant to the patient's condition. This is one reason why it is important that the laboratory should be told if any unusual infection is suspected. With specimens from sites that have a normal microbial population it is essential to have clearly defined aims—one of which is usually the discouragement of organisms normally present, in order to increase the chances of detecting those that are abnormal. There can be few hospital bacteriologists who have not at some time received a faecal specimen 'For organisms please'. Isolation and identification of all bacteria, viruses, fungi and protozoa in a single faecal specimen might well take years, whereas the question 'Does this specimen contain known intestinal pathogens?' can usually be answered by a few minutes' work spread over 2 or 3 days.

The presence of a normal microbial population creates problems of reporting. Clearly the report should include the names of any known pathogens found which might be responsible for the patient's illness or which it is undesirable that he should continue to carry. As a rule it will also include information about the sensitivity of such pathogens to appropriate antimicrobial agents. Organisms which are common commensals but also potential pathogens—e.g. pneumococci in the respiratory tract—need to be assessed in the light of clinical circumstances. An unexplained and unusual predominance of one of the normal commensals may be of some significance—e.g. an almost pure growth of viridans streptococci is sometimes obtained from a swab of an inflamed tonsil, and possibly indicates that this organism is in fact causing the tonsillitis. The correct wording of 'negative' reports is a subject of controversy. 'No pathogens isolated' gives the clinician a minimum of information, though it is probably the best formula for some specimens for which the range of normal findings is very wide. A list of pathogens which have been sought but not found is sometimes appropriate but is liable to be cumbrous. A report of the predominant organisms in the cultures—e.g. 'viridans streptococci and commensal neisseriae' from a throat swab—does not convey much useful information, and obscures the fact that this predominance was probably determined by the methods of culture. 'Normal flora' is defensible provided that it is taken to mean: 'The varieties and proportions of organisms identified appear to be within normal limits for such specimens when examined by the procedures in routine use in this laboratory.' The last point is important; an abnormality will be detected only if the procedures used are appropriate to its detection.

Infection hazards

Nearly every specimen sent to a diagnostic microbiology laboratory may contain pathogenic micro-organisms and is therefore potentially hazardous to the one who collects it, to those who carry it to the laboratory and to the laboratory staff. The degree of hazard varies greatly. Numerous reports have high-lighted the hazardous nature of blood and other fluids from hepatitis B carriers (p. 266); the special precautions indicated below should be taken when collecting these and other specimens from patients who are likely sources of readily transmissible and highly virulent pathogens. However, specimens from patients who are **unsuspected** sources of such organisms are equally dangerous, and so all microbiological specimens must be treated with due care. Furthermore, it is important to remember that a blood sample is no less dangerous when sent to a biochemistry, haematology or other laboratory than when microbiological investigations are requested! The following guide-lines should be observed:

(a) When collecting specimens from 'high risk' patients, disposable gloves should be worn and other special precautions may be appropriate.

(b) Specimens should be securely enclosed in protective containers, and fluid specimens should be in properly closed leak-proof bottles or tubes.

(c) Care must be taken to prevent contamination of the outside of the container.

(d) Ideally in all cases, and certainly when there is any special hazard, the container should be enclosed in a sealed plastic bag for transmission. The request form should **never** be wrapped round the container or enclosed in the same plastic bag; bags with two compartments, one for the specimen and one for the form, are best.

(e) Any known special hazard—e.g. of hepatitis B or AIDS—should be indicated by agreed colour markings on the specimen and the request form.

Collection of specimens

Most clinical specimens are collected in one of the following ways:

(1) Materials such as saliva, sputum, faeces, urine, crusts, and scabs and freely discharging pus can be collected **directly into suitable sterile disposable containers** with tightly fitting lids.

(2) Sometimes it is convenient to collect small quantities of fluid—e.g. vesicle contents for examination for poxviruses, or exudate from a suspected syphilitic chancre for dark-ground microscopy—into **capillary tubes**, which can then be sealed by heating their ends, so that the fluid does not dry up.

(3) For collection of material from skin and mucous surfaces, and also of exudates and discharges which are too small in amount for

direct collection as in (1), a **swab** can be used. This usually consists of a wooden or wire rod about 15 cm (6 inches) long, with a small quantity of cotton wool tightly twisted around one end and the other end inserted into the cork or stopper of the tube in which it is supplied. The swab and the inside of its container are sterile. For use, the swab is withdrawn from the tube, applied to the patient and then replaced in the tube for transmission to the laboratory, or else placed in a suitable **transport medium** (see below). Collection of organisms from dry skin is more efficient if the swab is moistened with sterile broth immediately beforehand. Use of serum- or albumin-coated cotton wool swabs may increase the chances that relatively delicate bacteria, such as *Str. pyogenes*, will reach the laboratory alive.

It needs to be emphasized that **swabbing is not a satisfactory substitute for the direct collection of such materials as pus or other exudates** when these could have been collected in a bottle or syringe; it should be used only when inadequacy of materials or other factors make direct collection impossible.

(4) **Washings** from cavities are used in certain circumstances. Throat washings, sometimes preferred to throat swabs for virus investigations, are obtained by asking the patient to gargle with physiological saline and then expectorate it. Gastric washings for examination for tubercle bacilli, from patients who cannot produce sputum, are obtained by running saline into the empty stomach and then withdrawing it through a Ryle's gastric tube. Washing out the maxillary antrum is a therapeutic procedure, but the washings may be sent for bacteriological examination.

(5) **Aspiration**, usually through a needle and often with the assistance of suction from a syringe, is used to collect materials confined within the patient's body, such as blood, cerebrospinal fluid, effusions into body cavities and joints, and closed abscesses. Organ biopsies can be carried out in the same way. Care must be taken to avoid contamination of the specimens, either from the apparatus used or by collecting organisms in the needle as it passes through the skin (see p. 343). Aspirated materials are either placed in a sterile container or added to suitable media directly from the syringe.

Transmission

Correct choice and collection of a specimen is of little use if the pathogens die on the way to the laboratory. The best way to prevent this is to take it there, or see that it is taken, without delay. (Inoculation of media at the bedside or in the clinic, while necessary in some cases, is unsuitable for general use because of administrative difficulties and the risks of contaminating the cultures.) If delay is inevitable, virtually all specimens other than blood cultures are better kept in a refrigerator than at room temperature; but it is usually better still to use suitable **transport media**, which are designed to keep the pathogens alive. This can be achieved in a general-purpose bacterial transport medium by incorporating reducing substances to protect the anaerobes from oxygen, and charcoal to neutralize components of the specimen that might be toxic to bacteria, and by excluding any nutrients to prevent overgrowth of fast-growing bacterial contaminants. Virus and chlamydial transport media often contain antibacterial drugs and are therefore inappropriate for transport of bacteriological specimens.

Special regulations govern postal transmission of microbiological specimens and other pathological materials.

Laboratory Procedures

Although there are overlaps, there are also some large differences between the procedures used for detection, identification and further study when dealing with bacteria, with viruses, with fungi, with protozoa and with helminths; they are therefore described below in separate sections. (Procedures for rickettsiae, etc., are outlined in Chapter 8.) However, there is considerably more overlap in the immunological approaches to infection with these different groups of organisms, and therefore immunodiagnosis of infection is dealt with in a single section.

Procedures for Bacteria (See also pp. 69–73).

Microscopy

Light microscopy

WET PREPARATIONS A drop of fluid from a specimen or a culture or any other suspension of organisms is placed on a glass slide, covered with a coverslip and examined with the high-power dry objective of the microscope, using a restricted amount of light. Objects such as bacteria that diffract light appear dark against a bright background. In this way it is possible to determine the size and shape of the organisms and whether they are motile. True spontaneous motility must be carefully distinguished from the Brownian movement to which all small particles in a fluid medium are subject, and from drifting due to currents in the fluid. Differentiation between organisms in such preparations cannot be carried very far, and some bacterial groups, notably the spirochaetes, are so feebly refractile that they cannot be seen at all in this way. When wet preparations are examined by **phase-contrast** microscopy, the diffracted light is also retarded by a quarter of a wavelength; consequently, details of the structure of organisms stand out sharply against a grey background. Wet preparations are also used for counting leukocytes and other cellular elements in such specimens as urine, cerebrospinal fluid and effusions, and are usually more satisfactory than stained smears for the identification of such structures.

STAINED SMEARS If material from a specimen or a culture is fixed to a glass slide (usually by heating), appropriately stained and examined with the oil-immersion objective of the microscope, using a bright light source, any bacteria present can be seen and to some extent identified. Even with only a simple stain such as methylene blue or carbol fuchsin, this procedure usually enables us to see more of the shape, arrangement and structure of the organisms than is visible in unstained preparations, though allowance has to be made for artefactual changes due to drying and staining. Additional valuable information can be obtained by using differential staining techniques such as **Gram's method** (p. 71). Gram-stained smears are made from a large proportion of specimens sent for bacteriological examination and often give valuable leads as to the most appropriate culture procedures. Such smears are of limited value, however, in the study of specimens from the alimentary tract and many of those from the respiratory tract, since even those from healthy subjects are likely to contain commensal bacteria of many different morphological types. Staining by the **Ziehl–Neelsen (ZN) method** (p. 103) is widely used for specimens which might contain tubercle bacilli.

Dark-ground microscopy

Some bacteria too feebly refractile to be seen by ordinary light microscopy—e.g. the spirochaetes of syphilis and leptospirosis—can be seen by dark-ground illumination, which also shows up fine details of their shape.

Dark-ground illumination is obtained by focusing a hollow cone of light from below on to the top surface of the microscope slide in such a way that, unless deviated from its path, the light will diverge again and miss the front lens of the objective. Thus the only light to enter the objective and reach the eye of the observer is that which has been deflected by striking bacteria or other objects on the slide. These objects shine brightly against a dark background. The process is essentially the same as that by which one sees fine dust particles when looking from the side at a shaft of bright sunlight. Dark-ground microscopy is usually applied to unstained wet preparations, with water or oil between the condenser and the slide to prevent total internal reflection of light within the condenser, and between the cover-slip and the objective to prevent scattering of light.

Fluorescence microscopy

It is sometimes possible to detect bacteria (or other micro-organisms) in clinical specimens or other material by staining fixed smears of the material with a dye which attaches preferentially to the organisms in question and which fluoresces when the smears are examined microscopically in ultra-violet light. One such dye, auramine, selectively stains tubercle bacilli when it is used in a ZN-like procedure in which it replaces the carbol-fuchsin (p. 103). This makes possible rapid detection of tubercle bacilli in sputum etc., though ZN staining may be necessary to confirm their morphology. In the **fluorescent-antibody technique (immuno-fluorescence**—p. 178) the fluorescent dyes are linked to antibodies specific for the bacteria or bacterial antigens sought, and so become selectively attached to these. This procedure has many applications to the recognition of bacteria or their antigens in clinical material and in cultures, and to determining their location in tissues, etc.

Electron microscopy

While this technique has added greatly to our knowledge of the structure of bacteria, its main diagnostic role is in virology (p. 168).

Other Methods for Rapid Detection and Counting of Bacteria

Microscopy, in any of its forms, is a time-consuming way of finding out whether bacteria are present in significant numbers in a specimen when it reaches the laboratory, or are multiplying in a broth culture too young to show naked-eye turbidity. The increasing work-loads of diagnostic laboratories call for rapid and preferably automated techniques for doing these jobs, and several have been developed in recent years—some of them to levels at which they can be taken into routine use. Automated screening of freshly collected urine specimens, making it possible to detect the many which do not contain bacteria in the numbers found when there is a true infection (p. 273), could save laboratories a great deal of work and materials. **Luminescence biometry** is one possible approach to this problem. It depends on the fact that the amount of bacterial adenosine triphosphate (ATP)—and thus the number of bacteria—in a given amount of liquid can be assessed by measuring the brightness of the flash of light emitted when the liquid is mixed with a standard preparation of luciferin–luciferase (the 'lighting system' of fireflies). This or some other sensitive procedure for detecting bacterial multiplication in young broth cultures, or its absence, can make it possible to determine within a few hours whether the bacteria present in a clinical specimen are capable of multiplying in a suitable liquid medium in the presence of one antibiotic or another; it is then possible to say which antibiotics should be appropriate for treatment of an infection well before the offender has been identified. One means of such sensitive detection is **electrical impedance monitoring**, which depends on the fact that the electrical resistance of the culture medium falls as the number of organisms increases; and a sensitive **turbidimetric** system can give comparable information. Early determination of whether bacteria are growing in a blood culture (p. 241) can be of great clinical importance, and older methods of doing this have taken time and have been liable to introduce confusing contaminant organisms into the culture; but these problems can be largely overcome by using appropriate equipment to monitor the **release of radio-active CO_2** from ^{14}C-labelled ingredients of the blood-culture broth or to measure electrical impedance. As problems of automating these and other comparable tests are overcome, laboratories will be able to handle still larger numbers of specimens, and this will doubtless lead to development of still more screening tests. The main effect of these may well be to make possible the rapid exclusion of some of the possible diagnoses in appropriate cases—a contribution which is often at least as useful to the clinician as the belated production of a positive report.

Culture

Preliminary microscopic examination, perhaps aided by special staining and by procedures such as capsule-swelling (p. 176), may allow a pathogenic bacterium to be identified in the original specimen with a high level of certainty—though this is often not so. In any case it is still usually necessary to obtain a pure culture of the organism (p. 67), not only to confirm any provisional identification but to acquire other needed information about it, such as its antibiotic sensitivities.

Plating out

The common procedure for obtaining a pure culture of a bacterium is to spread a little of the original material over the surface of a solid culture medium in such a way that individual bacteria, or small clumps of attached and so presumably related individuals, are well separated from one another. On incubation of the culture a distinct pile or **colony** of bacteria is formed by each individual bacterium or clump of attached individuals (in technical language, each colony-forming unit or CFU) that finds the conditions suitable for its multiplication. One of Robert Koch's valuable contributions to bacteriological technique was the use of **agar**, a seaweed derivative, as a means of solidifying media. Added in small amounts (1.5–2%) to a heated fluid culture medium, this substance causes it to set as a firm jelly when cooled to about 40°C. This low setting point allows heat-labile ingredients, such as red blood cells, to be added to a medium while it is still fluid. On the other hand, once set the medium will not melt until it is heated to nearly 100°C, and there is consequently no danger of its liquefying at ordinary incubation temperatures. Agar media are commonly used in circular, flat-bottomed plastic **Petri dishes**, 9 cm or so in diameter, with close-fitting lids. Media and dishes must be sterilized before use (see Chapter 26).

Culture in such a dish (plate culture) can be used to obtain pure cultures from a mixture of bacteria as follows:

(1) The material to be investigated is spread over a segment of the surface of the medium.
(2) A wire loop, sterilized by heating in a flame and then allowed to cool, is passed several times through the inoculated area on to a fresh area of medium. The bacteria will thus be less thickly spread on this second area than on the first.
(3) Similar transfers are made from the second area to a third and so on until the whole surface of the plate has been used.

In this way, as illustrated diagrammatically in Fig. 13.1, there is usually at least one area of the plate on which the bacteria are deposited sufficiently far apart to form separate colonies after incubation. In medical bacteriology 37°C is the usual incubation temperature and most organisms produce colonies large enough for recognition within 18 hours. It is then possible to see whether the original material contained a mixture of bacteria, and to select colonies for microscopic examination and subculture.

Figure 13.1 Diagram of plating out of a mixture of bacteria. Note that where the colonies are more widely separated they are larger and more obviously diverse.

Choice of culture media

Most specimens arriving in a diagnostic bacteriological laboratory are plated out on a selection of agar media, chosen in accordance with the nature of the specimen and the pathogens thought likely to be present. Simple **nutrient agar**—broth solidified with agar—is of limited use as a primary culture medium, but may be useful later as a means of distinguishing exacting pathogens, which cannot grow on it, from their more robust relations. **Blood agar**—i.e. nutrient agar to which has been added 5–10% of blood, usually horse blood—is used for most clinical specimens, as many of man's bacterial parasites will grow on it, and some (notably streptococci) cause characteristic haemolytic changes in it (p. 76). For many specimens both aerobic and anaerobic blood–agar cultures are set up, to ensure growth of anaerobes and to allow them to be distinguished from strict aerobes or facultative organisms (see p. 15; and see below for anaerobic culture techniques). Incorporation of an aminoglycoside antibiotic or nalidixic acid in the blood agar for anaerobic incubation often helps the isolation of anaerobes from specimens with mixed bacterial populations, as these agents suppress many facultative organisms. **MacConkey's agar**, which contains bile salts to select for intestinal organisms and lactose and an indicator to detect lactose-fermenters (p. 90), is valuable for specimens likely to contain enterobacteria—including, among others, faeces, urine, most pus samples and swabs from wounds and many ulcerative skin lesions, especially those on the lower half of the body. **Cooked meat medium** (boiled minced lean meat in peptone water—p. 90) supports the growth of most aerobic and anaerobic bacteria, and is commonly used in addition to the solid media for primary culture of pus and of swabs from many sites. Particularly when there has been delay in transport of the specimen or when the patient has been on antibiotic treatment, growth may occur in the cooked meat medium when there is none on the primary plate cultures; the cooked meat medium is then subcultured to further blood agar plates for aerobic and anaerobic incubation. Organisms from a patient receiving a β-lactam antibiotic (penicillin or cephalosporin) will sometimes grow in **broth containing β-lactamase** when they fail to grow in or on other primary culture media. They too can then be subcultured to blood agar. Means are available for neutralizing many other antibacterial drugs in specimens. Various broths are also used for primary culture of specimens unsuitable for plating out as described above—notably blood, in which the numbers of bacteria are likely to be far too small for that method.

The few media mentioned so far are sufficient for the examination of most clinical specimens except faeces, for which a highly selective approach is necessary (pp. 93 and 262). Specially nutritious media—notably **chocolate agar** (p. 81)—must be provided for unusually exacting bacteria. Various media containing selective inhibitors for the isolation of particular groups of bacteria have already been mentioned in discussing the relevant organism.

ANAEROBIC CULTURE This is commonly carried out in metal or polycarbonate **anaerobic jars**, which are large enough to hold piles of plate cultures; they can also be used for cultures in tubes or bottles, provided that these have cotton-wool stoppers or loose caps rather than airtight closures. The lid of the loaded jar is clamped down onto an airtight seal, and the air is then evacuated and replaced by an oxygen-free hydrogen-containing mixture, usually 80% nitrogen, 10% hydrogen and 10% carbon dioxide. The remaining traces of oxygen are removed by combination with some of the hydrogen, a reaction which is catalysed by palladium in a capsule fixed to the under surface of the lid. When its preparation has been completed the whole jar can be placed inside an incubator.

There are other ways in which anaerobes can be grown—e.g. reducing agents such as **glucose** and **sodium thioglycollate** can be added to the medium; and **cooked meat medium** is suitable for growing anaerobes because of the reducing activity of the pieces of meat at the bottom of the bottle. However, none of these methods is as reliable, or as suitable for the stricter anaerobes, as the use of anaerobic jars. Special anaerobic cabinets, in which all plating of specimens and all subculturing can be carried out as well as incubation, are a help to those doing a lot of anaerobic culture work, but are not essential for good routine anaerobic bacteriology.

CULTURE IN 10% CO_2 A simple and inexpensive way of achieving approximately the right atmosphere for isolation of CO_2-requiring bacteria, such as the meningococcus and *Br. abortus*, is to place the cultures and a lighted candle in any suitable container which can then be closed with an airtight lid. The candle goes out when the CO_2 level is of the desired order. An anaerobe jar containing 10% CO_2 in air instead of the anaerobic culture mixture is rather more reliable; but these simple methods are increasingly being replaced by use of special incubators in which the proportion of CO_2 in the atmosphere can be precisely controlled.

Further Identification

The use of growth characteristics, tests of metabolic function, product analysis and serological tests in the identification of bacteria has been outlined on pp. 69–73, and illustrated at many points in Chapter 7. There also we indicated the value of precise identification of strains causing disease—e.g. biotyping of brucellae (p. 101), serotyping of salmonellae (p. 92), phage typing of staphylococci (p. 75) and bacteriocine typing of shigellae (colicines, p. 94) and pseudomonads (pyocines, p. 96).

Antibacterial Drug Sensitivities

Procedures for determining these are discussed on pp. 284–5.

Procedures for Viruses

Lack of specific treatment for virus infections meant that for many years diagnostic virology was an unhurried and largely academic discipline, concerned with eventual understanding of virus infections rather than with immediate clinical and therapeutic problems. The tempo has now changed—not merely because of the development of antivirus drugs but because it is appreciated that early and precise diagnosis of virus infections may indicate appropriate action to be taken (in treatment of patients or in management of contacts), may save the patient from irrelevant treatment for other suggested diagnoses, and may allow more accurate prognosis. Rapid diagnosis depends on microscopic or serological demonstration that virus particles or antigens are present in material from the patient. Isolation of the virus by growing it in tissue culture, in eggs or in animals takes longer but may be more sensitive and more conclusive and may make possible more precise identification. Measurement of the patient's antibody responses (considered later in this chapter) may give indirect supporting evidence for the diagnosis or may be the only grounds on which it is based, but such responses take time (pp. 179–80).

Microscopy

Only rarely is ordinary **light microscopy** useful in virological diagnosis; examination of stained smears or tissue sections may show characteristic histological changes, or the presence of pathognomonic inclusion bodies in rabies, cytomegalic inclusion disease, etc., or even that of individual particles (elementary bodies) in poxvirus infections.

Fluorescence microscopy

Diagnostic virologists, like their bacteriological colleagues (p. 164), have found many uses for immunofluorescence techniques (p. 178). Specific antibodies, labelled with fluorescent dyes, can be used for the rapid detection and identification of viruses or virus antigens in clinical material or in tissue cultures.

Electron microscopy

This has applications in branches of microbiology other than virology—e.g. in determination of the finer structure of larger micro-organisms—but it is particularly valuable to the virologist. Much of the information given in Chapter 9 about the sizes, shapes and structures of viruses has been obtained, or at least confirmed, by electron microscopy; and it can also be used in recognition of viruses in clinical material—e.g. in distinguishing between varicella–zoster virus and poxviruses in vesicle fluid, or in detecting rotaviruses or other viruses of distinctive morphology in diarrhoeic faeces. In the electron microscope a beam of electrons, derived from an 'electron gun', is passed through a series of electromagnetic fields which correspond to the lenses in an optical microscope in that they bring about convergence of the beam. The material to be examined is mounted on a thin membrane of collodion, polyvinyl formal or carbon, which is supported on a metal grid, and the examination is carried out in a high vacuum which inevitably produces some distortion. Ultra-thin sections of tissue or of suitably embedded microbial or other cells can be examined, and so can films made from suspensions of organisms or from clinical material. To be clearly visible very small objects must differ from their surroundings in their opacity to the electron beam. This can be achieved by supplying a background of more electron-opaque material, such as sodium phosphotungstate ('negative staining'); or by 'positive staining' with some electron-opaque material that will selectively adhere to the particles (e.g. various heavy-metal compounds, or ferritin-labelled antibodies); or by 'shadow casting'—i.e. projecting a shower of metal atoms over the film obliquely, so that a thin layer of metal is formed all over it except in the 'shadows' of particles. After the electron beam has passed through the material to be examined it is made to produce a visible image on a fluorescent screen or to produce photographs that can usually be much enlarged. In **immune electron microscopy** the detection of virus particles or antigens under the electron microscope is facilitated by adding specific antisera to the material being studied; this turns them respectively into clumps of viruses or into antigen–antibody complexes. (Whereas ordinary electron microscopy can show outlines and structural details in a flat plane, the more sophisticated technique of **scanning electron microscopy** can produce clear 'aerial views' of particles such as microbial and other cells, showing the contours, irregularities and texture of their surfaces; but this has limited relevance to virology.)

Antigen detection

As well as being used in fluorescence microscopy and in immune electron microscopy as indicated above, specific antisera are used in CIE (p. 175), PHA (p. 176), RIA (p. 178) and ELISA (p. 178) tests for rapid detection of virus antigens in clinical specimens—e.g. of hepatitis B antigens in blood; of rotaviruses in faeces; of herpesviruses in vesicle fluid; and of influenza or respiratory syncytial viruses in respiratory tract secretions. Such methods can also be used to detect virus antigens in tissue-culture fluids long before infecting virus is detectable by other methods.

Virus isolation

The first essential is to ensure that the viruses survive the period between collection of specimens and setting up of cultures, since it is not usually practicable to set these up at the bedside or in the clinic. Some viruses are relatively robust; for example, poxviruses in material from lesions or polioviruses in faeces require no special precautions during transport so far as their own survival is concerned, though precautions must be taken to prevent infection of those who transport the specimens. Most viruses are unstable at normal room temperatures, and materials containing them should be kept at about 4°C (on ice or in a refrigerator) if the delay before culture is likely to be an hour or two, or at lower temperature (e.g. in an insulated flask partly filled with solid CO_2, or in a deep-freeze at $-20°C$, or better still at $-40°C$ or below) if a longer delay is anticipated. However, sub-zero temperatures should be used with discretion, as they are lethal to some viruses—notably respiratory syncytial virus. Repeated freezing and thawing must be avoided. Drying is another hazard to virus survival, especially when the specimens are swabs with only a little material on them; such specimens should be placed in a fluid transport medium (e.g. 0.2% bovine albumin in buffered salt solution).

Most clinical specimens contain bacteria which can interfere with virus isolation, and which therefore must first be killed (usually by incorporation of antibiotics into transport media and the nutrient media of tissue cultures) or removed (e.g. by filtration).

The methods of isolation appropriate to individual viruses are indicated in Chapter 9. If **animal inoculation** is used, viruses are detected and identified by the diseases and particular lesions which they produce. They may be identified in **hens' eggs** by the lesions which they produce on the chorio-allantoic membrane, or by the presence of their antigens in fluid from the cavities. However, because of the convenience and low cost of **tissue culture**, this method of isolation is used in preference to animal inoculation or hens' egg culture whenever the nature of the virus permits it.

Tissue culture

When it became possible to grow mammalian tissues in test tubes, a new method was available for artificial propagation of viruses. However, it was at first of limited value because the cells were in the form of tissue particles suspended in fluid and were therefore difficult to examine. Furthermore, their susceptibility to virus infection was hard to predict. It then became possible to grow tissues as **monolayers** (sheets of single-cell thickness) attached to the inner surfaces of the glass or plastic of culture tubes or bottles, where they can be examined microscopically at any stage without being disturbed. Furthermore, standard 'cell lines', mostly of human or simian origin and in many cases derived from neoplastic or fetal tissues, have been developed and distributed throughout the world, so that different laboratories can use tissue cultures of comparable and predictable virus susceptibilities. Monolayers need nourishment, and this is usually supplied by bathing them in nutrient fluid. However, it is then virtually impossible to separate out pure virus lines from a mixture, just as it was difficult to obtain pure bacterial cultures until solid media were introduced (p. 166). The virological equivalent of plating out bacteria to obtain separate colonies is to seed a monolayer with virus inoculum and then cover it not with fluid but with a nutrient agar. Any one virus particle and its progeny are then restricted to the cell originally entered and those in its immediate vicinity. If these cells are damaged (see below, CPE), a visible plaque of degeneration may appear in the monolayer. As with a bacterial colony, the appearance of the plaque may be characteristic of infection with a particular virus, and subculture from a single plaque is likely to yield a pure strain.

The replication of virus in tissue culture can be detected by:

(1) **cytopathic effect** (CPE)—i.e. degenerative changes in the infected cells which can be seen when a monolayer is examined microscopically. Viruses of many different groups produce such effects, and to some extent the nature of the virus can be deduced from the type of change which occurs in the cells;

(2) functional changes in the cells which can be detected by tests of such metabolic activities as acid production;

(3) the presence of detectable antigens, **haemagglutinins** (see below) or other virus components or products in the fluid bathing the cells;

(4) acquisition by the cells of the power to adsorb red blood cells on to their surfaces (**haemadsorption**). This results from infection by some of the myxoviruses, and the consequent formation of virus-containing buds on the surfaces of the infected cells;

(5) resistance of the cells to infection by other viruses (**interference**—see p. 49);

(6) attachment of virus-specific fluorescent antibodies to virus particles or virus antigens within the cells.

Though monolayer culture has many useful features, simple suspensions of cells in suitable nutrient fluids may suffice for virus propagation and for carrying out tests that depend on changes in cell metabolism.

Organ culture is another form of tissue culture that has given valuable results (e.g. with rhinoviruses, p. 119). Portions of intact ciliated epithelium from the respiratory tract of a human embryo can be kept alive on a culture medium for several weeks, and their cilia continue to beat. Some respiratory-tract viruses that cannot be grown in any other form of tissue culture (and also many that can) are able to infect such portions of epithelium and demonstrate their presence by damaging the cells—notably by stopping their visible ciliary activity. It seems probable that the development of similar cultures of other intact tissues will lead to the discovery of other new viruses. As well as being useful for primary isolation and subsequent maintenance of viruses that cannot be grown in other ways, organ cultures of respiratory epithelium have made possible some most elegant studies of precisely what viruses do when they infect tissue that was functioning normally prior to their arrival.

Haemagglutination

Orthomyxoviruses and members of several other virus groups (including poxviruses and togaviruses) have the ability to agglutinate red blood cells of various species. This property, which in some cases at least is related to the organism's method of attaching to and entering host cells (p. 42), has the following laboratory applications:

(1) Such viruses can be detected in fluids by their ability to agglutinate human or other suitable red cells.

(2) Some of them endow infected tissue culture cells with the power of haemadsorption, as indicated above.

(3) The haemagglutinins are antigens, and infected hosts are stimulated to form antibodies. These can be demonstrated and measured by their specific inhibition of the *in vitro* haemagglutinating activity of the responsible virus, and their presence in a patient's serum is evidence of infection with that virus (p. 177, haemagglutination-inhibition test).

Procedures for Fungi

This section deals only with fungi with which a diagnostic laboratory in Britain needs to be familiar.

Microscopy

Being large micro-organisms, fungi are in general easy to see by ordinary light microscopy. *Candida*, for example, can be highly conspicuous in a gram-stained smear of material from a mucous surface, as its yeast cells and pseudohyphae are strongly gram-positive and dwarf the surrounding bacteria. The mere presence of such organisms, however, does not mean that they are of clinical significance. In contrast, the demonstration of dermatophyte fungal elements in skin scrapings, hairs or nail clippings (p. 138) is diagnostic of ringworm; it is most readily achieved by treating such materials with 10–20% KOH, which dissolves the host cells and allows the highly refractile fungal elements to be seen by low-power microscopy with suitably reduced lighting. In skin or nails all of the dermatophytes form branching hyphae, some of which break down into short fragments (arthrospores); and all species look much the same. Hyphae also abound in and around the roots of infected hairs, but spores predominate further up the hairs, in distributions that allow a distinction between **ectothrix** infections (spores outside the hairs—*Microsporum* and some *Trichophyton* species) and **endothrix** infections (spores inside the hairs—other *Trichophyton* species). The microscopic appearance of the fungus in pityriasis versicolor, the other common fungal infection of skin, is described on p. 137, those of *Aspergillus* species on p. 140, and the use of microscopy in diagnosis of cryptococcosis on p. 141.

Culture

Candida species give visible growth on blood agar after overnight incubation at 37°C, but are more easily recognized on Sabouraud's medium (p. 24), on which they form moist creamy colonies. *C. albicans* can be differentiated from other members of the genus by its formation of germ-tubes within a few hours of inoculation into undiluted human serum, and by its carbohydrate fermentation reactions. In the case of suspected dermatophyte infection, material that has not been treated with KOH is inoculated on to Sabouraud's or other suitable medium and incubated at 30°C. Colonies take 1–3 weeks to appear and species recognition depends on the fluffiness, texture and colour of the colonies, on the liberation of pigment into the medium, and also on the microscopic appearances of the growth. Among the most important distinctive features are the multi-celled spore forms called macroconidia, which may be cylindrical (*Trichophyton*), spindle- or boat-shaped (*Microsporum*) or pear-shaped (*Epidermophyton*). The number, shape and distribution of the macroconidia and the presence or absence of other structures such as chlamydospores and spiral hyphae help in species identification (Fig. 13.2, p. 172).

Antigen Detection

Detection of fungal antigens in body fluids may allow a definite diagnosis to be made before the organism has been seen or grown. For example, a **latex agglutination** test (p. 176) provides a sensitive means of detecting *Cryptococcus neoformans* capsular antigen in the CSF of a patient with cryptococcal meningitis.

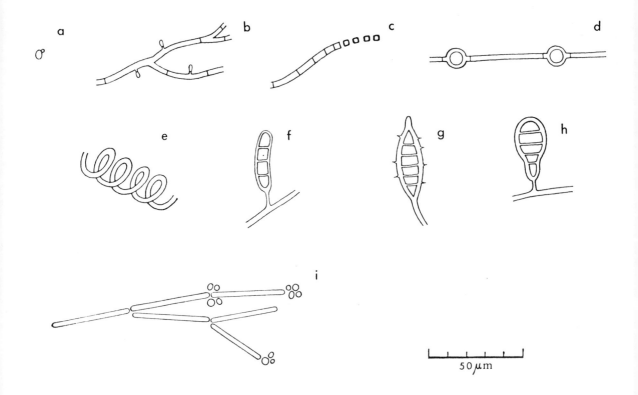

Figure 13.2 Some morphological features of fungi. (a) The same yeast as in Fig. 7.1 (q), to emphasise the difference of scale. (b) Branching septate mycelium with lateral microconidia. (c) A hypha breaking up to form arthrospores. (d) A hypha showing chlamydospores. (e) A spiral hypha. (f, g and h) Macroconidia of Trichophyton, Microsporum and Epidermophyton. (i) Pseudohyphae of a yeast-like fungus with yeast forms at junctions.

Procedures for Protozoa and Helminths

The laboratory diagnosis of most infections with protozoa and helminths is based mainly upon the detection of the parasites (or their cysts or ova) by microscopy in specimens of faeces (or urine for *Schistosoma haematobium*) or of blood, lymph-node fluid, bone marrow, liver, spleen or CSF. Methods of culture in artificial media or animal inoculation may be used in the diagnosis of trypanosomiasis and leishmaniasis. Immunodiagnosis is useful in some helminth and protozoan infections, particularly in long-standing, chronic infections; these tests are described on p. 183. Xenodiagnosis (exposure of the patient to the uninfected insect vector, then examination of the insect) is used to detect *Trypanosoma cruzi* infection (p. 246).

Microscopy

Parasites in the blood

Laboratory diagnosis of malaria, babesiosis, trypanosomiasis and filariasis is based on detection of the parasites in patients' blood. Blood samples can be taken at any time for diagnosis of any of the three protozoan infections; but in some forms of filariasis there is cyclical variation in the degree of parasitaemia and the blood should be collected when the microfilariae are most numerous—around midnight for most varieties of *Wuchereria bancrofti* (p. 247) and around noon for *Loa loa*. Blood can be collected by pricking the skin of a finger, an ear lobe or the heel of a small child (with minimal squeezing of the surrounding tissue because mixing of tissue fluid with the blood can cause artefacts), or by venepuncture. For diagnosis of the protozoan infections thick and thin films should be made from the blood and stained, usually by a Romanowsky method (e.g. Giemsa, Leishman, Wright or Field). Accurate diagnosis depends on care in the details of preparation and staining of the films. Staining must be carried out at pH 7.2 so that the parasite nuclei are stained red in contrast to blue cytoplasm; staining at pH 6.8 (as in routine haematological staining) fails to give this useful differentiation. If parasites are sufficiently numerous to be seen in thin films, they are most likely to be found at the edges and in the tails of the films. In thick films attention should be concentrated on areas where the leukocyte nuclei are purple (not blue or red), since this is where the differential staining of the parasites will be most marked. Finding the parasites may require 15–20 minutes of careful microscopy. Malarial parasites and babesiae are found inside red cells, whereas trypanosomes (and microfilariae) are extracellular. Observations should include estimates of the relative numbers of different parasite stages present, and in malaria the number of parasites per 100 red cells should be estimated in thin films. The possibility of multiple infection should be borne in mind. Microfilariae are not stained well by Romanowsky procedures. They are best looked for in thick blood films stained with a hot haemalum stain (e.g. Meyer's) in a technique similar to the Ziehl–Neelsen method (p. 103); the films should be washed in alkaline water to turn the nuclei blue.

Trypanosomes and microfilariae can be detected by microscopy of wet unstained blood films, in which they can be seen moving around and agitating the surrounding red cells; or by centrifuging blood in a haematocrit tube, placing the tube under the microscope and examining the plasma immediately above the packed cells. However, neither of these methods allows identification of the parasites. Trypanosomes too few to be seen in the routine thick and thin films may be found in some cases by triple centrifugation (p. 146) before making films. Trypanosomes can also be found by inoculating patient's blood into laboratory animals (in which *T. rhodesiense* and *T. cruzi* nearly always multiply and cause disease but *T. gambiense* does not—p. 146), or into tubes containing Novy–MacNeal–Nicolle or other suitable culture medium. After incubation of such a tube culture the water of condensation is examined microscopically for the presence of parasites—in their 'insect' phase, as trypomastigotes do not develop in culture.

Tissue fluids and biopsies

Aspirates of lymph nodes, for the diagnosis of leishmaniasis, and of chancres (p. 246), for trypanosomiasis in the early stages, can be prepared and stained in the same way as thin blood films. Smears of spleen, bone marrow or liver biopsies for detection of *Leishmania* are prepared by pressing the material between two slides; they are also stained in the same ways as thin films. Although *Leishmania* amastigotes are present only inside macrophages, the preparation of the film may result in the rupture of the parasitized cells and the amastigotes then appear to be scattered in groups between the cells as well as in them. The amastigotes are very small (c. 3 μm) and resemble platelets, but can be distinguished from them by careful examination. In cutaneous leishmaniasis the parasites may be seen in smears prepared from the base of the ulcer, but histological examination of fixed and stained sections of a biopsy taken from the edge of the ulcer is more reliable.

Parasites in faeces and urine

Protozoan and helminth intestinal parasites are detected in stool specimens by examining wet films of faecal suspensions (in 0.85% saline) on microscope slides with ×10 and ×40 objective lenses under reduced lighting.

Practice is required if accurate diagnoses are to be made, and a micrometer (measuring) eyepiece is helpful for the inexperienced observer. Most helminth ova are easily recognised by their size, shape and in some cases their colour (some have bile-stained shells). Protozoan cysts may be difficult to see if the film is too thick, or if the microscope lighting is not properly adjusted. Careful examination of *Entamoeba* cysts is needed to differentiate between *E. histolytica* (about 12 μm diameter) and *E. coli* (about 18 μm); patients with the former require treatment, those with the latter do not (p. 143). Apart from size, the two species can be distinguished by the numbers of nuclei in the mature cysts (*E. histolytica* 4, *E. coli* 8; but both have fewer in immature cysts). Nuclei are more easily counted in a fresh faecal preparation in which 1% iodine solution in potassium iodide has been substituted for saline. Preparations must be examined within 25 minutes because iodine is a progressive stain and overstaining may make observation difficult. The use of iodine is recommended only for differentiation between the two species of *Entamoeba*.

In protozoan infections the speed with which material passes through the intestine affects the development stage at which the protozoa are passed in the faeces; in constipated or normal stools, *Entamoeba* and *Giardia* are usually present as cysts, but in loose or diarrhoeic stools cysts may be few in number and trophozoites predominant. Trophozoites may be found in flecks of blood or mucus if specimens are examined promptly, but they die within 10–15 minutes after the stools have been passed. This is important if invasive amoebiasis due to *E. histolytica* is suspected; this diagnosis can be confirmed only by the identification of amoebic trophozoites containing ingested blood cells in freshly passed stools. It is of no value to keep the specimen warm in an incubator before examination; the trophozoites still die quickly.

More than one parasite is often found in one patient—e.g. multiple infections with *Ascaris*, hookworm and *Trichuris* are common in patients from the West Indies and the Indian subcontinent. The numbers of eggs and, particularly, cysts passed in faeces varies from day to day; therefore several specimens should be examined before a diagnosis is excluded. It may also be important to indicate the relative abundance of cysts or ova in a specimen. Some helminth parasites produce vast numbers of eggs/worm/day (e.g. *Ascaris* 250 000, *Ancylostoma* 10–20 000, *Necator* 5–10 000) and a simple saline film made with about 2 mg of faeces enables a semi-quantitative estimate of the number of eggs present (recorded as eggs/cover-slip or microscope field) to be given. More accurate quantitative methods are rarely necessary.

Several artefacts can mimic eggs or cysts and may mislead the unwary or inexperienced. Fat droplets, small air bubbles, yeasts and fungal spores may look like protozoan cysts; some undigested plant cells (especially some with thickened cell walls) may look like ova. Careful examination will usually reveal differences in size, shape, symmetry, colour or content, but the diagnosis should always be confirmed by examination of a repeat specimen. Some diagnostic or therapeutic measures may interfere with faecal examination: mineral oils, barium, bismuth, antibiotics, antimalarials and non-absorbable antidiarrhoeal preparations may greatly reduce the output of parasites for several weeks (without eradicating the infection), or may intereferre with accurate microscopical identification.

The examination of urine specimens for ova of *Schistosoma haematobium* is similar to the examination of faeces. A specimen of urine should be centrifuged and a wet film of the deposit examined with a ×40 objective lens for the 'egg-shaped' ova, usually coupled with the presence of red blood cells.

Immunodiagnosis

Antigen–Antibody Interactions in the Laboratory

Many experimental and diagnostic microbiological procedures make use of the fact that antibodies 'recognize' antigenic groupings with a precision that is often much finer than that of the most discerning chemical tests. This specificity of antigen–antibody interactions has a number of important practical applications, including the following:

(1) Two substances can be shown to be identical or closely similar by the fact that each reacts with antibodies produced in response to the other, and to some extent it is possible to estimate degrees of dissimilarity between substances by differences of reaction with the same antibodies.

(2) Consequently the identity of a microorganism in culture can be established and its antigenic composition analysed by testing its reactions with antisera of known specificity—that is, sera containing antibodies against known antigens.

(3) Current infection with a particular micro-organism can sometimes be detected by demonstrating that the patient's body fluids contain antigens which react with antisera specific for products of such organisms—e.g. toxins, capsular polysaccharides, hepatitis B virus antigens.

(4) The presence, in human or animal serum, of antibodies specific for components or products of a particular micro-organism is presumptive evidence of infection at some time with that organism or with one that is antigenically related; and if the amount of such antibodies is still increasing, the infection was a recent one.

Some of the techniques for demonstrating these and other interactions are outlined in this section. Many examples of their application to the problems of medical microbiology are to be found later in this chapter and throughout much of this book.

We have referred above, and shall continue to refer, to the use of specific antisera, which have in the past been produced in animals and refined by absorbing out unwanted antibodies. The introduction of monoclonal antibody preparations (p. 56) has largely overcome the major technical problem of immunodiagnosis—that of making antisera of such precise specificity that positive results can be interpreted with confidence as indicating the presence of the appropriate antigens.

Precipitation

A soluble antigen may combine with an appropriate antibody to form a precipitate. This happens maximally when the two meet in approximately **optimal proportions** and are both used up in the formation of large lattice complexes with alternating antigen and antibody layers. In the presence of substantial excess of either, there is a tendency to form much smaller complexes, each possibly consisting of just a single molecule of the scarcer component and enough molecules of the other to use up all of its binding sites; in antigen excess in particular such complexes are still soluble.

A precipitation reaction may be demonstrable by drawing up a small volume of antigen followed by a similar volume of antiserum into a capillary tube; the reaction then manifests itself as a layer of white precipitate near to the interface between the two fluids. This is one of the methods used for Lancefield grouping of streptococci (p. 77).

Alternatively, a **double-diffusion** (Ouchterlony) procedure may be used. Antigen solution and antiserum are placed in two suitably spaced holes cut in a layer of agar gel on a flat glass or plastic surface. Antigen and antibody diffuse out into the gel, and where they meet in optimal proportions a white line of precipitate is formed. Since the position of the line can be brought nearer to one hole by putting less of the relevant reactant into the hole in the first place, this procedure can be made quantitative. Multiple lines may appear if the two fluids contain several sets of antigens and corresponding antibodies. By using three holes it is possible to compare the reactions of, say, one serum and two antigen solutions, and so to arrange things that the lines formed by the two interacting systems meet at an angle. If the two antigens are the same, the two lines will merge into a curve—**the reaction of identity**; if the two are different, the lines will 'ignore' one another and cross. The Elek method for detecting toxin production by *C. diphtheriae* (p. 84) is based on the double-diffusion procedure.

In **immuno-electrophoresis** an electric current is used to separate out the components of an antigen mixture (or an antibody mixture) so that they are located at different points along a line in agar gel on a microscope slide or other suitable surface. Antiserum (or antigen solution) is placed in a trough cut in the gel parallel to and at an appropriate distance from the electrophoresis line. Diffusion is then allowed to take place, and since the electrophoresed components set out from point sources in their line and meet a linear front of the reactant from the trough, precipitate lines are arc-shaped.

Cross-over or **countercurrent immuno-electrophoresis (CIE)** is a rapid and very sensitive variant of double-diffusion, in which an electric field is used to hasten the diffusion of the reactants towards one another. This technique is widely used for detecting antibodies—e.g.

specific fungal precipitins in the sera of patients with fungal infections or allergies—and also for detecting microbial antigens—e.g. pneumococcal polysaccharide in the blood of a patient with pneumonia or in the CSF of one with meningitis. Detection of specific antigens can be particularly helpful when bacterial cultures are negative because antibiotics have already been given.

Agglutination

Bacteria are commonly identified by preparing aqueous suspensions of them on glass slides or in tubes and seeing whether these are agglutinated into visible clumps on the addition of antisera specific for known bacterial surface antigens. Conversely, suspensions of known bacteria can be used to detect and quantitate antibodies in sera. Quantitation is achieved by testing a series of dilutions of the serum against a standard bacterial suspension and determining the highest dilution of the serum which still gives definite agglutination (the titre of the serum, more fully defined on p. 180).

For many purposes agglutination tests are easier and more satisfactory than precipitation tests. For **indirect** or **passive agglutination tests** soluble antigens are made particulate and therefore agglutinable. This is done by allowing or persuading the soluble antigens to adhere to the surfaces of particles that are themselves immunologically inert, at least so far as the test system is concerned. For example, many polysaccharides adhere readily and firmly to the surfaces of washed red blood cells, and many protein antigens adhere similarly to red cells that have been treated with tannic acid or various other agents, or to particles of polystyrene or latex. This is the basis of **passive haemagglutination** (PHA) and **latex agglutination** tests.

Co-agglutination

Staph. aureus strains possessing surface protein A (p. 75) bind to the Fc regions of antibody molecules (p. 52), leaving their Fab regions free to attach to their specific antigens. If these antigens are on the surfaces of bacteria, mixing a suspension of those bacteria with a suspension of *Staph. aureus* carrying appropriate antibodies will result in clumping together of the mixed bacteria (co-agglutination) into visible particles—a procedure widely used for such purposes as Lancefield grouping of streptococci (p. 77) and the identification of gonococci and other organisms.

Capsule swelling

This phenomenon, alternatively known by its German name 'quellung', consists of a change in appearance of the bacterial capsule when it is exposed to a serum containing antibodies specific for the capsular polysaccharide. The change is probably due in part to actual swelling of the capsule and in part to an increase in its refractility. The phenomenon is highly type-specific—i.e. it occurs only if the organism being tested shares a polysaccharide antigen with that against which the antibodies were formed. Accordingly, capsulate strains of *Str. pneumoniae* can be divided into a number of types, differing in their capsular antigens; and a similar process can be applied to the typing of other capsulate species, such as *Haemophilus influenzae* and *Klebsiella pneumoniae*.

Complement fixation

It has been known since the end of the last century that certain antigen–antibody interactions which produce no visible result (such as a precipitate, agglutination or changes in the appearance of organisms) can be detected by the fact that they use up or 'fix' complement so that it is no longer available to take part in other interactions. (See p. 56 for a comment on the use of the word **complement** as though it referred to a single substance.) To demonstrate that such fixation has occurred in a tube which originally contained only a small known amount of complement, another antigen–antibody mixture can be added which will undergo a visible change only if complement is still present in adequate amounts. This is the basis of the **complement-fixation test** (CFT), which is most easily understood if considered in stages.

(1) Into a test tube are placed known amounts of antigen and of the serum to be tested for the presence of an antibody that will combine

with the antigen and fix complement. This serum has previously been heated to inactivate the unknown amount of complement that it contained.

(2) A known amount of complement is added to the mixture, usually in the form of guinea-pig serum. The amount is related to the known amount of antigen present, so that if the serum contains enough antibody all of the complement will be fixed.

(3) After the antigen–serum–complement mixture has been given an adequate opportunity to combine, an indicator system is added. This commonly consists of sheep red cells and rabbit serum containing lytic antibodies for such cells. The serum must of course be heated to inactivate its own complement before it is added to the mixture in the test tube. If the original guinea-pig complement is still present in adequate amounts, the red cells will be lysed:

(a) Antigen + serum containing no antibody — NO COMPLEMENT FIXATION
(b) Complement
(c) Red cells + haemolytic serum
} LYSIS

whereas if the complement has been fixed in the first reaction, no lysis will occur:

(a) Antigen + serum containing antibody
(b) Complement
} COMPLEMENT FIXATION
(c) Red cells + haemolytic serum — NO LYSIS

The result is described as 'positive' if lysis fails to occur (indicating that antibody was present in the serum being tested) and as 'negative' if lysis does occur.

Coombs' test

The presence of non-agglutinating antibodies on the surfaces of particles can be demonstrated by the technique developed by Coombs and his associates. This depends upon the fact that the antibody is of necessity a globulin of a type peculiar to the animal species in question. Particles coated with it are therefore agglutinated by serum of an animal of another species which has received repeated injections of globulin derived from the first species. Thus, for example, the presence of human globulin on particles can be shown by exposing them to the action of the serum of a rabbit which has been stimulated to produce anti-human globulin (AHG) antibodies. Furthermore, if the AHG has been prepared by using purified immunoglobulin of one class only (e.g. IgM or IgG), it will agglutinate the particles only if they are carrying antibody molecules of that class. Originally devised for detection of blood-group antibodies, this technique has found various applications in microbiology— e.g. the exposure of a suspension of brucellae to the serum of a patient with possible brucellosis and then to AHG to see whether they have acquired non-agglutinating antibodies.

Neutralization, blocking, inhibition

Many antibodies can be detected and quantitated by means of their ability to interfere with the actions of micro-organisms or their products. Neutralization of toxin, as tested by comparing the action of susceptible animals of untreated toxin and of toxin mixed with the serum being investigated, is a simple example. Similarly, virus neutralization tests measure the ability of a serum to protect animals or tissue against viruses. Other examples of this interference approach are metabolic inhibition tests, haemagglutination-inhibition tests and immobilization tests, which respectively test the ability of sera to prevent measurable metabolic activities of micro-organisms, to prevent the agglutination of red blood cells of various animal species which normally follows their exposure to certain viruses (p. 170) and to stop the spontaneous movement of flagellate or other motile organisms. Since such interference by antibodies is confined to particular species of organisms or even types within species, similar tests using antisera of known specificity can be used in identification of organisms.

Labelling

Attachment of recognizable labels—radio-isotopes, fluorescent dyes or enzymes—to antigen or antibody molecules is the basis of an increasing number of methods for immuno-diagnosis of infection.

RADIO-IMMUNO-ASSAY (RIA) The amount of radio-labelled antigen bound by a known amount of antibody is decreased in the presence of unlabelled antigen, which competes for binding sites. Quantitation of this competitive inhibition allows precise measurement of the amount of an unlabelled antigen—e.g. hepatitis B surface (Australia) antigen—in a sample of serum or other fluid. A similar system can be used to measure antibodies.

IMMUNOFLUORESCENCE When antibodies labelled with a fluorescent dye (e.g. fluorescein isothiocyanate) combine with antigens in a tissue section or otherwise fixed on a microscope slide, the complex is visible by ultraviolet microscopy (p. 164). Variations of this technique include:

The **direct test**, in which labelled antibody specific for the antigen in question is applied directly to the specimen.

The **indirect test**, in which the specific antibody is unlabelled but is itself the target for labelled antibodies specific for immunoglobulins of the animal species in which it was produced. For example, rabbit antibodies are allowed to attach themselves to their specific antigens and are then located by means of labelled guinea-pig anti-rabbit-immunoglobulin antibodies. Use of this procedure for detection of antibodies (and particular immunoglobulin classes) in human serum is illustrated by the account of the FTA test on p. 182.

The **sandwich test**, in which antibody is detected in a specimen by first treating the specimen with a solution of its corresponding antigen, then washing off any unbound antigen and finally locating the bound antigen as in the direct test.

ENZYME-LINKED IMMUNOSORBENT ASSAY (ELISA) Antigens or antibodies carrying enzyme labels can be located, after they have attached themselves to fixed antibodies or antigens of appropriate specificity, by adding a substrate that undergoes a visible change when exposed to the enzyme. At microscopic level this technique permits detection and localization of antigens in tissues, but on a larger scale changes visible to the naked eye may be used. The ELISA procedure for detecting and measuring antibodies in serum uses antigen firmly attached to a solid phase, such as plastic beads, the inner surface of a tube, or a well in a plastic tray. The serum to be tested is given an opportunity to react with the fixed antigen and is then washed off. Enzyme-labelled anti-immunoglobulin of appropriate specificity (as in the indirect immunofluorescence test described above) is then given a chance to attach itself to any antibody bound by the original antigen. Finally, after a further wash, retention of labelled anti-immunoglobulin by bound antibody is detected by adding enzyme substrate to the tube or well and looking for the visible change (usually of colour). Many modifications of the ELISA technique have been developed. By starting with specific antibody (instead of antigen) fixed to the solid phase, the procedure can be used to detect and measure antigen in serum or other fluids. **Antibody-capture** assays are particularly useful for measuring IgM antibodies; an anti-IgM (light chain) reagent is bound on the solid phase and this captures any IgM present, which is then assayed by allowing it to react with antigen.

Testing for Antigens in Body Fluids

In many infections the causative organisms enter the body and multiply during the prodromal phase to reach a critical concentration around the time of appearance of symptoms and signs referable to the infecting agent. During this early phase the concentration of antigen is high, and it is at this time that immuno-assay for antigen may be helpful in providing an early diagnosis. Although the host begins to produce antibody, it is quickly mopped up by the excess antigen and remains undetectable. However, as soon as antibody appears immune complexes are formed and can be detected in the body fluids (see below). By isolating and splitting these complexes of antigen and antibody it is possible to identify the antigen involved and consequently the organism responsible for the infection. Finally, as antibody production outstrips the supply of antigen, the situation is reversed and diagnosis, albeit retrospective, is based on the detection first of IgM antibodies and then of rising levels of IgG antibodies (see below).

Testing for Immune Complexes in Body Fluids

Methods are available for measuring levels of immune complexes in the body fluids. Serial measurement, particularly of IgG and IgA complexes, may be specially useful in substantiating the diagnosis of infective endocarditis (p. 237) and monitoring the efficacy of antibiotic therapy, since patients with endocarditis have elevated levels of circulating immune complexes, which fall on institution of appropriate therapy. However, many conditions are associated with elevated immune complex levels and normal healthy individuals have detectable circulating immune complexes in their blood intermittently, though not at the levels associated with infection.

Testing for Antibody in Body Fluids (Diagnostic Serology)

Since microbial infection commonly provokes the host to make specific antibodies, the demonstration of such antibodies in a patient's serum is often of considerable diagnostic value—particularly in conditions in which isolation of the causative organism takes a long time or is difficult or even impossible. However, important points to be borne in mind when interpreting the results of antibody determinations include the following:

(1) Sometimes antibodies are found which react with a particular organism or its products although they were formed in response to infection with a different organism. For example, antibodies that agglutinate various *Proteus* strains are formed as a result of infections with some of the rickettsiae (p. 113, the Weil–Felix test).

(2) The duration of detectable antibody responses to different organisms varies. After infection with some species, antibodies may be detectable for many years, and in such circumstances their presence is evidence only of infection **at some time**, not necessarily of **recent** infection. For example, the finding of rubella antibodies in the serum of a woman a few weeks after she had been exposed to rubella during early pregnancy may mean either that she became infected at that time, with the possibility that her fetus was severely damaged, or that she had her infection some years earlier and was not at risk. Two lines of evidence in favour of an infection being recent are the demonstration of a rising titre (see below) and the identification of some of the antibody molecules as IgM, since in many infections antibodies of this class are the first to appear in the serum after a primary antigenic stimulus but may persist for only a few weeks. In other infections IgM antibodies may persist much longer but their presence can be taken to indicate that the disease is still active.

(3) As a corollary of the preceding points about cross-reaction and persistence of antibodies, the significance of a given level of the antibody measured by a particular test cannot be assessed without knowing the levels to be expected in the blood of normal healthy individuals of the same age, habitat and social background as the patient.

(4) In general, antibody responses are not detectable for at least a week, and often not for several weeks, after the onset of an infection with an organism of which the host has no previous experience. Thus examination of a serum sample collected during the first few days of an illness cannot be expected to give useful direct information about the cause of the illness. However, in virtually all acute infections for which specific serological tests are available **it is desirable to collect a serum sample in the acute stage**. This is because demonstration of **a rising level of specific antibodies** is often the best way of identifying a recent infection. If antibodies are found in a serum sample collected 3 weeks or so after the onset of an illness, it may be impossible to say whether they have any connection with that illness and it may be too late to demonstrate a further rise (however, see p. 182). But the situation is much clearer if an acute-stage specimen is available to be tested at the same

time as the later specimen, and contains a significantly smaller amount of the antibodies in question. It is generally accepted that a 4-fold rise in titre is beyond the range of experimental error of most routine procedures, provided that the two specimens are tested at the same time, and is therefore indicative of a true increase in the amount of antibody in the patient's blood. (The **titre** of a serum is the highest dilution of that serum which gives a positive reaction in a given test for antibodies of a given specificity.)

(5) Even a rising titre of specific antibody is not unequivocal evidence that the organism in question was responsible for the illness that is being investigated. Contact with the organism may have occurred coincidentally at about the time of onset of the patient's illness—though this possibility can reasonably be ignored if the patient's condition is strongly suggestive of that which the organism commonly produces. The position may also be obscured by an **anamnestic reaction**—i.e. a rise in the level of a previously formed antibody in response to the non-specific stimulus of some quite different infection. This is particularly well recognized in connection with salmonella H agglutinins, which may increase in amount following a wide variety of unrelated febrile illnesses.

(6) When a patient is suspected of suffering from an illness in which detectable antibodies are usually formed, but no appropriate antibodies are found in his blood several weeks after the onset of his illness, considerable doubt is thrown upon the diagnosis. However, immunocompromised patients—e.g. those on immunosuppressive drugs or suffering from an immunodeficiency syndrome such as AIDS—may fail to make detectable antibody despite extensive infection.

(7) Detectable antibodies are not necessarily protective antibodies, and therefore the results of serological tests may have nothing to do with the patient's immunity or lack of it. This is the case in many protozoan and helminth infections.

The use of serological tests is referred to in many places in this book in relation to the diagnosis of many kinds of infection. Here we shall describe in some detail the serological tests used in the diagnosis of enteric fever (the Widal test), of brucellosis and of syphilis, since these account for most of the serological workload of many diagnostic bacteriology laboratories (though brucellosis is of rapidly decreasing importance in Britain) and illustrate many general principles. We shall also summarize the use of serological tests in diagnosis of infections due to viruses, fungi, protozoa and helminths, and in the elucidation of 'pyrexia of unknown origin'.

The Widal Test

In many cases of fever it is necessary to investigate the possibility of enteric fever or other salmonella infection. The serological part of such an investigation is measurement of the ability of the patient's serum to agglutinate various salmonella suspensions.

THE TEST The nature of the three types of salmonella antigen—H, O and Vi—has been indicated on pp. 92 and 93. Standard bacterial suspensions can be so prepared that each is of known agglutinability by antibodies to one of these antigen types only. In Britain, H and O suspensions of *S. typhi* and *S. paratyphi* B are routinely used in the test, sometimes accompanied by suspensions made from other salmonellae, especially when it is suspected that one of these is the cause of the patient's illness. Use of *S. typhi* Vi suspension as a means of detecting typhoid carriers has a long but unimpressive history, but purified Vi antigen preparations now available may provide the basis for a more reliable test for this purpose.

Serial dilutions of the patient's serum in saline are pipetted into special small tubes, one complete set of dilutions being prepared for each bacterial suspension that is to be used in the test. An equal volume of the appropriate bacterial suspension is then added to each tube. Control tubes are also set up, containing suspensions mixed with saline instead of serum, to ensure that the bacteria do not agglutinate spontaneously.

H and O agglutination tests are incubated in a water-bath at 37°C (or at 50–55°C). Floccular agglutination of the H suspensions becomes apparent within 2 hours or so, whereas the more granular agglutination of O suspensions takes from 4 to 24 hours. The results of the tests are expressed as antibody titres, as defined above.

INTERPRETATION Low levels of antibodies that agglutinate salmonella suspensions are common in the blood of patients with no history of relevant illness or of immunization. The picture is greatly confused by previous typhoid (or typhoid + paratyphoid) immunization, and it is important that all relevant information on this matter should accompany any request for a Widal test. Following such immunization, high levels of H agglutinins may be present for many years, and rising H agglutinin titres are of doubtful significance because they may represent anamnestic reactions (p. 180). Similarly, after natural infection the H agglutinin level may be high and variable for many years, but in this case it is only the H suspensions belonging to the appropriate species which is agglutinated or those which are antigenically related to it. O agglutinin levels, on the other hand, remain high for only a few months after immunization or natural infection, and are less liable to anamnestic rises. Furthermore, O agglutinins are formed more rapidly and more constantly than H agglutinins following typhoid or other salmonella infections, usually being detectable in the serum by about the tenth day of the illness. For these reasons the most reliable serological evidence of a salmonella infection is a fourfold or greater increase in the O agglutinin level between the first week and the second or later weeks of the illness. However, because there is considerable sharing of O antigens between salmonellae, a rise in the H agglutinin level may give more precise information as to the particular organism causing the patient's illness. Absence of agglutinin response does not exclude the possibility of typhoid, as occasionally patients fail to produce detectable levels of antibodies.

Serology tests for brucellosis

Where brucellosis is a possible cause of persistent pyrexia, agglutination tests using suspensions of the appropriate brucella species can be carried out in conjunction with the Widal test. The technique is essentially that of the Widal test for O agglutinins. Neither agglutination tests nor any of the other tests mentioned below can be relied upon to establish firmly the diagnosis of brucellosis, since antibody levels found in the blood of some apparently healthy people, notably of farm workers and others who have long been exposed to risks of brucella infection, may exceed those found in some patients with undoubted brucellosis. If the patient is seen early in the illness, it may be possible to demonstrate that his blood contains IgM antibodies, and such a finding strongly supports the diagnosis of brucellosis. Demonstration of a rising antibody titre may also be possible at this stage, and is a help to diagnosis. But because the onset of this disease is often insidious, the IgM antibodies may have disappeared and the initial IgG antibody rise may be over before the patient presents for investigation.

Serological tests for syphilis

NON-SPECIFIC OR REAGIN TESTS For well over half of this century the Wassermann reaction (WR) was the mainstay of the laboratory diagnosis of syphilis. It is a non-specific complement-fixation test using as 'antigen' a preparation of cardiolipin from ox heart which reacts with an IgG serum component called **reagin** (though it is not related to the similarly named IgE antibodies mentioned on p. 61). The WR has been replaced by the **Venereal Diseases Research Laboratory** (VDRL) test and its variant, the **Rapid Plasma Reagin** (RPR) test, in which the reaction between cardiolipin and serum is seen as a floccular precipitate on a slide. Reagin tests become positive within 2–3 weeks of infection, and revert to negative far more rapidly than the specific tests (see below) after effective treatment. 'Biological false positive' results may occur in many infections, notably in malaria, leprosy, tuberculosis, leptospirosis and hepatitis, in chronic diseases such as rheumatoid arthritis and disseminated lupus, after various immunizations and in pregnancy. Such false positives are usually transitory, except in certain chronic diseases such as rheumatoid arthritis and disseminated lupus.

TESTS FOR SPECIFIC ANTIBODIES The **fluorescent treponemal antibody** (FTA) and the ***T. pallidum* haemagglutination** (TPHA) tests are the most widely used of these. In the FTA test, serum from the patient is applied to a fixed smear of dead *T. pallidum* on a microscope slide and any antitreponemal antibodies are detected by treatment with rabbit serum containing fluorescent anti-human globulin (p. 177). The test is made specific for *T. pallidum* by first absorbing from the patient's serum antibodies that react with treponemes in general, not specifically with *T. pallidum* (FTA-Abs). The FTA becomes positive early in the disease. It can also be used to detect IgG and IgM antibodies separately. In an adult, IgM antibodies indicate active infection; in an infant, IgM indicates infection of the infant itself, because IgM does not cross the placenta. The TPHA test is a passive haemagglutination test in which red blood cells (from animals or birds) are coated with *T. pallidum* extract and are agglutinated by sera if they contain anti-*T. pallidum* antibodies. The ***T. pallidum* immobilization** (TPI) test is restricted to reference laboratories because it utilizes live treponemes, propagated by serial intratesticular passage in rabbits. If serum from a patient with syphilis, plus guinea-pig serum, is added to a suspension of motile treponemes, the organisms cease to move. This is the most sensitive and most specific serological test for syphilis, but is invalidated if the serum contains antibiotics or other substances that damage the treponemes.

Serological tests for virus infections

In virology, even more often than most other branches of microbiology, tests with known antigenic preparations are used to detect and measure antibodies in patients' sera, since it is often easier to establish a serological diagnosis than to isolate the virus. The principles outlined on p. 179–80 all apply, and it is particularly important to examine **paired serum specimens**, one taken at the beginning of the illness and one taken a few weeks later. This is commonly forgotten by those in charge of the patients. There is no difficulty about remembering the acute-stage specimen, when the patient is ill and the doctor is worried; but that specimen is worthless without a second, collected in most cases when all anxiety is passed and the patient may no longer be in hospital or under the doctor's care. A virologist's deep-freeze is often a repository for acute-phase sera which wait in vain for their partners! When no acute-stage specimen has been taken, a single specimen taken later may be of value, either by giving negative results and thus casting doubt on or excluding the suspected diagnosis, or by showing a level of antibodies (notably IgM—p. 179) high enough to suggest recent infection. In some diseases demonstration of an appropriate timed **fall** in antibody level may be confirmation of the nature of a recent illness.

The tests described below are among those most widely used in the serological diagnosis of virus infections.

COMPLEMENT-FIXATION (CF) TESTS (p. 176). As antigens, fluids from tissue cultures or from the allantoic cavities of hens' egg cultures of appropriate viruses are commonly used. Tests using such antigens are in general less precise in their specificity than are neutralization and haemagglutination-inhibition tests, and these latter methods are more valuable for distinguishing between related virus types.

NEUTRALIZATION TESTS (p. 177). The theory of these is simple. The activity of a virus suspension can be demonstrated, according to the nature of the virus, by inoculating it into animals, into hens' eggs or into tissue cultures. If a serum sample contains adequate amounts of neutralizing antibody for the virus in question, then a dose of virus suspension + serum will fail to produce the effect which is produced by the suspension alone. The amount of antibody can be determined by discovering how much the serum can be diluted before it ceases to neutralize the suspension. Antibodies in the blood of patients or of immunized animals will efficiently neutralize only those viruses that are identical with or closely related to the one that provoked their formation. Conversely, virus isolates can be identified by their susceptibility to neutralization by antisera of known specificity.

HAEMAGGLUTINATION-INHIBITION (HAI) TESTS
These are similar in principle to the neutralization tests, but the activity of the virus which is studied and which is inhibited by specific antibodies is the agglutination of red cells. Such tests can be used to detect antibodies against any of the assorted viruses that cause haemagglutination (p. 170); and, like neutralization tests, they can also be used to identify specific virus.

Serological tests for fungus infections

Detection and measurement of antibodies may be useful in the often difficult task of diagnosing systemic fungal infections (deep mycoses), but interpretation may be problematical. For example, in candida infection of a prosthetic heart valve it may be difficult or impossible to isolate the organism from the patient's blood and much easier to detect specific precipitins or agglutinins; but as *C. albicans* is part of the normal body flora the significance of these antibodies may be hard to determine. Antibodies to aspergilli, on the other hand, are always an abnormal finding. In the respiratory allergies due to inhalation of fungal spores the antibodies formed, whether IgE (measured by a radio-allergosorbent test, RAST) or IgG precipitins (detected by gel diffusion, CIE, RIA or ELISA) indicate both the nature of the offending fungus and the type of allergy. In extrinsic allergic alveolitis serial measurements of the circulating levels of immune complexes give an objective assessment of the progress of the disease and the efficacy of treatment, whereas without them the doctor is largely dependent on the patient's own impressions.

Serological tests in pyrexia of unknown origin (PUO)

The patient with persistent pyrexia and no clear indication as to its cause is a common diagnostic problem. Serological tests may provide vital clues. Tests which it is reasonable to carry out in such circumstances in Britain include the Widal and brucella-antibody tests already described in this chapter and those which we have mentioned in earlier chapters in connection with leptospirosis, toxoplasmosis and Q fever, with the addition, in influenza-like illnesses, of tests for influenza and other viruses capable of producing such illnesses, and for psittacosis and mycoplasma infection. Special circumstances or features of the illness may suggest other possibilities. (See also p. 245.)

Serological tests for protozoan and helminth infections

Serological tests are widely used in the diagnosis of invasive amoebiasis, toxoplasmosis, leishmaniasis, Chagas' disease, schistosomiasis, cysticercosis, hydatid disease, trichinosis and malaria. Complement fixation, indirect haemagglutination and immunofluorescence tests are used for the diagnosis of toxoplasmosis. Indirect haemagglutination and indirect fluorescent antibody tests can be useful in patients with chronic malarial infection. Napier's formol gel test is a useful screening test for leishmaniasis but what it detects is greatly increased levels of serum globulin, regardless of antigen-specificity, and it may give positive results in malaria, tuberculosis, leprosy and some other diseases. Indirect fluorescent antibody and direct agglutination tests are more specific, but the latter does not differentiate between the various *Leishmania* species. Complement fixation and agglutination tests are useful in *T. cruzi* infection (Chagas' disease); the complement fixation test is particularly valuable in the chronic stages of the disease, whereas direct agglutination provides a sensitive test in the acute stage and the indirect haemagglutination test is said to be both sensitive and highly reproducible.

Skin Testing

Tests for immunity

The **Schick test**, used extensively in the past but much less now, is a means of assessing immunity to diphtheria, and in particular of determining which people in a group about to be immunized do not in fact need this. It depends on injection of a standard minute amount of active diphtheria toxin into the skin of one arm and of heat-inactivated toxin (toxoid) into the other arm. In the non-immune the toxin causes an erythematous reaction in the skin in the next few days, whereas the toxoid fails to do so. The subject who has a protective level of circulating specific antitoxin does not develop such a reaction. One who is hypersensitive to the material injected and therefore liable to react adversely to immunization will react to both the toxin and the toxoid—but with a more rapid and in general more evanescent reaction than the true positive result. A similar procedure, the Dick test, was used in the past to test for immunity to the streptococcal erythrogenic toxin of scarlet fever.

Tests for hypersensitivity

THE TUBERCULIN TEST The theoretical basis of tuberculin testing will be discussed on p. 207. It depends on a Type IV hypersensitivity reaction.

It is important to appreciate certain features in which the tuberculin test is fundamentally different from the Schick test:

(1) A positive reaction, as described under (3) below, to the injection of tuberculin is due to the host's acquired hypersensitivity, not to the properties of tuberculin itself.

(2) Whereas a positive Schick reaction indicates lack of immunity and a negative reaction indicates immunity, it is the positive tuberculin reaction which is associated with immunity. This association is only an indirect one, however, for there is no clear evidence that tuberculin hypersensitivity is itself part of the mechanism of immunity. All that can be said is that those who have had and overcome tuberculous infections (natural or resulting from BCG vaccination) generally have some degree of immunity to further infection; and, therefore, since tuberculin hypersensitivity indicates past tuberculous infection, it also indicates probable immunity.

(3) The important component of a positive tuberculin rection is palpable induration, which is maximal two or three days after injection of the tuberculin. Erythema also occurs but is in part non-specific and is difficult to interpret.

Tuberculin testing of human patients can be carried out in various ways. In the **Mantoux test** a standard amount of tuberculin—Old Tuberculin (OT) or Purified Protein Derivative (PPD)—is injected intradermally by means of a syringe and needle. Since some patients, especially those with active tuberculosis, may give very strong reactions, it may be necessary to start with a smaller dose and to repeat the test using stronger solutions if indicated. The test is read after 72 hours, induration of greater than 6 mm in diameter being regarded as positive. The inconvenience of the Mantoux test for screening large numbers of people led to the introduction of speedier multiple-puncture techniques. The **Heaf test** employs a number of very short needles mounted on a spring-loaded device which drives them into the skin through a drop of tuberculin solution placed on the skin in the appropriate site. The needles are sterilized by flaming and allowed to cool before re-use. Although this procedure sounds formidable when described, it is in fact more acceptable to children than the Mantoux test and is less liable to produce excessive reactions. The test is read after four to seven days and reactions are graded from 0–4. Positive reactions (grades 2, 3 and 4) range from multiple small papules at the site of individual punctures to a zone of induration including all of the puncture sites and the surrounding skin. The **tine test** is a variant of the Heaf test that uses a disposable multiple-puncture instrument with a coating of dried tuberculin on its tines (prongs). It is readily portable and simpler to use than the Heaf gun but is less reliable and so not recommended for routine use.

Purposes for which tuberculin testing of human subjects is useful include:

(1) **Diagnosis** of individual patients. The test is positive in all cases of active tuberculous infection except those which are very early or of overwhelming severity. However, its diagnostic value is limited by the fact that it is also positive in a high proportion of older children and adults who are not suffering from active tuberculosis, these being people with healed primary infections (natural or from BCG vaccination). A negative reaction, contraindicating the diagnosis of tuberculosis, is often helpful to the clinician;

and a very strong reaction suggests active tuberculosis.

(2) **Detection** of foci of infection. In the absence of widespread vaccination, a high incidence of positive reactions in the children of any social unit—e.g. a family or a school—suggests the presence of an active disseminator of the disease.

(3) **Surveys** of population groups to determine the frequency of tuberculosis in the communities which they represent. This use also depends upon the situation not having been obscured by widespread vaccination.

(4) **Selection** of subjects (new-born infants excepted) for BCG vaccination. In the absence of any better criterion of immunity a negative tuberculin reaction is taken as indicating that vaccination is required. However, if there is a visible BCG scar (greater than 4 mm) from a previous vaccination, re-vaccination is unnecessary even if the tuberculin test is negative.

(5) **Assessment** of a patient's capacity for cell-mediated immunity.

Use of tuberculin testing in relation to BCG vaccination is discussed on p. 365.

Tuberculin testing of cattle, with elimination of those giving positive reactions, is the basis of the creation of tuberculosis-free herds, which has played an important part in eliminating the risk of human infection with tubercle bacilli of the bovine type.

OTHER TESTS Many other microbial antigens can be used to detect delayed-type hypersensitivity as an indication of infection, and also some helminth antigen preparations, such as hydatid fluid used in the Casoni test; but their general limitation is that they indicate only that the patient has at some time been exposed to relevant antigens, and may throw little light on his present condition. Their usefulness therefore depends very much on the frequency with which positive reactions, due to past exposure or subclinical infection, are found in healthy members of the community. For example, a positive histoplasmin test (p. 139) might be informative in some circumstances but is of no diagnostic value in those parts of Kentucky where it is positive in 95% of people!

Suggestions for Further Reading

Stokes, E.J. and Ridgway, G.L. 1980. *Clinical Bacteriology.* 5th edn. Edward Arnold, London.

Lennette, E.H. *et al.* 1985. *Manul of Clinical Microbiology.* 4th edn. American Society for Microbiology, Washington DC.

Also Section III of *Bacteriology Illustrated,* see p. 110.

PART III
Infections of Man

Mouth and Upper Respiratory Tract

Infections of the mouth and upper respiratory tract are common in all age groups and are responsible for a large proportion of general-practitioner consultations. Most are of little significance, but some are serious illnesses in themselves (e.g. diphtheria) and others may give rise to serious diseases by stimulation of the host's immune system (e.g. rheumatic fever or acute glomerulonephritis after a streptococcal sore throat), or may indicate the presence of some serious underlying condition (e.g. candidiasis in an immunosuppressed patient). From the microbiologist's point of view, the upper respiratory tract comprises all the air passages from the anterior nares to the larynx; this is not the definition used by respiratory physiologists and physicians and by pathologists, but is more appropriate for consideration of infection. This chapter deals with infections of the mouth, nose, nasopharynx and oropharynx, paranasal sinuses, middle ears and mastoids. Infections may be caused by bacteria, viruses (including those responsible for such well-known infections as the common cold and infective sore throat) or fungi. The upper respiratory tract is also the portal of entry of many organisms that cause generalized systemic infection—e.g. the meningococcus and the viruses of influenza, measles and rubella.

Normal Microbial Flora

All of the sites discussed in this chapter are heavily colonized by a normal commensal microbial flora. These organisms are generally harmless and often beneficial, but in appropriate conditions they may become opportunist pathogens and cause serious disease. Staphylococci are an important part of the normal flora of the anterior nares; these are usually coagulase-negative species ('*Staph. albus*') but about 30% of people in general and up to 60% of patients and staff in hospitals are carriers of potentially pathogenic *Staph. aureus*. Most carriers have no symptoms, but some suffer from recurrent and troublesome intranasal boils. Non-pathogenic coryneform gram-positive bacilli are also part of the normal nasal flora. Potential pathogens that may be carried include *Haemophilus influenzae* and *Streptococcus pneumoniae*, and also *Str. pyogenes*; nasal carriers of the latter are particularly important because they disseminate it in the environment more efficiently than do pharyngeal carriers.

Many species of micro-organisms are normal inhabitants of the mouth; saliva contains c. 10^8 bacteria/ml. The predominant species in the saliva, on the tongue and on the buccal mucosa are α-haemolytic streptococci ('*Str. viridans*') such as *Str. salivarius*. Lactobacilli are also found there and up to 50% of healthy people carry small numbers of *Candida albicans*. Anaerobic bacteria that may be found in the saliva—fusobacteria, bacteroides, veillonellae and gram-positive anaerobic cocci—are derived from their normal oral habitat in the gingival crevice, where they form a major part of subgingival dental plaque (p. 191). The species of gram-negative anaerobic bacilli present in the normal oral flora include members of the melaninogenicus–oralis group of *Bacteroides* (p. 102), *Fusobacterium nucleatum* and various other fusobacteria that have not been well characterized. (These anaerobes differ from the gram-negative anaerobic bacilli of the lower gastro-intestinal tract, where the fragilis group of *Bacteroides* predominates). Actinomycetes, including *Actinomyces israeli*, and spirochaetes are also integral parts of the subgingival flora; they can be seen by microscopy of films prepared directly from plaque samples, but are rarely cultured.

Many of these species are present also in the normal pharyngeal flora, together with commensal neisseriae, *Branhamella catarrhalis* (the gram-negative diplococcus formerly called *N. catarrhalis*), haemophili including *H. influenzae* (usually non-capsulate) and *H. parainfluenzae*, and *Str. pneumoniae*. Some healthy people are pharyngeal carriers of *Str. pyogenes* and some carry meningococci.

Oral Infections

Virus Infections

Herpes simplex

The mucocutaneous junction of the lip margin and the skin around the nares are common sites for the lesions of herpes simplex virus (HSV) infection (p. 130). Primary infection with this virus usually occurs in childhood and causes few symptoms other than a mild fever; but it may cause **herpetic (aphthous) gingivo-stomatitis**— an acute febrile illness in which vesicles form on the oral mucosa and rapidly break down to leave shallow bleeding ulcers. Despite antibody production and clinical recovery, the virus persists in a latent state within nerve cells, commonly those of the trigeminal ganglion, to re-appear as recurrent **herpes simplex** (alternatively called **herpes labialis** when it is in its commonest site, on the lips). This is a very painful condition in which vesicles form, break down and become crusted. The many factors that disturb the host–parasite equilibrium and induce the latent virus to form the active lesions include intercurrent febrile infections, such as pneumonia, and all forms of immuno-suppression. Herpes lesions usually recur at the same site in any individual. Almost all oral herpes infections are caused by type 1 virus, whereas most genital lesions (p. 282) are caused by type 2.

The diagnosis of herpes simplex is usually made on clinical grounds, but a confirmatory laboratory diagnosis may be required, especially in severely ill immunosuppressed patients with widespread oral ulceration. The virus may be identified by immunofluorescence microscopy of material from fresh vesicular lesions, or typical herpesvirus particles may be seen by electron-microscopy. The virus can be grown in tissue culture, producing intranuclear inclusions and a cytopathic effect. A rising antibody level may be of diagnostic value in primary infections, but most healthy adults already have antibodies. Vesicular fluid can be collected directly on to grids for electron-microscopy, and smears for immunofluorescence can be prepared either directly or from a swab placed in virus transport medium, which is also the appropriate specimen for the isolation of the virus. Topical acyclovir (p. 135) is the treatment of choice for moderately severe oral herpes.

Herpangina

This acute infection of the oral and pharyngeal mucosa, usually affecting children, is caused by group A coxsackieviruses (p. 119). Vesicles form over the palatal, buccal and pharyngeal mucosa and break down to leave shallow painful ulcers. The child is febrile and miserable, but the illness lasts for only a few days and requires no specific therapy. Laboratory diagnosis is rarely necessary, but the virus can be isolated by tissue culture and by intracerebral inoculation of new-born mice.

Hand, foot and mouth disease

This more generalised illness of young children is also caused by group A coxsackieviruses; it should not be confused with foot-and-mouth disease of animals. It is characterized by a febrile illness with vesicles on the soles of the feet, palms of the hands and the oral mucosa, particularly over the hard and soft palates. It occurs in epidemics amongst children in nursery and primary schools. Virus isolation helps to establish the particular strain causing the infection and to trace the spread of the virus, but for the individual child recovery is uneventful and there is no specific treatment.

Bacterial Infections

Dental caries

Although not often thought of as an infection, caries is in fact the result of bacterial action upon the host. The sequence of demineralization, destruction and cavitation of enamel and dentine results from acid production by oral bacteria, principally streptococci, in the build-up of dental plaque that occurs when oral hygiene is poor. The principal bacteria in supragingival plaque are streptococci of the viridans groups, especially *Str. sanguis* and *Str. mutans* (p. 78); these produce dextran, a highly adhesive glucose polymer which forms over the tooth surface an adherent coat containing bacteria, bacterial products and some food particles. Initially the occlusal surface and, later, the approximal dental surfaces inaccessible to cursory brushing are at particular risk of plaque formation, and are the common sites of caries. The streptococci produce large amounts of lactic acid from dietary carbohydrates, especially mono- and di-saccharides that diffuse readily into plaque. Demineralization of the enamel and dentine by this acid is the first stage of caries. Dental caries, therefore, is a result of bacterial metabolism; its development is encouraged by poor oral hygiene and by a diet rich in sugars, especially if high sugar concentrations are maintained for long periods. The avoidance of these predisposing factors is the key to its prevention.

Gingivitis and periodontal disease

Inflammation of the gingiva (acute gingivitis) and chronic periodontal disease are parts of a spectrum of diseases that affect the gums and tooth sockets, and are the most common reason for premature loss of teeth. Gingival inflammation with pain, ulceration, bleeding and the production of a purulent exudate was described by Plaut in 1894 and by Vincent in 1896; this condition, now called **acute ulcerative gingivitis**, is closely associated with the throat infection known as Vincent's angina (p. 196). Continuing neglect of oral hygiene results in chronic inflammation leading to destruction of the interdental papillae, recession of the gingival margin and deepening of the gingival crevice, with the formation of pus-filled gingival pockets and abscesses covered by a pseudomembrane; there is destruction of the collagen fibres that anchor the tooth to the jaw, and resorption of bone from the alveolar margins. The receding gingival margins are inflamed and irregular, purulent exudate can be seen in the subgingival pockets and the patient usually has a foul breath. Plaut and Vincent described two bacteria characterisically seen in smears prepared from gingival exudate—a coarse spirochaete (*Borrelia vincenti*) and a fusiform organism that has been variously named but is now classified as *Leptotrichia buccalis*. Both are anaerobes and neither is readily cultivated. Since the original descriptions it has been generally accepted that acute ulcerative gingivitis and periodontal disease are due to synergic fuso-spirochaetal infection, but there is evidence that oral bacteroides, in particular *Bact. gingivalis* (p. 102), are essential to the infection and may be the specific cause of tissue damage. However, laboratory diagnosis still depends upon the recognition of spirochaetes and fusiform organisms in smears prepared from the pseudomembrane or the periodontal pocket. The pathogenesis of the condition is only partly clear; poor oral hygiene allows subgingival plaque to build up, and this, unlike the supragingival plaque described in connection with caries, contains predominantly anaerobic bacteria. The anaerobes initiate local tissue damage and also stimulate the host phagocytic and immunological mechanisms. This combination of bacterial and host factors causes the clinical and pathological features of gingivitis and periodontal disease. Treatment consists of antimicrobial therapy to control the anaerobic bacteria (metronidazole is the drug of choice) and improved oral hygiene to prevent relapse.

Dento-alveolar abscesses

Oral bacteroides of the melaninogenicus–oralis group, fusobacteria and α-haemolytic streptococci are the species usually isolated from dento-alveolar abscesses; these may develop when infections of the dental pulp and root canal spread into peri-apical tissues and cause necrosis of surrounding tissue and bone. Such abscesses may be acute or chronic (pyogenic granuloma), and often involve several teeth; osteomyelitis and soft-tissue destruction may cause large abscesses that discharge to the outside or into the mouth or nose. Diagnosis is based on clinical and radiological findings. Treatment often necessitates tooth extraction for drainage. The post-extraction syndrome ('dry socket') may follow extraction of an infected tooth, or may develop in a clean socket. It is an anaerobic, possibly synergic infection that causes pain, clot destruction, cellulitis and a foul taste.

Bacteria from these oral infections may enter the blood stream, and cause abscesses in the brain (p. 225) or elsewhere; or as a result of inhalation they may cause necrotizing pneumonia, lung abscesses and empyema (p. 206). Anaerobes are of predominant importance in such complications.

Oral Candidiasis (Thrush) (p. 137)

Candida (usually *C. albicans*) infection of the oral mucosa is commonest in infancy. It is also common in adults when there are predisposing local factors, such as ill-fitting dentures; when they have underlying debilitating conditions such as diabetes or leukaemia or other malignant diseases; or when they are receiving steroids, immunosuppressive agents or broad-spectrum antibacterial drugs that suppress the normal flora and allow the yeasts to flourish. This painful and often distressing infection is characterized by the presence of numerous adherent white or creamy plaques on intensely red and inflamed oral or pharyngeal mucosa. In acute pseudomembranous candidiasis the soft plaques are easily rubbed off the mucosa. In acute atrophic candidiasis, which is usually provoked by broad-spectrum antibiotic therapy, the plaques are scanty and the mucosa is red and swollen, especially over the tongue. In chronic atrophic candidiasis (denture stomatitis), found underneath dentures, the mucosa is erythematous but plaques are rare.

Most of these infections are endogenous, but infants may acquire *Candida* from their mothers' vaginas at birth. The diagnosis is confirmed by seeing yeast cells and elongated pseudohyphae in stained films prepared from the mouth lesions and by growing *C. albicans* on Sabouraud's agar. Treatment requires measures to eliminate the predisposing factors, together with oral administration of one of the non-absorbed antifungal agents—nystatin, miconazole or amphotericin B.

Upper Respiratory Tract Infections

Common Cold (Coryza)

The commonest group of upper respiratory tract infections, responsible for an enormous amount of minor illness and short-term absence from work, is that encompassed by the term 'common cold'; local virus infection of the nasal mucosa is responsible for the characteristic irritation and watery nasal discharge. In the later stages, the discharge becomes thick and purulent but there is seldom evidence of specific bacterial superinfection and antibiotics are usually not beneficial. Most colds are caused either by rhinoviruses (p. 119), of which there are more than 100 serotypes, or by coronaviruses (p. 127), of which six serotypes have been described. Infection spreads by the airborne route and there is a high attack rate. There is no specific treatment for the common cold; prevention by immunization, though effective with individual strains, is of little value because immunity is strain-specific and the serotypes replace each other as immunity develops. A minority of cases of coryza are caused by other viruses from a wide range that include paramyxoviruses (p. 126), influenza B and C viruses (p. 125), respiratory syncytial virus (p. 127), some adenoviruses (see below) and in a few cases coxsackieviruses or echoviruses.

Laboratory diagnosis is rarely necessary, but the viruses may be isolated from nasal swabs or washings inoculated on to appropriate tissue cultures.

Sore Throat

Sore throat ('pharyngitis', 'tonsillitis'), often with associated laryngitis, is another common reason for medical consultation. Most cases are caused by viruses, with bacteria being primarily responsible for many of the others. Clinical differentiation between these two groups is unreliable, and a laboratory search at least for the likely bacterial causes is often desirable.

Virus infections (mainly of the throat)

The virus groups mentioned above as commonly causing coryza can also cause sore throats, but the rhinoviruses in particular rarely cause more than mild local discomfort with little or no systemic disturbance. In contrast, adenoviruses (notably types 3, 4, 7, 14 and 21—see p. 129) can cause more severe upper respiratory tract infections, particularly of the throat, often with generalized febrile illness ('febrile cold'). Adenovirus infection can be confirmed by isolation of the virus in tissue culture, and retrospectively by using complement-fixation tests to demonstrate antibody rises.

Infectious mononucleosis (glandular fever) (See also pp. 133 and 245)

Sore throat is a common presenting feature of this disease, which in about 80% of cases is caused by the Epstein–Barr (EB) virus (p. 133)—most of the other cases being due either to another herpesvirus (CMV, p. 132) or to *Toxoplasma gondi* (p. 151). The acute exudative tonsillitis or pharyngitis with which patients often present has features suggestive of a bacterial infection and may be treated with antibacterial drugs—with no benefit but possible unpleasant results (see below). Other main features are low to moderate fever with weakness and malaise, cervical and often generalized lymphadenopathy with splenomegaly, mild hepatitis (in some cases) and sometimes a rash. A quite different type of rash, florid and irritant, can be induced in almost all patients who are given ampicillin or related drugs, and may occur when other antibiotics are used. Serious complications of infectious mononucleosis are rare if the immune mechanisms are normal (p. 133), and the ultimate prognosis is good; but the illness can be tiresomely persistent and there is no specific treatment. Steroids are sometimes used to treat massive tonsillar enlargement. Prolonged excretion of EB virus in the saliva after clinical recovery (p. 133) accounts for infectious mononucleosis being known as the 'kissing disease', especially as it is most likely to occur in those who do not acquire their primary EB virus infection until they are 15-25 years old; it is rare in children and in later life.

Laboratory diagnosis of infectious mononucleosis depends on finding the characteristic abnormal mononuclear cells in the blood and on serological tests. The Paul–Bunnell test in its original form determined in tubes the presence and amount of heterophil antibodies (p. 133) in the patient's serum, and their pattern of absorption from the serum; those typical of infectious mononucleosis can be removed by ox red cells but not emulsified guinea-pig kidney. Rapid slide-tests based on similar principles are now used, at least for initial screening. Heterophil antibodies are detectable early in the illness, and their presence indicates fairly reliably that EB virus is responsible for the illness (p. 133). Early diagnosis is helpful in protecting the patient from antibiotic treatment. Specific serological tests can eventually establish EB virus, CMV or *Toxoplasma gondi* as the aetiological agent in nearly every case.

Streptococcal sore throat

The most important common bacterial cause of sore throat in developed countries is the β-haemolytic streptococcus of Lancefield's group A (*Str. pyogenes*). This organism typically causes an acute follicular tonsillitis, with flecks of pus seeping from the crypts of infected and enlarged tonsils; oedema and inflammation extend to the fauces and soft palate; and the cervical lymph nodes are enlarged and tender. The general systemic effects include fever, malaise and a neutrophil leukocytosis. The infection may extend to produce a peritonsillar abscess (quinsy), mastoiditis or sinusitis. Although the typical picture readily suggests a bacterial infection, some streptococcal sore throats are clinically indistinguishable from those commonly caused by virus infections, and some virus sore throats resemble typical streptococcal infections. Unfortunately it is impossible to tell from the severity or other clinical features of a streptococcal sore throat whether the patient is at risk of developing 2–6 weeks later the immunological sequelae that sometimes follow such an infection—rheumatic fever and acute glomerulonephritis.

Rheumatic fever has in recent times become uncommon in Britain and others of the more developed countries, but remains common elsewhere—notably in communities with poor living conditions and overcrowding. Although these factors facilitate streptococcal transmission, that is not an entirely satisfactory explanation of the epidemiological differences. The disease typically begins 2–3 weeks after an attack of sore throat, as a fever of acute onset with pain and swelling affecting one or more joints and 'flitting' from joint to joint during the illness; there is also a carditis which is the most serious part of the disease. All layers of the heart may be involved (pancarditis): myocarditis is invariably present and may cause acute heart failure; pericarditis may lead to a pericardial effusion; and endocarditis may damage the heart valves, causing serious long-term effects. The acute illness usually resolves spontaneously, and this process may be hastened by treatment with salicylates. Later, however, many patients are disabled by chronic valvular disease (stenosis and/or incompetence) as a result of the endocardial damage. The mitral valve is most commonly affected, followed by the aortic valve.

Rheumatic fever is a hypersensitivity reaction initiated by protein and polysaccharide cell-wall components of *Str. pyogenes* (which need not be of any particular Griffith type—p. 78); the antibodies formed cross-react with antigens in the myocardium and endocardium, especially the valvular endocardium. Histologically, rheumatic fever is characterised by Aschoff nodules—foci of hyaline material surrounded by lymphocyte and macrophage infiltration—in the affected tissues.

Acute glomerulonephritis is another result of a hypersensitivity reaction following *Str. pyogenes* infection; but it differs from rheumatic fever in that the initial infection is often in the skin (e.g. impetigo) rather than in the throat, and that particular Griffith types of *Str. pyogenes* are involved (notably type 12 in the throat and type 4 in impetigo). Streptococcal antigen and antibody and complement combine in an immune complex reaction (Type III—p. 61) in the glomerular basement membrane. The disease presents, 1–5 weeks after the streptococcal infection, with albuminuria, haematuria, oedema and hypertension. Recovery is usually spontaneous but in a few cases acute glomerulonephritis leads to permanent renal damage and failure.

Scarlet fever is a streptococcal throat infection with a strain of *Str. pyogenes* that produces an erythrogenic toxin. The toxin produces the characteristic skin rash and 'strawberry' tongue in non-immune patients. Toxin production is encoded in streptococcal genes carried by a lysogenic bacteriophage (p. 77). Scarlet fever is much less common than in the earlier part of this century, a change which may reflect changing epidemiological patterns of streptococcal disease during the antibiotic era.

LABORATORY DIAGNOSIS For a streptococcal sore throat this traditionally depends upon the isolation of *Str. pyogenes* from a throat swab. Alternatively, *Str. pyogenes* antigen may be detected by co-agglutination or latex agglutination test kits available commercially; results of these tests are available within 15–30 minutes. The swab must be collected carefully so that tonsils, fauces and pharyngeal mucosa are properly sampled. This must be done in a good light, with the patient's tongue depressed. In the laboratory, the swab is used to seed blood agar and, possibly, blood agar with crystal violet to inhibit most pharyngeal commensals. β-haemolytic colonies are seen best after anaerobic incubation, which retains the activity of haemolysins O and S; the Lancefield grouping of streptococcal isolates is determined either by a modification of Lancefield's precipitation method or by one of several commercial kits that use other methods, e.g. co-agglutination. When a patient presents with rheumatic fever or glomerulonephritis it may still be possible to grow the causative streptococcus from the throat, or from a skin lesion in a case of glomerulonephritis. Whether this is achieved or not, it may be possible to find evidence of recent streptococcal infection by examining the patient's serum for antibodies to streptolysin O (**ASO test**); an ASO titre of >200 units usually indicates a recent infection, but some patients with rheumatic fever may have low levels. Antibodies to other streptococcal products—hyaluronidase, DNA-ase—may also be detected. In glomerulonephritis, a reduced complement C3 level in the serum is evidence of immune-complex formation and may help to confirm the diagnosis.

TREATMENT The antibiotic of choice for *Str. pyogenes* infection is penicillin. Oral phenoxymethyl penicillin (penicillin V) is usually adequate, though an initial dose of parenteral benzylpenicillin may be appropriate for the acutely ill child. Treatment should be continued for 10 days to prevent the risks of rheumatic fever or glomerulonephritis and of further spread of infection. Erythromycin is a suitable alternative for patients who are hypersensitive to penicillin. Patients with rheumatic fever or glomerulonephritis should be given penicillin to eradicate any remaining streptococci. Children who have had rheumatic fever should be given long-term prophylaxis with oral penicillin to prevent reinfection leading to a recurrence of the rheumatic fever. Such prophylaxis is not indicated following acute glomerulonephritis, as recurrence is prevented by development of Griffith-type-specific antibodies.

EPIDEMIOLOGY Streptococcal infection is endemic throughout the world but epidemics also occur, particularly in closed communities such as boarding schools. Spread is airborne either from a patient who has an infection (a case) or from a carrier. Convalescent carriage is common and there are also chronic carriers who may or may not have had a previous streptococcal illness. The commonest carriage site is the throat, but the less common nasal carrier is more likely to be an important source of widespread dissemination.

Diphtheria

During the nineteenth and early twentieth centuries diphtheria was one of the most feared of infectious diseases, but in developed countries immunization has now reduced it to a rarity, though it remains common in other parts of the world. It is due to infection with a toxigenic strain of *Corynebacterium diphtheriae* (p. 84); the organism remains localized—usually in the upper respiratory tract but sometimes in a wound or elsewhere—and causes serious remote effects by means of the toxin. Nasal diphtheria is a milder disease than the far commoner throat infection. In the latter the tonsils and fauces are inflamed and covered with an exudate that forms a grey adherent membrane. This may cause a life-threatening respiratory obstruction, especially if the larynx is involved (laryngeal diphtheria). There is an intense local reaction involving the cervical lymph nodes and surrounding tissues; reactive inflammation and oedema give a 'bull-neck' appearance. The serious systemic effects are caused by the potent exotoxin, which has specific cardiotoxic and neurotoxic effects (p. 83). Most deaths of patients are due to heart failure following toxic damage to the heart. Neurotoxicity causes cranial and peripheral nerve palsies; paralysis of the XIIth cranial nerve leads to nasal speech and difficulty in swallowing—characteristic early signs of diphtheria.

LABORATORY DIAGNOSIS Treatment for diphtheria must be started without delay on the basis of clinical suspicion and without waiting for laboratory diagnosis. This comes from culture of a throat swab. *C. diphtheriae* cannot be reliably identified by microscopy of direct smears from the throat swab, but has characteristic appearances in films made from 18-hour cultures on Loeffler's serum medium and stained by Albert's or Neisser's method; so this is the earliest stage at which the laboratory can give an opinion. Selective culture on tellurite media (p. 84) allows recognition of the characteristic colonies of *C. diphtheriae*, but they take 2 days to form. Methods for confirming that the strain is toxigenic (and therefore capable of causing diphtheria) are described on p. 84.

TREATMENT Antitoxin must be given to the patient as soon as a clinical diagnosis of diphtheria is suspected, so that toxin can be neutralized before it is fixed to target tissues (but see p. 367 for precautions). Once it has been fixed, cell death is inevitable and neutralization impossible. *C. diphtheriae* is sensitive to penicillin and to erythromycin; either will eliminate it, but erythromycin may be preferred because it rapidly switches off production of proteins, including toxin. Erythromycin is also the treatment of choice for carriers.

EPIDEMIOLOGY *C. diphtheriae* spreads readily in non-immune populations, airborne from cases or convalescent carriers; but active immunization with diphtheria toxoid (heat-inactivated toxin) restricts this spread. This is because antitoxic immunity, as well as preventing damage by the toxin, greatly reduces the frequency of successful colonization of the human upper respiratory tract by the organism, and so reduces its spread and carriage rate in the community. It is not eliminated altogether, however, and a decline in immunization rate is liable to lead to the appearance of scattered cases of diphtheria. In Britain and other developed countries children are routinely immunized against this disease (for details see p. 372).

Infections with anaerobic bacteria

A combination of anaerobic bacteria that includes fusobacteria and spirochaetes causes **Vincent's angina**—an acute ulcerative tonsillitis with the formation of a necrotic pseudo-membrane. This is often associated with ulcerative gingivitis (p. 191) and is similar in its bacteriology and in the role of predisposing factors such as immunodeficiency, malnutrition and poor oral hygiene. The diagnosis is confirmed (as in ulcerative gingivitis) by seeing large numbers of the characteristic fusobacteria and spirochaetes in stained smears prepared directly from a throat swab. Treatment should include specific antibacterial therapy (with metronidazole or penicillin) and measures to correct the underlying factors.

A more severe form of necrotizing tonsillitis caused by *Fusobacterium necrophorum* was recognised in the pre-antibiotic era and is still occasionally reported in Britain. This destructive and invasive infection, with localized tissue damage and a high incidence of bacteraemia which may lead to metastatic abscesses, is known as **necrobacillosis**. Diagnosis is by isolation of *F. necrophorum* from a throat swab, or most convincingly from a blood culture, by anaerobic culture. Metronidazole is the most appropriate treatment, with penicillin a suitable alternative.

Candidiasis

Candida albicans infection of the mouth (oral thrush, p. 192) may also involve the throat, especially in debilitated patients.

Otitis Media, Mastoiditis and Sinusitis

The middle ears, mastoids and paranasal sinuses are integral parts of the upper respiratory tract, connected directly with the nose and nasopharynx. They are often involved in infections of the upper respiratory tract, and also have their own localized infections. Acute sinusitis is part of many virus infections primarily affecting the nose and pharynx. Virus otitis or sinusitis is usually mild and resolves quickly, but mucosal oedema and inflammation may obstruct the normal drainage, causing retention of secretions and providing ideal conditions for secondary bacterial infections with organisms from the nasopharynx.

Acute otitis media is common in children. The usual main symptom is ear-ache, but in bacterial otitis there may be fever, irritability and even meningism. On examination, the drum is seen to be inflamed and lacks the normal light reflex; if there is a build-up of pus, the drum will bulge and may perforate with discharge of pus. The bacteria most commonly involved in acute otitis media are *Haemophilus influenzae*, *Str. pneumoniae* and *Str. pyogenes*. Acute otitis is often recurrent. The infection may extend to the mastoid air spaces, with pain and mastoid tenderness.

Acute sinusitis is more often a disease of adults. Discomfort in the regions of the frontal and maxillary sinuses is common in virus infections of the respiratory tract; bacterial infection is indicated by more severe pain and local tenderness, with purulent nasal discharge. The bacteria are those involved in acute otitis media.

Chronic otitis media and **chronic sinusitis** are long-standing suppurative infections, usually with periods of quiescence followed by exacerbations, and with permanent pathological changes. These conditions may begin in childhood and persist into adult life. In chronic otitis media there are recurrent episodes of discharge with or without pain; there is often impairment of hearing; and cholesteatoma formation is common. Recurrent headaches, nasal obstruction and purulent discharge are characteristic of chronic sinusitis. The bacteria that may be responsible for these chronic infections include those that cause the acute infections, but the microbial flora is much more varied and usually mixed. *Staph. aureus* is commonly present as a pathogen; but enterobacteria (*Escherichia coli*, *Proteus* and *Klebsiella* species) and *Pseudomonas aeruginosa*, also commonly present, are of doubtful pathological significance. However, when anaerobic cultures are set up they often reveal that bacteroides, anaerobic gram-positive cocci and microaerophilic or carboxyphilic streptococci (e.g. *Str. milleri*) are present in large numbers, and these probably have important pathogenic roles.

These conditions are important not only because of the pain and discomfort experienced during attacks but because of more serious possible consequences. Chronic otitis is the commonest cause of hearing loss. Either otitis or sinusitis may spread to involve surrounding bone and soft tissue. Most importantly, spread from such primary sources is a possible source of meningitis and the commonest cause of brain abscesses, especially of the temporal lobe (otitis) and the frontal lobe (frontal sinusitis). Anaerobes and *Str. milleri* are particularly important in these abscesses (p. 225).

INVESTIGATIONS Suitable specimens from acute otitis are difficult to obtain unless the ear drum perforates or is incised (myringotomy). Any purulent exudate should be cultured aerobically and anaerobically. A persistent discharge is part of the clinical picture of chronic otitis. Antral lavage (washing) is part of the standard investigation of chronic sinusitis, but if pus is present in the sinuses it should be collected by aspiration before lavage, as it is more useful neat than when diluted by washing.

TREATMENT Antibacterial treatment of acute otitis media or sinusitis should be directed at the common causative organisms. Amoxycillin or ampicillin, with erythromycin as an alternative, are the most appropriate antibiotics in general practice. Amongst children admitted to hospital, possibly because the disease is more severe, there appears to be a higher incidence of pneumococcal infection and good results have been obtained with parenteral benzyl penicillin as initial therapy, followed by oral phenoxymethyl penicillin.

For chronic infections antibiotic treatment should include an agent effective against anaerobes (usually metronidazole), but this is only part of the appropriate management. Surgery may well be necessary, to provide adequate drainage and to remove any damaged bone or a cholesteatoma.

Acute Epiglottitis

This condition affects mainly young children, usually 1–5 years old, and is then almost invariably due to *H. influenzae* type b. (The bacteriology of the few cases that occur in adults is less certain.) Acute inflammation of the epiglottis (which is enlarged and cherry-like) and of the aryepiglottic folds may lead to sudden total respiratory obstruction; examination of the throat should not be carried out unless equipment for intubation or tracheostomy is immediately to hand. Within a few hours of the first symptoms the child becomes febrile and evidently ill because of developing septicaemia. The disease may be fatal within 12 hours or so of onset. Since blood culture, the best laboratory means of confirming the diagnosis, takes longer than that, treatment must be started on the basis of clinical indications, which include drooling, difficulty in swallowing, inspiratory stridor, muffling of the voice, a preference for sitting rather than lying (if old enough) and signs of septicaemia. Treatment must be directed simultaneously at ensuring an airway and adequate ventilation and at killing the haemophilus. For the latter purpose chloramphenicol is currently the drug of choice. Ampicillin and amoxycillin are inappropriate because the haemophilus strain may produce β-lactamase, and may kill the child before sensitivity tests can be carried out.

15

Lower Respiratory Tract

Microbiologists regard as the lower respiratory tract that part of the tract which is below the larynx. It comprises the large airways (trachea and bronchi), the small airways (bronchioles) and the alveoli. Unlike the upper respiratory tract, these parts do not have a normal resident microbial flora; the airways are not sterile, since micro-organisms are continually inhaled, but the mucus that lines them, the alveolar macrophages and the ciliary action entrap and remove inhaled particles, living or inanimate. Bacterial multiplication in, and colonization or infection of, the lower respiratory tract is a pathological condition.

Parasites of all classes—viruses, bacteria (including mycoplasmas, chlamydiae and coxiellae), fungi, protozoa and helminths—cause infections of the lower respiratory tract. Clinically, these can be divided into infections of the airways—the larynx, trachea, bronchi and bronchioles—and those of the lung alveoli and parenchyma (pneumonia). This chapter deals first with airway infections; then with pneumonia (including *Pneumocystis* pneumonia, the only protozoan infection included here); then with tuberculosis and with helminth infections; and finally with sputum examination and other laboratory procedures. Fungal diseases of this tract are dealt with in Chapter 10.

Bacterial and Virus Infections of the Airways

Laryngo-Tracheo-Bronchitis (Croup)

Virus infections of the large airways are common, and in children they are often manifest as croup—laryngo-tracheo-bronchitis—in which all of the large airways are involved. The name **croup** relates to the inspiratory stridor which is characteristic of this condition; it formerly included laryngeal diphtheria (p. 195) and epiglottitis (p. 198), but now is used solely for the much milder virus laryngo-tracheo-bronchitis.

Typically, croup affects young children. It may occur up to the age of 5 years, but it is seen mostly in those under 2 years. The child is febrile and unwell, with a harsh barking cough and marked inspiratory stridor. The respiratory obstruction is often sufficient to cause indrawing of the intercostal spaces, and in severe cases there may be evidence of respiratory failure. The virus infection of the respiratory mucosa causes oedema (leading to obstruction) and damages the ciliated epithelium. The viruses most commonly implicated are paramyxoviruses—usually parainfluenza viruses. The diagnosis of parainfluenza virus infection may be confirmed rapidly by fluorescent antibody staining of respiratory epithelial cells aspirated from the nasopharynx through a fine catheter (p. 211). Other respiratory tract viruses can cause similar clinical pictures. There is no specific antivirus therapy and antibiotics are rarely required because bacterial superinfection seldom occurs. The child may need to be nursed in 25–30% oxygen if there is evidence of respiratory failure; increasing the humidity (e.g. with a steam kettle) may give symptomatic relief from the drying effect of the respiratory effort.

Bronchiolitis

Airway infection with respiratory syncytial virus (p. 127) is common, but in adults or children other than infants it causes only 'colds' or other mild symptoms. In children under 1 year old, however, it causes a distressing and sometimes serious condition known as **acute bronchiolitis**. No doubt it causes bronchiolar infections in older patients, but the seriousness of the condition in infants is attributable to a combination of very small airways, which are readily obstructed by mucosal oedema, and a lack of protective antibodies. The baby is febrile and distressed; respiration is rapid with indrawing of the intercostal spaces, and high-pitched small-airway rhonchi are audible. In severe cases respiratory failure may necessitate artificial ventilation, usually for only a short period. The diagnosis is confirmed by immunofluorescent staining of respiratory epithelial cells obtained by nasopharyngeal aspiration, perhaps most reliably after physiotherapy has encouraged upward drainage of bronchiolar secretions. Treatment should be symptomatic and antibiotics are rarely necessary, since the great majority of patients do not develop secondary bacterial infections.

Bronchitis

Acute bronchitis can occur at any age, and is a common disease of people with previously healthy lower respiratory tracts. It is often part of a virus infection that also involves the upper respiratory tract, the causative agent being commonly a parainfluenza virus, an adenovirus or a coronavirus; however, *Mycoplasma pneumoniae*, which is best known as a cause of pneumonia (see below), is probably even commoner as a cause of acute bronchitis, though it escapes detection in this role unless suspected and tested for. Infection of the bronchial mucous membranes results in mucosal oedema and ciliary paralysis. The patient has a dry painful cough with 'tightness of the chest' and usually some fever and systemic upset. Sputum is absent or scanty in the early stages, but is later viscid and often purulent because of secondary bacterial involvement. Even so, acute bronchitis in patients with previously healthy lower respiratory tracts usually resolves in a few days without antibiotic treatment; the situation is markedly different in 'acute-on-chronic' bronchitis (see below).

Chronic bronchitis, a major cause of morbidity and mortality, is not primarily an infectious disease, but bronchi damaged in this way are liable to recurrent and persistent infections that may have serious consequences. The underlying damage is caused, often over long periods, by environmental factors such as atmospheric pollution and, most importantly, cigarette smoking. The important changes in the bronchi are mucosal thickening, an increase in the number of goblet cells and increased mucus production (causing persistent productive cough) and inflammation, fibrosis and destruction of distal small airways. Production of mucoid sputum is part of the underlying bronchitic process, but purulent sputum is a later development, associated with infection, and commonly occurring in acute exacerbations. In some patients purulent sputum may come to persist all winter or all year. Acute exacerbations may be precipitated by virus infections or by various environmental factors. Productive cough increases, and within a day or two the sputum becomes purulent and contains numerous bacteria—usually *Haemophilus influenzae* (p. 98), often accompanied by pneumococci (p. 79). Whereas the more serious haemophilus infections such as meningitis (p. 214) and epiglottitis (p. 197) are virtually always caused by type b capsulate strains of *H. influenzae*, the strains involved in exacerbations of chronic bronchitis are non-capsulate; like their pneumococcal companions, they are nasopharyngeal commensals that have become opportunist pathogens in the conditions prevailing in the bronchi. Such infections usually increase the patient's disability, and in many cases antibiotic treatment reverses this. Whether persistent or recurrent bacterial infection also adds to irreversible damage in the bronchi is not certain.

The diagnosis of an acute exacerbation of chronic bronchitis is made on clinical grounds, and sputum culture will confirm the involvement of *H. influenzae* in most cases and of pneumococci in many. Indeed, so predictable are the findings (unless the patient has had a recent course of antibacterial drugs) that it is reasonable to initiate vigorous treatment aimed at *H. influenzae*, without waiting for culture results, if the patient's clinical condition calls for this; such treatment is likely to be effective against pneumococci also. Purulent sputum alone, which may disappear spontaneously after a few days, is not necessarily an indication for treatment; it is important to treat the patient, not his sputum. Amoxycillin, tetracyclines and cotrimoxazole are the drugs most likely to be effective. Many patients with chronic bronchitis are given a supply of an antibiotic and sputum containers; as soon as exacerbations start, they collect specimens of sputum for culture and start to take the antibiotic. This approach is now more favoured than attempting to suppress exacerbations by continuous antibiotic treatment throughout the winter.

Bronchiectasis

In this condition the terminal bronchi and bronchioles are dilated to form sacs in which fluid accumulates; this becomes infected with bacteria, and persistent bacterial infection is the major feature of the disease. Unlike the underlying damage in chronic bronchitis, that in bronchiectasis is in most cases attributable to infection—e.g. to severe lower respiratory tract involvement in whooping cough or measles in childhood. The bacteriology of established bronchiectasis is initially similar to that of chronic bronchitis; *H. influenzae* is prominent. The sacs are then commonly colonized by *Pseudomonas aeruginosa* (p. 96) and at a later stage they become pockets of infection with mixtures of bacteria derived from the mouth and upper respiratory tract. Anaerobes similar to those found in lung abscesses (p. 206) are a major part of these mixed infections, and they contribute to the continuing tissue destruction and the production of large amounts of foul sputum that are characteristic of the late stages of this condition.

Pertussis (Whooping Cough)

This serious and common childhood infection is caused by *Bordetella pertussis* (p. 100). (The closely related *Bord. parapertussis* may be responsible for a few mild cases.) The site of infection is the mucosa of the trachea, bronchi and bronchioles, is an important cause of chronic chest disease (see above). *Bord. pertussis* spreads by the airborne route and the risk of infection is directly related to the length of time spent in the same room as an infectious patient. There is an incubation period of 1–3 weeks (usually about 10 days), followed by a catarrhal stage that lasts for 1–2 weeks during which time the patient is infectious. Only then do the characteristic paroxysms of coughing develop; this stage lasts for 2–4 weeks and a residual cough may persist for much longer. A violent paroxysm of coughing forces air out the lungs, sometimes to the extent that the patient becomes cyanosed, and is terminated by the typical inspiratory 'whoop' as a massive inflow of air is drawn into the lungs. This is often followed by vomiting. There is little sputum and the violent coughing results in the production of only small plugs of clear mucus.

Mortality from pertussis is now low in countries with high living standards and in which secondary pneumonia is rapidly controlled by antibacterial drugs. However, there were 13 deaths in the 1982 epidemic in Britain, at a time when fears about the safety of pertussis vaccine had led to low acceptance rates for it. (Pertussis vaccination is further discussed on p. 365.) This remains a serious infection for young children, with a significant risk of permanent damage—not only of bronchiectasis, as mentioned above, but also of brain damage, possibly caused by hypoxia during severe paroxysms of coughing.

Clinical diagnosis is difficult in the early stages, but easy once the child develops the classical cough and whoop. The isolation of *Bord. pertussis* is not easy in some cases and failure to grow it does not exclude the diagnosis, but attempts should be made in mild or atypical cases in order to confirm the diagnosis. The best specimen for this purpose is a fine pernasal swab passed along the floor of the nose to sample the nasopharynx. The alternative but less reliable method is the 'cough plate'—a plate of medium held in front of the child's face during a coughing paroxysm. Tests for antibodies to *Bord. pertussis* are unreliable and not generally used.

Antibiotic treatment does not affect the course of the illness once the coughing phase has begun. It may reduce the infectivity during the catarrhal stage; and treatment or 'prophylaxis' with erythromycin, or possibly ampicillin, during the incubation period or the catarrhal stage may abort or modify the infection.

Pneumonia

Infection of the lung tissue, with inflammatory exudate and consolidation leading to respiratory embarrassment, is a serious condition. It can occur as a **primary** infection in previously healthy people, or as a **secondary** complication of some other infection or disorder—often as a terminal event in the elderly or debilitated.

Primary Pneumonia

Pneumococcal pneumonia

The pneumococcus (*Str. pneumoniae*, p. 79) is the commonest cause of bacterial pneumonia in patients of all age groups. Primary pneumococcal pneumonia is typically **lobar** or **segmental**—i.e. infection is limited to one or more lung lobes or segments rather than being spread throughout the lung. The lung alveoli are filled with inflammatory exudate instead of air (consolidation). The infection spreads to the blood-stream, and in the absence of prompt and appropriate antibacterial treatment this is a serious infection, with a high mortality rate. The patient has a high fever, with a rapid pulse and rapid shallow breathing. The cough, initially dry, later produces purulent sputum which is often 'rusty' because it contains altered blood from the alveoli. Auscultation reveals bronchial breathing due to the lung consolidation. Extension of the infection to the pleura (pleurisy) is common unless prevented by antibacterial treatment; it is manifest initially by chest pain on breathing, and then commonly by development of a pleural effusion that may develop into an empyema. Extension to the pericardium may lead to septic pericarditis (p. 239). Since at first there are no pneumococci in the sputum and bacteraemia is common, blood culture is the most effective means of isolating the pneumococci in the early stages. (Alternatively, it may be possible to demonstrate the presence of pneumococcal capsular polysaccharide in the blood or in urine.) Of the 46 pneumococcal capsular types (p. 79) type 1 is the commonest cause of pneumonia, though type 3 tends to cause a larger number of fatal cases.

The treatment of pneumococcal pneumonia was revolutionized by the introduction of penicillin. In Britain, this is still the drug of choice but resistance does occur in several parts of the world (p. 80).

Pneumonia clinically and radiologically resembling primary pneumococcal pneumonia is sometimes caused by *H. influenzae*—nearly always by type b capsulate strains and mostly in children.

Klebsiella pneumonia

In some countries, *Klebsiella pneumoniae* (*K. aerogenes* type 3 or Friedlander's bacillus, p. 99) is a recognized cause of primary pneumonia. In one of its forms this is characterized by lobar involvement and the formation of multiple abscesses and cavities. Klebsiella pneumonia is rare in Britain.

Pulmonary anthrax (wool-sorter's disease)

This is a rare but severe haemorrhagic infection of the bronchi and lungs that follows inhalation of *Bacillus anthracis* spores (p. 86). Pulmonary infection is often accompanied by pleural and pericardial effusions and by septicaemia. Only a small proportion of workers who are exposed to the hazard of inhaling anthrax spores in wool or hide factories develop pulmonary infections. Infected bone meal is another possible source (p. 307). Treatment is as for cutaneous anthrax (p. 307).

Pneumonic plague

This is another rare but severe form of bacterial pneumonia. The septicaemia of bubonic plague (caused by *Yersinia pestis*, p. 95 and 244) may lead to pulmonary infection, which has a high mortality. A patient with pneumonic plague is a source of droplet spread to other humans who, since they acquire the infection by inhalation and not from a flea bite, also develop pneumonic plague.

Legionellosis

In 1976 the genus *Legionella* joined the list of causes of primary pneumonia. It contains at least 5 species, of which *L. pneumophila*—the cause of the severe outbreak of illness in American legionnaires (p. 102) from which the names **legionnaire's disease**, *Legionella* and **legionellosis** are derived—is responsible for the most serious infections. Outbreaks of legionellosis have been mainly associated with large buildings such as hotels and hospitals, and with growth of *L. pneumophila* in water in the domestic supplies or in the air conditioning/humidifying systems of such buildings in circumstances such that aerosols containing legionellae were released into the environment. Infection is acquired by inhaling droplets or dust particles derived from such aerosols; man-to-man transmission does not occur, and so this is not a highly infectious disease as was initially feared. Nor is it usually the deadly killer that news-media reporting has suggested; mild illness is common. In the severe form, pneumonia with fever, dyspnoea and a non-productive cough may be complicated by cerebral hypoxia, gastro-intestinal symptoms or renal failure; whereas at the other end of the spectrum the name **Pontiac fever** has been given to a much milder illness, with fever and few distinctive features, which occurred in epidemic form in the American town of Pontiac in 1967 and was diagnosed as legionellosis retrospectively by serological tests after the organism had been isolated in 1976. Diagnostic procedures are outlined on p. 102. The other *Legionella* species cause similar infections but these are rarely severe. Erythromycin is the antibiotic of choice for treating *L. pneumophila* infections, though it has given disappointing results in some series.

Mycoplasma pneumonia

Primary pneumonias that do not have the typical characteristics of bacterial (especially pneumococcal) pneumonia are called 'atypical' pneumonias but the term **primary atypical pneumonia** is reserved for that caused by *Mycoplasma pneumoniae* (p. 110). This organism is probably the commonest cause of pneumonia in previously healthy young adults, and accounts for 10–30% of all acute lower respiratory tract infection. A patient with mycoplasma pneumonia in its characteristic form is pyrexial and dyspnoeic, but there are fewer signs on auscultation than would be suggested by the widespread patchy consolidation seen on radiography. Penicillin is not effective against mycoplasmas because they do not have cell walls, and mycoplasma pneumonia therefore fails to respond to conventional treatment for pneumococcal pneumonia. *M. pneumoniae* is sensitive to tetracyclines and to erythromycin and the latter is the drug of choice if legionellosis is a possible alternative diagnosis. Isolation of *M. pneumoniae* from sputum is difficult and takes time, but the diagnosis can be established early in the illness by finding 'cold agglutinins' (p. 61) and subsequently confirmed by specific tests for *M. pneumoniae* antibodies. If such tests are not carried out because the possibility of mycoplasma pneumonia has been overlooked, the diagnosis may be delayed and the illness prolonged; but it responds promptly to appropriate therapy.

Chlamydia pneumonia (psittacosis–ornithosis)

Infection of man with *Chlamydia psittaci* (p. 114) causes an illness called psittacosis when the infection is derived from psittacine birds (parrots etc.), or ornithosis when the source is some other type of bird (e.g. pigeon, duck, gull). Some birds infected with *Chl. psittaci* may themselves be ill and so be obvious suspect sources, but others, especially amongst flocks of wild birds, may be healthy carriers. Man is usually infected by inhaling dust containing dried bird droppings, though occasionally transmission is from person to person by droplets. Infection in man may cause negligible symptoms or a mild febrile illness, or may produce a severe pneumonia which may be fatal if diagnosis and appropriate treatment are delayed. Diagnostic procedures are described on p. 116. *Chl. psittaci* may persist in the respiratory tract for months after recovery. Like *M. pneumoniae*, *Chl. psittaci* is insensitive to penicillin and should be treated with tetracycline or erythromycin.

Chlamydia pneumonia in neonates

Chl. trachomatis may cause pneumonia in infants of mothers who are vaginal carriers of the organism at the time of delivery (p. 326). *Chl. trachomatis* is a common cause of genital infection (pp. 280–1), and women who are found to be infected during pregnancy should be treated with erythromycin, which is also the drug of choice for the pneumonia in their babies.

Q fever

Human infection with *Coxiella burneti*, described on p. 113, is by inhalation of dust or droplets contaminated with the excreta, body fluids or secretions of infected animals. The organism can survive for many months in dust. Asymptomatic infection is probably common in rural communities; symptomatic disease ranges from a mild 'flu-like' illness to a severe pneumonia. A serious but uncommon complication of Q fever is endocarditis (p. 238). Like the other causes of atypical pneumonia, *Cox. burneti* is resistant to penicillin, and Q fever should be treated with tetracycline.

Virus pneumonias

When laboratory confirmation was less readily available the term virus pneumonia was loosely applied to most cases of atypical pneumonia, many of which were not caused by viruses. However, viruses do cause many mild cases of pneumonia and a small proportion of severe cases. Those that do so most commonly are the influenza A and B and parainfluenza viruses and, especially in children, respiratory syncytial virus. Measles virus can cause a severe pneumonia, characterized by the presence of giant cells, in debilitated or immunosuppressed children. Antibiotic treatment does not affect the course of a virus pneumonia unless there is bacterial superinfection. A particularly important and dangerous form of such superinfection is the secondary staphylococcal pneumonia that has been common in some influenza A epidemics. The virus damages the mucosa and depresses the immune responses, and the staphylococcal invasion of the lungs and blood-stream may kill even young and previously healthy patients in a few hours. Autopsy shows widespread destruction of respiratory tract mucosa and of the alveoli. Patients who survive such a staphylococcal infection are often left with lung abscesses. In epidemics in which such secondary infections are occurring flucloxacillin, fucidin, erythromycin or vancomycin should be given, singly or in combinations, as soon as there is any suggestion of a staphylococcal superinfection. Pneumococci can also cause post-influenzal pneumonia, and *H. influenzae* is so named because of an apparently similar role in influenza outbreaks between 1892 and 1919.

Secondary Pneumonia

In chronic bronchial disease (bronchitis, bronchiectasis)

In pneumonia that is secondary to chronic bronchial disease or to some underlying debilitating illness, infection is centred on small bronchi. It is called **bronchopneumonia**; the consolidation is widespread and patchy and the small bronchi are plugged with mucopus. The pneumococcus is the commonest cause of bronchopneumonia. Non-capsulate strains of *H. influenzae* may also be implicated, though their responsibility for causing bronchopneumonia is open to question because they are so frequently present in the lower respiratory tracts of patients with chronic bronchitis (p. 200). *Branhamella catarrhalis* (p. 82), another pharyngeal commensal, causes bronchopneumonia in some such patients.

In cystic fibrosis

In this common genetic abnormality (one case per 2000 babies in Britain) excessively viscid secretions cause blockage and dilatation of exocrine gland ducts and of bronchioles. The abnormal secretions in the bronchiectatic cysts formed in this way are colonized by bacteria—notably *H. influenzae*, *Staph. aureus* and *Ps. aeruginosa*. Chronic infection of surrounding lung, with acute exacerbations, is the most important complication of cystic fibrosis; it leads to a great deal of debility and is the commonest cause of death. Exacerbations of infection, which are associated with increased disability and with increased amounts of pus and numbers of bacteria in the sputum, can usually be brought under control by intensive antibiotic treatment appropriate to the pathogens present, but infection cannot be eradicated and lung damage is progressive.

Following operations

Patients who have had surgical operations are particularly susceptible to bronchopneumonia during the post-operative period. Anaesthetic gases and artificial ventilation during the operation irritate the mucosa and paralyse the cilia, preventing clearance of particles. After surgery, particularly to the thorax or abdomen, pain may restrict chest-wall movement and discourage coughing. Pneumococci and *H. influenzae* are the commonest causes of chest infection in these patients. *Staph. aureus* can cause a severe post-operative pneumonia. Many post-operative patients belong in the category of compromised patients discussed in the next paragraph.

In compromised patients

Bronchopneumonia in debilitated or immunocompromised patients is a common and important form of hospital-acquired infection (p. 352). The bacteria involved may be those mentioned above in connection with post-operative chest infections; but patients with impaired defences, and in particular those whose respiratory tract protective mechanisms have been by-passed by intubation and artificial ventilation, are particularly liable to opportunist infections. The organisms involved may be relatively 'orthodox' enterobacteria such as *Esch. coli* or *K. aerogenes*, or they may be *Ps. aeruginosa*; but an increasing proportion of such opportunist infections are caused by other enterobacteria, such as *Serratia* or *Enterobacter* species, or by other pseudomonads, and all of these are commonly resistant to many antibiotics and likely to be very difficult to treat. The use of antibiotics and immunosuppressive agents also predisposes these patients to colonization by fungi, especially *Candida albicans* and *Aspergillus*, and it is often difficult to distinguish between harmless colonization and significant invasive infection.

Pneumocystis pneumonia

The protozoan *Pneumocystis carini* is another opportunist cause of pneumonia in immunocompromised hosts. The common finding of low levels of antibodies to it in the blood of healthy people suggests that subclinical infection is common, and that when pneumonia does occur in adults it is a reactivation of latent infection because of loss of immunity. Patients most commonly affected are those on cytotoxic or immunosuppressive drugs and those with AIDS (p. 283), in whom *Pneumocystis* pneumonia may be the first indication of the syndrome. Premature or malnourished infants are also liable to develop this disease, and in them the infection may be transplacental, whereas spread to adults is airborne. The incubation period in infants is 2–6 weeks; it has not been established in adults because at least the great majority of adult disease is reactivation of latent infection. In both adults and infants the mortality is high. Diagnosis is difficult, as chest radiography shows only patchy infiltration and the protozoa are rarely detectable in the sputum (which is scanty). Needle aspiration may fail to find a focus of infection, and the most satisfactory ways of obtaining material in which the protozoa can be recognized by microscopy are open lung biopsy or alveolar lavage through a fine catheter passed through a bronchoscope. A high titre of *P. carini* antibodies may be diagnostic, but poor antibody responses are likely in the types of patient concerned. Cotrimoxazole is the drug of choice for treatment of *Pneumocystis* pneumonia and also for prophylaxis in patients undergoing treatment that puts them at risk of developing it; pentamidine is an alternative drug.

Aspiration pneumonia; anaerobic pleuropulmonary infections; lung abscess

Localized or widespread pulmonary infection may follow the aspiration of infected material from the upper respiratory or gastro-intestinal tracts. When acidic stomach contents are aspirated, infection commonly develops in the lung tissue damaged by the acid aspirate. The bacteria that cause these aspiration pneumonias are from the upper respiratory tract but, unlike other pneumonias, these infections are usually due to bacterial mixtures, with anaerobes as important components. Anaerobic pleuro-pulmonary infections are of four types. (1) **aspiration pneumonia** without necrosis or abscess formation, which is an acute illness and responds rapidly to antibiotic therapy; (2) **necrotizing pneumonia** (pulmonary gangrene), in which there are areas of tissue necrosis, suppuration and cavitation with the formation of foul pus, and which has a high mortality; (3) **empyema** (i.e. suppuration in the pleural cavity), which is usually associated with underlying lung parenchymal infection but may occasionally be secondary to a subphrenic abscess; and (4) **lung abscess**.

Lung abscess is the best recognized of these. The abscess develops in an area in which lung tissue has been damaged, usually as a result of bronchial obstruction by a tumour, enlarged lymph nodes or an inhaled foreign body. Secretions fail to drain and the obstructed segment is infected by organisms aspirated from the upper respiratory tract, in particular *Bacteroides* species, and anaerobic gram-positive cocci (p. 80). The patient is usually febrile and anaemic and loses weight; a cavity containing fluid (pus) can be seen by chest radiography. The copious sputum produced when the abscess breaks through into a bronchus and drains via the respiratory tract is putrid, giving the patient a foul halitosis.

The anaerobes found in these pleuro-pulmonary infections are mostly bacteroides from the mouth and upper respiratory tract—*Bact. oralis, Bact. melaninogenicus*, etc.—and fusobacteria and anaerobic gram-positive cocci; but *Bact. fragilis* is involved more commonly than would be expected of a faecal species. Treatment should be with antibiotics effective against the mixed flora, and it is usually appropriate to include metronidazole to deal with the anaerobes. Surgical drainage and removal of necrotic tissue may be necessary.

Lung abscesses may also arise as complications of staphylococcal (p. 204) or klebsiella (p. 202) pneumonias, or from haematogenous spread of septic emboli. These abscesses are usually multiple.

Tuberculosis

This chapter on lower respiratory tract infections is the appropriate place for a general account of tuberculosis because the lungs are the main sites for infection with *Mycobacterium tuberculosis* (p. 103).

In the 19th century, tuberculosis (consumption) was known as the Great White Plague and was one of the most feared diseases of the overcrowded city slums that grew out of the industrial revolution in Britain and Europe. Earlier, John Bunyan described it as 'captain of all the men of death'. Today it is a much smaller part of medical practice in most developed countries, though it is still a serious disease and there are approximately 7000 new notifications in Britain each year. World-wide, however, it remains the most important of the infectious diseases, causing more morbidity and a greater mortality than any other. The World Health Organization estimates that there are 15 million infectious cases in the world, 70–80% of them in developing countries; there are 2–3 million new cases and 1–2 million deaths each year.

Most tuberculosis in man, especially pulmonary tuberculosis, is caused by *Myco. tuberculosis*, a slowly growing bacillus with a specially resistant cell wall, one effect of which is to render it acid-fast and alcohol-fast in the Ziehl–Neelsen staining method (p. 103). The commonest form of *Myco. tuberculosis* infection is pulmonary. Involvement of other sites is usually the result of spread from a pulmonary source. Infection is acquired by inhalation of the bacillus in droplets, or in dust contaminated by respiratory secretions, and results in **primary** tuberculosis if the dose is adequate and there has been no previous infection. The site of such primary disease is in the periphery of the lung, usually in the upper lobe, particularly towards the apex. Inflammation develops here and along the draining lymphatics to involve the hilar lymph nodes. This complex of peripheral lesion and hilar lymphadenopathy is known as the

primary or **Ghon focus**. The response induced by *Myco. tuberculosis* infection is typically that of chronic inflammation. There is a brief initial acute response, but the polymorphonuclear leukocytes are unable to eliminate the bacilli, the area becomes infiltrated with lymphocytes and macrophages, and a granuloma forms, characterized by the presence of multinucleate (Langhan's) giant cells. This pattern depends on the resistance of *Myco. tuberculosis* to intracellular killing by phagocytes.

The primary infection is often asymptomatic and the primary lesion usually heals, or at least becomes quiescent, although the bacilli may remain viable. The affected lung is left with a scar which may calcify during a period of years and can be detected by radiography. Previous tuberculous infection can also be detected by demonstrating tuberculin conversion. As a result of primary infection, Type IV cell-mediated hypersensitivity to a component of the mycobacterial cell wall known as tuberculin develops and forms the basis of the tuberculin skin test used to detect previous infection (p. 184).

Not all primary infections are brought under control by the host. Some develop into serious and potentially fatal infections. If the primary focus involves the pleura, an effusion may form; tuberculosis is a major cause of pleural effusion in young people. Within the lung itself the infection may not be localized and may spread through the bronchial tree, causing tuberculous bronchopneumonia. The most serious sequel of primary infection, however, is the systemic spread of *Myco. tuberculosis* through the blood-stream to cause **miliary tuberculosis**. This is most readily detected by chest radiography, which shows scattered mottling of the lung field, representing many small granulomas (tubercles); or by ophthalmoscopy, which reveals such lesions on the retina. Untreated miliary tuberculosis has a high mortality. Even in the absence of generalized miliary tuberculosis, *Myco. tuberculosis* may be seeded through the body via the blood-stream and this systemic spread may lead to tuberculous meningitis (p. 218), which has a high mortality and a very high incidence of residual damage in those who survive. Organisms may also be seeded to other tissues (bone, kidney, adrenals, uterus, testis, etc.) where they may establish small unobtrusive lesions that may become evident only in later years when they turn into the destructive secondary form of tuberculosis.

The **secondary** (**adult**) form of the disease may result from such reactivation of a quiescent or dormant lesion, or in some cases may follow reinfection of a person who is already hypersensitive (tuberculin positive) as a result of an earlier primary infection. Reactivation is related to some reduction in immunity and in developed countries is often associated with age or other debilitating conditions—e.g. alcoholism, malnutrition, infections that depress the immune responses, malignant disease and its treatment, steroid treatment or (sometimes) pregnancy. In the lung the reactivated or new infection results in a destructive chronic inflammatory process for which tuberculin hypersensitivity is largely responsible. Large granulomas develop, with extensive destruction of lung tissue and the formation of a thick, cheese-like pus ('caseation'). A growing granuloma may erode into a bronchus, discharging its caseous centre full of *Myco. tuberculosis* into the airways and leaving a cavity. This is the most highly infectious form of tuberculosis, known as **open** tuberculosis, because the sputum is laden with *Myco. tuberculosis* from the discharging cavity. Spread of infection through the bronchial tree may cause tuberculous bronchopneumonia. Caseation and cavitation also commonly erode blood vessels causing coughing up of blood (haemoptysis), which may be dramatic if a large vessel is involved. (Similar processes elsewhere in the body are described in the appropriate chapters on bone, soft tissue, central nervous system and genito-urinary infection.)

A large proportion of tuberculosis seen in developed countries is secondary disease in the elderly and debilitated; but primary infection still occurs, and sometimes presents as miliary tuberculosis or tuberculous meningitis in children or young people. Infection in these cases can usually be traced to contact with an adult (e.g. an elderly relative) with open pulmonary tuberculosis. In Britain during the late 1960s and 1970s many cases of tuberculosis were seen in immigrants recently arrived from the Indian subcontinent, where there is a high incidence of infection, and in Asian children born in Britain but in contact with recent immigrants.

The diagnosis of pulmonary tuberculosis depends upon the demonstration of *Myco. tuberculosis* in sputum; the first morning sputum is the best specimen to examine as it is likely to contain the greatest number of tubercle bacilli. From children too young to expectorate sputum, gastric washings can be examined. The laboratory examination of these specimens is described on pp. 210–1. Testing for tuberculin hypersensitivity is useful for establishing past tuberculous infection, and particularly for screening contacts of open cases for evidence of primary infection. However, it must be remembered that tuberculin hypersensitivity does not develop until several weeks after infection and may be absent in the young, in overwhelming miliary tuberculosis, and in elderly or debilitated patients with secondary tuberculosis.

So far we have been dealing with *Myco. tuberculosis* infection. Tuberculosis in man can also be caused by *Myco. bovis*, the organism responsible for cattle tuberculosis. Man may become infected by direct involvement with live infected cattle or with carcasses, but since the organism is excreted in the milk of infected cows the great majority of human infections result from drinking such milk. The common sites of infection in man are the tonsils and cervical lymph nodes (p. 244) and the intestinal lymphoid tissue and mesenteric lymph nodes (p. 245), with subsequent spread to other parts of the body, notably bones and joints. *Myco. bovis* infection in man has been virtually eliminated from Britain and many other countries by pasteurization of milk (p. 358) and by tuberculin-testing of cattle and elimination of those with positive reactions (p. 185). (See also pp. 333–4.)

TREATMENT Chemotherapy of tuberculosis is described in Chapter 29 (pp. 397–8).

CONTROL The control of tuberculosis requires a combination of individual and public health measures (p. 333). The first requirement is the detection and treatment of cases, particularly open cases. In the past, patients were removed to sanatoria, but with modern drugs patients cease to be infective very soon after starting treatment and can then be managed adequately as out-patients. The prevention of further cases then depends upon contact-tracing. The high-risk groups are the immediate family and close personal contacts and the colleagues or classmates at work or school. A combination of tuberculin tests and chest radiography is used to detect infection in contacts so that they can be treated immediately, while the disease is at an early stage. Contacts, especially children, who have converted to tuberculin positivity but have no evidence of overt disease are given 'prophylactic' chemotherapy. Case finding by mass miniature radiography is no longer cost-effective and radiography programmes are now directed at high-risk groups, i.e. those suspected by general practitioners, miners (because of a long association with mining and pneumoconiosis), and people involved in health care or medical laboratory work (see Chapter 27; p. 353). Teachers and other public servants in close contact with large numbers of the public, especially with children, should be checked by radiography at regular intervals because they are a serious hazard if they become infected. These public health measures and improvements in nutrition and in living standards have been of major importance in the control of tuberculosis in the developed countries. In addition, however, vaccination with the live attenuated strain of *Myco. bovis*—Bacille Calmette–Guerin (BCG)—has made a significant contribution; its use is discussed in Chapter 29 (p. 365).

Other mycobacteria

The name 'tubercle bacilli' includes *Myco. tuberculosis* and *Myco. bovis*; but other mycobacteria, known collectively as 'atypical mycobacteria' (p. 105), can cause tuberculosis-like illness in man. Most of them cause superficial infections (p. 106) or mycobacterial lymphadenitis, but some cause pulmonary infection. The photochromogenic *Myco. kansasi* is the species that most commonly causes pulmonary infection resembling tuberculosis.

Strains of the *Myco. scrofulaceum/avium/ intracellulare* complex, which more often cause mycobacterial lymphadenitis, infect the lung on rare occasions. Laboratory investigations are as for tuberculosis, but treatment may be more difficult because many 'atypical' strains are resistant to the antituberculosis drugs.

Helminth Infections

Several parasitic infections have transient or more prolonged effects upon the lung. There are two groups of parasitic pulmonary infections: (a) those caused by parasites (usually nematode larvae) passing through the lungs; and (b) those for which the lung is the definitive site of infection. Several species of nematode found as adults in the intestine have immature stages that migrate through the lungs during development (*Ascaris*, p. 156, *Strongyloides*, p. 157, and the hookworms *Ancylostoma* and *Necator*, p. 157). *Ascaris* larvae hatch from ingested eggs in the small intestine and penetrate the mucosa to enter the blood or lymphatic vessels. The larvae are carried through the liver and heart to the lungs where they are filtered out of the cirulation by the alveolar capillaries and break out into the alveoli. They develop there for about 10 days before ascending the respiratory tree to the trachea to be swallowed. Small numbers of migrating larvae produce few obvious effects on the lungs, but large numbers may cause pneumonitis, which is usually short-lived and rarely causes serious disease. Radiographic changes are usually non-specific infiltrates and patchy areas of pneumonitis. Eosinophilia is common with *Ascaris*. Hookworm and *Strongyloides* larvae undertake a similar migration, but whereas a large number of *Ascaris* eggs may be swallowed at one time (e.g. on contaminated vegetables) giving rise to a large number of migrating larvae and hence to respiratory symptoms, infection with hookworm and *Strongyloides* larvae is by penetration of the skin and is therefore generally at a lower but more constant level, so that respiratory symptoms are less common.

Treatment of pneumonias caused by developing nematode larvae is hardly justified and probably rarely attempted. Symptoms are usually of short duration and diagnosis can only be presumptive until the adult parasites or their eggs can be demonstrated several weeks later. Symptomatic treatment of the respiratory symptoms if they are severe is all that is required; more important is the subsequent treatment for the adult parasites (see Chapter 18, p. 260).

In urinary schistosomiasis, some *S. haematobium* ova do not escape from the bladder wall into the urine but are carried in the bloodstream to the lungs, where granulomata form; these lead to blockage of the finer capillaries, increased pulmonary blood pressure, hypertrophy of the right ventricle and eventual right heart failure (cor pulmonale).

The cyst-like lung lesions caused by lung flukes (*Paragonimus* species) in people in parts of Asia, Africa and South America are due to presence of adult flukes in the lungs, as described on p. 154. Ingested larvae penetrate the gut wall to the peritoneal cavity and pass through the diaphragm and thoracic cavity to the lungs. The migratory stages cause few if any symptoms, but as the parasites grow in the lungs they provoke an inflammatory reaction which gradually changes to a fibrous cyst surrounding the parasite. Light infections may cause no evident host reaction but moderate infections may result in a cough with increased production of mucus (containing the eggs) and chest pain. Heavy and long-standing infections lead to increased fibrosis of the lung, the cysts showing as discrete nodules on radiographs. Rarely, individual immature parasites lose their way during migration and mature in other parts of the body, provoking symptoms depending on the sites affected. Cerebral paragonimiasis is not uncommon, particularly in Asia; such infection is commoner in people under 20 years old than in those older. Eggs may also be carried in the circulation from the lungs to almost any part of the body, where they provoke a granulomatous reaction similar to that produced by schistosome eggs. Diagnosis is based upon chest radiography and demonstration of the *Paragonimus* eggs in sputum or faeces. *Paragonimus* infection can be treated by oral bithionol; side effects are not uncommon but are usually transient.

Examination of Sputum and Other Laboratory Procedures

The microbiological diagnosis of the cause of lower respiratory tract infections is to a large extent based upon the examination of sputum coughed up by the patient and expectorated into a suitable container. This is suitable for many routine purposes, but there are occasions when either it is not the most appropriate specimen or it is not possible to obtain sputum; alternative methods must then be used. For example, pulmonary infection with anaerobes may be difficult to establish by examining a sputum specimen contaminated (as it virtually always is) with mouth material also containing anaerobes. A far better specimen for this purpose can be obtained by transtracheal aspiration—i.e. inserting a wide-bore needle through the skin into the trachea at cricoid level, passing a fine catheter through the needle, and using this to withdraw material from inside the trachea. However, many clinicians and microbiologists are loath to subject their patients to this procedure. In other refractory pulmonary problems, however, bronchoscopy is an essential investigation and incidentally provides the opportunity to obtain specimens of uncontaminated secretions from specific sites in the bronchi. In children, sputum may be difficult to obtain because they naturally swallow it. Gastric washings, sometimes useful for the diagnosis of tuberculosis (p. 208), are of little value for isolation of organisms less able to withstand gastric acidity.

However obtained, the specimen of sputum or respiratory secretions must be transported quickly to the laboratory to avoid overgrowth of organisms such as coliforms (in specimens from patients treated with broad spectrum antibiotics) or fungi (in those from patients on steroids). Expectorated sputum is not of uniform composition; flecks of obviously purulent material are often mixed with mucus and saliva. Random sampling gives unreliable results, and there are two alternatives: (1) careful selection of purulent portions for examination, or (2) homogenization of the sputum, either by chemical (N-acetyl cysteine) or by mechanical methods (shaking with glass beads), so that any sample from it represents the whole.

A gram-stained film of sputum may give useful information. As many of the organisms that cause lower respiratory tract infections are upper respiratory tract commensals, their presence in a culture may not tell us their role on this occasion (though the heaviness of the growth may be informative—see below). However, a massive predominance of one type of bacterium in the film (e.g. of the classical diplococcal forms of pneumococci, sometimes evidently capsulate) may give diagnostic help; and the presence of many polymorphonuclear leukocytes indicates infection. The occasions when *Branhamella catarrhalis* is present not as a commensal but as a pathogen are recognizable by finding it as many gram-negative cocci inside polymorphs; and *Staph. aureus*, to be discounted as a nasopharyngeal contaminant when the film shows many pharyngeal epithelial cells, should be taken more seriously if it is seen to be present in large numbers and accompanied by polymorphs.

A ZN-stained film should be made if tuberculosis is suspected (p. 103) and many laboratories do this, or use the auramine method (p. 165), for every patient with a chest infection, so that open tuberculosis is not missed. Tubercle bacilli are sometimes more readily detected in films made after concentration of the specimen (p. 104).

It is appropriate to culture most sputa and similar specimens on blood agar in air and on chocolate agar (p. 81) in air with increased CO_2 (5–10%). These media will grow pneumococci, haemophili, staphylococci and many of the other potential pathogens in such material. Because sputum resembles urine (p. 273) in coming from a site that has no normal bacterial flora but being almost inevitably contaminated during collection, a roughly quantitative approach to its bacteriology can be helpful; dilution of the homogenized specimen 1:2000 or so before plating out helps to distinguish between contaminants (usually present in only modest numbers) and pathogens (liable to be far more numerous). This is particularly true of enterobacteria, which are commonly present in the upper respiratory tracts and sputa of debilitated and antibiotic-treated patients, and which produce large colonies that obscure the presence of much larger numbers of the small colonies of organisms such as pneumococci. MacConkey agar is useful for preliminary identification of these enterobacteria. If fungal infection is suspected, Sabouraud's agar can be set up—though *Candida* and *Aspergillus* species grow well on blood agar. Anaerobically incubated blood agar, possibly with a selective agent such as kanamycin or nalidixic acid, should be used for isolation of anaerobes from patients with suspected bronchiectasis, lung abscesses or aspiration pneumonia. If tuberculosis is suspected, concentrated material should be cultured on appropriate media (p. 104).

Respiratory tract infection is an area in which the development of rapid methods for the specific diagnosis of virus infection has proved particularly valuable. Serological tests inevitably give only retrospective diagnoses and culture of respiratory viruses is time-consuming, sometimes requires prolonged culture and may not be successful. However, the development of immunofluorescence microscopy has placed rapid diagnosis of several important respiratory virus infections, especially in children, within the reach of most laboratories. It is particularly useful for the diagnosis of infections with respiratory syncytial virus and the para-influenza, influenza A and B and measles viruses. Cells infected with the virus and having virus antigens on their surface are collected by nasopharyngeal aspiration through a fine catheter. After being fixed to a slide they are treated with specific antivirus serum and then with fluorescein-conjugated antibody against the specific antibody. Microscopic examination under ultraviolet illumination shows brightly fluorescing virus-infected cells.

16

Central Nervous System

Infections of the central nervous system (CNS) include some of the most feared of infectious diseases. Despite modern antibacterial treatment, which has dramatically improved the prognosis in typical pyogenic meningitis, infection of the brain tissue itself or the surrounding meninges still carries a significant risk of death, and an even greater risk of permanent neurological damage, especially if diagnosis and treatment are delayed for any reason. The diagnosis of meningitis is a medical emergency.

CNS infections can be divided into three broad types—**meningitis**, in which only the membranous coverings of the brain and spinal cord are involved, **encephalitis** in which the neural tissue itself is infected, and localized lesions such as **brain abscesses**. In meningitis, there is no direct involvement of neuronal tissue, but inflammation over the surface of the brain and thrombophlebitis of blood vessels serving the brain or spinal cord may cause permanent neurological damage. Meningitis may also spread throughout the CSF to produce a **ventriculitis**. The damage in encephalitis is caused by direct infection of neuronal tissue leading to inflammation and cell death. The inflammation in encephalitis almost invariably involves the meninges, and the term meningo-encephalitis is appropriate.

Meningitis

Meningitis may be caused by a wide range of bacteria, including *Mycobacterium tuberculosis* and leptospires, by viruses and, much more rarely, by fungi or amoebae.

Bacterial Meningitis

Bacterial (often called 'pyogenic' = pus-producing) meningitis in children and in adults is caused predominantly by three bacterial species: *Neisseria meningitidis*, *Haemophilus influenzae* (usually type b) and the pneumococcus. The picture in neonatal meningitis, however, is quite different (see below). Although the three main bacterial species may cause meningitis in any age group the incidence

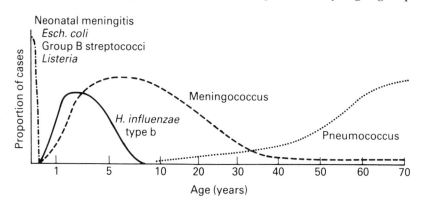

Figure 16.1 *Age-group incidence of the major types of bacterial meningitis.*

of the different types varies with age (see Fig. 16.1). In infants beyond the neonatal period and young children up to the age of 5 years, *H. influenzae* type b and meningococcus are both common causes of infection. In the USA *H. influenzae* is by far the commonest cause of bacterial meningitis in this age group, but in Britain the two organisms occur with about equal frequency. The pneumococcus rarely causes meningitis in young children. Over the age of 5 years haemophilus meningitis is rare, and throughout the remainder of childhood and into adult life the meningococcus is the predominant cause of bacterial meningitis. Primary pneumococcal meningitis is rare in children or young to middle-aged adults, but becomes increasingly common with advancing age and is the commonest form in the elderly. Its occurrence in the younger age group is commonly associated with fractures to the base of the skull (p. 216).

Meningococcal meningitis

The commonest cause of bacterial meningitis in children and in young otherwise healthy adults is *N. meningitidis* (p. 81), which colonizes the upper respiratory tract, particularly the nasopharynx, and is spread from person to person by the airborne route. Such spread is encouraged by bringing groups of people together in close proximity and by overcrowded living conditions. Most people who acquire *N. meningitidis* in this way carry it in their upper respiratory tracts for a period but suffer no ill effects. However, a small proportion of those who become colonized subsequently develop invasive infection. The reason why only some people are susceptible to infection is not known, but responsibility appears to lie with the immune system, and there may be a relationship with HLA type. Studies of American army recruits in the 1930s showed that the transmission of *N. meningitidis* increased as the density of recruits in sleeping quarters increased. When the distance between bed centres was reduced to less than 1.5 m (5 ft), epidemics of meningococcal meningitis occurred amongst the recruits. This finding has not been substantiated in subsequent studies.

In the early stages of invasive infection there may be mild upper respiratory tract symptoms, but the most serious presenting feature of meningococcal disease is septicaemia. It may be that meningococci sometimes gain direct access from the upper respiratory tract to the meninges, but their common route is via the blood-stream; bacteraemia, detectable by blood culture, is an almost invariable part of systemic meningococcal infection. There may be an associated septicaemia, apparent before the patient has evident meningitis and itself a life-threatening condition. *N. meningitidis* is a gram-negative organism and contains endotoxin which is released not only from dead meningococci as is classical endotoxin but also from the surface of living organisms. Meningococcal septicaemia has all the characteristics of an endotoxaemia and is a potent trigger of disseminated intravascular coagulation (p. 50). This is responsible for the widespread petechial rash characteristic of meningococcal disease, and for the widespread haemorrhages typical of the fatal form of meningoccocal septicaemia known as the Waterhouse–Friderichsen syndrome. In this syndrome (which also sometimes occurs in septicaemia with other species) death is due to haemorrhagic destruction of the adrenals. However, most patients with meningococcal disease are first seen when they develop the clinical signs of meningitis. Inflammation of the meninges causes meningism—headache and photophobia (which may be related to raised intracranial pressure), neck stiffness and Kernig's sign (inability to raise a straightened leg because this procedure stretches the nerves and irritates the inflamed meninges). As well as contributing to the headache and photophobia, the raised intracranial pressure resulting from the inflammation causes vomiting, and progressively interferes with the patient's level of consciousness, so that a patient who is initially only mildly drowsy may soon become deeply comatose. The specific diagnosis may be indicated by the presence of the typical meningococcal rash, which is rare in other forms of bacterial meningitis. It may consist of anything from a few isolated petechiae to huge areas of intradermal bleeding, particularly over the buttocks and lower limbs but also affecting the trunk and upper limbs.

Laboratory confirmation of the diagnosis is obtained by examination of the cerebrospinal fluid (CSF). The presence of intracellular gram-negative diplococci in a film of the centrifuged deposit of the CSF is diagnostic of meningococcal meningitis. However, small numbers of meningococci, particularly in patients who have received some antibiotic therapy, may be difficult to find. In such circumstances, latex agglutination (p. 176) or co-agglutination (p. 176) tests with specific antisera may detect the presence of meningococcal antigen. Culture of the CSF should always be done; it is more sensitive than microscopy but does not provide an immediate answer. In most cases the meningococcus may also be isolated from blood cultures taken before antibiotic therapy has started. An alternative method of demonstrating the presence of meningococci, particularly when the intracranial pressure is high and a lumbar CSF tap is contra-indicated because of the risk of coning, is the examination of tissue fluid from fresh petechiae. Gram-negative diplococci can be seen in films prepared in this way and the meningococcus may also be grown.

The treatment of meningococcal meningitis was revolutionized by the introduction of sulphonamides and then penicillin. Sulphonamides cross the blood–brain barrier easily and therapeutic concentrations are readily achieved in the CSF. However, as least 20% of meningococcal strains now isolated in Britain are resistant to sulphonamide and the proportion is higher elsewhere. Fortunately, the meningococcus is very sensitive to penicillin and this remains the drug of choice in the treatment of meningococcal disease. Benzylpenicillin should be given intravenously as soon as a diagnosis of meningococcal septicaemia or meningitis is made. There are strong arguments for the general practitioner giving penicillin on the basis of his clinical diagnosis. This may hamper bacteriological confirmation, but the earlier treatment can be given the better the chance of success. Children, especially, can deteriorate rapidly; the petechiae appear before the eyes of the attendants and the patient may die while arrangements are made for hospital admission. Benzylpenicillin should be given by intravenous bolus injections at least every 4 hours during the first few days of the illness. Recovery in uncomplicated cases is rapid and treatment can usually be stopped after one week. If the diagnosis is in doubt chloramphenicol can also be given to cover the possibility of *H. influenzae* infection. The meningococcus is sensitive to chloramphenicol and this drug has not been shown to reduce the efficacy of penicillin in this disease.

Meningococcal meningitis is infectious. Fortunately, epidemics do not occur in most temperate countries, but occasionally two or more related cases are seen in the same family or school class. There have also been instances of medical and nursing staff developing meningitis after being involved in mouth-to-mouth resuscitation or other close contact with a patient. Chemoprophylaxis may be given to close relatives, particularly other children who have been in close contact with the index patient, to prevent secondary cases. Although penicillin is the drug of choice for treatment, it is not effective as a prophylactic. Sulphonamide has been the most widely used prophylactic and should be given for 3 days. Rifampicin is also effective and is now regarded by many as the prophylactic of choice, though some prefer to restrict the use of this valuable anti-tuberculous drug to cases caused by sulphonamide-resistant meningococci.

True epidemics of meningococcal disease, some of them very large with thousands of cases, do occur, particularly in Africa and South America. Some of the worst epidemics have been in the sub-Sahara belt of West and Central Africa. These are usually caused by *N. meningitidis* of serogroup A, whereas the commonest serogroup in Britain is B. Thus, the patterns of infection vary in different parts of the world, with a marked contrast for example between the epidemic spread affecting broad age groups seen in Africa and the sporadic cases, about half of them in children under 5 years old, seen in Britain.

Haemophilus meningitis

The main diagnostic problem in dealing with children with bacterial meningitis is to distinguish cases caused by the meningococcus from the other large group caused by *H. influenzae* type b, which accounts for at least half of those in the under-5 age group in Britain. Haemophilus meningitis is rare above that age, probably because asymptomatic or undiagnosed *H. influenzae* type b infections are common in childhood and lead to immunity. Like the meningococcus, *H. influenzae* spreads from person to person by the airborne route and initially colonizes the nasopharynx. A small proportion of those colonized then develop bacteraemia and meningitis; the meninges are seeded from the blood-stream. Haemophilus meningitis occurs as sporadic cases, and although there is an increased incidence of the disease among family and other close contacts of the primary cases, there are no epidemics such as are seen in some countries with the meningococcus. This may be the result of the restricted age-range of susceptibility to haemophilus meningitis.

The clinical picture of haemophilus meningitis is usually indistinguishable from that of meningococcal meningitis in the individual patient, although there are some differences in the way that the disease develops. Whereas meningococcal meningitis has an acute onset, that of haemophilus meningitis may sometimes be more insidious. Rashes are far less common, and when they do occur are usually less obvious in haemophilus than in meningococcal infection. However, *H. influenzae* is a gram-negative bacillus containing endotoxin and may give all the signs and symptoms of endotoxaemia. The slightly less acute onset may lead to delay in diagnosis, so that a child may be in coma when eventually admitted to hospital. Such delay may be partly responsible for the somewhat higher incidence of permanent sequelae after haemophilus meningitis, which counter-balances the greater risk of fatal septicaemia when the infecting organism is the meningococcus. The permanent neurological damage is not caused by direct invasion of brain tissue but is the result of superficial inflammation and thrombophlebitis causing hypoxic damage to the underlying brain. As the dense inflammatory exudate resolves, fibrosis may cause further damage, sometimes leading to hydrocephalus. Permanent damage from pyogenic meningitis includes mental retardation, blindness, deafness and many types of motor and sensory deficit. Because haemophilus meningitis is, like meningococcal meningitis, a bacteraemic illness, other manifestations of the bacteraemia may be present—notably epiglottitis (p. 197) and suppurative arthritis (p. 302). (See also p. 234.)

The diagnosis of haemophilus meningitis is confirmed by the demonstration of small gram-negative coccobacilli with many polymorphs in the CSF or by rapid antigen detection techniques (p. 169) and by the isolation *H. influenzae* type b (typed by capsule swelling or by agglutination with specific antisera) from the CSF and from the blood. The same organism may also be isolated from the upper respiratory tract and, if there is evidence of middle ear infection, from any discharge. However, isolation of *H. influenzae* (untyped) from such sites is no evidence of haemophilus meningitis, since many people carry this species in their upper respiratory tracts, and non-capsulated strains are commonly involved in middle ear infections. An accurate diagnosis is essential because haemophilus meningitis will not respond to the penicillin therapy appropriate for the meningococcus. The drug of choice for haemophilus meningitis is chloramphenicol. High doses (100 mg/kg/day) provide bactericidal levels of chloramphenicol in the CSF against *H. influenzae*. Ampicillin, at one time preferred by many (especially in North America) because of fears about chloramphenicol toxicity, is now contra-indicated because too many *H. influenzae* type b strains produce β-lactamase and are thus resistant to it. So far the incidence of chloramphenicol resistance remains low. Where it does cause concern, cefotaxime or latamoxef or one of the other new cephalosporins may be a suitable alternative. Even when optimum treatment is given, recovery may be slow and it may be several days before the child begins to respond. During the recovery phase, and particularly after antibiotic treatment has been stopped, relapse may occur, presumably a result of loculated pockets of infection that have not been cleared. There is also a greater risk of a subdural effusion as a result of haemophilus meningitis than with meningococcal meningitis, and focal neurological signs may be the result of localized collections of fluid.

There has been less pressure for chemoprophylaxis for contacts of patients with haemophilus meningitis than in connection with meningococcal infection, because secondary cases have been thought to be less common (though it now seems that this is not true) and because of the absence of epidemics. Siblings under 5 years old are at increased risk, but there is no reliable prophylactic regimen; some trials have suggested that rifampicin is the best drug available.

Pneumococcal meningitis

The third of the 'big three' causes of bacterial meningitis is the pneumococcus. In most countries it is the least common (but see below). As we have seen, pneumococcal meningitis is mainly a disease of adults, especially of the elderly. When it does occur in younger patients, it is often secondary to a skull fracture, usually affecting the base of the skull and providing the pneumococcus with direct access from the patient's upper respiratory tract to the meninges. If this access remains open, repeated attacks of meningitis may occur (p. 216). In contrast, pneumococcal meningitis in the elderly is usually secondary to bacteraemia and is commonly associated with pneumonia due to the same organism. Immunosuppressed patients of all ages are also particularly liable to this type of infection. A third pattern of incidence occurs in some developing countries, particularly in tropical West Africa. Here epidemics of pneumococcal infection, including pneumococcal meningitis, affect all age groups, including young children. These are devastating outbreaks of disease with high mortality.

Even with modern antibiotic treatment the mortality from pneumococcal meningitis remains high at around 30%. This may, in part, reflect the type of patient who suffers from it, but the virulence of the organism itself must be largely responsible. The drug of choice for pneumococcal meningitis is benzylpenicillin, given intravenously in high doses and at frequent intervals (4-hourly, or even 2-hourly at first). As with other types of bacterial meningitis the severity of the condition demands immediate diagnosis. A film of the CSF will show elongated gram-positive cocci in pairs or in very short chains, usually amongst many polymorphonuclear leukocytes. Invasive pneumococci are invariably capsulate and evidence of the capsules may be apparent on the gram-stained film. The pneumococcus may be isolated from the CSF and also from blood cultures taken before antibiotic therapy is started.

Because pneumococcal meningitis in developed countries is not an epidemic infectious disease, the question of prophylaxis for contacts does not arise. It may arise, however, for patients who have repeated episodes of pneumococcal meningitis as a result of an unhealed skull fracture. Such patients may be protected by continuous treatment with amoxycillin or an oral cephalosporin, but will need surgical treatment to effect a permanent cure. Prophylaxis with penicillin and sulphonamide is also appropriate for patients with fractures of the base of the skull, particularly when the sinuses, middle ear or mastoid are involved and when there is leakage of CSF. Where epidemics of pneumococcal disease occur, the question of widespread vaccination against the pneumococcus must be considered as the only feasible protective measure (p. 80).

Neonatal meningitis (See also p. 325)

The pattern of meningitis in neonates differs markedly from that seen in older children and in adults. The bacteria that cause meningitis in neonates are more varied (the big three are rarely implicated) and the clinical signs and symptoms are often far less dramatic than in older patients. The organisms encountered most commonly in neonatal meningitis are *Escherichia coli*, group B β-haemolytic streptococci and *Listeria monocytogenes*.

Esch. coli is the commonest cause of neonatal meningitis, and other enterobacteria are responsible for a few cases. Such infections are examples of the general increased liability to infection of high-risk babies of low birth weight, or with some other underlying problem, who are nursed in special care baby units. Babies with congenital abnormalities of the central nervous system such as spina bifida and hydrocephalus are at greatest risk of enterobacterial meningitis, but it is by no means restricted to this group. The strains of Esch. coli that cause meningitis are those prevalent in the hospital in question, and are often resistant to several antibiotics. The clinical diagnosis of neonatal meningitis may be difficult. The babies rarely show specific clinical signs (neck stiffness and photophobia are difficult to demonstrate in a neonate) and a high index of suspicion is necessary for early diagnosis. Meningitis should be considered in any neonate who is pyrexial, vaguely unwell, lethargic and floppy and, in particular, irritable and resistant to handling. The diagnosis is confirmed by examination of a sample of lumbar, cisternal or ventricular CSF. The polymorph response in the CSF may not be great but gram-negative bacilli will usually be visible. The treatment of neonatal meningitis is difficult. Ideally, it demands the use of an agent effective against gram-negative bacilli and able to cross the blood–brain barrier readily. Chloramphenicol is an attractive candidate for these tasks, but for the risk of toxicity, notably of causing the 'grey baby' syndrome. The choice of treatment lies between an aminoglycoside and one of the newer cephalosporins (p. 392). Aminoglycosides such as gentamicin and, more recently, amikacin have been used successfully but they do not cross the blood–brain barrier well, and intrathecal or intraventricular injection is therefore necessary throughout the initial course, backed up by intravenous treatment. It is clear that some of the newer cephalosporins (e.g. latamoxef, cefotaxime) achieve acceptable CSF levels following intravenous administration, without the need for intrathecal injection. Intrathecal therapy is always necessary if the meningitis progresses to the most serious form of infection—ventriculitis. In this event antibiotics must be injected directly into the ventricles. Despite modern treatment, there remains a high risk of permanent damage from neonatal enterobacterial meningitis; in particular, the associated inflammation may result in a secondary hydrocephalus.

In recent years a great deal of attention has been focused on two other causes of neonatal meningitis, of which the epidemiological features are remarkably similar. Both group B β-haemolytic streptococci and *L. monocytogenes* are carried asymptomatically in the vagina. Reported studies have given widely different figures for the proportion of women who carry group B streptococci in their vaginas late in pregnancy (p. 325). Those who do so at term may transmit them to their infants during birth. As with *Staph. aureus* and *Esch. coli*, infection may also be transmitted from baby to baby in the nursery. These two routes of transmission account for the two types of group B streptococcal infection in neonates—early and late (see p. 325). Meningitis, almost invariably associated with septicaemia in the early form, may occur in either of these types, but is commoner in the late form. The signs of meningitis caused by group B streptococci are not distinctive and infection must be suspected in any infant who shows the general signs of infection outlined above. Confirmation of the diagnosis depends upon the isolation of the streptococcus from the CSF or from blood culture. The treatment of choice is benzylpenicillin, given by intravenous bolus injections. Prevention of this disease by examination of vaginal swabs from pregnant woman and eradication of streptococcal carriage may not be feasible, but examination of a vaginal swab at delivery may show which infants are at highest risk and indicate those who should be given penicillin if there is any clinical evidence of developing infection.

L. monocytogenes (p. 325) is the least common of these three causes of neonatal meningitis. This small gram-positive bacillus is carried in the vagina and may be transmitted to the neonate during birth. A few cases have resulted from the transmission of the organism either from one infected baby to another or from a colonized mother to another mother's infant whom she has handled. The clinical picture is that of a septicaemia and meningitis without distinctive signs, as described above. The treatment of choice is with ampicillin or amoxycillin and an aminoglycoside. Listerial meningitis is not restricted to neonates and can occasionally occur in immunosuppressed patients and in the elderly. The source of infection in these patients is not known.

Tuberculous meningitis

Mycobacterium tuberculosis infection of the meninges used to be a common and serious form of tuberculosis throughout the world. It is now uncommon in western countries, although most large centres still see a few cases each year, but in the world as a whole it remains a major cause of death and disability. Tuberculous meningitis may arise either as part of the initial spread of infection in primary tuberculosis or as part of secondary (adult) tuberculosis. In children, spread via the blood-stream (miliary spread, p. 207) deposits *Myco. tuberculosis* directly in the meninges and initiates meningitis. For this reason tuberculous meningitis is thought of primarily as a disease of children. However, the elderly or immunosuppressed are also at risk; in these patients initial spread of *Myco. tuberculosis* during primary infection has seeded organisms in the meninges, leaving small quiescent tubercles. These may reactivate under any of the conditions that lead to the development of the secondary (adult) type of tuberculosis (p. 207). The clinical presentation of tuberculous meningitis usually differs from that of acute pyogenic meningitis. The onset is more gradual and protracted and the initial signs are much less clearly those of meningitis. In the juvenile form, a child who has been generally unwell may develop focal neurological signs or become comatose; the differential diagnosis in these patients usually includes intracranial tumours or other space-occupying lesions. In the elderly, a patient who has become increasingly lethargic and dull may slip into coma before the nature of the illness is recognised. The diagnosis depends upon the examination of the CSF and the detection of tuberculous lesions elsewhere in the body. It is particularly important to examine the optic fundi carefully for the presence of tubercles. Unlike other bacterial causes of meningitis, *Myco. tuberculosis* causes a mainly lymphocytic response in the CSF. Although in the early stages of the infection there is a predominance of polymorphonuclear leukocytes, by the time of clinical presentation this has, in most cases, been replaced by a lymphocytic response with between 200 and 600 lymphocytes/mm^3 of CSF. The differential diagnosis lies between tuberculous meningitis and virus meningitis or meningo-encephalitis. In tuberculous meningitis there is a very high protein concentration and usually a low glucose concentration in the CSF. Much of the excess protein is fibrinogen, and CSF from a patient with tuberculous meningitis tends to clot in the form of a spider's web. Ziehl–Neelsen-stained films of the CSF deposit, and in particular of the clot that may have trapped many organisms, may reveal the acid–alcohol-fast bacilli of *Myco. tuberculosis*; but their examination requires perseverance because the bacilli may be only scanty. The treatment of tuberculous meningitis is based upon the same principles as the treatment of pulmonary tuberculosis (p. 397). Three drugs must be given initially and intrathecal treatment may be necessary to ensure adequate levels in the CSF. Streptomycin must be given by the intrathecal as well as by the intramuscular route; but fortunately rifampicin, which should always be part of the initial regimen, is well absorbed from the gastro-intestinal tract and crosses the blood–brain barrier easily, and can therefore be given by mouth. One of the most important complications in tuberculous meningitis is the formation of fibrous adhesions as a result of the very thick exudate that forms over the meninges. These may cause permanent brain damage, and also obstruct the flow of CSF, causing hydrocephalus. Corticosteroid treatment is often used in an attempt to reduce this inflammatory exudate and so prevent these complications and help the penetration of the antibiotics. However, the use of steroids in infection is always debatable. Recovery from tuberculous meningitis is a very long process and most patients are left with some residual damage.

Leptospiral meningitis

Meningeal involvement may be part of the clinical picture of any type of leptospirosis (pp. 109 and 267). If present in Weil's disease, it is usually only a minor component, but canicola fever is predominantly a lymphocytic meningitis. It occurs most commonly in children, who acquire their infections through close associations with infected dogs, usually young puppies. In the puppy the leptospires cause nephritis and are shed in the urine. Children are infected by contact of mucosa or skin surface with infected urine. The early signs of infection are typical of an acute meningitis and similar to those of virus meningitis, but they are not followed by the severe, life-threatening stage characteristic of a bacterial meningitis. CSF examination reveals a lymphocytic response and a raised protein concentration, and leptospiral antibodies can be eventually detected in both serum and CSF. Benzylpenicillin is the drug of choice in the treatment of canicola meningitis and recovery is usually uncomplicated.

Recurrent bacterial meningitis

Most causes of meningitis result in life-long immunity to reinfection with the same organism, and recurrent meningitis (which must be distinguished from relapse after inadequate treatment) is rare except when some anatomical abnormality, congenital or the result of trauma, provides direct communication between the meninges and skin or mucosal surfaces. Apart from neonates with hydrocephalus (p. 226) such a pattern is seen most commonly in patients with unhealed skull fractures. If the defect is not corrected, there may be recurrent episodes of meningitis (up to 20 in one patient have been recorded). The pneumococcus is the most common species causing these episodes and pneumococcal meningitis in a young person should initiate an examination for a communicating skull defect. Other upper respiratory tract commensals—e.g. *H. influenzae, Branhamella catarrhalis*—can also cause this type of meningitis.

Otogenic meningitis

Chronic infection of the middle ears, mastoids or sinuses can lead by direct penetration to infection involving the intracranial tissues. The most common complication is brain abscess (p. 225), but meningitis without involvement of brain tissue may occur. The bacteria found in such cases reflect the bacteriology of chronic otitis—which usually has a mixed flora of upper respiratory tract species, including a large proportion of anaerobes (bacteroides, fusobacteria and anaerobic cocci).

Virus (Aseptic) Meningitis

Virus meningitis is commoner than bacterial meningitis, but is usually a benign self-limiting infection that resolves spontaneously without residual damage. Treatment is symptomatic—analgesia and mild sedation. Meningitis may be the presenting feature of the virus infection, or may be a complication of a virus infection affecting other organs—e.g. of mumps.

When meningitis is the major presenting feature the commonest causative agents are the enteroviruses—particularly echoviruses and, less commonly, coxsackieviruses. These are transmitted by the various faecal–oral routes, or maybe at times by air, to the mucosae of the gastro-intestinal and upper respiratory tracts; they multiply there and reach the meninges during the viraemic phase of the infection. There is often a biphasic illness; a mild gastro-intestinal disturbance (probably due to initial infection and sometimes accompanied by a fever) is followed 10–12 days later by meningitis. The symptoms and signs of the meningitis itself are similar to (but often less severe than) those of bacterial meningitis, but the accompanying systemic disease is more suggestive of influenza than of septicaemia. There may be a mild rash (especially with some echoviruses) but not the petechial haemorrhages of meningococcaemia. The meningitic signs do not progress to coma.

Echoviruses cause sporadic cases of meningitis, but nationwide surveys reveal extensive epidemics due to single serotypes. These are apparent only when data from wide areas are correlated, because only a few of the people exposed to the virus develop meningitis; many are already immune from previous exposure and most of those infected have only mild subclinical infections, usually confined to the gastro-intestinal mucosa and lymphoid tissue (cf. polioviruses, below). A similar pattern is seen with coxsackievirus infection, but outbreaks tend to be more localized.

Polioviruses are also enteroviruses, and in general resemble echoviruses and coxsackieviruses in their epidemiology and clinical manifestations. In unimmunized populations subclinical or mild infection is common; it is confined to the gastro-intestinal tract, or after preliminary multiplication of the viruses in lymphoid tissue there is a light viraemia causing a mild febrile illness with no distinctive features. In a minority of cases, however, the viruses enter the CNS, and the febrile phase is followed by aseptic (i.e. non-purulent) meningitis or by paralysis due to infection of the anterior horn cells of the spinal cord or the brain-stem (poliomyelitis—p. 221).

Mumps virus may well be the commonest cause of mild aseptic meningitis, but this is only sometimes the presenting feature of the infection and is more commonly a minor complication of an attack of mumps (p. 221). Lymphocytic choriomeningitis virus, which causes epidemic infection in wild mice, is occasionally transmitted to man, causing an aseptic meningitis with a high CSF lymphocyte count (up to $1000/mm^3$).

Diagnosis of virus meningitis is by examination of the CSF. The appearance of the fluid is rarely turbid; usually it is either clear or opalescent because of the relatively small cellular exudate. The cell count can be anything from 5–10 (normal ≤ 4) up to $200–300/mm^3$, and almost all of the cells are lymphocytes (hence the terms lymphocytic or aseptic meningitis). The glucose concentration is normal and the protein either normal or slightly raised. The diagnosis may be confirmed by isolating the virus in tissue culture, but it is seldom isolated from CSF and the best specimens to send to the laboratory are faeces (since the intestine is the primary site of enterovirus infections) and a throat swab in virus transport medium. Serological examination of acute-phase and convalescent-phase specimens of serum may be of use in outbreaks due to a known enterovirus or to mumps virus, but serological screening for enteroviruses in general is not feasible because of the large number of serotypes.

Fungal Meningitis

Fungi are rare causes of meningitis in the immunocompetent host. *Cryptococcus neoformans* (p. 141) can cause a subacute or chronic meningitis which is liable to be mistaken for a brain abscess or tumour. Immunocompromized patients, especially those with Hodgkin's disease or AIDS, are particularly liable to develop this disease and with increasingly aggressive chemotherapy and improved survival, cryptococcal meningitis has become more common. The yeast cells can be seen by microscopic examination of the CSF and their characteristic large capsules are well demonstrated if indian ink is first added to the fluid to provide a dark background. Cryptococcosis is best treated with a combination of amphotericin B and flucytosine.

The dimorphic fungus *Blastomyces dermatitidis* (p. 139), which occurs in North and South America and in Africa, may also cause meningitis as part of a chronic systemic infection. Amphotericin B is the most effective treatment but hydroxystilbamidine is a less toxic and often successful alternative.

Protozoan Meningitis

Two species of amoebae which are normally free-living can occasionally infect the brain, causing acute and potentially fatal infections. *Naegleria fowleri* is usually responsible for primary amoebic meningo-encephalitis; *Acanthamoeba* is more rarely involved. Symptoms progress quickly; headache and fever, then nausea, vomiting and meningism, are followed by coma and death, the entire clinical course being 3–6 days. The CSF is usually purulent and may be blood-stained. White cell counts range from a few hundred to 20 000 or more (mostly neutrophils)/mm^3. Motile amoebae, which may contain red cells, may be seen in unstained CSF preparations; explosive formation of pseudopodia is characteristic of *Naegleria* and is best seen in fresh warm preparations. The epidemiology and pathology of these infections are described in Chapter 11 (p. 143).

Encephalitis, Meningo-Encephalitis and Myelitis

Infection of the brain tissue itself (encephalitis) is often a life-threatening disease; the meninges may also be involved, in which case the condition is meningo-encephalitis. A wide range of viruses can infect the neuronal cells of the

CNS, causing changes in personality and level of consciousness, as well as focal signs that depend upon the part of the brain affected. Progressive damage leads to coma and death. Protozoa and *Treponema pallidum* can also attack the brain and cord.

Herpes Simplex Virus Encephalitis

Encephalitis is the least common manifestation of primary herpes simplex virus infection (p. 130). Nevertheless this virus is the commonest cause of sporadic encephalitis. Any part of the brain may be affected and infection is often widespread, but involvement of the temporal lobe is particularly characteristic. The clinical signs and symptoms are those of progressive disturbance of brain function: changes in personality and confusion leading to coma. It may be difficult to differentiate between herpes encephalitis and other causes of confusion and coma. There is usually, but not always, some evidence of herpesvirus infection elsewhere in the body. Patterns of brain-cell activity shown by an electro-encephalogram (EEG) and changes detected by computerised axial tomography (CAT scan) are characteristic but cannot confirm the diagnosis. This can be achieved by electron-microscopic examination of the brain biopsy, especially from the temporal lobe, in which herpesvirus particles can be seen within neurones. Serology may also be helpful; detection of rising or very high titres of anti-herpesvirus antibodies in the serum suggest currrent herpesvirus infection, and the presence of such antibodies in CSF is diagnostic of herpes encephalitis—the most useful CSF finding in this condition. There may be a small excess of lymphocytes in the fluid if there is meningeal involvement, but often the fluid is entirely normal; herpes simplex virus cannot be grown from the CSF.

It is important to confirm the diagnosis of herpes encephalitis without delay, as prompt and appropriate treatment can substantially reduce the high mortality and the high incidence of residual damage in survivors which otherwise occur. Various antiviral agents (e.g. idoxuridine, vidarabine) have been used in different combinations, often with steroids to suppress cerebral oedema and the damaging host responses; some success was achieved with vidarabine, but the most effective treatment is with acyclovir (p.135).

Mumps and Measles Encephalitis

In various systemic virus infections, notably mumps (p. 126) and measles (p. 126), the acute phase of the illness may include a mild encephalitis, which is often masked by other more prominent features of the illness and usually resolves without complications. In mumps it is likely to take the form of a meningo-encephalitis (distinct from the simple aseptic meningitis described above), whereas the usual pattern in measles is an encephalitis with little meningeal involvement. Of greater concern is **subacute sclerosing panencephalitis** (SSPE), a rare but fatal disease of children or adolescents who have had measles some years ealier. It is apparently due to reactivation of measles virus latent in the brain, and is inevitably fatal.

Poliomyelitis

One of the most damaging infections of the CNS is caused by polioviruses (three types—p. 118). These are enteroviruses, and in general show the epidemiological behaviour of that group. Their normal habitat is the human intestine, and they spread by the faecal–oral route—the amount of spread being an index of the sanitary standard of the community. In most people, poliovirus infection is subclinical, producing little more than a mild intestinal or systemic disturbance. The initial site of infection is the lymphoid tissue of the intestinal tract. In some people this leads to viraemia and in a small proportion to aseptic meningitis (p. 219) or to poliomyelitis—a destructive infection of motor neurones of the anterior horn of the spinal cord, or of the brain stem, which causes paralysis. Whereas recovery from poliovirus meningitis is rapid and complete, paralytic poliomeylitis may be fatal (particularly if the brain stem is involved), or may leave severe permanent disabilities.

It used to be thought that paralytic poliomyelitis was a disease of developed countries, and that in conditions of poor hygiene infants came into regular contact with poliovirus while still protected by maternal antibody, and thus became immunized. While general spread is certainly an index of low sanitary standards, it is now clear that there are many paralysed children in developing countries whose condition was caused by unrecognized poliomyelitis. Control has been achieved in developed countries by improved sanitation and by vaccination, first with a killed (Salk) vaccine and more recently by widespread use of the live attenuated (Sabin) vaccine (p. 368).

Certain coxsackievirus and echovirus types sometimes cause paralytic disease which may be clinically indistinguishable from that due to polioviruses.

Rabies

Rabies is a distinctive type of virus encephalitis. All mammals are susceptible to this grim disease, which is endemic in many parts of the world though not in Britain. It is transmitted chiefly through bites inflicted by its victims upon other animals; vampire bats are exceptional in that they can carry and pass on the virus without themselves having the disease. Human infection is usually the result of a bite from a rabid dog. However, even a lick may be sufficient to transmit the virus, which multiplies in the salivary glands and may be present in large amounts in the saliva. The incubation period of the disease varies from less than 2 weeks to several months. The virus travels from its point of entry along the nerves or perineural lymphatics to the brain, and the incubation period is to some extent dependent on the length of this journey, being short if the bite is on the face. After multiplying in the brain, where it does extensive damage, the virus travels to the salivary glands and other parts of the body, probably again along the nerves. The outstanding clinical feature of the disease is a violent and painful spasm of the throat on attempting to swallow, with a consequent fear of drinking (**hydrophobia**). Once established, the disease is incurable and invariably fatal.

A bite or an area that has been licked by a possibly rabid animal should at once be thoroughly scrubbed with soap and water and then flooded with 70% ethyl alcohol or tincture of iodine if available. In hospital it may be thought necessary: (1) to infiltrate rabies immunoglobulin into the area of the wound and also to give an intramuscular dose; and (2) to carry out more thorough debridement of the wound, and to give other treatment to cover the possibility of bacterial infection from an animal bite—notably with *Clostridium tetani* or *Pasteurella multocida*. The other matter for urgent attention is to trace the dog (or other animal), find out whether it has been vaccinated against rabies, and ensure that its subsequent health is carefully observed (see below).

DIAGNOSIS If the animal responsible for the bite is available, it is important to determine whether it really has rabies. If it is dead, microscopic examination of its brain tissue is likely to reveal the typical Negri cytoplasmic inclusion bodies in the nerve cells, and the virus (p. 127) can also be demonstrated and identified by immuno-fluorescence or by electron-microscopy. Mice inoculated with infected brain tissue or saliva develop typical nervous system signs in about a week. If the animal is still alive and fails to become paralysed within 10 days, it is not rabid. The diagnosis can be made by mouse-inoculation of specimens from the human patient, but by the time this is possible it is too late to do anything about treatment. Patients do not live long enough for serological tests to be useful.

IMMUNIZATION Specific neutralizing antibodies confer protection against rabies. Passive immunization, begun soon after a bite or other exposure and consisting of injection of human rabies-specific immunoglobulin (p. 363) intramuscularly and into tissues around any wound, may prevent development of the disease; but it cannot be relied on to do so on its own. The long incubation period leaves time for the exposed person also to be actively immunized. This was first done by Pasteur. Although he did not know the nature of the rabies organism, he so modified it by drying infected rabbit spinal cords that it was harmless and yet an effective antigen when given by subcutaneous injection. Patients could be given further injections of such cord preparations dried for progressively shorter periods, until they were able to tolerate a preparation containing fully virulent virus. Both Pasteur's live vaccines and inactivated vaccines such as that introduced by Semple had the disadvantage that repeated injections of tissue from the central nervous system were liable to cause allergic encephalitis. The vaccine now used is prepared in human diploid cell culture, and has the advantages that fewer injections are required (six, on days 1, 3, 7, 14, 30 and 90) and that the adverse reactions are minimal. It is therefore suitable for routine prophylaxis of those likely to be at risk, as well as for treatment of those who have been exposed.

CONTROL In countries where there is rabies, stray dogs should be destroyed and all others actively immunized. Quarantine regulations and some other aspects of control of this disease are discussed on p. 334.

Post-infectious Encephalitis

A wide range of common virus infections, including influenza and some of the exanthemata (e.g. mumps, measles) that may cause encephalitis as part of the acute infection can cause post-infectious encephalitis. This is a late allergic phenomenon, triggered by the virus infection but without this infection itself reaching the brain tissue. The onset may be several weeks after the initial illness, and the symptoms and signs may be indistinguishable from those of infective encephalitis. They include headache and photophobia, changes in behaviour and in sensation, and then alterations in conscious level. There is cerebral inflammation, which may be focal, and intracranial oedema; and one of the most characteristic features is demyelination. The disease is usually self-limiting and most patients recover, although there may be some permanent damage. The role of steroids in the management is uncertain; they may help to reduce the inflammation and its deleterious effects.

'Arbovirus' Encephalitis

This group of virus encephalitides is caused by togaviruses (p. 121) and bunyaviruses (p. 122). There are more than 300 of these viruses, but not all infect men and not all of those cause encephalitis (e.g. yellow fever, p. 122, and dengue, p. 121).

Togavirus diseases are most common where animal hosts and insect vectors are plentiful—notably in tropical forests. Individual species show quite sharp geographical localization—e.g. eastern and western equine, Venezuelan, Japanese B and Murray Valley encephalitis viruses are found respectively on the east and west sides of North America, in Central and South America, in East Asia and in Australia. Among the many animals and birds known to act as hosts for togaviruses are monkeys, deer, horses, cattle, sheep, pigs, poultry and pigeons. One virus species may have more than one major host—e.g. birds as well as horses are commonly infected with the equine encephalitis viruses. Only for a few species—e.g. dengue—is man the sole known host. He plays an important part in the cycle of transmission of some others—e.g. yellow fever—but most are zoonoses (animal diseases with occasional human infections). The arthropod vectors are blood-suckers, and their infection depends on the occurrence of viraemia in the vertebrate hosts. The viruses multiply within the bodies of the vectors but cause no disease in them.

Infection in man may be subclinical or may produce clinical pictures ranging from a mild generalized febrile illness to a severe encephalitis. The localized illness is often preceded by a more generalized febrile illness, presumably resulting from virus multiplication at some site other than the point of entry or the central nervous system.

Spongiform Encephalitides

These rare forms of encephalitis are caused by 'slow viruses'—agents quite different in structure from normal viruses and much more resistant to inactivation by physical agents such as heat. Two human diseases come within this group. **Kuru** was described only in the highlands of New Guinea where it was transmitted by cannibalism—the ritual consumption of the brains of deceased relatives. The clinical signs were ataxia, progressive dementia, lethargy, coma and death. The suppression of the rituals has broken the chain of infection. **Creutzfeld-Jacob encephalitis** is a more widespread but still very rare form of pre-senile dementia, with softening and destruction of brain tissue leading ultimately to coma and death. The normal route of transmission is not certain, but the disease has been transmitted to laboratory animals and great care must be taken in handling tissues (especially nervous tissue) and secretions from such patients. Probably the most widely studied slow-virus infection is a sheep encephalitis known as scrapie.

All slow-virus infections share the characteristic that no clinical signs appear until many years after the initial symptomless infection; the disease is then slowly and inexorably progressive.

Progressive multifocal leukoencephalopathy is a rare form of encephalitis that occurs in the terminal stages of neoplastic disease, or sometimes in association with other diseases in which the reticulo-endothelial system is extensively involved. Virus particles similar to the papovaviruses that cause warts have been demonstrated by electron microscopy in (but not yet grown from) brain tissue of these patients.

Protozoa in the Brain

(1) Toxoplasmosis

Infection of the fetus with *Toxoplasma gondi* (p. 151) *in utero* (transplacental transmission from the mother) may cause widespread damage; CNS involvement may be manifest as hydrocephalus, microcephaly, intracerebral calcification, choroidoretinitis and psychomotor retardation (p. 321). Infection acquired in early childhood may also cause a fatal encephalitis. Cerebral toxoplasmosis is an increasingly common and serious condition in patients with AIDS (p. 283).

(2) Malaria

Mental disturbances, confusion and coma occur in the more severe cases of malaria (p. 242). These neurological actions may be due either to the effects of toxins and metabolites released from the erythrocytic schizonts when these burst, or, in cerebral malaria, to oxygen deprivation of the brain tissue following blockage of capillaries. The malarial paroxysms that begin with a sudden rigor are followed by the hot stage, during which the patient may be restless, disorientated or delirious, with severe frontal headache and pains in the limbs and back. Coma may follow, gradual or rapid in onset. In fatal cases of cerebral malaria, post-mortem examination shows that the brain capillaries are dilated, grossly congested and blocked with parasitized erythrocytes. The brain may be heavily pigmented and greyish in colour and there may be petechial haemorrhages.

(3) Trypanosomiasis

The neurological effects in African trypanosomiasis (sleeping sickness, p. 246) are probably the result of invasion of the CNS by trypanosomes and not due to interference with the blood supply as in malaria. The inoculation of *Trypanosoma rhodesiense* or *T. gambiense* by an infected tsetse fly during a blood meal results in the invasion of the CSF by trypomastigotes. This usually occurs 1–3 months after the bite in the Rhodesian form but later (6–12 months) in the Gambian form; but the disease may progress very much more rapidly, death occurring within 2–3 weeks of infection. The invasion of the CNS produces a chronic meningo-encephalomyelitis characterized by severe headache, progressive mental dullness, disturbances of co-ordination and reflexes, and lassitude and apathy increasing as the number of parasites in the CSF increases. There is a steady progression of confusion and somnolence; without good nursing care, lack of interest in food contributes to extreme emaciation. In the final stages of sleeping sickness there is profound mental deterioration; the patient becomes more and more difficult to arouse and finally comatose. Examination of CSF shows increasing turbidity, and many trypomastigotes may be seen in the centrifuged deposit. Polymorphs may be present in early cases, being replaced by plasma and morula cells in older infections. CSF protein levels are raised; the higher the level, the longer the infection has been present and hence the more advanced is the disease. Treatment of sleeping sickness is by a course of injections of suramin, pentamidine or melarsoprol; the last is the only drug effective once the CNS is involved.

Neurosyphilis (See also pp. 278–9)

Invasion and destruction of the central nervous system by *Treponema pallidum* in the late (tertiary) stage of syphilis used to be common. The treponemes are present in the brain and spinal tracts and cause progressive loss of function. Destruction of the posterior spinal tracts results in loss of sensation from the limbs and the characteristic stamping gait and loss of balance of **tabes dorsalis**. Destruction of cells in the higher centres of the brain causes personality and behavioural changes and neurological changes, followed by progressive dementia and paralysis known as **general paralysis of the insane** (GPI). With better detection of early syphilis and adequate treatment, these late manifestations are now rare.

Brain Abscesses

Localised infection of brain tissue with pyogenic bacteria causes a brain abscess. The acute inflammatory characteristics are similar to those elsewhere in the body, but the site means that they are very dangerous. If an abscess affects a vital centre directly, it may be rapidly fatal. More commonly, the local damage causes localized neurological signs, which may include epilepsy, and the surrounding inflammation and cerebral oedema cause raised intra-cranial pressure which can lead to coning and death. There are two main types of brain abscess, haematogenous and otogenic, differing in their routes of infection.

Haematogenous abscesses arise by metastatic spread from foci of infection elsewhere in the body, and usually occur in the area of distribution of a main artery such as the middle meningeal artery. The common sources of infection are lung abscesses and periodontal infections.

Otogenic abscesses arise by direct spread of infection from the middle ear, and characteristically affect the temporal lobe of the brain. They are usually associated with chronic otitis and mastoiditis (often with cholesteatoma formation). Similar abscesses may develop from paranasal sinus infections, usually chronic and most commonly in the frontal sinuses with consequent location of the abscesses in the frontal lobes.

The bacteriology of brain abscesses reflects their origins. A pure culture of a single species (e.g. of the pyogenic cocci *Staph. aureus* or *Str. pneumoniae*) is sometimes obtained, but mixed growths are far more common; anaerobes are often a major component. This reflects the importance of such organisms in the primary infections. Thus in brain abscesses secondary to lung abscesses there may be fusobacteria and oral bacteroides, anaerobic and micro-aerophilic gram-positive cocci and some viridans streptococci; abscesses secondary to periodontal disease may contain actinomycetes, notably the micro-aerophilic *Actinomyces israeli* or the aerobic *Nocardia asteroides*; and in those secondary to chronic otitis or sinusitis, bacteria of at least four groups are common—bacteroides (usually of the melaninogenicus-oralis group, but sometimes asaccharolytic species or *B. fragilis*), viridans streptococci (notably the carboxyphilic *Str. milleri*), anaerobic gram-positive cocci and various enterobacteria.

The management of such an abscess requires aspiration or surgical drainage to remove pus and to relieve pressure, together with high dosage of antibiotics effective against the known and suspected pathogens. It is a serious condition, with a high incidence of permanent neurological damage among survivors. Treatment must also be directed at the underlying pathology in the lungs, mouth, middle ear or sinuses.

Brain abscess similar to those already described may result from penetrating injuries. These may carry in exogenous pathogens, but penetration through the frontal sinus (which is the commonest type of injury) will result in abscess bacteriology similar to that associated with chronic sinusitis.

Amoebic Abscess, Hydatid Disease and Cysticercosis of the Brain

Brain abscesses caused by *Entamoeba histolytica* (p. 253) are rare but often fatal; clinical signs are those of any brain abscess (see above). They do not have a defined capsule, and when an amoebic aetiology is suspected a diagnostic biopsy is contra-indicated because of the risk of spreading the infection to other parts of the brain. There are invariably invasive lesions in the gut wall and careful examination of the faeces should reveal cysts or trophozoites. Serological tests may be helpful (p. 183). Treatment is with metronidazole followed by diloxanide furoate to eliminate cysts from the bowel (p. 254).

Hydatid cysts and cysticerci, the larval stages of the tapeworms *Echinococcus granulosus* (p. 156) and *Taenia solium* (p. 155) respectively, may occur in the brain. These cysts are enclosed in well-defined capsules which are partly of host and partly of parasite origin. Tapeworm cysts in the brain are relatively rare, probably occuring in not more than 5% of cases of infection with these parasites. The embryos develop slowly into fluid-filled cysts. *Echinococcus* cysts elsewere in the body may contain several litres of fluid after a number of years, but such unrestricted growth is not possible in the brain. *Taenia* cysticerci are smaller, measuring 1–2 cm in diameter when fully developed. Neurological signs of infection depend upon the site, size and number of cysts and the amount of brain tissue damaged by their presence. Epilepsy is one of the more common effects. The diagnosis of intra-cerebral hydatid cysts or cysticerci depends upon clinical and radiological findings and biopsy should not be attempted. There is no effective drug therapy, but surgical removal of the cyst may be helpful.

Extradural and Subdural Empyemas

Localised and often loculated collections of pus either between the inner surface of the skull and the dura (extradural) or between the dura and the pia-arachnoid or the brain itself (subdural) are rarely primary events. They are usually secondary to some other pathological process which may or may not be primarily infective. A head injury that causes a fracture of the skull often leads to an extradural or subdural haematoma, depending upon which vessels are damaged. The haematoma may then become secondarily infected, particularly if the fracture results in a connection between an extradural haematoma and a sinus or the middle ear, providing a direct route of infection. A haematoma also provides ideal conditions for blood-borne bacteria to lodge and become established, even when the bacteraemia is itself clinically insignificant. Also either extradural or subdural empyemas may develop by direct extension of chronic infections of the middle ears, mastoids or sinuses, as do brain abscesses. This process involves osteomyelitis of the skull, and the underlying meninges are readily involved also.

Subdural empyema may be a complication of pyogenic meningitis, particularly when it is caused by pneumococci or haemophili, and in such an infection it is not uncommon to find localized pockets of pus continuing to cause abnormal clinical signs during treatment.

The bacteriology of these empyemas reflects their different origins. Upper respiratory tract flora can be expected following fractures involving the middle ear or sinuses or when infection has extended from these sites, with bacteroides and other non-sporing anaerobes if the infection was chronic; whereas the bacteria of post-meningitic empyemas are those responsible for the initial meningitis.

CSF Shunt Infection

An important part of the management of hydrocephalus is the provision of an artificial system of CSF drainage by inserting a catheter into the ventricular system and connecting it through a one-way valve to the heart or superior vena cava or to the peritoneal cavity. The catheters and valves are usually made of silastic, with a few metal parts. Like all artificial implants they are readily colonized by bacteria, most commonly by coagulase-negative staphylococci but in a few cases by skin diphtheroid bacilli or viridans streptococci. Most infections are the result of implantation of small numbers of the bacteria, derived from the patient's skin, at the time of operation. It may be several months before infection is apparent, but during that time the bacteria have established themselves in the implant and proliferated until they are sufficient to cause serious trouble. This is usually manifest as shunt blockage or malfunction, and the patient develops signs of raised intracranial pressure and the general features of low-grade infections—a raised white cell count, grumbling fever, raised ESR and in long-standing cases an immune-complex type of nephritis. Antibodies to the staphylococci or other bacteria may develop, together with acute-phase reactants such as C-reactive proteins. The diagnosis is confirmed by culturing the causative organisms from the blood (where they may be present only intermittently) and from CSF taken from the shunt system. Spread of the bacteria may cause a persistent ventriculitis. Treatment is difficult; antibiotics may suppress the infection but they are rarely curative. This difficulty may be due to the protected condition of the bacteria when associated with the foreign implant, which is linked with the bacterium's ability to produce a sticky carbohydrate that may have an important role in shunt colonization. Treatment often entails replacement of the whole valve system. A great deal of effort is therefore directed towards preventing infection, by careful surgical techniques and by prophylactic measures such as extensive skin preparation (e.g. with povidone iodine) and the use of local prophylactic antibiotics at the time of valve insertion.

Laboratory Diagnosis of CNS Infections

CSF examination

The investigation of most CNS infections includes the examination of a specimen of CSF collected aseptically, either by inserting a wide-bored sterile needle between the spines of two lumbar vertebrae and through the dura into the spinal CSF below the termination of the cord (**lumbar puncture** or LP) or by a ventricular tap, for which a needle is inserted through the skull (through fontanelles in babies or through surgical bore holes) and through the brain into the ventricles. From the bacteriological point of view, the most important technical problem is to avoid introducing contaminant organisms either into the subdural space or into the specimen. This calls for rigorous aseptic technique and for antiseptic treatment of the skin - e.g. by applying povidone-iodine or chlorhexidine in 70% alcohol (p. 343). The CSF is collected into a sterile container, and examined by cytological, biochemical and microbiological methods; the examination must be done as soon as possible after collection.

(1) *General appearance.* Normal CSF is crystal clear and colourless; any turbidity indicates either infection or the presence of blood (possibly as a result of trauma during collection of the specimen). In virus meningitis the fluid may be clear, or more commonly faintly opalescent, because it contains small numbers of cells. In acute bacterial meningitis, the fluid is milky or turbid, and in severe cases may be almost too thick to pass through the needle. In tuberculous meningitis, however, the cell count is less and the fluid is usually opalescent, but it also contains a great deal of protein and a 'spider's web' clot often forms in it shortly after collection (p. 218). CSF collected 1 day or more after an intracranial bleed may be yellow ('xanthochromic') because of breakdown products of blood.

(2) *Cell count.* A drop of the well-mixed CSF is placed in a counting chamber for a total white and red blood cell count and for a differential white-cell count. Mixing a drop of CSF with an equal volume of white-cell diluting fluid will lyse red cells and stain the nuclei of white cells to aid the differential count. Red cells in the CSF are the result of trauma during collection or of intracranial bleeding. Normal CSF contains no red cells and not more than 4 white cells (all lymphocytes)/mm^3. In virus meningitis, the cell count is raised; there may be as few as 10 and rarely more than 200 cells/mm^3, and they are predominantly (usually > 90%) lymphocytes, though the proportion of polymorphs may be higher in the very early stages. In encephalitis the cell count may be normal or there may be a slight increase in the number of lymphocytes. In bacterial meningitis, however, the turbidity of the CSF is due to large numbers of white cells; the count is rarely < 1000 and may be 20 000/mm^3, the most common counts being 2000–10 000, virtually all polymorphonuclear leukocytes. Tuberculous meningitis gives a mixed picture. The cell count is usually several hundred (200–500, rarely as many as 1000/mm^3); early in the illness many are polymorphs, but in the later stages of the infection, the more usual time of investigation, lymphocytes predominate.

(3) *Biochemistry*. After the cell count, the CSF is centrifuged and biochemical estimations of the protein and glucose in the supernate are made. Normal CSF contains about 4 mmol glucose (80% of the blood level) and 0.2–0.4 g protein/litre. In virus meningitis these figures are little changed, but in bacterial meningitis the protein is increased, often to several g/litre, and the glucose is low, often undetectable. The protein increase is a result of the inflammatory response, and the glucose is lowered, probably by the metabolic activity of the polymorphs. The biochemical tests are particularly important for differentiation between virus and tuberculous meningitis, both of which produce a lymphocytic response. In tuberculous meningitis the protein is raised (usually very high) and the glucose reduced, in contrast to the virtually normal levels in virus meningitis. The CSF lactate concentration is raised in bacterial meningitis and its estimation (usually by GLC—p. 73) may help with difficult diagnostic problems.

(4) *Microscopy*. Films of centrifuged CSF deposit are stained by Gram's method and by the Ziehl–Neelsen method if tuberculosis is suspected. The characteristic appearance of the common pathogens is readily seen by the experienced eye in gram-stained films: the gram-negative, often but not necessarily intracellular, diplococci of the meningococcus; the small gram–negative cocco-bacilli of *H. influenzae*; and the gram-positive extracellular diplococci of the pneumococcus. Because of the urgency of achieving a precise diagnosis, this immediate microscopic examination is of vital importance for the clinical management.

(5) *Bacteriological culture*. The centrifuged deposit is also used for making cultures for bacteria. Generous loopfuls are spread on chocolate agar for incubation in air with additional CO_2 and on blood agar plates for aerobic and anaerobic culture. With specimens from unusual patients (e.g. those with hydrocephalus) or when bacteria may have been damaged by prior antibiotic therapy, culture of the remaining deposit in broth (e.g. brain–heart infusion broth) may increase the chances of growing the infecting organism.

(6) *Virus culture*. In virus meningitis it is occasionally possible to isolate the virus from CSF that has been placed in virus transport medium and then seeded on to cell cultures. This procedure is, however, unreliable and as most of the viruses that cause meningitis (without other systemic signs) are enteroviruses, throat swabs and specimens of faeces should also be submitted for virus culture.

(7) *Antigen detection*. Because the identity of the pathogen has an important bearing upon the treatment of acute bacterial meningitis, initial investigations applied directly to the CSF may include procedures for the identification of bacteria seen by microscopy or for the detection of bacterial antigens; the latter approach is particularly important if bacteria cannot be seen in the gram-stained film, e.g. as a result of prior antibiotic therapy. The bacteria may be identified by capsule-swelling (p. 176) or by immunofluorescence (p. 178), and the bacterial antigens may be detected by crossed immunoelectrophoresis of CSF with specific antibody, or by co-agglutination, in which the CSF (containing antigen) is mixed with staphylococcal protein-A preparations or latex particles coated with antibody (p. 176). It may also be possible to detect bacterial antigen by these means in serum or even in urine.

Blood cultures

Because bacteraemia is an integral part of most meningitic infection, blood cultures should always be taken. In meningococcal infection the meningococcus may be isolated from blood taken during the bacteraemic phase, before it has progressed to meningeal involvement. Blood cultures taken from febrile children have also shown that some cases of meningococcal, haemophilus or pneumococcal bacteraemia resolve without treatment and without progression to meningitis.

Pus from brain abscesses

Pus aspirated or drained surgically from a brain abscess should be examined by microscopy and by aerobic and anaerobic culture. Microscopy of gram-stained films may give immediate help in diagnosis—e.g. by revealing the mixed flora characteristic of otogenic abscesses, or the characteristic morphological appearances of an actinomycete. Because anaerobes are common in cerebral abscesses, direct examination of the pus by GLC may be useful in confirming their presence.

Diagnosis of virus encephalitis

Routine examination of the CSF may be of little help in encephalitis; there may be no meningeal involvement and the fluid often appears normal in routine tests. Serological studies on both CSF and blood may eventually show raised antibody titres to herpes simplex virus, and this is useful confirmation of a clinical diagnosis, often in retrospect. Serology is also used in the diagnosis of the togavirus encephalitides. However, active treatment—e.g. of herpes encephalitis with acyclovir—may be called for early in the illness, before antibody rises are detectable, and immediate confirmation of the diagnosis is therefore desirable. A CAT scan may indicate the nature of the disease, but the only reliable early confirmation comes from brain biopsy. A needle biopsy is taken from the area of brain known to be affected, and is examined by electron microscopy for typical herpesvirus particles.

17
Blood and Lymphoid Tissue

The blood-stream is the main transport mechanism connecting different parts of the body, and acts as such for many micro-organisms. However, it has no 'normal flora', and presence of micro-organisms in it always represents failure of the defence mechanisms to maintain its sterility. Even though in many cases such failure is transitory and of no clinical importance, in others it is a serious matter and may be life-threatening. Lymphoid tissue, as we have seen in earlier chapters, is an important part of the defence system, being both an elaborate filter to intercept potentially invasive pathogens and the headquarters of the lymphocytes on which immunity is heavily dependent. However, the filter function means that the lymphatic system is liable to clinically significant infection by intercepted pathogens; and it is also the specific primary target for some forms of infection. Involvement of the blood, the lymphatic system or both occurs in many infections discussed in other chapters of this book. Here we deal with infections in which such involvement is such a major feature that they fit more easily into this chapter than elsewhere; and we also deal with infections of the heart because of its obvious close association with blood, even though in some of the heart infections included there is no involvement of the blood-stream.

Some Bacteraemic Illnesses

Bacteraemia and Septicaemia

Bacteraemia is the presence of bacteria in the blood, which they may have entered via the lymphatic system or directly from the tissues or even by direct introduction into damaged or punctured blood-vessels. Transient bacteraemia, with small numbers of bacteria, is a common event, induced by many different activities or procedures (p. 236), and is usually eliminated by the reticulo-endothelial system without any trouble. In other circumstances, however, the virulence of the organism or the impaired resistance of the host or both may result in serious illness. The term **septicaemia** is often used in describing severe bacteraemic infections. It has been defined as meaning 'pus in the blood' or 'bacteria multiplying in the blood'; but there is no practicable means of demonstrating the existence of either of these situations in a particular patient, and the term is best regarded as a clinical description of the form of severe illness that is commonly seen in overwhelming bacteraemic infections (p. 60). (When it is likely that actual pus from an existing abscess has entered the blood and been transported to other parts of the body, the term **pyaemia** is more appropriate.) The existence of bacteraemia is demonstrable by blood culture (p. 240), a procedure that should be carried out whenever a patient has a febrile illness suggestive of a bacterial infection but has no readily accessible localized lesion from from which material for culture can be collected. Such an investigation can be of great value not only in acute and obviously dangerous illnesses but also when there is persistent low-grade fever (see PUO, p. 247). The nature of the organisms grown from blood cultures often suggests the site of an underlying lesion from which they have entered the blood—e.g. pneumococci are likely to have come from a lung infection (p. 202), enterobacteria from the urinary tract (p. 274) and a mixture of *Esch. coli* and *Bact. fragilis* from a gut-associated lesion (p. 349).

For convenience we have divided bacteraemic illnesses into two groups, though in practice they frequently overlap. The first group contains illnesses that commonly present as bacteraemias—i.e. as febrile illnesses with the general features of bacterial infection but with no obvious localization of the infection. In the second group are some illnesses (our list is by no means comprehensive) that commonly present with evidence of local infection, but in which bacteraemia is already present or develops subsequently. In some of those illnesses allocated to our second group the bacteraemia precedes local infection but not usually for long enough to be itself the reason for seeking medical attention.

Illnesses That Commonly Present as Bacteraemias

Bacterial endocarditis

This is an important example of our first group of infections, often presenting as a nondescript febrile illness in which blood culture reveals the diagnosis; but further discussion of it is deferred to the part of the chapter dealing with heart infections.

Typhoid and paratyphoid

Enteric fever is the term used to encompass infections with *S. typhi* or *S. paratyphi* A or B (or occasionally C), which differ from most other salmonella infections (p. 256) in that their main presenting feature is fever rather than intestinal symptoms. The organisms come from other cases or carriers (p. 93), having been excreted in their faeces or urine and transmitted usually in food (p. 361) or water (p. 255). After entering the body of the new host through the mouth they are thought to reach the blood-stream via the intestinal lymphatics. The bacteraemic phase of **typhoid** is followed by localization, chiefly in Peyer's patches of the small intestine, in the gall bladder and in the kidneys, but also sometimes in the bone marrow (possibly leading to osteitis) and in other sites. The clinical picture in the first week of the illness usually consists of progressively mounting fever, headache and severe malaise. Myalgia, cough and sore throat may increase the possibility of confusion with influenza or similar virus infections; but typhoid is suggested in many cases by the severity of the illness, the height of the fever, and the presence of mental confusion and other signs of endotoxaemia (p. 40). Diarrhoea is not common at this early stage, and indeed the patient may be constipated. In the second week the nature of the infection is likely to become more obvious, with the appearance in many cases of the characteristic 'rose-spot' skin eruption on the trunk, the onset of diarrhoea (which may be profuse) and often palpable enlargement of the liver and spleen. Ulceration of Peyer's patches may lead to intestinal perforation or haemorrhage, which are common causes of death in untreated cases. Relapse during convalescence and persistent carriage (p. 269) are common following typhoid that has not been treated or in which the infection has not been eradicated. **Paratyphoid** is in general similar to but milder than typhoid. Some of the other salmonellae can cause bacteraemic illnesses on occasions, particularly in very young, elderly, debilitated or otherwise compromised patients, and abscesses may result, notably in bones. Salmonella osteitis is particularly common in patients with sickle cell haemoglobinopathy.

LABORATORY DIAGNOSIS OF TYPHOID *S. typhi* can be isolated from the blood of about 80% of typhoid patients during the first week of the disease. This can be done by standard blood-culture techniques (p. 240) or by clot culture. In the latter, a blood sample is allowed to clot and the clot is cultured in broth, leaving the serum available for the Widal test for salmonella antibodies (p. 180). Positive blood cultures are less common as the disease progresses; bone marrow cultures may continue to be positive later than those of peripheral blood, particularly when antibiotic treatment has already been started. Faecal culture, which may be positive at any stage, is more often so in the second and third weeks. Urine culture may be positive from the second week. Repeated examinations of the faeces and urine of all convalescents allow early detection and treatment of carriers.

Antibacterial treatment of typhoid is discussed on p. 93, epidemiology on p. 225 and prophylaxis on p. 366.

Brucellosis

Brucellosis in man is alternatively known as **undulant fever** because its clinical pattern often consists of periods of bacteraemia and fever alternating with afebrile periods. Fatigue, drenching sweats, malaise, headache, anorexia, weight loss, joint and muscle pains, splenomegaly and persistent depression are other common features of a clinical picture that is often difficult to recognize. In some patients the illness is mild and in others it is disabling and may persist for years if not recognized and correctly treated. Of the three brucella species (p. 100), *Br. abortus* is the one that was endemic in Britain until recently. It is readily transmitted in unpasteurized milk, but routine pasteurization of milk restricts the incidence of infection to those in direct contact with infected cattle. The organism becomes widely distributed in the body of the human patient, and causes multiple small granulomatous nodules and micro-abscesses in infected tissues. The organisms are predominantly intracellular (pp. 49 and 58). Their isolation from blood is seldom achieved in *Br. abortus* infection, though success is somewhat greater on the rather rare occasions when the diagnosis is suspected, and blood cultures in the necessary special media and under special conditions (p. 101) are set up, soon after the onset of symptoms. *Br. melitensis* is easier to isolate from blood, in part because it tends to cause a more acute and dramatic early illness and so to be investigated earlier. Serodiagnosis of brucellosis is discussed on p. 181, and treatment and prevention on p. 101.

Relapsing fever

Another kind of infective illness in which bacteraemic febrile bouts alternate with afebrile periods is caused by *Borrelia* spirochaetes (p. 110). **Louse-borne relapsing fever**, caused by *Bor. recurrentis*, is a disease mainly of cold and temperate climates, which is transmitted from man to man by the human body louse and has caused major epidemics on many occasions in association with wars and other disasters; it is currently endemic, with seasonal epidemics, in Ethiopia. **Tick-borne relapsing fever**, caused by *Bor. duttoni*, occurs in warmer climates (the Mediterranean area, Africa, southern USSR, southern USA, Central and South America), and is transmitted by soft-bodied ticks to man from other humans or from rodents; cases are mainly sporadic, but sometimes occur in groups of people all exposed at the same time. The spirochaetes of either species are taken up in the blood meal by the tick and multiply in its body; but the means of transmission to man differ. That of *Bor. recurrentis* requires that the louse be crushed on the human skin, allowing the borreliae from its body cavity to penetrate the skin through the louse's feeding puncture or through an abrasion. *Bor. duttoni*, in contrast, infects the tick's salivary and coxal glands and is transmitted to man either by injection of saliva during feeding or by secretion of coxal fluid onto the skin and subsequent penetration as by *Bor. recurrentis*. Infection with either species results in an illness of acute onset, characterized by high fever, rigors, headache, dizziness, nausea and vomiting, muscle pain, conjunctivitis, a rash and hepatosplenomegaly, sometimes with jaundice. The fever lasts 3–5 days and then subsides rapidly, only to recur after a 4–8 day afebrile period. This cycle is repeated several times, each febrile bout being less severe than its predecessor, and finally the patient recovers. Borreliae are plentiful in the peripheral blood during the febrile episodes but scanty in the afebrile phases. Their presence can be demonstrated directly by examining the blood by dark-ground microscopy, or indirectly by injecting some of the blood into mice and examining their blood 3–4 days later. The explanation of the recurring bouts of fever in this disease is that after each bout the borreliae undergo antigenic variation while in the host's organs and then re-emerge into the blood in their new and temporarily less manageable form.

Bacteraemic Illnesses That Commonly Present With Evidence Of Local Infection

These conditions are arranged below in the order in which their causative agents are described in Chapter 7, and page references for those descriptions are given.

Staphylococcal infections (p. 74)

Staph. aureus is one of the commonest causes of bacteraemia, and such infection with it often takes the severe form known as staphylococcal septicaemia. The staphylococci usually enter the blood-stream from a superficial lesion, which may be anything from a small pustule to an infected wound or intravenous catheter site or a boil, carbuncle or abscess. Bacteraemia may be light and undetected during the initial seeding phase but become much heavier with the subsequent development of internal infections such as osteitis (p. 299) or deep abscesses (p. 285). An acute and destructive endocarditis (p. 236) is a particularly serious complication. *Staph. aureus* bacteraemia is readily detectable by blood culture in most cases; but problems may arise when patients have extensive skin lesions with heavy colonization by *Staph. aureus*, since it is then difficult to obtain uncontaminated blood cultures. Treatment of *Staph. aureus* bacteraemia with a β-lactamase-resistant penicillin (e.g. cloxacillin) may be more effective if another suitable antibiotic such as fusidic acid is also used; and it is often essential to deal with local lesions by draining pus or removing infected catheters, etc. Some cases are fatal, usually because treatment was started too late or because of serious underlying disease.

Staph. epidermidis bacteraemia is a rather common and tiresome complication of the use of intravenous lines or prostheses (pp. 351 and 319), especially in neonates and compromised patients. The organisms come from the skin, probably at the time of insertion of the devices in most cases, and colonize the devices—which almost always have to be removed to bring the bacteraemia to an end. The bacteraemia is usually fairly light, and may be difficult to confirm because *Staph. epidermidis* is a common skin contaminant of blood cultures. Repeated cultures, taken with care to avoid contamination and preferably from different venepuncture sites, usually reveal whether there is a true bacteraemia.

Pneumococcal infections (p. 79)

Pneumococci are another common cause of bacteraemia. In many cases the patients have pneumonia (p. 202), and this probably preceded the bacteraemia. In pneumococcal meningitis secondary to skull fracture (p. 219) the bacteraemia is also presumably secondary to the local lesion, as it is when associated with peritoneal infection that entered via the vagina and uterus (p. 79). However, in other cases of pneumococcal meningitis it is probable that the organisms entered the body from somewhere in the respiratory tract without causing any local lesions, and that at least some degree of bacteraemia preceded the development of meningitis; and the same applies to other presumably haematogenous infections such as suppurative arthritis (p. 302) when they are not secondary to pneumonia. Liability to pneumococcal bacteraemia, with or without pneumonia, is high in patients with various predisposing conditions including chronic bronchial disease, alcoholism and leukaemia; and in those whose spleens have been removed surgically or have been severely damaged, as by the multiple infarcts that may occur in sickle cell disease. Such patients may be helped by immunization with polyvalent pneumococcal vaccine (p. 80) or by long-term penicillin prophylaxis. Benzylpenicillin remains the drug of choice for infections with pneumococci, apart from strains (very rare in Britain) that are resistant to it (p. 80).

Meningococcal infections (p. 81)

Probably all meningococcal infections begin with entry of the cocci from the nasopharynx into the blood-stream, and consequent initial bacteraemia. This commonly results in meningitis (p. 213) as the main feature of the clinical presentation, but even so much of the picture is attributable to the bacteraemia itself (meningococcal septicaemia or meningococcaemia). This can also occur without meningitis—i.e. as an illness which by itself would have qualified for inclusion in the previous section of this chapter. Meningococcal septicaemia can be a fulminating infection, sometimes including the Waterhouse–Friderichsen syndrome (massive DIC and haemorrhagic necrosis of the adrenal glands—p. 40) and fatal within a few hours. More often it is less dramatic, with a petechial rash as its most characteristic feature. These petechial haemorrhages, which occur mainly on the trunk and lower limbs but in dark-skinned patients are most easily seen under the conjunctiva, are due to thrombocytopenia, which itself reflects the underlying DIC triggered by meningococcal endotoxin. Larger haemorrhagic lesions and mucosal bleeding are indications of more serious DIC, and may be accompanied by internal haemorrhages such as those into the adrenal mentioned above. The diagnosis can be confirmed by culture of blood, CSF or the skin lesions. Treatment consists of parenteral benzyl penicillin and measures to combat DIC and shock. On occasions meningococcal bacteraemia has an insidious onset and becomes chronic, causing intermittent fever, rashes and arthritis. (See also under Gonococcal Infection.)

Gonococcal infections (p. 82)

Gonococcal bacteraemia occurs in 1–2% of patients with gonorrhoea (p. 279), most of them women, and is secondary to the gonorrhoea. It causes flitting polyarthritis, mostly of large joints, and a pustular erythematous skin rash on the limbs. Gonococci can be grown from the blood and the skin lesions, and later in the illness from joint fluid. Patients with disseminated gonococcal infection, and also those with persistent or recurrent meningococcal bacteraemia, should be screened for evidence of complement deficiency (p. 57).

Enterobacterial and pseudomonal infections (pp. 90–96).

Enterobacteria, pseudomonads and other similar gram-negative bacilli are between them responsible for more cases of significant bacteraemia than any other group of bacteria. *Esch. coli* is the species most often responsible, because its frequent involvement in intra-abdominal sepsis and in infections of the urinary tract (p. 274), the biliary tract (p. 286) and abdominal wounds (p. 349) gives it many opportunities to enter the blood-stream. The result can range from a transient mild febrile episode (e.g. 'catheter fever', p. 352) to septicaemia with DIC and severe shock. The latter condition, when due to organisms of this group, is often known by the unfortunately compressed term 'gram-negative septicaemia', which underlines the important role of the endotoxin of gram-negative organisms (cf. meningococcal septicaemia, above). *Pseudomonas aeruginosa, Klebsiella aerogenes, Serratia marcescens* and similar organisms have often caused bacteraemic illnesses in patients on ventilators, having first colonized the apparatus and then the patient's respiratory tract, but such episodes have been made far less common by attention to design, sterilization and proper management of such apparatus. These and other opportunist bacteria are responsible for many episodes of bacteraemia in compromised hosts, as indicated in Chapter 24, where preventive measures and treatment are also discussed.

Haemophilus influenzae type b infections (p. 98).

Haemophilus meningitis (p. 214), epiglottitis (p. 197), pneumonia (p. 202), suppurative arthritis (p. 302), cellulitis (p. 98) and related acute infections have in common that they are virtually always due to capsular type b of *H. influenzae*, that they occur predominantly (and often in combination) in young children, and that there is an associated bacteraemia. The form of association is almost certainly that the bacteraemia is primary, initiated by haemophili that enter the blood-stream from the nasopharynx and are carried to their target sites in the blood; though epiglottitis may be an exception, in that it seems probable that the epiglottis is invaded before or at about the same time as the blood. Blood culture is an important aid to diagnosis of these infections, and treatment must be directed at the bacteraemia as well as the local infections. The occasional occurrence of petechial lesions like those of meningococcal septicaemia, and even of the Waterhouse–Friderichsen syndrome, are reminders that this too is a gram-negative organism producing endotoxin.

Bacteroides and other anaerobe infections (p. 102)

Increasing efficiency in growing *Bacteroides* species and related anaerobes in recent years has shown that they are common causes of bacteraemia—though still less common than the enterobacteria, with which they share some of their main portals of entry into the blood and so may cause mixed bacteraemias. The anaerobes come most often from abdominal or pelvic sepsis associated with intestinal surgery or perforation. The growth of *Bacteroides* species from a blood culture is sometimes the first indication that the patient has an intestinal neoplasm. Other sources of anaerobe bacteraemia include superficial necrotic ulcers, pressure sores and other gangrenous lesions, necrotizing tonsillitis (necrobacillosis, p. 196), lung abscesses (p. 206) and chronic ear infections (p. 197). Metronidazole is usually the drug of choice for treating infections with these organisms.

Mixed infections

Polymicrobial bacteraemias, including those due to mixtures of enterobacteria and anaerobes which we have just mentioned, appear to be increasing in frequency. This may be merely a reflection of the tendency of bacteriology laboratories not to be content with isolating a single organism from a blood culture but to reincubate primary cultures and carry out further subcultures so as to detect other slower-growing or more fastidious species. However, such mixed bacteraemias are commonest in debilitated and compromised patients, and some of the increase in frequency may be due to the increased numbers of such patients in hospitals (p. 313). The importance of detecting such mixed infections is that antibacterial treatment aimed only at the species that was easy to grow may leave the patient with an untreated infection due to a second or perhaps several more species.

Viraemia

Transmission in the blood is an integral part of the pathogenesis of many virus infections, and may occur at more than one stage of a disease (p. 42). It is the means by which the viruses reach their targets. These targets may be cells in particular systems or organs (e.g. anterior horn cells of the CNS in poliomyelitis, liver cells in virus heptatitis or yellow fever) or widely scattered (e.g. in measles or chickenpox). Viraemia commonly causes general and non-specific symptoms and signs—e.g. fever (often mild), malaise, anorexia, muscle aches—often known as **prodromal** symptoms and signs because they precede the specific and recognizable manifestations of the disease. In other conditions the effects of the viraemia are far more serious—notably in the viral haemorrhagic fevers (p. 124), in which the viruses activate the complement, coagulation and fibrinolytic systems, with consequent DIC, shock and haemorrhage. The methods of isolation of viruses from blood are those used for their isolation from other specimens (p. 169); in severe infections that might be due to highly virulent viruses (e.g. those of the viral haemorrhagic fevers) such isolation should be attempted only in laboratories equipped and designated to handle dangerous pathogens (p. 124).

Infections of the Heart

Infective Endocarditis

Pathogenesis

Many different species of bacteria can cause infection of the endocardial lining of the heart (endocarditis), as can *Coxiella burneti*, *Chlamydia psittaci* and fungi. There is usually, but not always, some underlying abnormality at the site of infection. In most cases one or more heart valves are involved—notably the aortic or mitral valve or both—and were susceptible because of congenital abnormality or damage by rheumatic fever or degenerative disease. Replacement valves, especially prostheses, are also liable to infection—though with different organisms from those that commonly cause endocarditis of natural valves (see below), and with perhaps more infection of the prosthesis itself and overlying fibrin than of endocardium. Endocarditis can also occur on the margins of septal defects, on scars from previous infarcts or almost anywhere where there is an abnormality of the inner surface of the heart. Irregularity of the surface, together with blood turbulence, leads to platelet and fibrin deposition on the endocardium, and it is to these **vegetations** that passing organisms attach themselves. (The sources of these organisms are discussed below.) The organisms that cause endocarditis (many of them streptococci) often have marked ability to adhere to surfaces, which suggests an analogy with the ability of the dextran-producing *Str. mutans* to adhere to tooth enamel and so initiate formation of dental plaque and consequent development of caries (p. 191). Further platelet–fibrin deposition on infected endocardial vegetations provides the adherent and multiplying organisms with protection against the body's cellular and humoral defences and also makes them, as the vegetations grow, increasingly inaccessible to antimicrobial drugs.

The organisms and their sources

The common recognized source of the 'passing organisms' mentioned above is a transient bacteraemia following some disturbance of a heavily colonized mucosal surface—notably dental extraction, or surgical or instrumental procedures involving the intestine or the urinary tract (p. 352). The range of bacterial species put into the blood-stream by such activities is very wide, and many of them have occasionally caused endocarditis; but streptococci do so far more often than any other bacteria, perhaps because of their adhesive capacity (see above). Viridans (α-haemolytic) streptococcal species are particularly prominent—*Str. sanguis*, *Str. mutans* and *Str. mitior* presumably coming usually from the mouth and *Str. bovis* endocarditis being particularly associated with carcinoma of the colon or other gastro-intestinal disease and presumably entering the blood-stream via intestinal capillaries. Prostatectomy or other genito-urinary tract procedures in elderly men with degenerative disease of their heart valves may result in enterococcal endocarditis. Some cases of endocarditis are secondary to illnesses that involve invasion of the blood by bacteria or other organisms. For example, *Staph. aureus* endocarditis, which commonly attacks previously healthy valves and rapidly destroys them, is often secondary to an existing staphylococcal septicaemia; and *Cox. burneti* endocarditis is a complication of Q fever (p. 238). Despite the high incidence of enterobacterial bacteraemia (p. 234), organisms of this group seldom cause endocarditis. The patients in whom they are most liable to do so are those with cirrhosis and drug addicts. The latter group, because they inject contaminated fluids into themselves intravenously and because of their lowered resistance to infection, are at risk of endocarditis (particularly of the tricuspid valve) from organisms of many kinds, including *Staph. aureus*. Microbial colonization of intravenous catheters and feeding lines and other prostheses can provide organisms capable of colonizing damaged endocardium, and is one of the ways by which candida endocarditis can be acquired. This type of infection also occurs in drug addicts, and both *Candida* and *Aspergillus* species (particularly the former) can cause endocarditis following heart surgery—notably after valve replacement—when the lack of any reliable and acceptably non-toxic drug for their treatment means that further surgery is certainly required and that the prognosis is poor. Prosthetic valve endocarditis is most commonly due to *Staph. epidermidis* (p. 76), but many other organisms have been incriminated, including saprophytes rarely encountered by medical microbiologists and presumably introduced at the time of operation. *Staph. epidermidis* infection of the patient's own valve is rare but not unknown.

Clinical features

Acute infective endocarditis is relatively rare. It is caused by aggressive pathogens—most commonly *Staph. aureus*, but sometimes *Str. pyogenes*, *Str. pneumoniae* or *N. gonorrhoeae*—and presents with high fever, general toxaemia and rapidly deteriorating cardiac function due to valve destruction. It is rapidly fatal unless promptly treated with appropriate antibiotics (bactericidal—p. 377), and even when the infection is eliminated the valves are often so badly damaged as to require immediate replacement. The commoner **subacute** endocarditis, such as that due to viridans streptococci, typically develops insidiously, with low fever, night sweats, weight loss and general ill-health. It used to be a disease largely of young patients with valves that were congenitally abnormal or had been damaged by rheumatic fever; but since the advent of antibiotics and the decline in incidence of rheumatic fever the picture has altered greatly in Britain and other developed countries, so that by now more than half of the patients are over 50 years old and the underlying damage in most of them is due to degenerative disease. The clinical presentation of subacute endocarditis is variable, particularly in the elderly. As well as the general indications of an infective illness already mentioned, there are more specific features. These include heart murmurs that are liable to change from day to day as the lesions alter; petechial haemorrhages in the skin, under the conjunctivae and under the nails ('splinter haemorrhages'); raised, hot and usually transient tender lumps, most often in the pulps of fingers or toes (Osler's nodes); and sometimes macular plaques, also transient, on the palms and soles (Janeway lesions). Congestive heart failure may develop, particularly if the infection is not controlled in time to prevent valve damage sufficient to result in incompetence or stenosis. Various complications may result from small fragments of infected vegetations breaking off into the blood and forming micro-emboli. These are responsible for the Janeway lesions described above, but can also have more serious consequences that may include septic infarcts in the spleen, kidneys, brain or other organs. They can also lodge in small arteries, or even in the arterioles of the walls of large arteries, and set up infections that lead to aneurysmal dilatations of the vessels (mycotic aneurysms) which may then rupture; such lesions occur commonly and most seriously in the branches of the cerebral arteries, and may cause intracerebral or subarachnoid haemorrhages.

Investigations

Echocardiography is increasingly used to demonstrate the presence and location of vegetations. It may also give an indication of the infecting organism, in that very large vegetations are suggestive of fungal endocarditis. Typical laboratory findings in infective endocarditis include anaemia, ESR elevation, proteinuria and microscopic haematuria. High levels of immune complexes are found in the blood, and it is useful to monitor these during treatment, since they fall rapidly if it is appropriate. These complexes may be responsible for various non-cardiac manifestations of infective endocarditis, including Osler's nodes and the glomerulonephritis which underlies the presence of protein and red cells in the urine (though gross haematuria may be due to septic infarction of the kidney). However, by far the most important investigation of suspected infective endocarditis is culture of the blood. The general procedure is described in p. 240, but investigation of endocarditis requires special considerations. It is particularly important to set up several blood cultures before treatment is started, because it must last several weeks (see below) and even so may be ineffective if it is not fully appropriate to the causative organism and only suppresses the infection instead of eradicating it. The need for several blood cultures is particularly clear when infection of a prosthetic valve is suspected, because of the difficulty of establishing that coagulase-negative staphylococci (or other skin organisms, such as diphtheroid bacilli, which can also infect prostheses) have come from the blood and not from the skin (p. 240). When initial blood cultures are negative after a few days, incubation should be continued for several weeks in the hope of finding slow-growing organisms; but other special cultures may also be indicated (provided that the patient is not already on antimicrobial treatment) aimed at growing such organisms as streptococci with special nutritional requirements, brucellae, anaerobes or fungi. Fortunately the cases of infective endocarditis that present the greatest problems in isolating their causative agents are seldom acute enough to demand urgent treatment. *Cox.*

burneti endocarditis is identified by serology (p. 113).

Treatment

To be effective, antimicrobial treatment for infective endocarditis must be cidal and sustained—presumably because of the privileged position of the organisms, as indicated above under Pathogenesis. The aim should be to maintain drug levels in the blood such that an eight-fold dilution of the serum is cidal to the causative organism *in vitro* (p. 386). This clearly calls for close collaboration between the clinician and the microbiologist, who needs not only to identify the organism but to determine what drug (or in many cases what synergic mixture of drugs—p. 378) is suitable for killing the organism, and to monitor the level of cidal activity in the blood (p. 386). Antimicrobial treatment is as a rule only part of what the patient needs. Many need medical treatment for heart failure, for example, and some need valve replacement. This may be an emergency procedure because of valve destruction (e.g. by *Staph. aureus* infection); or necessitated by post-infection scarring and contracture, which has made the valve inefficient; or the only way of eradicating the infection when the organism is one against which it is not possible to devise a suitable cidal regimen (e.g. *Cox. burneti* or a fungus). However, the remainder of this section on treatment concentrates on antimicrobial drugs.

STREPTOCOCCAL ENDOCARDITIS For many of the streptococci that cause endocarditis benzylpenicillin is a highly effective cidal drug. High levels can be achieved and maintained by giving it intravenously or intramuscularly and also giving probenecid—otherwise after larger doses the kidneys work hard to keep the levels down! After the first 2 weeks of treatment it is usually possible to substitute an oral penicillin, to be continued for the next month, but the blood levels on such a regimen should be checked to see that they are adequate. For enterococci and other streptococci which are less sensitive to penicillin, a penicillin–aminoglycoside mixture is in many cases synergic *in vitro*, and such results are likely to be reproduced in the patient. Sometimes it is necessary to carry out elaborate tests of a number of drugs for bactericidal effect and synergy before a promising combination is found. Vancomycin is a possible drug for treatment of streptococcal endocarditis in penicillin-hypersensitive patients, and it may be synergic with aminoglycosides against these organisms—though unfortunately the nephrotoxicity of such a mixture is high.

STAPHYLOCOCCAL ENDOCARDITIS When *Staph. aureus* endocarditis is diagnosed or suspected, antibacterial treatment must be started at once, without waiting for laboratory guidance, and a mixture of cloxacillin and gentamicin is appropriate. If the organism turns out to be cloxacillin-resistant, or if the patient is hypersensitive to penicillins, vancomycin is again a possible alternative, subject to the qualification given above. *Staph. epidermidis* infection, usually of a prosthetic valve as we have indicated, presents a far more complex therapeutic problem because the sensitivity patterns of such organisms are varied and unpredictable and multiple resistance (including cloxacillin) is common. Vancomycin may be the most appropriate drug, but giving it for the necessary period of several months may be difficult or impossible (p. 396). Other appropriate forms of treatment have to be determined by *in vitro* tests of cidal activity against the strain in question, but it is hardly ever possible to avoid removal and replacement of the prosthesis.

Q FEVER ENDOCARDITIS *Cox. burneti* infection is best treated by prolonged administration of a tetracycline, with rifampicin as an alternative; but these drugs rarely achieve eradication of the infection, and excision of the infected valve is nearly always necessary.

FUNGAL ENDOCARDITIS Once again the drugs available for treatment of such infections— amphotericin B and the imidazoles (p. 142)— cannot eradicate infection from a valve, and excision is required.

Prophylaxis

A history of recent dental extraction or dental surgery is given by some 15–20% of patients with infective endocarditis; and enterococcal endocarditis has a similar association with recent prostatectomy and some other urinary tract procedures. These are thus good theoretical grounds for believing that it is wise, when carrying out operations of this kind on patients with damaged heart valves or other heart lesions at risk of infection, to ensure that at the time of operation the patient's blood contains antibacterial drugs at levels adequate to kill the likely potential pathogens on their way to the heart. Such an approach has strong support from studies of experimental endocarditis in rabbits, but the unpredictability of occurrence of endocarditis has made it difficult to determine how effective such prophylaxis is in man. Starting 'prophylaxis' too early has the highly undesirable effect of encouraging overgrowth of antibiotic-resistant bacterial strains on the relevant mucosal surfaces. A single high dose (3 g for an adult) of oral amoxycillin, given 1 hour before dental extraction or other major dental work, provides at the time of operation and for some hours afterwards a blood level adequate to deal with oral streptococci; erythromycin is a suitable alternative drug if the patient is hypersensitive to penicillin or has recently received a penicillin which may have altered the oral flora, but at least two doses are recommended. Although minor dental work or even vigorous brushing of the teeth may induce a transient bacteraemia (p. 60), there is no evidence that such procedures carry an increased risk of endocarditis. Bacteraemia following intestinal or genito-urinary tract (including obstetric and gynaecological) surgery or instrumentation is usually enterobacterial, and this carries little risk of endocarditis (p. 236); but the risk of enterococcal endocarditis in adults with damaged valves undergoing prostatectomy or urinary tract instrumentation is clearly established, and a recommended prophylactic regimen consists of two doses of parenteral amoxycillin and gentamicin, 1 hour before and 6 hours after operation. It may also be wise to give some prophylaxis to at-risk patients when any surgery in the abdomen or pelvis involves an infected or heavily contaminated or colonized area; and it is commonly given in such circumstances if the patient has a prosthetic heart valve, because of the high risk and serious consequences of infection of these devices.

Parenteral prophylaxis (e.g. with cloxacillin and gentamicin) is generally used perioperatively in valve-replacement surgery.

Myocarditis

Infective myocarditis can be caused by any of a wide range of viruses, but is most commonly due to group B coxsackieviruses—in Britain, usually in the summer, when these viruses are most prevalent. (They also commonly cause pericarditis—see below.) The diagnosis can be confirmed by isolating the virus from the throat or the faeces, or from the myocardium itself or pericardial fluid if any, or by demonstrating a high or rising antibody level for the virus in question.

Allergic myocarditis is the major life-threatening component of rheumatic fever, induced by group A β-haemolytic streptococcal infection usually of the throat (p. 194). *Trypanosoma cruzi* myocarditis (p. 246) probably also has an immunological basis.

Toxic myocarditis is one of the main effects of *C. diphtheriae* toxin, which stops protein synthesis in heart-muscle cells, so impairing the function of the heart and causing acute circulatory failure. The actual infection is usually in the pharynx or larynx (p. 195).

Pericarditis

Inflammation of the pericardium commonly leads to a **pericardial effusion**, which may be large enough to interfere with heart function. If the effusion is caused by a pyogenic organism it is likely to become purulent and to need surgical drainage. Following infection (particularly if it is tuberculous) the pericardium may become thickened, fibrous and contracted. Such **constrictive pericarditis**, in which the heart is enclosed in an inelastic sac that increasingly restricts its movements and output, eventually requires surgical removal of some of the sac. Most patients with early infective pericarditis have (in addition to the general signs of significant infection), chest pain and dyspnoea, and if there is not already enough fluid to separate the inflamed pericardial surfaces, movement of the heart will cause a pericardial rub, audible on auscultation. The presence of any substantial amount of pericardial fluid can be demonstrated by radiology or echocardiography.

Pericarditis of acute onset in a previously healthy patient, particularly a young adult, is likely to be due to infection with a virus, most commonly a group B coxsackievirus. Bacterial pericarditis may be (1) secondary to a nearby lung lesion—it is then usually pneumococcal; or (2) the result of a haematogenous spread from septic lesions elsewhere, usually caused by common pyogenic bacteria such as *Staph. aureus*, *Str. pyogenes* or enterobacteria; or (3) a manifestation of a primary bacteraemia—e.g. with *H. influenzae* type b, in young children (p. 234). However, most adults who develop bacterial pericarditis without evidence of infection elsewhere have underlying systemic diseases such as diabetes or chronic renal failure. Tuberculous pericarditis may be a component of miliary tuberculosis, or may result from post-primary activation of foci seeded earlier (p. 207). Fungal pericarditis occurs in histoplasmosis (p. 139), and also in some immunocompromised patients (notably those on cytotoxic drugs) as a component of invasive candidiasis or aspergillosis. Aspirated pericardial fluid is often the most useful specimen for identification of the organism causing pericarditis. Treatment includes use of systemic antimicrobial agents as appropriate to the organisms involved, and surgery is necessary in many cases, as indicated above.

Blood Cultures

Culture of venous blood is the most important microbiological investigation in many forms of infection, including most of those (other than virus infections) discussed in this chapter. The wise clinician is quick to have blood cultures set up whenever a patient has a febrile illness with no obvious explanation or accessible local lesion and without features typical of a virus infection (see PUO, p. 247). Unexplained leukocytosis is another reason for doing so, as is neutropenia in some circumstances, and as is hypothermia in neonates or the elderly, whose response to serious infection may be paradoxical in this respect. Even poor feeding by a neonate may be a sufficient reason for suspecting a bacteraemic illness. Delay in diagnosis and appropriate treatment of a serious infection can often be avoided by early and careful collection of blood cultures.

Blood for culture should be collected by venepuncture, with scrupulous care to avoid contamination from the skin of the patient or the operator, since the laboratory procedures are calculated to grow bacteria present in only small numbers and growth of skin-type organisms can be very difficult to interpret (pp. 236 and 237). The skin over the chosen vein—usually in the antecubital fossa or the fore-arm—should be cleaned and treated with a suitable disinfectant (p. 343), which should be allowed a few minutes to work. The vein, distended by means of a tourniquet, should then be punctured with a sterile needle attached to a sterile syringe or to a specially designed sterile evacuated ampoule. It may be best (if circumstances permit) to take blood when the patient's temperature is rising, as this is likely to be the time of heaviest bacteraemia. When the patient is desperately ill a single blood culture at least can be taken before antibacterial treatment is started, and is likely to give a useful result because in such a patient the organisms are probably numerous and easily grown. When time permits, however, it is best to take three samples of 10–15 ml each over a period of a few hours, before antimicrobial treatment is started. This allows the sorting out of problems due to contamination from the skin (p. 237) and covers the possibility of accident to one of the specimens; and it also makes some allowance for the possibility of intermittent bacteraemia. There is seldom any point in taking more than three samples initially, though if these give negative results in the first few days (which is most likely to happen in low-grade infections that have not demanded immediate institution of treatment) further samples should be taken (p. 101). As soon as blood has been collected (before it has had time to clot) it is introduced into various containers (preferably sealed, with diaphragms through which the blood can be injected) containing broth media suitable for aerobic, anaerobic and any appropriate special forms of culture. 3–5 ml of blood per broth culture is suitable for most systems. (Alternatively, all of the blood may be added to a small amount of broth containing sodium polyanethol sulphonate—'Liquoid'—which prevents coagulation and allows distribution into appropriate media to be carried out in the laboratory.) A long-standing problem in laboratories has been whether to subculture blood-culture broths daily—which maximises the chances both of detecting pathogens early and of introducing contaminants that will confuse the interpretation of later subcultures—or to subculture only when the broth shows naked-eye evidence of microbial growth. A solution to this dilemma now widely available is to use broths that include bacterial nutrients containing radio-active carbon; and to examine them by means of a machine that can sample the atmosphere in the tops of culture bottles daily, or more often, without introducing contaminants and can detect the arrival there of radio-active CO_2 released by the metabolic processes of the multiplying bacteria. As soon as the machine records a positive finding from a bottle it can be subcultured.

To avoid tedious repetition we have described blood culture as though it referred only to culture of bacteria. What we have said applies also to culture of fungi from blood, except that some of them require special media.

Protozoan Infections of Blood

Malaria and babesiosis are protozoan infections predominantly of the blood. The blood and cardiovascular system are also importantly involved in trypanosomiasis, but this disease is described in the next section of the chapter, since lymphoid system involvement is even more prominent.

Malaria

To understand the description that is given here, it is necessary to be familiar with the life cycles of *Plasmodium* species as described in Chapter 11 (pp. 148–51).

The clinical features of malaria depend largely on the parasites' invasion of and multiplication in the host's red blood cells (the erythrocytic schizogony cycle) and the consequent destruction of the red cells; and the clinical differences between the forms of malaria caused by the four *Plasmodium* species reflect differences in their erythrocytic cycles. Whereas invasion by *P. vivax* and *P. ovale* is largely restricted to reticulocytes (early red cell forms) and that by *P. malariae* to the older among the red cells, *P. falciparum* can invade red cells of any age; and the severity of the illness caused by *P. falciparum* is in part due to the fact that it customarily invades a much larger proportion of the red cell population than do the other species. The incubation period between infection and onset of clinical symptoms depends on the speed with which red cells become parasitized, and so depends to some extent on the species of parasite. With any of them, parasitaemia is usually greatest at about 2–3 weeks after the first invasion of red cells, as after that host immunity decreases the number of red cells invaded. The main presenting features of malaria are paroxysms of chills, fever and sweating, occurring (except early in the disease) at intervals that are determined by the length of the erythrocytic cycle of the species responsible and the consequent frequency of release of merozoites from erythrocytic schizonts. *P. falciparum* completes its erythrocytic cycle in 36–48 hours (subtertian malaria), *P. vivax* and *P. ovale* do so in 48 hours (tertian malaria) and *P. malariae* in 72 hours (quartan malaria)—the terms tertian and quartan reminding us of the antiquity of the disease, since they are derived from the Roman method of counting days. The length of the intervals is a less reliable guide to species in the early stages of the disease because several unsynchronized broods of parasites may at that time be acting independently; but before long they are coordinated, and the bouts of fever become regular and more pronounced. The fever of malaria is thought to be caused by the release of toxins and metabolites of parasite origin when the erythrocytic schizonts burst. Anaemia in malaria is in most cases largely due to red cell destruction, and in falciparum malaria may rapidly become severe. The situation is complicated in some cases by a host auto-immune response that destroys large numbers of uninfected red cells. This results in heavy excretion of haemoglobin in the urine, a condition known as **blackwater fever** and seen most often in falciparum malaria. Another serious complication seen when parasitaemia is particularly heavy (usually in falciparum malaria) is the blocking of blood capillaries in deep tissue, notably in the brain, by accumulations of erythrocytic schizonts. This, adding to the tissue hypoxia already existing because of anaemia, combines with circulating parasite toxins to cause serious metabolic disturbances in the brain (**cerebral malaria**) and elsewhere, with death of the patient unless appropriate treatment is started promptly (or even in spite of such treatment). Acute glomerulonephritis is another complication sometimes seen in falciparum malaria, but it is more characteristic of *P. malariae* infection. It leads to lowering of the albumin level in the plasma and elevation of the globulin (particularly γ-globulin) level, with little change in the total plasma protein concentration.

Chemotherapy and chemoprophylaxis

The chemotherapy and chemoprophylaxis of malaria is complicated by the fact that *Plasmodium falciparum* exhibits increasingly widespread resistance to several drugs in various parts of the world, including SE Asia, eastern India, Central and South America, and East and Central Africa. Resistance is chiefly and most importantly to chloroquine but in some areas there is resistance to other antimalarials as well. The 4-aminoquinoline drugs—chloroquine and related drugs such as amodiaquine—are the most rapidly effective in destroying the erythrocytic schizonts of *P. falciparum* and are the first choice for treatment. However, the extent and degree of resistance are increasing. Chloroquine is generally given by mouth but in patients who are seriously ill or who cannot retain the drug because of vomiting it may be given intravenously or by intramuscular injection, although the latter is less satisfactory.

Where there is chloroquine resistance alternative drugs, often in combination, must be used. In areas such as Thailand, where there is a high level of resistance to several drugs, a seriously ill patient would require prompt treatment with intravenous quinine or quinidine. Where resistance is to a narrower range of drugs, 'Fansidar' (pyrimethamine + sulphadoxine) may be used, although quinine would still be the best choice for seriously ill patients. *P. falciparum* multiplies in the blood much more quickly than the other species of *Plasmodium*; parasitaemias of 30–40% may be reached within a few days. Such levels are life-threatening and failure by the patient to respond quickly may require a change of treatment.

Chloroquine resistance does not occur in *P. vivax*, *P. ovale* and *P. malariae*, and acute episodes of malaria due to these species are best treated with chloroquine. However, *P. vivax* and *P. ovale* persist in the liver as hypnozoite stages which are not susceptible to chloroquine and may give rise to relapses. Hypnozoites should be treated with primaquine after chloroquine treatment has eliminated the erythrocytic schizonts. Hypnozoites are not known to occur in *P. malariae*, and chloroquine alone is adequate treatment for this type of malaria.

Prophylaxis of malaria is also complicated by the emergence of drug resistance, and up-to-date advice should always be sought. Prophylaxis is recommended for three groups of people:

(1) Non-immune visitors to malarious areas; these need not necessarily be from the temperate areas of the world (malaria has been eradicated from some areas of the tropics and was never endemic in others). This group includes pregnant women, for whom an attack of malaria is particularly dangerous.
(2) Local people who are returning to a malarious area after long residence (a year or more) in a non-malarious area and who have, therefore, lost the immunity induced by repeated exposure to infection.
(3) In some cases, local children under the age of 5 years living in an endemic area.

Prophylactics must be taken with strict regularity and at the doses recommended. Treatment should start one week before entering the malarious area, continuing for the duration of the stay and for a further four weeks after leaving the area. However, although prophylaxis considerably reduces the chances of infection, it does not guarantee absolute protection.

Chloroquine, given as a single weekly dose, is a suitable prophylactic against *P. vivax*, *P. ovale* and *P. malariae*, and against *P. falciparum* in areas where chloroquine-resistant *P. falciparum* does not occur. In areas where chloroquine resistant *P. falciparum* occurs, alternative prophylactic drugs have been recommended, such as Fansidar (see above) or Maloprim (pyrimethamine and dapsone) or amodiaquine. However, Fansidar causes severe skin reactions and Maloprim causes bone marrow failure in some patients, particularly when given in conjunction with chloroquine. An alternative regimen now favoured by some physicians is a weekly dose of chloroquine plus daily proguanil (Paludrine), with strong advice that any case of fever in a patient taking these drugs should be referred immediately to a suitably experienced clinician for careful investigation in case malaria has developed despite the prophylaxis.

Babesiosis

Babesiae are tick-transmitted animal parasites (p. 151). Some species are on rare occasions transmitted to man, directly invading his red blood cells and multiplying there. Haemolysis is a major feature of the type of human babesiosis reported from Europe, which is due to *B. bovis (divergens)*, occurs almost exclusively in patients who have had splenectomies and is virtually always fatal. Human infection with *B. microti*, a rodent parasite, occurs in eastern USA and is a milder illness; it begins 10–20 days after a tick bite, with gradually increasing malaise, fever, chills, sweats, joint and muscle pains, fatigue and weakness, and lasts for weeks or even months. Chloroquine appears to afford some symptomatic relief in babesiosis but not to affect the extent or duration of parasitaemia.

Lymphoid Tissue Infections

Lymphangitis

The lymphatic system is a common route for passage of pathogens from peripheral portals of entry or primary lesions to the blood-stream and so to other parts of the body. As a rule the lymph vessels themselves are not evidently affected by such passage, but sometimes they become inflamed (acute or chronic lymphangitis). *Str. pyogenes* infection of, say, a cut or abrasion or puncture wound on the hand can provide the classic example of **acute lymphangitis**; the primary lesion is soon linked to enlarged and tender lymph nodes in the arm and axilla by clearly visible red lines on the skin that mark the positions of acutely inflamed lymph vessels. In such circumstances the streptococci can often be grown from the blood as well as from the local lesion, and prompt penicillin treatment must be given to prevent development of streptococcal septicaemia. *Pasteurella multocida* infection of an animal bite can produce a somewhat similar picture, except that the sequence is less acute, and local cellulitis and suppuration precede the appearance of lymphangitis. Penicillin is again the appropriate drug. **Chronic lymphangitis** can occur in mycobacterial infections (p. 106), in filariasis (p. 246) or in fungal infections, as typified by sporotrichosis. In this infection the fungus, found in soil and on some timbers and plants, enters the skin through a penetrating injury such as a prick from a rose thorn, and forms an erythematous nodule. The nature of this lesion becomes clear when it fails to respond to antibiotic treatment and when a nodular thickening of the local lymph vessels develops. Potassium iodide is the recommended treatment.

Lymphadenitis

Microbial infection causes lymph-node infection (lymphadenitis) far more often than lymphangitis. Infected lymph nodes are recognizable by their enlargement and, when the infection is acute, their tenderness. The infection may involve a single node or a local group, or sometimes nodes throughout the body. (The term lymphadenopathy can be used in description of the clinical findings, but it is non-committal as to the cause of the lymph-node abnormalities and so is inappropriate to this discussion limited to infective disease of lymph nodes.) Temporary enlargement and even tenderness of one node or a small group is often the only evidence of an otherwise subclinical and insignificant illness—part of the continuing battle between the body's defences and potential invaders. More marked swelling and tenderness of local lymph nodes, often accompanied by fever and polymorphonuclear leukocytosis, are common, especially in children, in association with local infections with *Staph. aureus* or *Str. pyogenes* (especially tonsillitis or pharyngitis), and may be more prominent than the primary lesion and a useful indication of its whereabouts. The cervical and inguinal lymph nodes are the groups most often palpably enlarged by infection, and mesenteric lymphadenitis also has special importance but is far more difficult to detect. Lymphadenitis in these sites will now be discussed, then generalized lymphadenitis. After that we shall deal with particular infections involving the lymphatic system.

Cervical lymphadenitis

Some enlargement of cervical lymph nodes is common in virus infections affecting the upper respiratory tract; and posterior cervical lymph-node enlargement is a useful diagnostic sign in rubella. More marked enlargement is common in infectious mononucleosis (see below). The commonest bacterial infections causing cervical lymphadenitis are staphylococcal infections of the face, scalp or neck and streptococcal tonsillitis; in the absence of antibiotic treatment, infection of tonsils and the tonsillar lymph nodes can become chronic. Mycobacterial chronic cervical lymphadenitis was common in Britain when milk containing *Myco. bovis* was widely available, as infection with this organism was often initiated by entry through the pharyngeal mucosa; and spread from a pulmonary infection with *Myco. tuberculosis* can sometimes involve cervical lymph nodes. Today, however, such cases of mycobacterial lymphadenitis as do occur in this country are usually due to atypical mycobacteria (p. 106). The lesions develop insidiously into large fluctuant swellings, and as atypical mycobacteria are commonly resistant to antituberculous drugs surgical removal of the affected nodes is usually the best treatment.

Inguinal lymphadenitis

Unilateral inflammation of the inguinal lymph nodes is usually associated, particularly in children, with septic lesions on the appropriate lower limb—sometimes obvious, but in other cases (especially when they are between or under the toes) unsuspected until the lymph nodes attract attention. Bilateral inguinal lymphadenitis is common in some of the sexually transmitted diseases. In primary **syphilis**, if the chancre (p. 278) is at one of the usual genital sites it is accompanied by painless enlargement of inguinal lymph nodes. (Lymphadenitis is also part of the picture of secondary syphilis, but is then generalized.) Bilateral inguinal node tenderness is common in primary or recurrent **genital herpes**, and in **gonorrhoea** but not in non-gonococcal urethritis. The transient primary lesion of **lymphogranuloma venereum** (p. 280) often passes unnoticed, and the patient may present some 10–30 days after exposure complaining of fever, malaise and an inguinal lump. This unilateral lesion, a **bubo**, is a swollen tender lymph node that later suppurates, becomes fixed to the skin and may rupture and develop a persistent sinus. (For treatment see p. 114). A similar but more painful bubo develops in **chancroid** (p. 280) about a week after the primary lesion.

However, the infection that classically produces buboes is not a venereal disease but **bubonic plague** (p. 95). This *Yersinia pestis* infection is spread by flea bites. The resultant illness begins with fever, malaise, headache and tender swelling of lymph nodes draining the bitten area; and as the rat-flea most often bites humans on their legs, the affected nodes are commonly inguinal. The mass of oedematous and matted lymph nodes suppurates and is liable to discharge through the skin, but a more serious development—likely to be fatal in the absence of appropriate antibiotic treatment—is septicaemia. Suspected plague should be promptly treated with tetracycline. The organism can be isolated from the bubo pus, and from the blood in cases sufficiently far advanced. (Pneumonic plague is mentioned on p. 199). **Tularaemia**, caused by *Francisella tularensis* (p. 101), is a plague-like illness that commonly presents with an ulcerative skin lesion and painful lymphadenitis, often inguinal.

Mesenteric lymphadenitis

Acute mesenteric lymphadenitis is commonest in children aged 5–14 years, and is usually caused by adenoviruses or by *Yersinia enterocolitica* or *Y. pseudotuberculosis* (p. 95). Its presenting features are often fever and right iliac fossa pain and rebound tenderness—in other words, it mimics acute appendicitis. When it is due to adenoviruses, or perhaps sometimes other viruses, there is often an associated infection of Peyer's patches in the intestinal wall, and those swollen masses of lymphoid tissue may be carried along inside the intestinal lumen by peristalsis, causing intussusception. Support for the diagnosis of adenovirus or yersinia infection may be provided by serological tests, or by isolating the virus from the throat or the faeces or yersiniae from faeces by 'cold enrichment' (p. 95). Chronic mesenteric adenitis occurs in intestinal tuberculosis. The tubercle bacilli also localize in the intestinal lymphoid tissue, particularly in the caecum; caseation produces transverse ulcers that are typical of intestinal tuberculosis.

Generalized lymphadenitis

Widespread lymphadenitis occurs in many blood-borne infections, including infectious mononucleosis (see below), secondary syphilis (p. 278), leptospirosis (p. 267) and miliary tuberculosis (p. 207). Indications of the aetiology may come from a history of recent travel to endemic areas (dengue, filariasis, histoplasmosis) or of occupational risk of exposure (leptospirosis, brucellosis); from the presence and nature of a rash (rubella, measles, secondary syphilis); from the presence of abnormal cells in the blood (infectious mononucleosis); or from various other sources. Generalized lymphadenopathy persisting for more than three months in a homosexual male may well be due to infection with HIV virus and may be a prelude to the development of AIDS (p. 283).

Infectious Mononucleosis (Glandular Fever)

The first, and now more commonly used, of these two names for this disease emphasizes the haematological findings; whereas the second refers to the clinical prominence of enlargement of lymph nodes (formerly called lymph glands) and requires it to be mentioned in this chapter. However, it has already been described in Chapter 14 (p. 193).

Toxoplasmosis

The most serious infections caused by *T. gondi* (p. 151) are those acquired *in utero* or during the neonatal period (described in Chapter 25, p. 321), and choroidoretinitis (p. 297). Here we are concerned only with non-ocular infections of older children and adults. Infection is acquired by ingestion either of oocysts, usually from the faeces of domestic cats, or of encysted stages in meat. Trophozoites develop in cells of the reticulo-endothelial system. In many cases this initial infection is asymptomatic or causes only a mild nondescript illness, but some patients develop fever, chills, headache, fatigue and cervical lymphadenitis. Illness of this type usually resolves within 2–3 months. However, toxoplasmas disseminated through the lymphatic system and the blood to all parts of the body during the primary infection (asymptomatic or symptomatic) form tissue-cysts in which they can persist. Many years later lowered host immunity may result in their reactivation (p. 319). Histological examination of lymph nodes enlarged by *T. gondi* infection shows characteristic changes. Serodiagnostic tests available include those using immunofluorescence, ELISA or agglutination, and a dye test that depends on the fact that exposure to specific antibodies renders the parasites resistant to staining with alkaline methylene blue; but when interpreting such tests allowance must be made for the frequency of *T. gondi* antibodies in blood from healthy people.

Trypanosomiasis

The lymph nodes are an important site of parasite development in the early stages of both

African and South American trypanosomiasis. The epidemiology, causative organisms and pathology of these diseases are described in Chapter 11 (pp. 145–7).

In **African trypanosomiasis** (sleeping sickness) infection from a tsetse fly bite is followed by an incubation period which may last several weeks in Africans but may be only a few days, particularly in non-Africans. A painful weeping sore with surrounding swelling, the **trypanosomal chancre**, forms at the site of the bite in about 25% of patients. At this stage trypomastigotes are to be seen in aspirates from the chancre and there may be a few in the blood. The most characteristic early feature of the disease is tender enlargement of lymph nodes, particularly of those in the posterior cervical group (Winterbottom's sign). From the time of development of this adenitis there may be irregular episodes of fever lasting a few days and associated with the appearance of more numerous parasites in the blood. Trypomastigotes are also often visible in aspirates taken from the enlarged cervical nodes. The infection may terminate spontaneously at this stage, but more typically there is progressive involvement of the nervous system (p. 224), with trypomastigotes present in the CSF. The laboratory diagnosis is based on demonstration of trypomastigotes in the various fluids mentioned above, including both thick and thin films of blood (cf. malaria, p. 173). To find scanty parasites in blood it may be necessary to concentrate them by such methods as triple centrifugation (p. 146). The treatment of African trypanosomiasis depends upon the stage of the infection. If there is no abnormality in the CSF examined before treatment and no clinical evidence of neurological involvement, suramin should be given by slow intravenous injection. A freshly prepared test dose of 200 mg is given first, and if there are no adverse reactions a course of five doses is given over a period of three weeks. A number of side-effects are associated with suramin treatment and the patient requires careful monitoring.

If the CSF protein level is raised or there are > 5 leukocytes/ml of fluid, neurological involvement is assumed to have occurred and melarsoprol (Mel B) is given instead of suramin, again by the intravenous route. Melarsoprol is a highly toxic trivalent arsenical drug. It is given as a 4-week course with a precisely controlled schedule. The patient must be carefully monitored for adverse side effects which may include neurological changes and shock, which can be fatal.

In **South American trypanosomiasis** (Chagas' disease—p. 146) high fever and generalized lymphadenitis are characteristic of the acute phase, and there may also be generalized CNS involvement at this stage. The most severe illnesses are usually in young children, in whom death is common within a few weeks or even a few days. Survival from the acute phase may be followed by complete recovery or by a chronic phase lasting months or years. In long-standing infections there is myocarditis (p. 239), with considerable enlargement of the heart and thinning of its wall at the base, and congestive cardiac failure is common, especially in the elderly; electrocardiography may help in diagnosis. Digestive tract dilatation (p. 45) is less common. *T. cruzi* trypomastigotes are present in blood and in lymph nodes in the acute illness and can be seen by microscopy, but in the chronic phase it may be necessary to use **xenodiagnosis**; this consists of allowing uninfected bugs of the vector species to bite the patient and then examining their faeces for trypanosomes 7–10 days later. The only drug available for treatment of Chagas' disease, nifurtimox, inhibits the intracellular development of the parasites and is reported to be effective in both acute and early chronic infections when given by mouth daily for 3–4 months. Administration of immunosuppressive drugs to patients with quiescent Chagas' disease results in exacerbation and should be avoided.

Filariasis

Wuchereria bancrofti and *Brugia* species (p. 158) are the only helminths that are important in relation to the lymphatic system, which is their headquarters in the host. From the infecting mosquito bite the larvae travel through the lymphatic system to settle in regional lymph nodes and in the large lymph vessels, where maturation occurs. Fever, lymphangitis and lymphadenitis are the early manifestations of filariasis. Lymphangitis commonly affects the limbs, especially the legs, but may occur in other sites including the genitalia (*Wuchereria* only) and the breasts. It usually spreads outwards along the vessels from a regional node. The vessels become enlarged, hot and tender, the overlying skin is also hot and is tense and erythematous, and there is often surrounding oedema. Episodes of lymphangitis may recur at irregular intervals, and abscesses may form in the lymph nodes or along affected vessels. The common sites for lymphadenitis are the femoral and epitrochlear nodes. The nodes become enlarged, firm, well-defined and rather tender, and the enlargement is usually permanent. In males *Wuchereria* infection of the scrotal lymph vessels may be accompanied by orchitis and spermatic cord inflammation, and lymph varices may rupture into the scrotal sac and cause hydrocoeles. Alternatively they may rupture through the skin, and lymph then oozes out persistently. In some cases of filariasis rupture of abdominal lymph vessels into the urinary tract results in passage of lymph in the urine, which then appears milky, or possibly has a pink tinge because some blood is also present, and may contain microfilariae, the immature forms normally found in the blood (p. 158). In many patients with chronic filariasis obstruction of the lymphatic drainage of a limb or other part of the body leads to elephantiasis. This consists of oedema, due to accumulation of undrained lymph, and progressive proliferation of fibrous tissue. It is a disfiguring and disabling condition, and once established it is irreversible.

The laboratory diagnosis of filariasis due to *Wuchereria* or *Brugia* is made by demonstrating the presence of microfilariae in films of peripheral blood. Blood samples usually have to be taken at night, as the microfilariae show nocturnal periodicity, appearing in the peripheral blood in highest numbers around midnight, though the difference between daytime and nocturnal numbers are less marked with Pacific area strains. A daytime peak can be stimulated by giving a dose of diethylcarbamazine but only reaches about one third of the nocturnal level. Diethylcarbamazine is also the drug for treatment of filariasis, and is usually given two or three times a day for 2–4 weeks—often in reduced amounts at first in an attempt to minimize the allergic reaction which may result from the killing of microfilariae in the early stages of treatment. The adult parasites take longer to kill than the microfilariae, and failure to complete a course of treatment may allow them to recover and produce more microfilariae.

Pyrexia of Unknown Origin (PUO)

This term is used to describe an illness in which persistent fever is the principal feature and remains unexplained after careful clinical examination and routine radiological and laboratory investigations. Clearly many widely different diseases can present in this way, and some of them, including lymphomas and other forms of malignant disease, are outside the territory of this book. However, most such illnesses are due to infection; and our justifications for dealing with PUO in the present chapter are that some of those infections are bacteraemic and eventually have their nature revealed by blood cultures, and that there is no other chapter in which this important but complex topic can be more logically placed.

The inability of young children, unconscious patients and some of the elderly to describe symptoms may be coupled, particularly in the young children, with a shortage of the classical physical signs of particular infections which are to be expected in other patients. Thus respiratory tract virus infections that cause no obvious nasal discharge or abnormalities on auscultation may escape detection, and urinary tract infections may be difficult to recognize when there are major problems in collecting uncontaminated urine specimens. Even bacterial meningitis in an infant may offer very little in the way of suggestive signs. At any age there may be difficulty, especially in obese patients, in finding clinical evidence of deeply placed indolent abscesses—e.g. in the liver or under the diaphragm; and the same may be true of osteitis of the pelvis or other well-protected parts of the skeleton. Pulmonary tuberculosis should be shown up by routine radiography, but non-pulmonary forms are more likely to remain undetected for long periods. Subacute endocarditis due to a 'hard-to-grow' organism may not cause such evident changes in heart sounds as to be easily distinguished from the underlying valvular disease. *S. typhi*, *Listeria monocytogenes* and *Brucella* species are examples of bacteria that may be widely disseminated inside host cells and cause fever for quite long periods without localizing signs or symptoms. Histoplasmosis can similarly cause fever while infecting only the reticulo-endothelial and lymphatic systems; and other systemic fungal infections, notably candidiasis, can present similar problems in immunocompromised patients.

The patient's recent history may give important clues as to the possible aetiology of his pyrexia—e.g. whether he has been in an area where histoplasmosis or malaria or some other agent capable of causing such an illness is endemic, whether he has occupational exposure to infection from animals, whether he has been bitten by possible insect vectors, and whether any of his recent contacts have had similar illnesses. The pattern of his temperature chart may be helpful. Persistent but climbing fever is characteristic of the early stages of typhoid, and in other conditions in which the organisms are continually present in the blood—e.g. typhus—the temperature is likely to be persistently raised. In contrast, hectic swings of temperature, with chills and rigors, may be caused by large deep abscesses or various other conditions. Regular periodicity of fever may indicate malaria, though alternation of periods of fever with periods of a few afebrile days can occur in other infections, such as relapsing fever (as well as being a well-recognized pattern in Hodgkin's lymphoma).

It is obvious that a case of PUO calls for very thorough clinical examination, with particular attention to the lymphatic system and to the possibility of skin rashes, and that such an examination should be carried out repeatedly so that recognition of developing signs is not delayed. Microbiological investigations should in nearly all cases include blood cultures, as well as cultures from the throat and of urine, sputum if any, faeces and (in young children or if any other circumstances suggest it) CSF. Culture of biopsies from the liver, bone marrow or any palpable lymph nodes may be indicated. Blood films will already have been examined to see whether there is a neutrophilia or lymphocytosis and for the presence of abnormal cells such as those of infectious mononucleosis; but in PUO it may be appropriate also to examine the blood for the organisms of malaria, trypanosomiasis or relapsing fever. A wide range of serological tests (p. 183) may be appropriate to the particular patient, covering many different possible aetiological agents (particularly those that are difficult to isolate rapidly) and aimed to detect antibody rises (provided that appropriately timed specimens are available—p. 179) or IgM responses suggesting recent infection. It is probable that the virologists' well-established habit of providing 'serological screens' for particular types of infection will be the prelude to even more sophisticated screens, each of which uses a single serological technique (and so can be readily mass-produced and even automated) and a large assortment of microbial antigens.

18
Gastro-intestinal Tract

The gastro-intestinal tract, in particular the large intestine, is the body's major site of bacterial colonization; the normal bacterial flora amounts to some 10^{12} bacteria/g (wet weight) of colonic contents and faeces. The stability of this flora is continually challenged by regular additions of food and drink, and, inevitably, by bacteria which are either carried in with the food and drink or introduced into the mouth in other ways, e.g. on fingers. Much of this challenge is harmless and goes unnoticed, but pathogenic organisms may be introduced and can lead to infection. The commonest manifestation of intestinal infection is a disturbance of function, obvious to the patient as diarrhoea or vomiting or both. The size of the challenge and the incidence of infection are direct reflections of the standards of hygiene and sanitation in a community. In large areas of the world these standards are low and infective diarrhoeal diseases are common. In many such countries these infections are responsible for a very high childhood mortality rate; and the debility that they cause compounds the problems of malnutrition. Visitors are liable to develop 'traveller's diarrhoea'. However, diarrhoeal disease is not only a problem of poor countries; it remains common in those with supposedly good standards of public hygiene. Part of this persistence is due to ignorance, part is inevitable because faecal–oral transmission occurs so easily, and part results from large-scale food production and mass catering practices that fly in the face of bacteriological knowledge.

The gastro-intestinal tract may be infected by a wide range of bacteria and viruses, some fungi and many protozoa and helminths. The routes of transmission and avenues of infection are many and varied, but the final common route for most is ingestion (some helminths that penetrate skin are exceptions, p. 45), and intestinal pathogens can be divided into three broad groups according to their means of reaching the mouth.

(1) Faecal–oral or hand-to-mouth spread, in which a breakdown of simple hygiene allows ingestion of faecal materials (often in only small amounts) from a human case or carrier or an animal.

(2) Simple contamination of food or water from the same sources, the food or water then being merely a vehicle of transmission.

(3) Food poisoning (not always clearly distinct from the previous category), with multiplication of the pathogens in the food and therefore the possibility of ingesting a large number of bacteria or a large dose of their toxic products.

Diarrhoeal infections that spread readily from person to person by the faecal–oral route can be caused by bacteria, viruses or protozoa. The bacterial species involved are the shigellae (bacillary dysentery), campylobacters and certain serotypes of *Escherichia coli*. The viruses clearly established as intestinal pathogens are rotaviruses and the Norwalk agent; the role of others is less clear. The protozoa are *Entamoeba histolytica*, *Giardia intestinalis* and *Cryptosporidium*.

Bacterial Enteritis

Bacillary Dysentery (Shigellosis)

The name bacillary dysentery is usually given to the diseases caused by any of the four species of

Shigella (p. 94). It has been the scourge of armies throughout history. The advance of Alexander the Great through the Middle East was hindered by it; during the 100 Years War, there were more deaths from it in the armies than there were from the fighting; and more recently, it was a major problem for the British Army in India and Burma during the Second World War, and was responsible for many deaths in the prisoner-of-war camps in East Asia. Major outbreaks have also occurred in civilian populations, especially following social upheavals and breakdown of life patterns produced by wars.

Typical bacillary dysentery is an acute diarrhoeal illness characterized by the passage of frequent (often very frequent) loose stools that tend to be small in volume and to contain blood, mucus and pus. (Dysentery can be defined as passage of stools with these three components.) There is often pain on defaecation and persistent tenesmus (a constant feeling of the need to pass a stool even when there is little there). However, the severity of the illness varies, being in large part determined by the *Shigella* species involved. *Sh. dysenteriae* (Shiga's bacillus) tends to cause the most severe illness, which may be disabling and may well be fatal, especially in debilitated patients or those who are malnourished and living in poor conditions. *Sh. boydi* is next in virulence. Neither of these species is endemic in Britain or other developed temperate countries, but they remain important causes of illness in developing countries in the tropics and sometimes cause dysentery in visitors or in immigrants recently arrived in Britain. *Sh. sonnei*, the commonest cause of bacillary dysentery in Britain and in most developed countries, is endemic throughout the world. The disease that it causes may be quite trivial, with only a few loose stools and little inconvenience, but the commoner manifestation is moderately severe diarrhoea lasting for several days. *Sh. flexneri* is also endemic in some major cities in Britain (e.g. Glasgow, Liverpool) but is much less common than *Sh. sonnei*. It is of world-wide distribution and tends to be intermediate between *Sh. sonnei* and *Sh. dysenteriae* in the severity of the illness that it causes.

Shigella infection is confined to the intestine. There is no invasion of deeper tissues and no bacteraemia; fever is rare, except in the severest forms. Only very small numbers of organisms need to be swallowed to establish infection—i.e. the infective dose is low. In the intestine, the bacilli adhere to the mucosal surface to establish infection; surface components that probably include fimbriae (pili) are important for this attachment. The bacilli invade the superficial layers of the mucosa, causing ulceration and inflammation; and this results in bleeding, mucus secretion and production of a purulent exudate.

The source of infection is always a human case. Shigellae do not affect animals and even in their only host, man, chronic carriage does not occur, though convalescent carriage may continue for several months. Bacillary dysentery is a disease of poor personal hygiene and is common where those who have a relatively low standard of personal hygiene are brought together and share toilet facilities—e.g. among young children in play-groups and nursery schools, and in institutions for the mentally handicapped. Transmission may be by direct person-to-person contact, or via fomites such as toilet fittings or taps. Food contaminated by someone excreting shigellae may be a vehicle for infection, but this is not regarded as 'food poisoning' because the infective dose is low and there is no requirement for multiplication of the organism in the food. When the source is a patient with profuse diarrhoea, there is clearly a high risk of contamination and of spread to contacts, but such a patient is usually confined to home and removed from general circulation. A milder case may cause wider spread of infection by remaining at school or work, in contact with more people and using communal toilet facilities.

Sh. sonnei has only a single biotype and serotype, but epidemiological tracing of outbreaks of sonnei dysentery is helped by colicine (bacteriocine) typing (p. 94). In this, the types are distinguished by the range of activity of their colicines (antibiotic-like products that inhibit other shigellae and related enterobacteria) against a set of indicator strains. Antibiotic treatment is not usually necessary for bacillary dysentery, and is generally contra-indicated. It may prolong symptoms and the excretion of shigellae. The exceptions are illnesses caused by *Sh. dysenteriae*, in which antibiotic therapy may be necessary because of the severity of the illness, and severe illnesses due to other species in very young or debilitated patients.

Campylobacter Enteritis

This is a relatively 'new' disease. *Campylobacter jejuni* and *Camp. coli* (p. 97) were described only in 1976–77, but they are now recognized as major causes of diarrhoeal disease. These micro-aerophilic vibrio-like organisms cause enteritis in man and in a wide range of animal species. After an incubation period of 2–10 days, campylobacters produce a dysentery-like illness; the patient has frequent loose stools that may contain blood and/or mucus. In addition, and in contrast to shigellosis, headache, malaise, fever and abdominal pain are common. These may be the presenting features of the disease and can cause confusion with appendicitis and other causes of an acute abdomen, with the risk that the patient may be subjected to a laparotomy before the true nature of the infection becomes manifest. Most campylobacter infections resolve spontaneously within about 7 days, but some run a prolonged course; the diarrhoea may persist for weeks, or it may improve and then relapse. In these cases, treatment with erythromycin is usually beneficial. Infection is usually confined to the intestine, but bacteraemia may occur in severe infections or in debilitated patients.

Man may acquire his campylobacter from another human, but far less frequently than in the case of shigellosis. Individual human infections and large outbreaks commonly result from transfer of campylobacters from animals. Infection is widespread among these, and cows are particularly important sources of human infection because, as well as developing scours in response to infection, they may become long-term symptomless excreters. Man is infected as a result of drinking untreated milk (unpasteurised; 'green top' in England and Wales) that was contaminated with faeces at the time of milking (campylobacters are **not excreted** in milk). Several large epidemics affecting hundreds of people have occurred in Britain in this way.

Chickens also carry campylobacters, and there has been epidemiological evidence to link some human outbreaks to chicken. Dogs also develop campylobacter infections and may be part of the pattern of transmission between members of human families.

Campylobacters in the stomach

A motile curved bacillus, *Campylobacter pyloridis*, has been seen in close association with the mucosal epithelium of the gastric antrum in biopsy specimens from patients with active chronic gastritis and from some with peptic ulcers; and it has been grown from some such specimens. Its relevance to the aetiology of these two diseases is not yet established.

Escherichia coli Enteritis

Although *Esch. coli* (p. 91) is a normal commensal of the large intestine of man, some strains are pathogenic and can cause diarrhoea. These strains are probably the commonest cause of diarrhoeal disease in the world as a whole. The capacity of *Esch. coli* to cause enteritis was first recognized in the late 1950s in babies. There were explosive outbreaks of enteritis in hospital nurseries and in maternity homes, with a high mortality. These outbreaks were due to *Esch. coli* strains of particular O-serotypes, which became known as epidemic or enteropathogenic *Esch. coli*. Subsequently it was found that adults could develop *Esch. coli* enteritis and that this species was the main cause of traveller's diarrhoea (p. 359). It had been known for a long time that travellers arriving in new places often suffer episodes of diarrhoea. This is often only a relatively mild inconvenience, but can be severe and debilitating; it has been given many colourful epithets – Delhi belly, Tokyo trots, Montezuma's revenge (a particularly vicious form in Mexico) etc. Traveller's diarrhoea is generally associated with visits to places with dubious sanitation, but it can occur on arrival in any new environment because, despite normal hygiene measures, there is a general exchange of intestinal organisms, including *Esch. coli* strains, between neighbours in any locality. It is now known that *Esch. coli* enteritis can affect people other than infants and travellers and that it is responsible for a large proportion of diarrhoeal illness, especially in developing countries.

Three groups of *Esch. coli* strains cause enteritis:

(1) **Enteropathogenic (EPEC)** strains belong to particular O-serotypes (e.g. O111, O127) and cause enteritis in infants and young children. Their pathogenic mechanism is not known.

(2) **Enterotoxigenic** strains produce either or both of two toxins—one heat-labile (LT) and the other heat-stable (ST). LT is an enterotoxin very similar in structure and in mode of action to cholera toxin. One subunit attaches to the mucosal cell membrane, and the active subunit then stimulates adenylcyclase activity; the cyclic adenosine monophosphate (cAMP) produced inhibits the flux of sodium and potassium ions across the cell membrane which leads to ion loss and an associated outpouring of fluid. This is manifested as a profuse watery diarrhoea without blood, mucus or inflammatory cells. Dehydration may occur rapidly, especially in children. The mode of action of ST is unclear, but it also produces a watery diarrhoea.

(3) **Entero-invasive** strains are taxonomically related to shigellae and are similar to them in action. After attaching to the mucosal surface by means of fimbriae, they invade the mucosal cells and multiply within them. This destruction of mucosal cells causes superficial ulceration and diarrhoea, with blood, mucus and pus in the stool (i.e. this is in fact a bacillary dysentery).

The relationships of these pathogenic mechanisms to the epidemiological patterns of *Esch. coli* enteritis are complex. Some of the strains that cause infantile gastro-enteritis produce enterotoxin, but many do not, neither are they entero-invasive. Some that cause traveller's diarrhoea are toxin producers and may be the cause of severe, watery diarrhoeal illness, but others are entero-invasive. However, in many episodes of traveller's diarrhoea no strains can be found with either pathogenic mechanism, and it is suggested that mere acquisition of a new bacterial strain on arrival in a new area may be sufficient to cause an intestinal disturbance as it establishes itself in the new host. There is debate about the role of antibiotics in the treatment of *Esch. coli* enteritis, but the general opinion is that they are of no benefit.

Yersiniosis

Two species of *Yersinia* (p. 95) cause infections of the intestinal tract—*Y. enterocolitica* and *Y. pseudotuberculosis*. Both are pathogenic in animals and can also cause disease in man. *T. enterocolitica* causes an enteritis characterized by mild-to-moderate diarrhoea. *Y. psuedotuberculosis* infection, however, is usually localized to the mesenteric lymph nodes and causes mesenteric adenitis (p. 245); the consequent abdominal pain and mild fever may be mistaken for appendicitis.

Clostridial Enteritis

Enteritis necroticans is a rare condition characterized by diffuse, sloughing necrosis of the mucosa of the large and small bowel; it is caused by type F strains (and possibly others) of *Cl. perfringens* (p. 89). Type C strains of this species are responsible for **pig-bel,** a similar necrotizing jejunitis that occurs in Papua-New Guinea following pig-meat feasts. (Necrotizing enterocolitis of neonates, p. 327, resembles pig-bel in some respects but has no certain clostridial aetiology.) The association of *Cl. perfringens* with food-poisoning is discussed on p. 257, and that of *Cl. difficile* with pseudomembranous colitis on p. 259.

Virus Gastro-Enteritis

Rotaviruses

It has long been recognized that many cases and outbreaks of enteritis with no discoverable bacterial agents must be caused by viruses; and in 1976 a group of viruses capable of doing so were discovered and were named rotaviruses (p. 120). They have since been recognized throughout the world and, although human rotaviruses attack only man, closely related members of the group have been found to cause enteritis in many animal species. Rotaviruses are probably the commonest cause of enteritis in children but can also affect adults; in Britain they are the most commonly reported intestinal pathogens overall. The wheel-like virus particles can be seen by electron-microscopy in the stools of those with the infection, and the capsid antigen can be detected by ELISA. The viruses spread by the faecal–oral route, or are perhaps sometimes airborne. An incubation period of 48–72 hours is followed by vomiting and diarrhoea. Rotavirus enteritis is an acute self-limiting disease. The watery diarrhoea may lead to a moderate degree of dehydration. In children, the diarrhoea may be preceded by mild upper respiratory tract symptoms. Infection is commonest in the winter months. There is no specific treatment, and recovery usually takes only a few days. Malnourished and debilitated children in developing countries may be more severely affected.

Norwalk Agent

An outbreak of enteritis, with diarrhoea and vomiting, that was clearly of virus origin occurred in the town of Norwalk, Connecticut in 1968. Many local epidemics have been recognized since then, but the causative virus(es) has(have) not been clearly characterized, and they retain the designation of Norwalk or Norwalk-like agent (p. 121). They are readily transmissible, causing a relatively mild self-limiting illness.

Other Viruses

Various other virus particles can be demonstrated in the faeces of patients, particularly children, who have enteritis. Adenovirus particles are frequently seen, but they are not easy to grow and appear to differ from the upper-respiratory-tract adenoviruses. Small round virus particles, some called caliciviruses and some astroviruses (p. 121) because of their electron-microscope appearances, are also common, but the relationship of any of these to diarrhoeal disease is not proven.

Protozoan Infections

Many protozoa can be present in the faeces without causing disease, but of those frequently found in the intestine *Entamoeba histolytica* and *Giardia intestinalis* most commonly cause diarrhoea. Both are transmitted from man to man mainly by faecal contamination of food or water. The use of human excreta (night soil) as fertilizer for vegetable crops is an important means of infection in the Far East. Outbreaks due to contamination of water supplies have been described, and normal levels of chlorination do not kill cysts. Flies and cockroaches can transfer infective cysts from faeces to food either by ingesting the cysts, which pass undamaged through their digestive systems, or by transmitting cysts on their feet or bodies.

Amoebiasis (p. 143)

In some 80% of cases amoebic infection is non-invasive. Even in the many cases in which it causes no symptoms for many years, the host is at risk of invasive disease and the excretion of cysts is a hazard for those around. It is not clear whether invasion in a person with long-standing infection results from a new infection or from activation of the existing non-invasive infection. In invasive infections, the trophozoites (amoebae) destroy tissue cells, generally starting with those in the base of the crypts of Lieberkuhn in the large intestine, producing flask-shaped ulcers in the intestinal wall. Trophozoites are concentrated at the edges of these ulcers where they feed on tissue debris and blood cells. Cysts are not found in the ulcers. The surface layers of the mucosa may become undermined by a series of adjacent ulcers and extensive areas of it may be sloughed off. Both trophozoites and cysts are found in the faeces, and the presence of trophozoites containing red blood cells is diagnostic of invasive amoebiasis; faeces must be examined within about 15 minutes of being passed for the trophozoites to be recognized (p. 174). Trophozoites penetrate deep into the wall of the intestine and reach the blood, and they may then be carried to the liver (p. 269), lungs (p. 269), brain (p. 225), etc. where secondary amoebic abscesses may be established. Cutaneous amoebic infections can occur in the peri-anal area, or in the abdominal or the thoracic wall when a hepatic or a lung abscess tracks to the body surface.

In the treatment of amoebiasis metronidazole and diloxanide furoate are effective against trophozoites in the intestinal lumen; these and emetine hydrochloride are effective against amoebae in ulcers and abscesses. Metronidazole is used for treatment of amoebae in liver or lung abscesses (often with emetine), and it is also usual to aspirate the pus, tissue debris and parasites from such abscesses to speed recovery if they are large or about to burst. Under no circumstances should an amoebic abscess be opened surgically, because generalized invasive amoebiasis will be almost inevitable (p. 269). Patients with invasive amoebiasis (i.e. those passing trophozoites containing erythrocytes) require treatment for their own benefit, and both they and the asymptomatic cyst passers should be treated for the benefit of the community. In particular, any cyst passers who are food handlers must be effectively treated before resuming work, since many outbreaks of amoebiasis have been traced to such sources.

Giardiasis (p. 144)

In *G. intestinalis* infections the mucosa of the small intestine may be lined with large numbers of the parasites; and though not invasive, these impair the host's general health by interfering with digestion and absorption. Symptoms in the acute infections range from mild diarrhoea, abdominal pain and flatulence to acute discomfort and the passage of copious fatty stools. Children are more frequently affected than adults. Heavy chronic infection may interfere with food absorption to such an extent that clinical malabsorption may result. Infection is by the faecal–oral route, either on food or more commonly as a result of drinking contaminated water (p. 355). The incubation period is 6–22 days. The diagnosis is made by microscopy of appropriate specimens. In acute giardiasis, motile trophozoites are seen in wet preparations of faeces, and cysts may be present; the trophozoites encyst as the temperature of the specimen falls. In chronic infections, trophozoites are not present in the faeces and even the cysts may be difficult to detect, and may fluctuate in numbers widely from day to day; cysts are more likely to be seen in aspirated duodenal contents, and trophozoites in histological slides from a duodenal mucosal biopsy. Metronidazole is usually an effective treatment for giardiasis.

Cryptosporidiosis (p. 144)

The protozoan parasite *Cryptosporidium* causes enteritis in immunocompromized patients (in whom it is often fatal), but can also affect previously healthy people, especially children. In immunocompromized patients it causes a profuse, watery, greenish diarrhoea. *Cryptosporidium* cysts can be detected in faeces by microscopy; films made from faecal specimens after concentration for parasite ova are stained by a modified Ziehl–Neelsen or auramine–phenol method and the cysts are readily seen (p. 145). There is no specific treatment and the disease is self-limiting in patients with a normal immune system. What is known of its epidemiology is outlined on pp. 144–5.

Water-Borne Infection

Water is essential to human life, both for drinking and for the vegetable crops and animals on which man depends for food; but water is also a major vehicle for the spread of infection, and a safe unpolluted water supply is essential for health (p. 355). The major feature of a safe water supply is the separation of sewage (human excreta) from drinking water, because most of the serious infections spread by water are intestinal infections in which the water forms a link in the faecal–oral chain of transmission. Some of these infections are diarrhoeal diseases in which infection is confined to the intestine; others are systemic infections with an intestinal portal of entry.

Two of the most important diseases that spread by water are cholera and enteric fever (typhoid and paratyphoid).

Cholera

This is a purely intestinal infection, but a serious and potentially fatal one. It is commonly acquired by drinking water contaminated with faeces from infected cases; but direct faecal-oral spread from patients occurs readily. *Vibrio cholerae* (the classical organism—p. 97) or its eltor biotype colonizes the intestine and produces a toxin that causes a profuse watery diarrhoea. The toxin comprises five B subunits that attach to the mucosal cells and one A subunit that activates adenylcyclase, which generates cAMP, resulting in a massive loss of electrolytes and water as described in connection with enterotoxigenic *Esch. coli* strains (p. 252). The typical rice-water stools of cholera result—copious nearly colourless fluid teeming with the cholera vibrios. A patient may lose up to 20 litres in a day, and this rate of dehydration can be rapidly fatal. The main component of treatment is rehydration, by intravenous infusion or by oral administration of electrolyte solution with glucose; oral treatment with tetracycline also helps to reduce the severity of the disease.

There have been seven pandemics of cholera; the most recent one, caused by the eltor biotype, began in the mid-1960s in SE Asia and spread across Asia to the Middle East, into Africa and to the east coasts of South and Central America. Local outbreaks may occur wherever the normal social order breaks down—in refugee camps as people flee from wars or from drought; in areas suffering after natural disasters such as earthquakes and floods; and annually in India and Bangladesh (where cholera appears to have originated) where the monsoon causes widespread flooding.

The principal means of spread of cholera has been known since John Snow's investigations of the outbreaks in London in 1853 and 1854. He showed the link between cholera and drinking sewage-contaminated water—first in Golden Square, Soho, where a broken sewer leaked into the water supply of the Broad Street pump; and later in south London, where cholera occurred in people whose water supply came from the Pool of London near to the mooring places of ships containing cholera patients brought back from the Crimean War, but did not occur in those whose water supply came from further up the Thames, above the area of contamination. Prevention still depends upon providing a safe water supply, and this has been given top priority by the World Health Organization. The eltor biotype is somewhat more difficult to control than the classical *V. cholerae* because it causes less severe illness and therefore more patients are ambulant but vigorous excreters.

Typhoid

Unlike other *Salmonella* species, which usually cause diarrhoeal illnesses as described below under Food-poisoning (p. 256), *S. typhi* and *S. paratyphi* A, B and C (p. 92) all typically cause febrile bacteraemic illnesses. They have therefore been described in the previous chapter (p. 231), but typhoid in particular deserves mention here because it resembles cholera in that most large outbreaks have been water-borne, and that it is a disease of poor sanitation. The most important epidemiological difference from cholera is that recovery from typhoid may be followed by carriage of the bacilli, often for many years, in the gall-bladder or kidneys, and by their intermittent excretion in faeces or urine. Symptomless excreters, especially if they are unrecognized, may be important sources of infection for others because their excreta can contaminate water supplies. (Such people can also cause much trouble if employed as food-handlers, since they can easily contaminate food — p. 360.)

Typhoid is common in all areas of the world where standards of hygiene and sanitation are poor and is a major cause of death in those areas. Travellers to such places should be cautious about their drinking-water supplies, and also about the type of food they consume (freshly cooked food and fruit that can be peeled immediately before eating are safest).

Other Water-Borne Infections

Polluted water is important in the spread of *Esch. coli* enteritis (see above), which may indeed be the commonest form of water-borne enteritis. Water can be important also in the spread of organisms that are excreted in faeces and enter the next patient via the mouth but then cause diseases elsewhere than in the intestine. Examples are polioviruses (p. 221) and hepatitis A virus (p. 264), both of which spread readily in areas with poor sanitation.

Bacterial Food-Poisoning
(see also pp. 358–61)

A useful definition of bacterial food-poisoning is 'an intestinal disturbance (diarrhoea and/or vomiting) caused by bacteria or their products when the bacteria have multiplied in food before it was eaten'. Bacterial multiplication in the food before ingestion greatly increases the challenge dose, and therefore the food is not just a vehicle for infection but is an important factor in increasing its likelihood. The two main types of food-poisoning are: (1) infective, when the disease depends on ingestion of adequate numbers of live bacteria; and (2) toxic, when the illness is caused by toxin produced in the food by growing bacteria and does not depend on ingestion of live organisms. Almost any bacteria, when present in food in very large numbers, may cause spoilage of the food and intestinal disturbances in those who eat it despite its spoiled state; but far more important are those bacteria that can cause illness without making the food obviously unfit for consumption.

Salmonella Food-poisoning

Food-poisoning is the usual form of illness caused by many different salmonellae (about 200 serotypes) other than *S. typhi* and *S. paratyphi* (p. 255). In Britain these are the commonest causes of food-poisoning, in terms of cases and of outbreaks; most outbreaks involve only a few people, but large banquets can involve hundreds. Some serotypes—notably *S. typhimurium*—are persistently responsible for numerous outbreaks each year, but the frequency of many others varies from time to time in accordance with farming practices, importations of infected animal foods and many other factors.

Salmonella contamination of food can sometimes be traced to human cases or carriers, but animal sources are commoner. Salmonella infection is widespread among domestic and wild animals and birds, and is often asymptomatic in them. It is particularly common among chickens and turkeys — often introduced in poultry feed, with a high incidence of infection maintained by keeping the birds under crowded conditions. A somewhat similar situation occurs frequently among cattle. Cross-contamination of carcasses happens readily during plucking and dressing of poultry and in abattoirs. The housewife or caterer is wise to assume, under present conditions of mass-production, that all fresh or frozen poultry or meat may be contaminated and should be handled accordingly. Adequate cooking will kill all salmonellae (as well as most of the other food-poisoning organisms mentioned below), but it is adequate only if the heat has penetrated all through the bird or joint. The risk of inadequate cooking is increased if a large bird is tightly stuffed, if a joint is too large, or if a frozen bird or joint is not adequately defrosted before cooking. Even properly cooked food can be subsequently contaminated, either by a food-handler who is an excreter or by contact with contaminated raw food or with surfaces or utensils that have been contaminated and not subsequently cleaned. Other foods besides meat—notably dairy foods—can also become contaminated in such ways.

Contamination alone, unless it is very heavy, is unlikely to cause any trouble. Food-poisoning, as defined above, requires that the bacteria be given a chance to multiply in the food. This can be avoided by immediate consumption of even inadequately cooked meat, or of other contaminated food, or by its immediate and continued refrigeration. Long cooling periods after cooking, holding food at room temperatures, or gentle and prolonged warming of food after removing it from a refrigerator can all provide opportunities for bacterial multiplication.

Salmonellae cause an infective type of food-poisoning. After their ingestion in food there is an incubation period of 12–36 hours (depending upon challenge dose and host resistance) during which the salmonellae colonize the intestinal mucosa and begin to damage it. This mucosal damage causes increased fluid secretion and increased intestinal motility, and these result in the major clinical feature of salmonella food poisoning—diarrhoea. The presenting features include headache and abdominal pain and the patient may be febrile. Vomiting may occur but is usually a minor component. In otherwise healthy people, the disease is usually self-limiting and treatment is supportive; antibiotics are not beneficial, and may increase the incidence of long-term carriage. A small proportion of patients—mostly very young, old or debilitated—have more serious illnesses in which salmonella pass through the intestinal mucosa to the blood-stream and cause bacteraemia. (Some *Salmonella* serotypes are particularly liable to behave in this way, even in previously healthy adults.) This does require antibiotic treatment, usually with chloramphenicol. After recovery, which may take 1–2 weeks, some food-poisoning patients become long-term carriers of the salmonella, excreting the organism for months or years. They may be a hazard to others, especially if the carriers handle food (see above).

Clostridial Food-Poisoning

Outbreaks of food-poisoning due to *Cl. perfringens* are fewer than those due to salmonellae but tend to be larger because they depend on bulk food production, as in residential institutions, hospitals and catering establishments. *Cl. perfringens* is a gram-positive anaerobic sporing bacillus (p. 89); the spores of the antigenically distinct type A strains associated with food-poisoning are particularly heat-resistant and are able to withstand prolonged boiling. This species is a normal intestinal commensal of man and animals; thus meat, the food most commonly implicated, is often contaminated with the spores. High-temperature cooking of small amounts of meat kills the spores, but they can survive prolonged low-temperature cooking of bulk meat products, as in commercial or institutional preparation of stews, broths and meat pies. In these circumstances, heat penetration is poor and the spores are protected in the protein-rich mixture; in fact, the heating may be just the shock necessary to stimulate subsequent outgrowth. If the cooked product is not either eaten or refrigerated immediately but is allowed to cool, cooling in the centre of the large mass of food is slow and the temperature remains for some time between 50°C and 20°C, at which level the spores will germinate and the vegetative cells multiply rapidly. The conditions in cooked meat products, equivalent to the cooked meat broth used in the laboratory, are ideal for growth of anaerobes. If such a rich culture of *Cl. perfringens* (up to 10^7/ml can easily be attained within a few hours) is not thoroughly reheated to kill the vegetative cells, a large challenge dose will be available for ingestion. In the intestine the clostridia sporulate (which they rarely do in culture), and as a by-product of sporulation they produce an enterotoxin that acts on the intestinal mucosa to cause fluid loss and consequently a watery diarrhoea. *Cl. perfringens* food poisoning is therefore a combination of infective and toxic types. The incubation period after eating the meal is at least 8 hours, because this is the length of the sporulation process and the toxin produced early in the process is not released until the final stage of lysis; it may be up to 20 hours. The onset of symptoms after this incubation period is usually rapid and precipitate. There is abdominal pain and diarrhoea, which is often violent. Prostration is common but there is no fever and seldom any vomiting. The illness lasts for several hours, and recovery is usually complete by the next day. This is rarely a serious illness, except that profuse diarrhoea in elderly or debilitated patients may lead to a degree of dehydration that can be severe and even fatal. Therapy is supportive only, because the illness is self-limiting.

Staphylococcal Food-Poisoning

If food is contaminated by an appropriate *Staph. aureus* strain (p. 75) and then held at room temperature, the staphylococci multiply and produce a toxin in the food. This toxin causes the illness. Being moderately heat-stable, it is not inactivated by gentle reheating sufficient to kill the vegetative organism; it will even withstand 100°C for 30 minutes. There are several types of staphylococcal enterotoxin, but most food-poisoning episodes are associated with type A. Its mode of action is uncertain, but it probably acts on receptors in the intestine, from which sensory stimuli travel to the vomiting centres in the brain via the sympathetic nerves and the vagus. The predominant effect is vomiting, sometimes followed by diarrhoea. The vomiting begins suddenly, a few (usually 2–6) hours after eating the contaminated food. The patient may be prostrated and usually has abdominal pain. Complete recovery takes place within a day of onset without specific therapy. Many of the *Staph. aureus* strains that have been shown to cause food-poisoning belong to a small number of types of phage group III (p. 75).

The source of infection is usually a food-handler, because *Staph. aureus* is essentially a parasite of man. A food-handler who is a nasal carrier (p. 189) or has a staphylococcal septic lesion can easily contaminate food unless there is scrupulous attention to good handling practices (in particular, **no** direct handling of food that is to be eaten without subsequent cooking). The foods commonly affected are those handled a great deal during serving and preparation—cooked meats (especially ham, which provides suitable conditions for the salt-tolerant staphylococcus), cheese, cream cakes and other dairy produce.

Staphylococcal food poisoning is an unpleasant illness but is self-limiting. Prevention depends upon: (1) ensuring that catering staff do not work when they have septic lesions, or at least have them covered by an impervious dressing and do not touch cooked food; (2) adopting no-touch methods for food handling; (3) keeping all cooked foods, dairy produce, etc. in refrigerated display cases. The same measures should be applied in the home.

Bacillus cereus Food-Poisoning

Certain strains of *B. cereus*, a widely distributed saprophytic gram-positive aerobic sporing bacillus (p. 87), cause food-poisoning of the toxic type, similar to staphylococcal food-poisoning but particularly associated with one food (rice) and with Chinese restaurants in which it is the practice to boil rice and then keep it warm and moist throughout the day ready for frying. Rice is often contaminated with *B. cereus* spores, which are not killed by the boiling and germinate when the rice is kept warm, producing their relatively heat-stable toxins that survive light frying. Two distinct forms of *B. cereus* food poisoning are recognized. In most of the rice-associated outbreaks nausea and vomiting come on 1–5 hours after eating the rice. The disease is self-limiting with recovery within a day. Strains of serotype 1 are most frequently isolated from outbreaks, probably because their spores are more heat resistant than those of other types. The other form of *B. cereus* food poisoning is less common and is characterized by abdominal pain and diarrhoea coming on 8–16 hours after eating the food responsible; vomiting is uncommon in this type.

Vibrio parahaemolyticus Food-Poisoning

A few outbreaks of food-poisoning in Britain and many more elsewhere—notably in Japan—are caused by the halophilic (=salt-loving) *V. parahaemolyticus*, which has been ingested in raw shell-fish; in Britain these have often, but not always, been imported from Asia. After an incubation period of about 12 hours, there is acute diarrhoea with moderate-to-severe abdominal pain and mild fever. The symptoms subside after 24–48 hours.

Botulism

The toxic effects of the highly potent toxin of *Cl. botulinum* (p. 87) are not usually classified as food-poisoning, because none of the symptoms relate directly to the intestinal tract, but botulism is the result of ingestion of a toxin produced by bacteria growing in food. *Cl. botulinum* is strictly anaerobic, and botulism is associated only with foods that provide suitable anaerobic conditions. It is rare in Britain, but more common in Europe, where it is associated with pâtés and sausages, and in the USA, where it is associated with home canned fruit and vegetables, particularly beans; it is sometimes associated with fish. In all cases, the food is contaminated with *Cl. botulinum* spores (which are common in soil and in the mud of lakes, rivers and estuaries) and has not been heated sufficiently to kill them. Commercial canners are required to apply sufficient heat (by means of steam under pressure) to kill *Cl. botulinum* spores and give a safety margin. This standard does not apply to cured meat and to home canning. In the last outbreak in Britain (1978) four people were affected, two fatally. The tin of salmon which they ate had been suitably heated, but during subsequent cooling the spores had been drawn in from the contaminated exterior through minute defects in the tin. Germination of the spores in the food leads to production of the potent neurotoxin without any perceptible deterioration of the food. The toxin causes motor paralysis. Characteristically, neurological symptoms appear some 12-36 hours after toxin is ingested. They include hoarseness, visual disturbance, headache, nausea and vomiting, and progress to difficulties in swallowing and speaking and often to death from respiratory or cardiac arrest. A different and usually much less severe form of botulism occurs in infants (p. 327).

Scombroid poisoning

This is not a true bacterial food-poisoning but is the result of bacterial spoilage of food. If scombroid fish (mackerel, tuna, etc.) are poorly preserved, bacteria degrade histidine in the fish to a histamine-like compound. If the fish is eaten, this toxin causes nausea and vomiting, flushing, headache, abdominal pain, diarrhoea and other 'allergic' phenomena such as dysphagia, thirst, pruritus and urticaria. The onset is rapid (often within a few minutes) and the symptoms subside within 12 hours.

Antibiotic-Associated Diarrhoea

That antibiotic treatment could lead to colitis and diarrhoea was recognized in the late 1950s, and at that time the usual pattern of the condition was that the patient had received a tetracycline or other broad-spectrum antibiotic which had seriously disturbed the bacterial flora of his intestine and allowed colonization by *Staph. aureus*; this led to **staphylococcal enterocolitis**—a condition quite unlike the staphylococcal food-poisoning described on p. 257, and having a high mortality. For reasons perhaps associated with increasing antibiotic-resistance of the human intestinal flora this condition virtually ceased to exist, but other forms of antibiotic-associated diarrhoea are common. In some cases the diarrhoea is merely inconvenient; no particular bacterial or fungal infection can be demonstrated, and there may well be nothing more than a disturbance of balance of the normal intestinal flora—perhaps in particular a disturbance of the homeostatic function of the anaerobes. In the 1970s, however, a more clearly defined entity was recognized, with characteristic histopathological and bacteriological features—**pseudomembranous colitis (PMC)**. In fully developed PMC, treatment with antibiotics has resulted in a severe colitis; sigmoidoscopy shows flattened areas of pseudomembrane and histological examination of a mucosal biopsy shows typical 'exploding volcanoes' disrupting the superficial layers. These changes occur because the antibiotic disturbance of the intestinal flora allows colonization by *Cl. difficile* (p. 90), which produces two toxins (enterotoxin and cytotoxin) that are responsible for the pseudomembrane formation and the colitic symptoms. Initially a particular association with clindamycin or lincomycin treatment was reported, but it is now recognized that PMC can follow any fairly broad-spectrum antibiotic treatment (in occasional cases there is no history of antibiotic treatment). It is not certain whether *Cl. difficile* is carried in small numbers in the intestinal flora of most people, but the incidence of PMC varies from place to place and there is evidence in several outbreaks that the organism is transmitted from patient to patient. PMC is best treated by prompt oral administration of vancomycin.

Disturbance of the intestinal bacterial flora by antibiotics can also lead to overgrowth by *Candida albicans* and candida enteritis. This causes a severe diarrhoea and should be treated with oral nystatin or amphotericin B.

Diarrhoea in Systemic Infections

Diarrhoea is a common symptom of many systemic infections, particularly in young children. Most febrile infectious diseases, especially the common virus infections, induce some loose and often watery stools, although the intestine is not the primary infection site. In well-nourished otherwise healthy children the diarrhoea is usually mild. In malnourished and debilitated children, however, the enteritis may be severe, leading to dehydration and in some cases to death. This is particularly true in measles in many developing countries.

Helminth Infections

Most of the helminth parasites that infect the intestines of man cause few symptoms if present in fairly small numbers in otherwise healthy individuals, but a heavy infection in a patient with malnutrition or some underlying chronic disease may have serious consequences.

Nematodes (Roundworms)

Infection with hookworms (*Ancylostoma duodenale* or *Necator americanus*, p. 157) can result in anaemia, of the microcytic hypochromic type. The degree of anaemia depends on the number of parasites, on their species (blood loss due to *Ancylostoma* is several times that due to *Necator*) and on the nutritional status of the host. Patients on a good diet can tolerate a reasonably heavy infection with little adverse effect, while in those on a poor diet even a modest infection can lead to a chronic, severe or even fatal anaemia. Apart from areas which have been damaged by the cutting mouthparts of the worm, histological changes in the mucosa are small. Abandoned feeding sites may continue to bleed for some time after the worm has moved elsewhere, because of the secretion of an anticoagulant by the parasite. Eggs are produced about 7 weeks after exposure, and the infection may last for several months or even years but is usually self-limiting. However, hookworm infections should be treated with bephenium or pyrantel to avoid the consequences of the blood loss. *Necator* is more difficult to treat than *Ancylostoma*, and repeated courses may be necessary. In an endemic situation, reinfection of the treated patient is almost certain to occur unless improved sanitation is provided and used; and agricultural workers in particular should use simple but effective footwear.

Infection with *Strongyloides stercoralis* (p. 157) is less common than with hookworms. *Strongyloides* does not suck blood, but it causes systemic illness in the host because the larvae migrate through the tissues (**systemic larva migrans**—p. 157). Infection can last much longer than with hookworms and is much more difficult to treat because larvae develop rapidly in the lumen of the intestine and are capable of reinfecting the host (auto-infection) without being passed out in the faeces. Treatment must therefore be prolonged. Depression of the immune response, e.g. by immunosuppressive drugs or in AIDS, can lead to a flare-up of *Strongyloides* infection, and in massive infections can be fatal, especially in young children. Severe and even fatal perinatal infections have been described (e.g. from New Guinea). Thiabendazole is the preferred drug for *Strongyloides* infections, although minor side effects are common; repeated treatment may be necessary and auto-infection has been known to continue for 40 years.

Although *Ascaris lumbricoides* (p. 156) does not suck blood and light infections are usually symptomless, the presence of even a few worms can be dangerous. The larvae are fairly large, and migration of large numbers through the lungs can cause a pneumonitis which begins about 5–6 days after infection and lasts for 10 days or so. Eosinophilia may develop (up to 40% of the white blood cells about 3 weeks after infection). The usual symptoms of *Ascaris* infection in the intestine are digestive disorders, vomiting and abdominal discomfort. In children *Ascaris* infection may cause acute intestinal obstruction. Individual worms attempt to migrate up the bile and pancreatic ducts, causing obstruction and secondary bacterial infection. *Ascaris* infection should be eliminated before intestinal surgery, because the adult worms can burst open any surgical incisions. The drugs effective against *Ascaris* are pyrantel and bephenium. Unfortunately in the endemic areas successful treatment is usually followed by reinfection; only major improvements in sanitation and personal hygiene and avoidance of the use of night soil as a fertiliser will improve matters.

Trichuris trichiura infection (p. 158) is neither as common nor as severe as *Ascaris* infection, although heavy infection may contribute to failure to thrive in children in the tropics.

Considerable numbers of worms may be present but they are much smaller than *Ascaris* and their effects generally are correspondingly less. Most infections are asymptomatic and only occasional eggs are seen in the faeces. Diarrhoea, abdominal discomfort and weight loss are common in heavily infected individuals, particularly children; and anal prolapse is not uncommon in infected children. Patients with over 200 worms may be anaemic; the degree of anaemia depends on diet, but more than 1000 worms can cause severe hypochromic anaemia. An eosinophilia of up to 25% is common. Although *Trichuris* infection is typically found in areas with high relative humidity and with temperatures between 22 and 28°C, it can reach high prevalence rates in temperate areas of the world; there it is found in institutions, particularly those for mentally handicapped children, where the standard of hygiene is poor and where there may be promiscuous defaecation and coprophagy. In the tropics *Ascaris* and *Trichuris* infections often occur together (with hookworm infection also in many areas) because the worms flourish in similar environmental conditions and have similar epidemiology. Treatment of *Trichuris* infection is the more difficult; mebendazole is the drug of choice.

Enterobius vermicularis (threadworm; pinworm) infection (p. 157) is usually asymptomatic, although an allergic response to the ova may cause peri-anal irritation, giving rise to sleeplessness in sensitive patients. Adult worms are frequently found when the appendix is examined histologically but whether there is any relationship between *Enterobius* infection and appendicitis is unproven. Diagnosis is made by sampling the peri-anal skin with a strip of self-adhesive tape (Sellotape; Scotch tape etc.) early in the morning before the patient has had a bath; the strip is then applied to a microscope slide and examined under a × 40 objective. Treatment is straightforward; piperazine or pyrantel salts are the drugs of choice (the latter is also effective against *Ascaris* and hookworms). Since all people sharing the accommodation are likely to be infected, all members of the household should be treated at the same time. All bedding and clothes should be thoroughly washed, and furniture should be washed and exposed to sunlight. However, *Enterobius* is often reintroduced into the family by young children.

Cestodes (Tapeworms)

Infection with *Diphyllobothrium latum* (p. 156) is generally not serious and can be readily treated with niclosamide, but some patients develop an anaemia which is similar to pernicious anaemia and is probably the result of vitamin B_{12} deficiency caused by the parasite. Similarly, light intestinal infection with *Taenia* (p. 155) is usually of little consequence; there may be vague abdominal pains or discomfort. Treatment with niclosamide is again effective. However, tissue infection with the cysticercus stage of *Taenia solium* (cysticercosis, p. 293) is of greater concern. Even here, light infections may cause no symptoms unless the cysticerci are lodged in some vital area such as an eye. Cysticercosis of the muscles is the usual manifestation, frequently detected only when calcified cysticerci are found incidentally in radiographs taken for some other purpose. With heavier infections there is a greater risk of nervous system involvement (p. 226).

Schistosomes (see p. 155)

Intestinal schistosomiasis is caused by *S. mansoni* in Africa and the Middle East and by *S. japonicum* in China, Japan and the Philippines. The adult worms live in the mesenteric veins of the colon where they cause granulomas to form. The gut mucosa becomes congested and nodules form, which may develop into large polyps causing intestinal obstruction. There is heavy protein loss from this damaged mucosa. Infection is often asymptomatic in the early egg-laying stage but may cause abdominal discomfort and irregular bowel action, with mucus and blood in the stool. Diarrhoea is uncommon except in heavily infected children or in visitors to endemic areas. Eggs that do not escape into the gut lumen may be carried in the portal veins to the liver, where they cause nodular fibrosis which may result in portal hypertension and ascites.

Praziquantel is the drug of choice for intestinal schistosomiasis.

Fluke Infections

Fasciolopsis is the commonest intestinal fluke that infects man. Infection is endemic in East Asia. The adult flukes attach to the small intestine mucosa and erode the gut wall. The life cycle is described in Chapter 12 (p. 153).

Examination of Faeces

The diagnosis of intestinal infection depends largely upon microbiological examination of specimens of faeces. In food-poisoning, any available specimens of suspect food are examined in much the same way as the patients' faeces.

Microscopy

Microscopy of faeces has limited value in the diagnosis of bacterial infections, but it is the mainstay of the investigation of protozoan and helminth infections. A wet film of a faecal emulsion examined with a ×40 objective will show red and white blood cells, as evidence of intestinal pathology. Microscopic examination of gram-stained films is generally unhelpful; the pathogens are similar in appearance to the normal commensals. One exception to this generalization was staphylococcal enterocolitis (p. 259) in which characteristic replacement of the normal flora by clumps of gram-positive cocci provided a presumptive diagnosis; and another exception is candida enteritis (p. 259), since it is abnormal to see numerous yeasts in faeces. Microscopy of wet films, either unstained or stained with iodine, is the basis of the laboratory diagnosis of infection with protozoa and helminths. The vegetative cells (trophozoites) and cysts of *Entamoeba* and *Giardia*, the ova of the intestinal round worms and schistosomes, and the larvae of *Strongyloides* are large when compared with bacteria and can be recognised with the ×40 objective (p. 173). The proglottides of tapeworm are visible by the naked eye.

Culture for bacterial pathogens

In the search for infective causes of diarrhoea, specimens of faeces are inoculated on to a series of selective culture media for aerobic incubation in an attempt to separate the possible pathogens from the large number of commensal bacteria. **MacConkey's agar** and **deoxycholate citrate agar** (DCA) distinguish the pink lactose-fermenting *Esch. coli* from salmonellae and shigellae (pale, non-fermenters). DCA is more inhibitory towards *Esch. coli*. **Xylose lysine deoxycholate (XLD) agar** is an alternative medium, especially for the isolation of shigellae. Media for the isolation of campylobacters contain antibiotics and are incubated in a micro-aerophilic atmosphere with additional CO_2 at 43°C. Cholera and other vibrios may be isolated on **thiosulphate – citrate – bile salt – sucrose (TCBS) agar**. Cycloserine cefoxitin fructose agar incubated anaerobically at 37°C is selective for *Cl. difficile* and **mannitol–salt agar** for *Staph. aureus*. All of these media are seeded directly from the faecal specimens. Broth enrichment may be used when specific pathogens are suspected. **Tetrathionate** or **selenite broth** are used for the detection of small numbers of salmonellae, particularly in faeces samples from suspected carriers. **Alkaline peptone water** incubated at 37°C for 18–24 hours encourages the growth of cholera vibrios and **cooked meat broth**, heated to destroy vegetative cells after seeding with faeces or food, is used for isolation of *Cl. perfringens*.

Some pathogens may be identified immediately by their colonial and microscopic appearance on primary isolation, but the identity of most must be confirmed by further biochemical and/or serological tests of the purified isolates. For lactose non-fermenters which might be salmonellae or shigellae, a series of biochemical tests (p. 91) will confirm the identification to genus level (or to species level with some shigellae), and agglutination tests with specific antisera will determine the serotype or species. *Esch. coli* is similarly identified by its biochemical profile, and the enteropathogenic serotypes are detected by agglutination tests with O-antisera. Biochemical tests will also identify the pathogenic yersiniae. The typical appearance of *V. cholerae* on TCBS medium is confirmed by agglutination with *V. cholerae* O1 antiserum (p. 97), and that of *Cl. perfringens* on anaerobic blood agar by a Nagler test (p. 89).

Tests for enterotoxin production

The advent of PMC (p. 259) as an important clinical problem re-awakened interest in the detection of enterotoxin in faeces. Culture is an insensitive way of finding *Cl. difficile* in faeces, and is in any case of little value because mere presence of the organism does not prove its pathogenicity. Detection of toxin is of far greater diagnostic value. Of the two toxins produced by this species—toxin A, a cytotoxin, and toxin B, an enterotoxin—the former can be detected by its cytopathic effect on certain cell-cultures and the neutralization of this effect by an appropriate antiserum. This is the basis of a routine diagnostic test. Toxin B has been detected, far less conveniently, by its effects on intestinal loops from laboratory animals; but it is now possible to detect both toxins by an ELISA technique applicable directly to faecal suspensions from patients.

Neutralizable cytopathic effects and ELISA techniques (p. 178) can also be used to detect *Esch. coli* LT and ST (p. 252) and the enterotoxins of *Staph. aureus* and *Cl. perfringens*.

Detection of viruses

With the recognition of the pathogenicity of rotaviruses, a definitive diagnosis of this type of virus enteritis became possible. There are two common approaches to the detection of rotavirus in faeces—demonstration of virus particles by electron-microscopy, and detection of rotavirus antigen by an ELISA technique (p. 178). For electron-microscopy, a filtrate of a faecal suspension is centrifuged at high speed and negatively stained, and the deposit is loaded on to grids. Not only rotavirus particles but also adenovirus, astrovirus, calicivirus and other small round viruses can be detected by this method; but the significance of the latter group is not clear. In the ELISA technique, microtitration plates are first coated with anti-rotavirus antibody and then exposed to a faecal filtrate. Enzyme-labelled antibody is added, followed by the substrate. The colour change produced by the action of enzyme on substrate indicates the presence of rotavirus. Electron-microscopy gives more information than ELISA, but the value of this information is unproven. ELISA is a more sensitive means of detecting particular viruses, and is far easier to apply to a large number of specimens because making preparations for electron-microscopy is laborious.

19
Liver and Biliary Tract

By virtue of its rich endowment of phagocytic cells and its strategic position in the circulation, the liver is continually filtering out micro-organisms from the blood and also inactivating endotoxin absorbed from the gut. It is therefore vulnerable to infection and to damaging effects of infection elsewhere. However, by far the most important form of infective liver damage is virus hepatitis caused by organisms that are primary liver pathogens.

Virus Hepatitis

This term is used to describe inflammation and cellular necrosis of the liver caused by viruses. There are at least five different hepatitis viruses (p. 134); and the liver can also be involved in virus infections that are not primarily hepatic (p. 267).

Virus hepatitis has a spectrum of severity ranging from subclinical infection through to acute liver necrosis with death of the patient. Intermediate degrees of liver damage may be followed by complete recovery as in hepatitis A, or by chronic hepatitis sometimes leading to cirrhosis in hepatitis B and in non-A and non-B hepatitis. Two major epidemiological patterns were recognised in the 1940s—**infective** or **infectious hepatitis,** often occurring in epidemics, spread mainly by the faecal–oral route, and having an incubation period of 2–6 weeks; and **serum hepatitis,** transmitted in blood or blood products, usually occurring as sporadic cases but sometimes in outbreaks associated with receiving particular batches of blood products or with use of one syringe (and sometimes one needle) to inoculate a group of people, and having an incubation period of 2–6 months. Individual cases were often difficult to classify, because of lack of any relevant history. The situation has now become clearer if somewhat more complex with the discovery of the **hepatitis A, hepatitis B** and **delta viruses** and with accumulated evidence that there are at least two **non-A non-B viruses.** There may be epidemiological clues as to which virus is responsible for a given case, but clinical findings, biochemical tests and liver biopsy all fail to provide reliable differentiation. Marked rises in serum bilirubin and aminotransferases, reaching their peaks early in the icteric phase and gradually returning to normal over the next few weeks, are common to all of these types of infection. Recognition that the infection is due to A, B or delta virus depends on specific serological tests (see below).

Hepatitis A

An RNA enterovirus (p. 119) is responsible for the pattern of disease formerly called infective hepatitis, but can now be shown also to cause many of the sporadic cases that would previously have been unclassifiable. Patients are mainly children and young people, and outbreaks are common when groups of young people are brought together to live, as in schools, colleges and military camps. Faecal excretion of the virus occurs for 7-10 days before onset of the illness and for a few more days after that (about 2 weeks in all), and nearly all transmission takes place during this period—by the faecal-oral route (p. 249), but with flies, food (particularly shellfish, p. 359), drinking water and other vectors often involved and accounting for the higher incidence of the disease in warm countries with poor standards of hygiene. Transmission in blood is also possible, but there is no persisting viraemia such as occurs in hepatitis B.

The symptoms of hepatitis A, which develop after an incubation period of about 30 days (range 2–6 weeks), are largely non-specific—fever, malaise, anorexia, nausea, vomiting and dull abdominal pain. Jaundice may develop 3–10 days later, but many patients (1 in 4 adults, 1 in 12 children) are anicteric. During the icteric phase the urine is dark because of bilirubin excretion. Patients usually begin to feel better as soon as the jaundice appears. The diagnosis can be confirmed by demonstrating the appearance of antibodies for hepatitis A virus in the serum. IgG antibodies may relate to past infection, but the presence of specific IgM antibodies indicates that the current illness is hepatitis A. Active immunization against this infection is not possible yet, but an injection of human normal immunoglobulin (p. 363) provides passive protection for 4–6 months. Such passive immunization is a valuable means of controlling an outbreak in a closed community, such as a school, as it usually prevents the disease even if not given until a few days after exposure. It is also recommended for travellers visiting countries in which they are at increased risk of acquiring hepatitis A.

Hepatitis B

In 1964 the blood of an Australian aborigine was found to contain an antigen that gave a precipitin reaction with the serum of a much-transfused haemophiliac. This antigen, which became known as **Australia antigen**, was also found in the sera of many patients with hepatitis of the long-incubation type (serum hepatitis, as described above). It is now known to be a capsid antigen of a DNA virus called hepatitis B virus (p. 135), and it has been renamed hepatitis B surface antigen or HBsAg. HBsAg can also be detected in the urine, saliva, tears and semen of infected individuals.

Hepatitis B corresponds roughly to the old definition of serum hepatitis. Before its aetiology was known it was often transmitted by medical procedures such as blood transfusion or injections (see above). Such routes of transmission can now be closed, by testing blood for presence of hepatitis B antigen before allowing its use for transfusion and by using disposable syringes and needles; but other routes remain open. These including sharing of syringes and needles by drug addicts, and such procedures as tattooing and acupuncture carried out by non-sterile techniques. Doctors, dentists, nurses, laboratory workers and others may be infected when patients' blood gets on abrasions in their skin, or is splashed in their faces (the conjunctiva being a possible portal of entry for the virus), or when they scratch themselves with needles after inserting them into patients. The virus can also be transmitted during sexual intercourse (particularly among homosexual males), or by sharing a common tooth-brush, and probably in many other ways not yet detected.

The onset of hepatitis B is insidious, with increasing malaise and usually with some degree of jaundice, though many cases are anicteric. There may be a skin rash (often itchy), urticaria, arthritis and in children glomerulonephritis; these extra-hepatic manifestations are probably due to immune complex deposition. HBsAg and HBeAg (p. 135) are detectable in serum during the last few weeks of the long incubation period, before the onset of clinical hepatitis, and as a rule for some weeks afterwards, HBsAg persisting longer than HBeAg and each of them disappearing at about the time that its corresponding antibody becomes detectable. The presence of HBeAg indicates that the patients' blood is highly infectious, and that chronic infection may well develop (see below). The finding of HBe-antibody indicates that the infectivity and risk of transmission are extremely low. Antibodies to HBsAg and HBeAg are long-lasting, but HBc-antibody (p. 135) appears only during the early stages of the clinical illness, and high titres of anti-HBc IgM indicate recent hepatitis B virus infection.

In round figures 80% of patients with acute hepatitis B recover completely, 15% develop chronic persistent hepatitis (a benign condition) and 5% develop chronic active hepatitis, which in about half of them will progress to cirrhosis. By definition chronic hepatitis is the persistence of symptoms or of abnormal liver function tests for more than 6 months. Paradoxically, complete recovery is the rule for patients who have acute florid hepatitis with deep jaundice, and even for those with fulminant but non-fatal hepatitis B; whereas those whose illness is insidious in onset and causes mild symptoms with minimal jaundice are more likely to develop chronic hepatitis—as are those who are in any way immunocompromised. Chronic hepatitis B is commonest in middle-aged men, and is associated with persistence of HBeAg in the blood and absence of HBe antibody. Steroid treatment may halt the progress and reduce the mortality of chronic active hepatitis.

Not only in patients with chronic hepatitis but in some who have made complete clinical and biochemical recoveries and in others who have become carriers without ever having clinical hepatitis, hepatitis B virus persists for years, as indicated by positive tests for HBsAg and HBeAg and by infectivity of their blood to any who may be exposed to it. All potential blood donors must therefore be screened to exclude the possibility of their being carriers of this virus. Furthermore, all samples of human blood should be regarded and treated as potential sources of hepatitis B infection for medical and nursing staff, laboratory workers and all others exposed to them, until tests for HB antigens have excluded this possibility—though clearly the risk is greater if the samples come from jaundiced patients. The assortment of people among whom persistent carriage of hepatitis B virus is particularly common includes children with Down's syndrome, patients with lepromatous leprosy, patients undergoing haemodialysis and drug addicts. The carriage rate in the community in general is of the order of 1:1000 in Britain, but as high as 5–10% in some countries, notably in the tropics. In South-eastern Asia a correlation has been established between HBeAg carriage by young adults and liability to develop hepatoma (a malignant liver tumour).

Transmission of hepatitis B from mother to baby in the perinatal period is discussed in Chapter 25 (p. 328).

Active and passive immunization for hepatitis B are discussed in Chapter 28 and the problems of hospital-acquired infection in Chapter 27.

Delta Virus Infection

Delta virus (p. 135) can cause infection simultaneously with hepatitis B virus, or it may be responsible for an episode of acute hepatitis in a known chronic hepatitis B virus carrier. The hepatitis resulting from delta virus infection is frequently fulminant. It is commonest in intravenous drug abusers, in haemophiliacs, in others who have had multiple blood transfusions, and in the contacts of these groups. The incubation period varies from 2 to 12 weeks and infection can be diagnosed by finding delta virus antigen or anti-delta antibodies in the serum. Since delta virus depends on hepatitis B virus for replication, measures that prevent hepatitis B will also prevent delta virus infection.

Non-A Non-B Hepatitis

Now that it is possible to show which patients with hepatitis are infected with hepatitis A or B viruses or with one of the other viruses (such as CMV or EB virus—see below) that can cause hepatitis, there remain some who have similar illnesses for which no agent can be identified. There are two clinical forms of non-A non-B hepatitis. One resembles hepatitis A in its faecal–oral transmission and in causing epidemics (particularly water-borne outbreaks in India). The incubation period appears to be about 40 days and the clinical presentation is that of any virus hepatitis. It is usually a mild illness, causing jaundice in only 25% of those infected. Chronic infection appears to be rare.

The other type of non-A non-B hepatitis, which includes nearly all cases of post-transfusion hepatitis seen in the USA, resembles hepatitis B in its parenteral transmission. However, not all sporadic cases of non-A non-B hepatitis follow known parenteral exposure, and there are probably other routes of transmission. The incubation period is usually from 6 to 10 weeks. Fulminant hepatic failure is rare, but chronic infection is more common than in hepatitis B.

At present there is no serological test for positive identification of infection with non-A non-B hepatitis viruses. However, the incidence of post-transfusion hepatitis can be significantly reduced by measuring the serum transaminase levels of all donors and rejecting donations from those with raised levels.

Liver Involvement in Other Virus Infections

Hepatitis, usually mild (often detectable only by biochemical tests) and self-limiting, may be a component of a more generalized virus infection—notably of EB-virus infectious mononucleosis (p. 143), in which slight elevation of serum transaminase is common but hepatitis severe enough to cause jaundice is rare. The diagnosis of EB virus infection can be established by looking for its typical clinical, haematological and serological features.

In immunocompromised patients, particularly those with defective cell-mediated immunity, herpesviruses (herpes simplex virus, p. 130; cytomegalovirus, p. 132; VZ virus, p. 131) may cause disseminated infections with liver necrosis and failure. Cytomegalovirus in particular has been associated with acute hepatitis in patients who are on immunosuppressive drugs because they have received organ-transplants; and it can be found in the blood and the urine of such patients.

Yellow fever (p. 122) is a virus infection particularly associated with jaundice resulting from liver damage, which is often severe and commonly fatal. This diagnosis should be suspected in non-immunized jaundiced patients from areas where it is endemic—in Central and South America and Central Africa—and can be confirmed by isolating the virus from blood or in due course by demonstrating a specific antibody rise.

Bacterial Infections of the Liver

Leptospirosis

Leptospiral infection, particularly with *L. interrogans* serotype *icterohaemorrhagiae* (p. 109), causes Weil's disease, in which there is fever, icteric hepatitis, haemorrhage, meningitis and renal failure. In the past this disease occurred typically in fish workers, sewer workers and miners who were exposed to water contaminated with infected rat urine, the organisms gaining entry via cuts and skin abrasions. With modern pest control measures, the wearing of protective clothing and the extensive use of detergents which are lethal for leptospires, the epidemiology of leptospirosis has changed (p. 356). Icterohaemorrhagiae infections are now less common and principally affect farm workers, bathers in fresh-water streams and ponds, and those involved in water sports such as canoeing where immersion occurs frequently.

Syphilis

Hepatitis may occur in either the primary or the secondary stage of syphilis (p. 278), but it rarely causes jaundice and the main evidence of its occurrence is usually elevation of the serum alkaline phosphatase level. The diagnosis depends on clinical findings (a chancre or a typical rash), confirmed by positive treponemal serology. Hepatic gummas are part of the picture of late syphilis.

Q Fever

Coxiella burneti (p. 113) causes a generalized infection which sometimes includes hepatitis. Persistent infection with granuloma formation may follow (see below).

Hepatic Granulomas

The agents responsible for granuloma formation in the liver include certain bacteria that can survive inside host cells—tubercle bacilli, brucellae, *List. monocytogenes* and *Cox. burneti*. The persistence of these organisms or their products inside liver macrophages triggers their transformation into epithelioid cells that in turn fuse to form multinucleate giant cells. A delayed-type hypersensitivity response to microbial antigens, mounted by T lymphocytes and macrophages in concert (p. 62), may be involved. (Non-bacterial agents capable of initiating similar granuloma formation are the yeast *Histoplasma capsulatum* and the helminths *Schistosoma mansoni* and *Toxocara canis*). In chronic granulomatous disease, a rare condition in which there is a genetically determined defect in the microbial killing capacity of host cells (p. 315), a wide range of micro-organisms can survive inside phagocytes and initiate granuloma formation, particularly in the liver. Tissue for laboratory diagnosis of hepatic granulomas can be obtained by percutaneous needle biopsy and examined by appropriate special cultural and histological techniques; serological and skin tests may also help to identify the causative agent.

Liver Abscesses (For amoebic liver abscesses, see p. 269)

Liver abscesses, single or multiple, may develop when pyogenic bacteria reach the liver either from intra-abdominal septic lesions via the portal vein, or by extension of an infection of the biliary tract (see below), or as part of a septicaemic illness. The classical presentation—swinging pyrexia, prostration and rigors, liver tenderness and signs of pleural or pulmonary involvement at the right base—is easily recognized, but not all liver abscesses produce such florid signs and the diagnosis is often delayed. Abscesses can be identified and located by radiological or ultrasound scanning techniques. Blood culture may reveal the bacteria involved, though it is unlikely to reflect the complexity of the bacteriology of abscesses secondary to intra-abdominal sepsis. The mixture of bacteria involved in such lesions commonly includes anaerobes (bacteroides and anaerobic cocci) and enterobacteria, and *Str. milleri* (p. 80) has been increasingly recognized as making an important contribution. Pus from a liver abscess, acquired either by percutaneous aspiration via a fine needle or by open surgery, can be checked rapidly by gas-liquid chromatography for the presence of the volatile fatty acids produced by anaerobes; this procedure gives a guide to appropriate antibacterial treatment long before culture results are available. Bacterial liver abscesses should be drained surgically as soon as suitable antibacterial therapy has been started, and that therapy should be prolonged.

Biliary Tract Infections

The bile ducts and gall bladder are normally sterile. However, obstruction of the biliary tract (e.g. by gallstones) interferes with the flow of bile and allows intestinal bacteria to invade the stagnant bile and cause infection of the bile ducts (**cholangitis**) and of the gall bladder (**cholecystitis**). General symptoms of bacterial infection—malaise, fever, etc.—are accompanied by nausea and (sometimes) vomiting, and by right hypochondrial pain and marked hypersensitivity to even gentle palpation in that area. Irritation of the abdominal surface of the diaphragm may cause referred pain in the right shoulder. Infection of a gall bladder behind an obstruction to its outlet may turn it into a swollen and distended sac full of pus under pressure—an **empyema**—and rupture of this may follow, leading to peritonitis. Infection of the upper biliary tract readily gives rise to seeding of bacteria into the blood-stream, and cholangitis is one of the commonest predisposing causes of bacteraemia; when this occurs, swinging fever and rigors and sometimes the signs of endotoxic shock (p. 40) are superimposed on the picture of biliary tract infection. Liver abscess (see above) is another possible complication of cholangitis. Enterobacteria, notably *Esch. coli*, are the bacteria responsible for most biliary tract infections; they are commonly accompanied by *Bacteroides* species of the fragilis group, especially in infections of obstructed gall bladders. The organism commonly responsible for bacteraemia secondary to biliary tract infection is *Esch. coli*; the anaerobes rarely reach the blood-stream. However, blood culture is still the most useful means of investigating the bacteriology of these infections, apart from surgical exploration of the tract.

Most patients with biliary tract infections require surgery for removal of stones or relief of other forms of obstruction. It is usually best to delay the operation until the acute infection has been brought under control by antibacterial treatment. This should be aimed at enterobacteria and anaerobes. A special problem in selection of suitable drugs in the presence of obstruction is that the penetration of most of them into the gall bladder and bile ducts depends on free flow of bile. Common regimens include an aminoglycoside or one of the newer cephalosporins, in either case accompanied by metronidazole. Even when there has been no recent acute episode or one has been brought under control by antibacterial treatment, biliary tract surgery should be covered by prophylactic administration of an appropriate antibacterial agent—usually a cephalosporin—during the operative period, to prevent both bacteraemia and wound infection.

Typhoid and the gall bladder

The gall bladder is the principal site for chronic carriage of *S. typhi* (p. 93), which is excreted in the bile and the faeces, often intermittently. Numerous faecal cultures may be needed to identify such carriers, and the Widal test (p. 180) is an easier approach, but has not proved reliable; the status of the Vi-antigen version of this test is considered on p. 180. If the gall bladder is healthy apart from colonization by *S. typhi*, treatment with amoxycillin or cotrimoxazole may eradicate the organism; but a diseased gall bladder often has to be removed to achieve this end.

Protozoan Infections of the Liver

Amoebiasis (p. 253)

Where intestinal amoebiasis due to *Entamoeba histolytica* is endemic, the amoebae also cause liver abscesses in a small proportion of cases. Vegetative amoebae reach the liver from the gut lumen by penetrating the mucosa and entering the portal circulation. They may cause diffuse hepatitis, but the commoner result of such an infection is the formation of one or, more rarely, several liver abscesses, with resultant enlargement and tenderness of the liver, malaise, irregular fever and leukocytosis. Amoebic abscesses do not have well-defined walls but enlarge gradually as the amoebae, which are mostly at the periphery, destroy the surrounding liver parenchyma. The centre of such an abscess is filled with tissue debris and blood; this material may contain some vegetative amoebae (trophozoites), but amoebic cysts are never found there and the pus is bacteriologically sterile. Abscesses are more often in the right lobe of the liver than in the left. Penetration through the diaphragm may occur, leading to an empyema that may drain via the bronchi, so that the patient coughs up the pinkish-brown 'anchovy sauce' pus that is characteristic of an amoebic abscess. Alternatively, penetration of the abdominal wall may occur.

Laboratory diagnosis

Trophozoites may be seen on microscopy of aspirated or expectorated pus, and stool specimens may contain both cysts and trophozoites (p. 174). Serological methods (latex-agglutination, complement-fixation, indirect haemagglutination and fluorescent antibody tests) are useful in the diagnosis of this and other forms of non-intestinal amoebiasis.

Treatment

Large amoebic abscesses, like bacterial liver abscesses, require both appropriate antimicrobial treatment and effective drainage; and the antimicrobial treatment, usually with metronidazole and emetine (p. 152) should be started before the abscess is drained. However, in this case drainage should be by aspiration through a wide-bore needle, not by open surgery, as the latter almost inevitably allows trophozoites to escape into the abdominal cavity where they cause a generalized abdominal amoebiasis that is often fatal. A large amoebic abscess may contain several hundred millilitres of pus and may require repeated aspiration.

Malaria (p. 241)

The liver is involved in malaria in several ways. In the early stages of the infection the parasites enter liver parenchyma cells and there develop into liver schizonts, but this probably has little immediate significance for the host or his liver function. Later, when the parasites move into the blood and destroy red cells, the liver and spleen (which have to deal with the red-cell remains) become enlarged during the acute episodes. In rare cases, mostly of *P. falciparum* infection, a severe form of liver infection is seen, with fever, abdominal pain, nausea, persistent vomiting and enlargement and tenderness of the liver. Jaundice in malaria can be due either to haemolysis or to liver damage. In chronic malaria, congestion and enlargement may become more or less permanent.

Leishmaniasis (p. 303)

Leishmaniae invade and multiply inside the reticulo-endothelial cells of the liver and spleen. Enlargement of these organs, abrupt in onset, is a typical feature of leishmaniasis, particularly of *L. donovani* infection; and they may be greatly enlarged in the later stages of the disease, with accompanying progressive anaemia and emaciation. The enlarged organs are soft and as a rule are not tender.

Giardiasis (p. 254)

Giardia sometimes invades the gall bladder, causing jaundice and biliary colic. It is also sometimes found in gall bladders that have been removed because of gallstones, but it is not clear whether it is responsible for the formation of these.

Helminth Infections

Schistosomes (p. 155)

Although schistosomes undergo the final (pre-adult) stages of their development in the hepatic blood vessels, this rarely causes any serious pathology. However, a significant proportion of the eggs laid by the female worms (perhaps 10–20%) do not pass through the intestinal or bladder wall to be excreted in faeces or urine respectively, and some of them (especially those of *S. mansoni* and *S. japonicum*, of which the adults live in the portal veins) are carried in the blood-stream to the liver. There they become trapped and provoke immunological responses that result in fibrosis around them. Heavy *S. mansoni* infections and lighter *S. japonicum* infections provoke initial enlargement of the liver, followed by gradual reduction in its size as it becomes progressively more fibrosed. The fibrosis causes obstruction of the hepatic circulation, portal hypertension, and the development of a collateral circulation including oesophageal varices, and rupture of these may lead to fatal haemorrhage. The more rapid and severe response of the liver to *S. japonicum* infection is due to its inhabiting more anterior branches of the portal circulation than does *S. mansoni* and producing more eggs per female per day. The less important role of *S. haematobium* in liver disease is the result of its eggs being carried to the lungs far more than to the liver.

Liver Flukes

Fasciola (p. 153)

Human infection with the sheep and cattle liver fluke *F. hepatica* is rather rare; it follows ingestion of contaminated aquatic vegetation, usually wild watercress. The immature stages of the fluke burrow through the duodenal wall, cross the peritoneal cavity, penetrate the liver capsule and migrate through the liver parenchyma to reach the bile ducts. There may be some minor symptoms, such as liver tenderness, at this stage; but it is the adult flukes that cause the major effects. They are large worms (up to 3×1.5 cm) and can cause mechanical irritation to the bile ducts and physical obstruction even to the larger ducts; and their toxic action on neighbouring host tissue can result in further partial or total biliary obstruction. Portal cirrhosis is the final outcome of serious infections. The diagnosis can be confirmed by finding *Fasciola* eggs in the patient's faeces, as long as allowance is made for incidental passage of such eggs, without infection, after eating liver from a parasitized animal; repetition of the investigation several days later may be necessary to confirm true infection. Treatment is with bithionol; emetine hydrochloride has also been used successfully.

Opisthorcis (p. 153) **and Dicrocoelium**

Infection of the distal parts of the human bile ducts with either of these small worms results from ingestion of the immature stages and their subsequent migration from the intestine up the biliary tree. When established in the bile ducts the adults provoke epithelial hyperplasia and fibrosis; heavy infection may cause portal cirrhosis. *Opisthorcis* infection of man is fairly common in East Asia; it is acquired by eating raw or inadequately cooked fresh-water fish. *Dicrocoelium* is a common parasite of herbivorous mammals in many parts of the world, but only rarely infects man; most apparent cases of infection, diagnosed by faecal examination, are due to incidental passage of eggs after eating liver from an infected animal (cf. *Fasciola*).

Ascaris (p. 156)

Young *Ascaris* larvae migrate to the lungs via hepatic blood vessels, causing tissue destruction and, if their numbers are large, enlargement and tenderness of the liver. This hepatitis is usually short-lived. The diagnosis cannot be established until a few weeks later, when the worms are mature and their eggs are to be found in the host's faeces. On occasions an adult *Ascaris* migrates from the intestine up the common bile duct and causes obstructive jaundice.

Echinococcus (p. 155)

Hydatid disease, a condition caused by the development of the larval (hydatid cyst) stage of the dog tapeworm *E. granulosus* in human tissues, occurs most frequently in the liver, with the lungs as the other common site. As the dog is the definitive host and sheep and goats are among the common intermediate hosts, most of the humans affected are closely involved with these animals—in Britain, mostly in parts of Wales, Cumbria and Scotland; they acquire their infections when their hands or food become contaminated with eggs from dog faeces. After ingestion the egg hatches and the embryo bores into the gut wall and is carried by the bloodstream to the liver or some other organ. There it develops in the course of a few years into a fluid-filled hydatid cyst, which after 10 years or so may have a capacity of several litres. The cyst wall consists of a thin layer of parasite germinal cells and a thicker laminated layer which merges with the surrounding host tissue. The germinal layer produces many immature tapeworms, some of which become detached and are free in the hydatid fluid. Hydatid cysts in the liver are generally more or less spherical; elsewhere their shapes are determined by resistance from the host tissue.

The diagnosis of hydatid disease can be confirmed radiologically or by immunological tests. The time-honoured **Casoni** test, consisting of the intradermal injection of sterilized and standardized hydatid fluid, may cause either an immediate wheal-and-flare or a delayed-type hypersensitivity reaction; but it has been largely replaced by more reliable specific antibody tests.

Surgical removal is the only reliably effective treatment for hydatid cysts, but the lack of a defined capsule makes it difficult. Furthermore, leakage or spillage of hydatid fluid may cause anaphylactic shock; and escape of any of the immature tapeworms or failure to remove all of the germinal layer is followed by further cyst development, which occurs in up to 50% of cases. The best approach to the problem is first to aspirate the fluid carefully and replace it with formaldehyde solution or some other fluid that will kill the germinal layer and immature tapeworms; the cyst is then carefully excised. Chemotherapy with mebendazole has been successful in some cases, but high levels of the drug must be maintained for long periods; the use of more soluble drugs or of slow-release preparations may prove more effective.

20

Genito-urinary Tract

Infections of the urinary and reproductive tracts are common and varied; some are relatively trivial and self-limiting, but others can have serious consequences. The two tracts are considered together in this chapter because of their proximity and inter-relations, but the infections are divided into two major groups—urinary tract infections, affecting any part of that tract from the kidneys to the bladder; and genital tract infections, which include the classical sexually transmitted (venereal) diseases and a much wider range of infections that may be related to sexual activity but are not necessarily the result of sexual transmission of a single microbial pathogen. The urethra belongs to both tracts, but urethritis is usually considered as a genital tract infection.

Urinary Tract Infections

Bacterial Infections

The only part of the urinary tract with a resident bacterial flora is the urethra. Normally, the tract above the vesico-urethral valve is sterile and any bacteria that do enter the bladder will be removed by the mucosal defence systems or washed out by the flow of urine. The urinary tract is infected when bacteria gain entry to the bladder, with or without involvement of ureters and kidneys, and begin to multiply. Attempts are made to distinguish between infections that involve the kidneys (pyelonephritis) and ureters and those in which only the bladder is affected (cystitis). This may be important in assessing the long-term prognosis of recurrent infection, because repeated infection of the kidney itself may be related to the condition of permanent renal damage with scarring known as chronic pyelonephritis (a pathological diagnosis, not a microbiological one); but the distinction between upper and lower tract involvement is often indistinct and initially it is better to regard the urinary tract as a single entity.

The diagnosis of urinary tract infection (UTI) is established bacteriologically by the demonstration of large numbers of the infecting organism in a freshly voided specimen of urine. Although there are signs and symptoms that typify UTI, many patients have an infection without symptoms (asymptomatic bacteriuria), whereas others may have symptoms in the absence of bacterial UTI. In the most obvious cases of infection, the patient complains of frequency of micturition, urgency and dysuria (pain or a burning sensation on passing urine). Together with low abdominal pain these are the cardinal symptoms of cystitis. The urine is passed in small amounts, and is often cloudy and offensive in smell. Although sufficient bleeding to make the urine red is rare, presence of red cells may give it a smokey appearance. If the kidneys are infected (pyelonephritis) there may be loin pain and fever, often with rigors that indicate bacteraemia; pyelonephritis is one of the commonest causes of bacteraemia. In most of these patients, large numbers of bacteria of a single species will be found in the urine, usually with many polymorphonuclear cells (pus cells—**pyuria**); these patients have symptomatic bacteriuria. However, many patients, especially women, have the large numbers of bacteria (i.e. an infection) without symptoms—asymptomatic bacteriuria—or with only minimal non-specific symptoms. There are also other patients who have dysuria, frequency and urgency without bacterial infection. The reason for this is not clear but it is sometimes attributable to infection and inflammation in the urethra—the urethral syndrome (p. 276).

Because symptoms are such unreliable indicators of UTI, bacteriological examination of the urine is important as a means of confirmation. All per-urethral specimens of urine will be contaminated to some degree by organisms from the urethra and the peri-urethral area (particularly the vulva and introitus in women). This contamination must be distinguished from true infection, and quantitative urine bacteriology is useful for this purpose. To maintain an infection in the bladder despite dilution by the continual flow of fresh urine, bacteria must be able to multiply rapidly and keep their numbers high. It has been established that in general the numbers of viable bacteria in freshly passed and carefully collected urine samples from patients with UTI (other than those with localized intra-renal infections) are 10^5/ml or more—usually many times more—and that these bacteria will usually be nearly all of one species (though often mixed with a few contaminants). In contrast, comparable urines from patients without infections usually give counts of less than 10^4/ml, and the organisms are mixed contaminants. More than 10^5 bacteria/ml in a suitable specimen is commonly referred to as 'significant bacteriuria', and it was the establishment of this criterion which allowed recognition of asymptomatic bacteriuria (see above). The measures necessary for collection of a satisfactory specimen are described on p. 274–5; and rapid transmission of the specimen to the laboratory is essential, since species commonly associated with urinary tract infections are also commonly present in small numbers of contaminants of non-infected urine, and can multiply fast at room temperatures to give a misleading impression of true bacteriuria.

Epidemiology

Urinary tract infection is essentially a disease of women, or of men with underlying abnormalities or pathology of the urinary tract.

Infections can be divided into primary and secondary types. **Primary** infections occur in those who have normal urinary tracts, and are uncommon in men. The main factor that makes primary UTI almost exclusively a disease of women is the anatomy of the urethra. The female urethra is short (c. 5 cm); it provides little barrier to the entry of bacteria into the bladder and opens to the surface at the vulva, an area heavily colonized by bacteria, whereas the much longer male urethra is a better protection for the bladder and is well away from the perineum. Moreover, the anatomical relationship of the female urethra to the vagina makes it liable to trauma during sexual intercourse (when bacteria may be massaged up the urethra into the bladder) and during child-birth. Primary UTI in women is therefore commonest in association with sexual activity and with child-bearing. These aspects are exemplified by (1) the frequent occurrence of 'honeymoon cystitis' in young women beginning sexual activity; and (2) the very low incidence of UTI in nuns.

UTI is one of the commonest infective diseases. Numerous surveys of women aged 15–50 years in various countries have shown that at any one time 4–6% of them have significant bacteriuria, which means that a large proportion of them must have at least single episodes of UTI during this age-period—possibly many episodes. About half of the cases of significant bacteriuria detected in this way are symptomatic, and in the majority of women with infections (symptomatic or asymptomatic) there is no evidence of urinary-tract abnormality, even in those who have repeated episodes of infection.

Secondary UTI results from some abnormality or instrumentation, and occurs in both men and women. The predisposing factor may be anything that causes disturbance of function—i.e. obstruction of flow or failure of emptying. The list of possibilities includes urethral strictures or valves, bladder diverticula, urinary calculi, prostatic enlargement, carcinoma of the bladder, renal cysts, duplex ureters, horse-shoe kidney, ureteric reflux, neurogenic bladder and therapeutic procedures such as bladder catheterization or cytoscopy. Secondary UTI is a common complication of the management of many other conditions of hospital patients, particularly when catheterization is necessary.

About 1% of children have UTI in the first year of life. This is rather more common in boys and is usually secondary to congenital abnormalities of the urinary tract, often minor.

Bacteriology

The bacteria that cause UTI are shown in Table 20.1. In simple uncomplicated primary UTI, 80% of infections are caused by *Escherichia coli*, 7–8% by *Proteus mirabilis* and a similar proportion by *Staphylococcus saprophyticus*. The remaining 5% are mostly caused by enterococci. The *Esch. coli*

TABLE 20.1 Bacteria that Cause Urinary Tract Infection

Primary UTI	Species	Secondary UTI
80	Esch. coli	60
8	Staph. saprophyticus	1
8	Proteus mirabilis	15
3	Enterococci	10
< 1	Ps. aeruginosa	5–10
	Klebsiella species	5–10
	other enterobacteria	5
	other staphylococci	5–10

*Secondary infections are not uncommonly mixed.

strains most commonly involved are of the serotypes most commonly found in the intestine, which suggests that their frequent involvement is due to their being most often in the right place at the right time. However, certain properties possibly related to pathogenicity are commoner among strains that have caused urinary tract infections than among otherwise similar intestinal strains; these properties include the ability to adhere to mucosal cells, resistance to serum killing and to phagocytosis, and haemolysin production.

Staph. saprophyticus is a skin commensal found on the perineal surface, but less commonly elsewhere on the body. It causes UTI mostly in young women; these infections are usually symptomatic, with pronounced lower tract symptoms (frequency and dysuria), haematuria and heavy pyuria.

Esch. coli is not so predominant in secondary UTI; many cases are due to other gram-negative bacilli—particularly the more antibiotic-resistant hospital-associated ones, such as *Proteus, Klebsiella, Enterobacter* and *Serratia* species and *Pseudomonas aeruginosa*. There is also a greater variety of gram-positive cocci, but *Staph. saprophyticus* does not feature in these infections. Mixed infections are also much more common in secondary UTI.

Most of the bacteria that cause UTI (other than *Staph. saprophyticus*—see above) are faecal organisms that contaminate the perineum, enter the urethra and then cause an ascending infection of the urinary tract. In secondary UTI, although the commonest source is still the patient's faecal flora there is greater opportunity for cross-infection and the introduction of hospital strains on hands or on materials used, e.g. in the manipulation of catheters.

Tuberculosis of the urinary tract is a bacterial infection distinct from all those considered above. It is part of the secondary (adult) type of tuberculosis (p. 207) that occurs when tuberculous foci in the kidneys, seeded via the blood stream during the primary phase of the disease, are reactivated, causing destruction and caseation in the renal parenchyma. There is progressive renal damage, and pus cells (polymorphs) and *Mycobacterium tuberculosis* are shed in the urine. The bladder may also be infected, with chronic inflammation leading to thickening of the bladder wall and poor emptying.

Laboratory diagnosis

The laboratory diagnosis of non-tuberculous UTI depends upon the demonstration of significant bacteriuria ($\geq 10^5$ organisms/ml; p. 273) by quantitative culture of a fresh, properly taken specimen of urine. The quality of the bacteriological examination is directly related to the quality of the specimen. In men, collection of a mid-stream specimen of urine (MSSU) is easy; after cleansing of the glans and urethral meatus with mild soap, the first part of the urine stream is allowed to flow away, carrying with it urethral contaminants, and a sample from the middle part of the stream is then collected into a sterile container. Collection from women requires more attention to detail. In comfortable conditions, the patient should cleanse the vulva and peri-urethral area with 3-4 swabs (without any antiseptic), then spread the labia and collect a MSSU in a wide-mouthed sterile container.*

In the laboratory, the urine is commonly examined by microscopy, and by semi-quantitative culture on a medium that supports the growth of most potential pathogens—e.g. CLED (cystine-lactose-electrolyte-deficient) agar. Several semi-quantitative methods of culture, accurate in the 10^3–10^6 range that is important in diagnosis, are used—e.g. standard streaking of a plate with a 0.01 ml loop, or sampling by means of a small filter-paper strip that transfers a fixed volume of urine on to the agar. Once a sample has been taken for culture, the specimen is centrifuged and the deposit examined by microscopy. Pus cells, red blood cells, epithelial cells, crystals, casts and bacteria may be seen in wet films examined with the × 40 objective. The morphology of the bacteria can be determined if necessary by examining a gram-stained film. Microscopy is also useful in investigations of renal disease other than UTI, in which, for example, the presence of red cells or casts might be significant. The presence of pus cells does not always indicate infection, although UTI is the commonest cause of pyuria. It does indicate some abnormality of the urinary tract and may aid the diagnosis of renal tuberculosis, calculi or tumour. If sterile pyuria is repeatedly found and tuberculosis is suspected, three consecutive complete early-morning specimens of urine should be collected, since bacilli are scanty in the urine and early-morning specimens are the most concentrated. In the laboratory these specimens are pooled and centrifuged. The deposit is treated with acid or alkali to destroy other bacteria, and then films are stained by the Ziehl-Neelsen method for acid-alcohol-fast bacilli (AAFB). The deposit is inoculated on to Lowenstein-Jensen or similar media. The presence of AAFB in the film is suggestive of renal tuberculosis but not diagnostic, because other saprophytic mycobacteria—e.g. *Myco. smegmatis* from the genitalia—may contaminate the specimen.

Bacteriological examination of urine is a very large component of the work-load of almost any clinical microbiology laboratory, and full examination is costly in time and materials. There have therefore been many attempts to find methods that screen out specimens with low bacterial populations and give early indications of those likely to give significant growths. Methods based on detecting abnormalities in the chemical composition of infected urine (the presence of nitrite or leukocyte esterase) have lacked specificity. Automatic methods for detecting large bacterial populations in the specimens (or early growth in cultures) by light-scattering (nephelometry), ATP-production (luminometry), electrical impedance changes or heat-production (calorimetry) may prove more useful in suitably equipped laboratories. Meanwhile, simple methods have been developed which allow screening for bacteriuria when there is no urgent need for the answer and microscopy is not required—e.g. during routine investigation of asymptomatic pregnant women. Plastic strips coated with a suitable culture medium ('dip slides') can be dipped into freshly passed urine, drained and incubated (in the clinic or in the laboratory); or the layer of medium can be at the bottom of a plastic pot, which is filled with urine, emptied and drained and then incubated. In either case the number of bacterial colonies on the incubated medium reflects the bacterial population of the urine.

Management

Generally, the prognosis in uncomplicated UTI is good, but pyelonephritis is one of the commonest sources of enterobacterial bacteraemia. The risks and the approach to management differ with different groups of patients.

(1) An otherwise normal healthy woman with symptomatic UTI should be treated with a suitable antibiotic. Many would recover spontaneously, but recovery is quicker with treatment and any risk of pyelonephritis and bacteraemia is avoided. Only short treatment is needed—3-5 days is effective, and more recent studies suggest that one day may be sufficient.

---

*Urine specimens should be collected through a catheter **only** if catheterization is required for other purposes (p. 352).

The antibiotic restricts bacterial multiplication and allows the normal clearance mechanisms to act; this is helped by ensuring a large fluid intake.

(2) A normal non-pregnant woman with healthy kidneys and asymptomatic bacteriuria does not need treatment. The kidneys will not be damaged, even by repeated episodes of infection, and if symptoms develop she will move to group (1). Treatment of asymptomatic bacteriuria does not affect the rate of relapse or reinfection. On this basis, there is no need to screen non-pregnant women for bacteriuria.

(3) In patients with damaged (scarred) kidneys, however, even asymptomatic infections may lead to further damage. They should be screened for infection regularly and all episodes treated, regardless of symptoms. If infections recur frequently, long-term prophylactic treatment with an antibiotic (once a day, or every second day) should be considered. Such prophylaxis is also effective in giving relief to women with normal kidneys who have repeated episodes of symptomatic UTI.

(4) In pregnant women UTI is very important. About one-third of those in whom asymptomatic bacteriuria is detected will subsequently develop clinical pyelonephritis. This predisposes to toxaemia of pregnancy, premature delivery, small-for-dates babies and increased perinatal mortality. To prevent these complications, all pregnant women should be screened for bacteriuria and those with UTI should be treated, whether symptomatic or not.

(5) Men and boys who develop UTI should be treated and thoroughly investigated, because primary UTI is much less common in them than in women and bacteriuria often signifies some underlying pathology.

Population screening

Cheap and easy methods for detecting bacteriuria enable screening of high-risk groups. These groups include pregnant women and men or women with damaged kidneys. Because infection early in childhood, especially when associated with ureteric reflux or minor anatomical abnormalities, may initiate damage and scarring in the developing kidneys, the more difficult exercise of screening infants might be worthwhile.

Urethral syndrome

This term is used to describe a common condition in which the symptoms of dysuria and frequency are not accompanied by significant bacteriuria. It probably represents inflammation or infection of the urethra without extension to the bladder. There is some overlap with urethritis (see below), in which there is in addition a purulent urethral discharge.

Schistosomiasis

Of the three important species of schistosome which infect man (p. 155), only *Schistosoma haematobium* normally affects the urinary tract. The worms reach the veins of the vesical plexus about 2 months after cercariae penetrate the skin and initiate infection. As in all species of *Schistosoma*, the sexes are distinct but male and female live as a pair; the female *S. haematobium* lays eggs within the venules in the bladder wall. To continue the life cycle of the parasite these eggs must escape from the body. Most of the eggs are liberated through the bladder wall into the lumen, and their passage through the wall is often accompanied by bleeding—the source of the characteristic haematuria. They are excreted in the urine. Eggs excreted within 2 weeks or so of laying are viable, but some are retained longer in the bladder wall and, when eventually passed, are dead and coated with a fibrin layer resulting from the body's immunological response. Others never escape into the bladder lumen, and immunological responses to substances secreted by the miracidia developing within them result in a fibrous reaction around them. At its full extent this produces a fibrous nodule two or three times the size of the egg, which itself is eventually absorbed. In a heavy infection, the large number of these nodules eventually reduces the muscular tone of the bladder.

The adult worms live for several years in their hosts (some for 10 years or over, probably more for only 4–5 years), but egg production gradually falls. During the most active phase each female *S. haematobium* lays several hundred eggs each day, and the passage of so many through the bladder wall causes various changes: formation of papillomata containing eggs at various stages of encapsulation; eroded areas of bladder lining ('sandy patches'); fibrosis of the bladder wall, with eventual distortion and radiological filling defects of the bladder; and calcification. Transitional-cell carcinoma is common in such damaged mucosa. Obstruction of the openings of the ureters into the bladder leads to raised intra-ureteric pressure, which causes hydro-ureter and eventually hydronephrosis, and these complications are associated with reduced life expectancy. Effective treatment of the infection may allow improvement in the condition of the urinary tract.

Not all of the trapped eggs remain in the bladder wall; some are carried in the blood stream to the lungs, where they may cause considerable damage (p. 209).

Epidemiology

Urinary schistosomiasis is a disease of tropical Africa and the Middle East. Infection is contracted by occupational, recreational or domestic immersion in or contact with water (lakes, rivers, streams, or irrigation systems) containing the infective stages (cercariae) and the snails in which they have developed (p. 155). Humans aged 10-15 years, especially males, are most frequently infected and infection rates may reach 80–90% in some areas; older people, particularly agricultural workers, may also be affected, often severely. The immunological response to infection is directed against the eggs, not the adult worms. However, a condition known as concomitant immunity develops in which a response induced by the adults prevents or limits the acquisition of further schistosomes but does not affect the existing infection; this situation continues only while the adult worms are alive. Thus curing the infection removes the concomitant immunity, and re-infection can then occur.

Laboratory investigations

Although haematuria is a characteristic presenting feature of urinary schistosomiasis, and is sometimes heavy enough to produce blood clots in the urine, it is in other cases detectable only by microscopy or may be absent. Other diseases can cause haematuria, but in endemic areas the simultaneous presence of blood and excess protein in the urine (readily detectable by use of urinalysis test strips) is a strong indication of a moderate or heavy *Schistosoma* infection.

The concentration of eggs in urine is much higher around mid-day than at other times, and urine samples should therefore be collected within 1–2 hours of mid-day whenever possible. Urine should be allowed to sediment for 20 minutes in a conical urine glass, or centrifuged; the deposit is then examined for eggs by microscopy. The method can be made roughly quantitative by counting the eggs present in the sediment from 10 ml of urine. For accurate work, three daily urine samples are taken and the mean egg count determined. Care is required to differentiate live from dead eggs; the latter are black, in contrast to the golden-brown viable eggs. Dead eggs may be passed in small numbers by all schistosomiasis patients, and may be passed for several months after successful treatment. In areas where both *S. mansoni* and *S. haematobium* are endemic, double infection occurs. In about 5% of infected patients *S. mansoni* eggs may be found in the urine and *S. haematobium* in the faeces.

Immunological methods for the diagnosis of schistosomiasis are available (e.g. ELISA, complement fixation and fluorescent antibody tests and intradermal skin tests), but are generally more appropriate for surveys than for diagnosis of the individual patient. They are of no use in evaluating cure, because antibodies persist for at least several months after successful treatment.

Cystoscopy, bladder biopsy and radiography are also used in diagnosis, particularly in long-established cases, in whom the consequences of infection are as a rule advanced but the low egg output makes it difficult to confirm the diagnosis by examination of urine.

Treatment

Urinary schistosomiasis can be treated successfully, safely and cheaply with two or three doses of metrifonate given at two-week intervals.

Alternatively, praziquantel is effective in a single dose but is considerably more expensive; it is the drug of choice in cases of mixed schistosome infection because metrifonate has little or no effect on schistosomes other than *S. haematobium*.

Urinary Tract Complications of Malaria

Haemoglobinuria (known in this context as **blackwater fever**) used to be a frequent complication of falciparum malaria (pp. 241–2), especially in patients with recurrent malaria treated with quinine; it usually occurred during a bout of malaria, but sometimes in an otherwise asymptomatic phase. The underlying acute intravascular haemolysis, which leads to the excretion of haemoglobin sometimes in amounts sufficient to make the urine very dark, may be due, in part at least, to antibodies induced by and capable of lysing the infected red cells; but renal anoxia may make a contribution. The resultant acute anaemia is sometimes fatal, and even when it is less severe the onset of haemoglobinuria is an indication for immediate treatment. Though blackwater fever may still occur, it is now rare.

An association has been established in young children between malaria (particularly quartan malaria) and the **nephrotic syndrome**—proteinuria, hypoproteinaemia and oedema. There is renal tubular degeneration, glomerular fibrosis and thickening of the glomerular basement membrane. The malarial nephrotic syndrome may be due to an excess of malarial antigen in chronic low grade quartan malaria. Treatment is difficult and the condition does not respond to the administration of steroids.

Urinary Filariasis

Occasionally a patient with long-standing filariasis (p. 246) passes urine that is mixed with lymph containing microfilariae. The mechanism of this begins with blockage of lymph vessels by adult filariae, and consequent development of lymph varices. When these occur in the scrotum they may give rise to a lymphocoele, in which there may be microfilariae; but rupture of such varices into any part of the urinary tract may result in a milky appearance of the urine (chyluria), commonly of sudden onset, lasting a few days, and liable to recurrence at long intervals. Microscopy of the centrifuged deposit of such urine may show microfilariae. (Microfilariae have occasionally been seen in cervical smears, presumably reaching this site by a comparable mechanism.)

Genital Tract Infections

The great majority of infections of the genital tract are sexually transmitted diseases (STD). They include the classical venereal diseases which have long been recognized—syphilis, gonorrhoea, chancroid, lymphogranuloma venereum and granuloma venereum. However, other infections also largely or entirely transmitted by sexual intercourse are now recognized as being much commoner than any of the above, including gonorrhoea. These conditions include chlamydial and other forms of non-gonococcal urethritis, other chlamydial infections of the female genital tract, trichomoniasis, genital herpes and anaerobic vaginosis ('non-specific vaginitis'). Candidiasis of the genital tract is probably less dependent than any of these other infections upon sexual transmission.

Syphilis

This disease, with its widely diverse manifestations, is caused by the spirochaete *Treponema pallidum* (p. 108). The first sign of infection is a painless ulcerating lesion known as a primary **chancre**, which appears several weeks after exposure, usually on the skin or mucosa of the genitalia, but sometimes around the mouth or anus. This soon heals, to be followed in a few weeks by the secondary manifestations—dull red macular skin lesions, oral and ano-genital **condylomata** (warty growths) and 'snail track' mouth and throat ulcers. These also heal, but slowly, and there follows a quiescent phase, lasting several years before the multi-system involvement of tertiary syphilis becomes evident. In this, **gummata** (destructive granulomatous lesions) are widespread in both superficial tissues and deep tissues, including bone, and are the result of an increase in cell-mediated immunity against *T. pallidum*. At this stage the spirochaetes are widely disseminated in the body. As immunity wanes, the damage that they have done to the cardiovascular and central nervous systems becomes apparent in the characteristic manifestations of late syphilis—aneurysm of the aorta (resulting from damage to its wall), general paralysis of the insane (p. 225) and tabes dorsalis (p. 225).

Laboratory diagnosis

In the primary or secondary stage it may be possible to recognise *T. pallidum*, by its characteristic shape and movements, when exudate is collected from a lesion (with care to protect the collector from infection) and is examined under a dark-ground microscope. However, in most cases of syphilis laboratory confirmation of the diagnosis comes from serological tests, of which there are two groups—**non-specific (reagin) tests** and **tests for specific (anti-treponeme) antibodies**. These tests are described on pp. 181–2. A combination of a reagin test (VDRL or RPR) and a specific test (the TPHA) enables early diagnosis, monitoring of treatment and detection of old infection, without undue cost in time or materials. The reagin test provides early diagnosis and, in quantitative form, reflects the progress of treatment; the TPHA test identifies false-positive reagin tests and has a longer 'memory' for past infection. The FTA-Abs test is a valuable confirmatory test and is the most useful serological test for investigating patients who might have very early syphilis. Early treatment may prevent any serological response. During the secondary stage, treatment results in the disappearance of detectable reagin 6–12 months later, but specific tests remain positive for years. If treatment is further delayed, all tests may remain positive for years, despite eradication of the infection.

N.B. The serological responses to syphilis are indistinguishable from those to the other treponematoses—yaws, pinta etc. (p. 108)—and this may cause confusion in patients from endemic areas.

Treatment

A special problem of treatment in clinics for sexually transmitted diseases is that certain types of patients are unlikely to complete treatment if it requires repeated visits or faithful taking of oral drugs. Penicillin is the drug of choice for treatment of syphilis, and a single injection of a long-acting preparation is effective. Most treatment regimens for gonorrhoea will eradicate *T. pallidum* as well, and treatment for gonorrhoea may have helped in the control of syphilis in countries with effective STD services.

Gonorrhoea

This is the commonest of the classical venereal diseases; its incidence has reached epidemic proportions during the last 20 years. It is caused by *Neisseria gonorrhoeae* (p. 82), which is nearly always transmitted by sexual intercourse. The initial infection is an acute suppurative urethritis. This is usually accompanied in women by infection of the cervix uteri, with possible spread by direct extension to Bartholin's glands, the uterus and fallopian tubes and ovaries, and the peritoneal cavity. In men spread may include the prostate, seminal vesicles, epididymis and testes. The rectum is often colonized by direct extension in women (some of whom develop mild proctitis), and is a common site of initial infection in homosexual men. Colonization of the throat is recognized with increasing frequency, and there may be a mild pharyngitis. Blood-stream spread may lead to suppurative arthritis and to tenosynovitis (infections that can seldom be confirmed by isolation of the gonococcus from these sites), and occasionally to endocarditis and other diseases. Non-venereal transmission of gonococci is responsible for neonatal ophthalmia of babies born to infected mothers (p. 295), and for vulvovaginitis of young girls; the latter may occur in institutions, being transmitted on towels and other fomites.

Laboratory diagnosis

Presence of gram-negative and predominantly intracellular diplococci in smears of pus from a urethral or cervical discharge is strong presumptive evidence of gonorrhoea. Confirmatory isolation of the organism is best achieved by using a sterile wire loop to transfer such purulent material directly to suitable culture media. Alternatively, swabs can be used. Specimens should be collected from the urethra of every patient; from the cervix and rectum of every female and the rectum of every homosexual male patient; and from the pharynx when appropriate. Transport medium—e.g. Stuart's or Amies'—should be used for sending the swabs to the laboratory if they cannot be plated out immediately. Culture media made selective by incorporation of antibiotics inhibitory to other bacteria should be used, especially for specimens likely to give heavy mixed growths. In contrast to syphilis, there is no reliable serological test for gonococcal infection.

Treatment

Penicillin remains the drug of choice for most gonococcal infections. Adequate treatment may be achieved by a single injection of a long-acting penicillin preparation, or by two doses of an oral penicillin with probenecid to delay excretion. Where β-lactamase-producing gonococci are prevalent (p. 82), alternative regimens are needed and spectinomycin is widely used.

Chancroid

Classically chancroid is a painful ulcer of the skin or mucosa of the genitalia (known as a **soft chancre** in distinction from the hard chancre of syphilis), often complicated by enlargement and suppuration of the inguinal lymph nodes (buboes). It is caused by *Haemophilus ducreyi* (p. 100). It is common in many third-world countries but rare in Britain and other developed countries. However, on specially enriched media *H. ducreyi* may be isolated from the mixed flora of less specific forms of genital ulceration (p. 100) and it may have a role as a secondary pathogen.

Lymphogranuloma Venereum (LGV)

This infection is caused by certain serotypes of *Chlamydia trachomatis* (p. 114), transmitted as a rule by sexual intercourse, though non-venereal transmission can occur—e.g. through the conjunctiva. LGV is largely confined to tropical and subtropical countries. In its common form it produces ulcerative genital lesions with regional lymph-node suppuration (called 'climatic bubo' when the inguinal glands are involved). Generalized dissemination follows, with fever and diffuse aches and sometimes with conjunctivitis, arthritis or encephalitis. Chronic infection leads to anal and genital strictures and elephantiasis (a result of lymphatic obstruction—cf. filariasis, p. 247).

Granuloma Venereum (Granuloma Inguinale)

This slowly progressive ulcerative disease of the genitalia is widespread in tropical countries but rare elsewhere. It is caused by a pleomorphic, micro-aerophilic and very demanding bacillus known as *Calymmatobacterium granulomatis* (formerly *Donovania granulomatis*). There is some doubt about whether it is truly a sexually transmitted disease.

Non-Gonococcal or 'Non-Specific' Urethritis (NGU, NSU)

Urethritis with dysuria and purulent discharge but no gonococcal infection is a very common form of STD, especially in men. (A similar condition, though with a slightly different pattern of aetiological agents, sometimes persists after gonococcal urethritis has been effectively treated; it is called post-gonococcal urethritis—PGU—and is due to double or multiple infection.) Over 80 000 cases of NGU are diagnosed annually in England and Wales—about twice the frequency of cases of gonorrhoea. About half of the NGU cases are caused by *Chl. trachomatis* (of serotypes distinct from those causing LGV—pp. 114–5); others are caused by *Trichomonas vaginalis* (see below), which may also be carried asymptomatically in the male urethra. *Ureaplasma urealyticum, Mycoplasma hominis* and certain bacteroides strains may each be responsible for some cases.

In chlamydial NGU there is usually a 7–10 day incubation period before urethral irritation and dysuria develop. The exudate contains polymorphs but no gonococci. The diagnosis is confirmed by the isolation of *Chl. trachomatis* from the urethra by tissue culture. As the exudate itself is not a good specimen for chlamydial isolation, a swab should be taken from within the urethra, so that infected epithelial cells are collected. It should be placed in antibiotic-containing transport medium to preserve the chlamydiae and eliminate contaminating bacteria. Liquid expressed from the swab is used to seed cell cultures, and after 4–8 days characteristic inclusion bodies of *Chl. trachomatis* are evident in the cell monolayers. Rapid diagnosis is possible by an immunofluorescent test, in which a smear of exudate is treated with fluorescent anti-chlamydial antibody and then examined by ultra-violet microscopy, or by ELISA for chlamydial antigen. Chlamydiae are sensitive to tetracycline and erythromycin, which are the treatments of choice for NGU in general because they are also effective against ureaplasmas and mycoplasmas.

Chlamydial Cervicitis and Associated Infections

Certain non-LGV serotypes of *Chl. trachomatis* (p. 114) cause infections of the cervix uteri in women, ranging from asymptomatic colonization of an apparently normal cervix to cervicitis in which the cervix is inflamed and oedematous and there is a purulent discharge. Cervicitis may be followed by ascending infection (*Chl. trachomatis* is the commonest cause of salpingitis), and may lead to acute perihepatitis. One of the common complications of chlamydial infection in women is eye infection in their babies; neonatal ophthalmia is far more often chlamydial than gonococcal (p. 326).

The diagnosis of chlamydial infection in women is by isolation of *Chl. trachomatis* in tissue culture from a specimen collected by a fine swab from within the cervical canal after wiping away the exudate or by immunofluorescence or ELISA tests (p. 178). Treatment is by tetracycline or erythromycin.

Trichomoniasis

Infection with the flagellate protozoan *Trichomonas vaginalis* (p. 145) is one of the common causes of a vaginal discharge. It produces a frothy, grey and often copious discharge, and causes vulval and vaginal irritation. The vaginal mucosa may be mildly inflamed. The diagnosis may be made immediately by microscopic examination of a wet film of the discharge, in which the actively motile trichomonads are easily seen; motility is most active when the preparation is kept warm on a heated microscope stage. It is also possible to grow *T. vaginalis* in broth, but this is usually unnecessary except in light infections or when the carrier stage is suspected. Whereas women infected with *T. vaginalis* are usually symptomatic, their male consorts, who are usually also infected, are commonly asymptomatic and may serve as unsuspected reservoirs of infection for women. Some men, however, do have symptomatic urethritis. Trichomoniasis responds rapidly to treatment with metronidazole, but to prevent equally rapid reinfection the male consorts of female patients should also be treated.

Anaerobic Vaginosis ('Non-Specific Vaginitis', NSV)

One of the commonest forms of vaginal discharge was known for many years as non-specific vaginitis (NSV) because there was no obvious aetiological agent. In recent years it has been recognized that various anaerobic and micro-aerophilic bacteria are associated with the condition and it has been renamed **anaerobic** (or, by some workers, just 'bacterial') **vaginosis**. The change to the -osis ending reflects one of the characteristic features of this condition, that there is no inflammation of the vaginal mucosa and no inflammatory exudate (pus cells, etc.) in the discharge. There is a foul-smelling ('fishy') discharge that may be distressing to the patient, but no pain or irritation. The pH of the vaginal secretions is above 5.0, and addition of a drop of KOH to the secretions on a glass slide releases a strong fishy smell, due to amines (the amine test). The bacteria associated with this vaginosis are *Gardnerella vaginalis*, a gram-variable CO_2-dependent bacillus (p. 86), *Bacteroides* of the melaninogenicus–oralis group (p. 102) and micro-aerophilic, motile, curved rods known as *Mobiluncus* (p. 98). Combinations of these in large numbers are usually present in the discharge, although not all are present in every case. The foul smell and the presence of amines in the discharge indicate that the anaerobes have a significant role. A diagnosis made on the basis of symptoms and a positive amine test can be confirmed by microscopy. The normal lactobacilli are absent and are replaced by masses of gram-negative (bacteroides) and gram-variable (*G. vaginalis*) rods and gram-negative curved rods. A characteristic feature is the presence of 'clue cells'—vaginal epithelial cells coated with gram-variable bacilli. Anaerobic culture gives a very heavy, mixed growth.

It is not certain whether this is a sexually transmitted condition. Bacteroides and *G. vaginalis* can usually be found in small numbers in the urethra of the male consort, but the epidemiology is difficult to confirm because there is no specific pathogen. It is likely that the introduction of alkaline seminal fluid into the vagina during intercourse is an important precipitating factor.

The condition can be treated by a short course (even one large dose) of metronidazole, but relapse or reinfection is common whatever the length of the course.

Candidiasis

The yeast-like fungus *Candida albicans* (p. 137) causes a vulvovaginitis known as **vaginal thrush.** In this condition there is inflammation of the vaginal mucosa with pain and irritation and often an intense pruritus. There is a white, curd-like discharge and white flakes characteristic of candidiasis can be seen adhering to the inflamed mucosa. *C. albicans* is present in small numbers in the vagina in about 10% of healthy women. Symptomatic infection may be induced by many factors such as diabetes, steroid therapy, debilitating diseases, immunosuppression and antibiotic therapy which suppresses the normal bacterial flora; thrush is particularly common during pregnancy and in women taking oral contraception.

The clinical diagnosis of thrush is confirmed in the laboratory by seeing yeast cells and elongated forms known as pseudohyphae in gram-stained films of the discharge and by growing *C. albicans* on Sabouraud's agar. Treatment is by correction of any underlying abnormality, if possible, and by local application of nystatin or oral treatment with an imidazole antifungal agent (p. 142).

In men, *C. albicans* causes superficial infections of the penis ranging from an irritant erythematous rash to a severe balanitis in which the surface of the glans is red and inflamed with superficial white pustules and plaques. These infections are commonest in male consorts of women with vaginal candidiasis.

Genital Herpes

Herpes simplex virus (HSV) infection of the genitalia occurs both in men and in women, but may be associated with more serious consequences in women. HSV is highly infectious and spreads readily by sexual contact from either primary or recurrent lesions. Most genital infections are caused by type 2 viruses (p. 130), but type 1 strains are responsible for a few cases. In primary herpes, small vesicles at the site of infection rapidly break down to shallow, weeping and very painful ulcers. In men, the glans, prepuce or shaft of the penis may be affected. External genital lesions are also common in women but in addition many of them have cervicitis. This is perhaps the most serious aspect of the infection because there is an association (not yet proved to be a causal relationship) between herpes cervicitis and the development of carcinoma of the cervix. Many patients with cervical carcinoma have evidence of herpes infection, and herpesvirus DNA has been detected in many of the tumours, but these findings could be mere reflections of the fact that genital herpes is common in the epidemiological groups with a high risk of carcinoma of the cervix (women with early onset of sexual activity and multiple partners).

Primary herpes lesions of the genitalia resolve in 7–10 days but, as with herpes infections elsewhere (p. 131), the virus is not eliminated but becomes latent in the dorsal root ganglia serving the affected area. Various stimuli (intercurrent infection, immune changes, hormonal changes, etc.) may precipitate reactivation and recurrent herpes lesions. Genital herpes is probably responsible for about half of all cases of genital ulceration; some deep necrotic ulcers may be initiated by the virus and then superinfected by anaerobic bacteria.

The diagnosis of genital herpes is often evident on clinical grounds, but it may be confirmed by the isolation of HSV-2 in tissue culture from swabs of the ulcers. These swabs must be placed in virus transport medium for transfer to the laboratory. A rapid diagnosis may be obtained by the indirect immunofluorescent antibody technique. Cells from suspect lesions are fixed on slides and treated with anti-herpes antibody and then with fluorescent anti-IgG antibody; when examined under ultraviolet illumination, herpes-infected cells fluoresce.

Once established, genital herpes cannot be cured, but the duration of the intensely painful and distressing symptoms can be markedly shortened by local or systemic treatment with acyclovir (p. 135). The period of virus shedding (i.e. infectivity) is also shortened.

Genital Warts

The moist skin and mucocutaneous junction of the genitalia are common sites for warts caused by papovaviruses (p. 308). Infective virus particles are shed from the warts and infection spreads to sexual partners. The diagnosis is made on clinical grounds; warts must be distinguished from the condylomata of syphilis.

Acquired Immune Deficiency Syndrome (AIDS)

The history, epidemiology and pathology of this largely sexually transmitted disease are described in Chapter 9 (p. 128). It is the result of infection with human immunodeficiency virus (HIV). Most people known to have been infected have remained healthy, though it appears that some of them carry the virus in their blood for long periods (cf. hepatitis B, p. 266). Of those who do become ill, some develop persistent lymphadenopathy with the lymphocyte abnormalities described on p. 128—a condition known as persistent generalized lymphadenopathy (PGL) syndrome or pre-AIDS. The fully developed AIDS syndrome is characterized by severe infections with opportunist pathogens. The commonest of these is *Pneumocystis carini* pneumonia (p. 205), but other organisms commonly involved include the protozoa *Cryptosporidium* and *Toxoplasma*, cytomegalovirus and herpes simplex virus, *Candida* and other fungi (causing systemic infections) and mycobacteria. These are mostly difficult or impossible to treat quite apart from the special problem of the patients' immune deficiency, and once the full AIDS syndrome is established death is inevitable. A formerly rare malignant neoplasm, Kaposi's sarcoma, is a surprisingly common complication of AIDS; in this setting it differs somewhat from the classical form of this sarcoma, as seen in Central Africa, which is not associated with the presence of HIV.

Since AIDS is incurable, urgent attention is being given to its prevention. A more responsible, less promiscuous life style among male homosexuals is essential to the control of spread. Transmission by blood transfusion or in blood products can be prevented by serological screening of all blood donors to detect and exclude all carriers of HIV, and by heat-treatment of blood products to be given to haemophiliacs—though there are worries about the resistance of the virus to such treatment. Protection of health workers is as for hepatitis B (p. 354).

Superficial Parasites

Ectoparasites (e.g. scabies mites and lice) can seldom be said to cause infections of the genital tract, but they are spread from person to person by close contact and sexual intercourse produces particularly suitable conditions for this. In scabies the characteristic tracks are often seen on the external genitalia and in the inguinal skin folds. Of the various types of louse, the pubic ('crab') louse specifically infests the hair of the genital region. Diagnosis of these infections is usually by clinical observation. The identity of the ectoparasites can be confirmed in the laboratory by macroscopic or low-power microscopic examination. The treatment is given on p. 311.

Diagnosis and Management of Sexually Transmitted Diseases

Because of the social connotations of many of these infections, a precise and confirmed diagnosis is essential. As detailed above, this is usually achieved by microscopic examination and bacteriological or virological culture of exudate or discharge from the local lesion.

The treatment of the individual patient depends upon the infection (see above), but management of this group of infections has some special features.

(1) When the disease itself is not serious—e.g. in trichomoniasis or ectoparasite infestations—it is not necessary to investigate the patients' sexual partners but they should be treated, because they are almost certainly infected.

(2) Most of the other diseases are so serious (and, in the case of gonorrhoea, antibiotic-resistant bacterial strains are such a problem) that all sexual contacts should be traced and persuaded to attend clinics for investigation and any necessary treatment. Such contact tracing is an essential part of the control of syphilis and gonorrhoea and is increasingly applied to other sexually transmitted diseases.

21
Soft Tissue and Eye Infections

This chapter brings together a group of infections of widely different aetiology. Some are of minor inconvenience, others are severe infections with a high morbidity and, in some cases, mortality. It is most logical to consider some of these infections in terms of the causative organisms, regardless of anatomical site, whereas others are grouped according to the site and the type of infection.

Bacterial Infections of Soft Tissues

Infections with Aerobic or Micro-Aerophilic Bacteria

Staphylococcal abscesses

Staphylococcus aureus (p. 74) is the major cause of a wide range of localised acute pyogenic infections. It is an aggressive pathogen that is carried asymptomatically in the anterior nares of at least 30% of the population, and in the skin folds of a smaller proportion, is readily transmitted by direct or indirect contact and is the pathogen most commonly isolated from septic lesions. It is an invasive organism that needs only minor damage or disruption of function in the host to establish infections. These range from superficial pustules or boils, which arise when hair follicles or sweat glands become infected with the staphylococcus, to large deep-seated abscesses at varied sites. Infections of eyelash follicles are known as styes. If subcutaneous tissues are involved, a carbuncle may develop; this is typically a large multilocular abscess that drains to the surface through several sinuses. All of these infections develop from the outside, the staphylococci coming directly from the skin surface. Deeper infections result either from direct spread of superficial lesions to adjacent structures or to the regional lymph nodes, or from haematogenous spread in which staphylococcal bacteraemia (originating from a superficial site) leads to the deposition of staphylococci at sites, often multiple, around the body. Suppuration of lymph nodes, such as cervical or inguinal nodes, is commonly staphylococcal, as are other deep abscesses such as perinephric abscesses or those between muscle planes. Abscesses may develop in previously healthy tissues, but are more likely to occur at sites where there has been some previous accidental or other damage.

Staphylococcal abscesses have a common general pattern. There is an acute inflammatory response, with a localized collection of thick creamy pus containing bacteria, necrosed tissue and polymorphonuclear leukocytes, mostly dead. The abscess is surrounded by a dense capsule of fibrin and lined with granulation tissue containing large numbers of polymorphonuclear and mononuclear phagocytic cells (macrophages).

The diagnosis is usually suspected clinically and confirmed by microscopy and culture of the pus; gram-positive cocci in clusters giving rise to colonies that are coagulase- and DNAase-positive (p. 75) confirm a staphylococcal aetiology. Treatment is by drainage of pus and, in other than minor infections, the use of an antistaphylococcal antibiotic—usually flucloxacillin or erythromycin, with fusidic acid and vancomycin reserved for severe, deep infections.

Streptococcal cellulitis

β-haemolytic streptococci of Lancefield's group A commonly colonize the pharynx and nose and are readily spread from person to person. They are invasive organisms that usually enter the body through some minor superficial trauma but do not cause localized abscesses. The infections are not confined by capsules and granulation tissue, but cause spreading cellulitis. There is intense acute inflammation; the infected tissue is red and tense but there is no local collection of pus. Damaged tissue and phagocytes are broken down by enzymes produced by the streptococcus (p. 77), so that the pus is thin and watery. Infection spreads rapidly, both directly along tissue planes and also via lymphatics. An ascending lymphangitis in which the lines of the lymphatics are seen as red inflamed streaks under the skin, particularly along the limbs, is characteristic of streptococcal infection. The regional lymph nodes are acutely enlarged and tender (lymphadenitis, p. 243), and infection may also spread to the blood stream. Streptococcal septicaemia is a serious condition and has a high mortality if not recognized and treated early. Probably because of this, distant secondary sites of streptococcal infection are much less common than when *Staph. aureus* is disseminated in this way.

The clinical diagnosis of streptococcal cellulitis is confirmed by microscopy and culture of the thin exudate to show streptococci that are β-haemolytic and of Lancefield's group A. Treatment is by administration of penicillin; in severe cases when septicaemia is suspected, this should be given intravenously with the dose repeated every 3–4 hours. Streptococci are highly sensitive to penicillin, and treatment may initially appear to worsen the patient's condition, as the streptococci are killed and lysed releasing many toxic components into the circulation, before improvement begins. No attempt should be made to incise and drain the streptococcal lesion. There is no pus to release and the incision does further damage which does not heal well. Antibiotic treatment alone is sufficient.

Actinomycosis

Infection with the branching higher bacterium *Actinomyces israeli* (p. 106) causes a subacute or chronic soft-tissue abscess. *A. israeli* is a commensal of the gingival crevice and is also found in the lower gastro-intestinal tract. The most common sites of infection are around the mouth, usually in the submandibular tissues, particularly at the angle of the mandible. The origin of infection is the gingival flora, and the predisposing factor is usually some trauma to the teeth or gums. This may be obvious—e.g. a broken tooth, or bone damage or a residual piece of root after a difficult tooth extraction. In other cases there may be only minor gingival trauma. Infection is most likely in people with poor oral hygiene; there are large numbers of *A. israeli* in their enlarged and inflamed gingival pockets, and local damage provides routes of entry.

Actinomycotic abscesses are typically large and indurated, with thick capsules and vigorous granulation-tissue responses around them. They are multiloculate and spread extensively and destructively across tissue planes. The mandible and the roots of several teeth are often involved in the infection. Such an abscess points to the surface, usually in several places, but natural drainage is slow because only small amounts of pus are extruded from the multiple sinuses. The diagnosis of actinomycosis is confirmed by examination of the pus. It is yellow and, although thicker than streptococcal pus, is less thick than that of a staphylococcal abscess. On naked-eye examination it is seen to contain small yellow particles, 'sulphur granules', which are micro-colonies of *A. israeli*. These are best seen if a drop of pus is shaken up with a few ml of saline in a tube; the pus disperses but the solid granules remain and sink to the bottom of the tube. These washed granules are the most suitable material for microscopy and culture. When such a granule is crushed on to a microscope slide and stained by Gram's method, the tangled branching filaments of *A. israeli* are easily recognized. Although this species is micro-aerophilic, it grows satisfactorily in the conditions routinely provided for the growth of anaerobes.

Treatment of actinomycosis requires a combination of surgical drainage and prolonged antibiotic therapy. Extensive exploration may be necessary to break down all the loculated parts of the abscess, and teeth may have to be removed if there is extensive involvement of the mandible and roots, since antibiotics cannot penetrate well through the thick indurated walls of actinomycotic abscesses. Penicillin is the antibiotic of choice, but success has also been achieved with erythromycin or lincomycin. Because *A. israeli* is a slowly growing organism and because of the nature of the abscesses, treatment must be continued for at least 6 weeks to ensure eradication of the infection. Initial parenteral treatment with benzylpenicillin may be followed by a combination of an oral penicillin with probenecid and a weekly injection of procaine penicillin to help maintain adequate levels.

Actinomycosis may also occur in the abdomen, usually involving the appendix and caecum in a typical actinomycotic abscess that may fill the right iliac fossa and cause symptoms of subacute or chronic appendicitis. Treatment is difficult because of the amount of loculation and fibrosis, and after drainage a persistent sinus may develop along the track of the surgical incision. Secondary abscesses develop in the liver as a result of metastatic spread through the portal system. The rectum can be involved in a similar actinomycotic abscess, and this is a relatively uncommon but well recognized form of peri-rectal or peri-anal abscess. One of the major complications of rectal actinomycosis is the formation of persistent, sometimes multiple, faecal fistulae.

Nocardiosis

The other genus in the actinomyces group that causes infection in man is *Nocardia* (p. 107); the two pathogenic species are *N. asteroides* and *N. brasiliensis*. A typical nocardial infection is a localized but deep and destructive abscess. It is similar to an actinomycotic abscess in being subacute or chronic, multiloculate, extensive and surrounded by a dense capsule of fibrin and granulation tissue. Nocardiosis occurs in tropical and subtropical areas where the pathogenic *Nocardia* species, like other non-pathogenic members of the genus, are environmental saprophytes found principally in soil. Infection occurs by implantation of the organisms in the tissues, usually as a result of some minor trauma. The common site of infection is the foot, and it is commonest in people who go barefoot or at least with little protection for their feet. A large abscess develops slowly, without much acute inflammation or severe tenderness, and discharges through multiple sinuses. This is one form of **mycetoma**. Other causes of similar subacute or chronic abscesses of the foot in similar circumstances are fungi (p. 45).

Either species of *Nocardia* can cause mycetoma. *N. asteroides*, however, can also cause systemic infection. The initial site of infection may be a superficial abscess, or in the lung as a result of infection by inhalation. Haematogenous spread may then occur and metastatic abscesses develop at multiple sites, including liver, kidney, peritoneum and lung. Such systemic infection occurs in previously healthy people in areas where the disease is endemic, and also sometimes in immunosuppressed hospital patients in non-endemic areas.

The diagnosis of nocardiosis is made by examination of the pus obtained from the abscesses by aspiration or drainage. A gram-stained film shows tangled gram-positive filaments which are often beaded and may be mistaken for chains of streptococci. A modified Ziehl–Neelsen stain decolourized with 0.5% sulphuric acid shows tangled acid-fast filaments (N.B., *A. israeli* is not acid-fast). Unlike actinomyces, nocardia are strict aerobes that do not grow anaerobically. They are very resistant to many antibiotics, but successful treatment has been reported with surgical drainage and a mixture of cotrimoxazole and amikacin given for 6-8 weeks.

Leprosy

The chronic and destructive disease caused by *Mycobacterium leprae* (p. 105) is mainly restricted to tropical countries but in endemic areas there are more than 10 million cases.

Contrary to popular belief, human infection does not spread readily, and depends on prolonged exposure; skin-to-skin contact is probably less important than was once thought, infection being mainly by inhalation of bacilli shed from the noses of patients with lepromatous disease (see below). The name leprosy is today restricted to *Myco. leprae* infections; but as used in the Middle Ages and in English translations of the Bible it must have included a range of other diseases.

The diverse clinical manifestations of infection with *Myco. leprae* depend on differences in immunological response, both between patients and at different times in the same patient (cf. tuberculosis, p. 207). There is a spectrum of disease, running from **lepromatous** through **borderline, dimorphous** or **intermediate** to **tuberculoid** leprosy. In lepromatous disease bacilli are numerous in the granulomatous skin lesions and along sensory nerves and are shed in large numbers in nasal and other discharges; the host seems to offer little resistance to the infection, which is progressive and has a bad prognosis in the absence of treatment. At the other end of the spectrum, tuberculoid leprosy is characterized by cell-mediated immunity and by a Type IV hypersensitivity response in the **lepromin test** which is somewhat analogous to the tuberculin test. Bacilli are scanty in the raised erythromatous skin lesions and the hypertrophic and sclerotic sensory nerve lesions of tuberculoid leprosy, and there is a strong tendency towards spontaneous healing as a result of the cell-mediated response, but at the cost of a good deal of tissue and nerve destruction. The loss of sensory nerve function leads to repeated and ultimately crippling damage to the areas served by those nerves, particularly the feet and hands.

Infections with Mixed Anaerobic Bacteria

Although the next group of infections, pilonidal and sebaceous abscesses, synergic gangrene and peripheral gangrene, have distinct clinical presentations, they are linked together by similar bacteriological findings. In each, mixtures of anaerobic bacteria that include anaerobic cocci and bacteroides appear to act synergically to cause tissue damage.

Pilonidal abscess and sebaceous cysts

These two types of superficial abscess are considered together because they are similar in their pathogenesis and in the bacteria that contribute the infective element to their development.

Pilonidal abscesses develop at the upper end of the natal cleft at the base of the spine. They are deep and multi-channelled, and spread extensively in the subcutaneous tissues. They are common in people with sedentary occupations, especially drivers who have pressure and repeated trauma at the base of the spine, and are thought to begin as small sinuses resulting from a combination of local trauma and malfunction or malformation of a hair follicle. Hairs characteristically protrude from the sinuses. Once such a sinus has formed, it becomes infected and develops into an indurated spreading abscess.

Sebaceous cysts develop in any area of skin that is well supplied with sebaceous glands. If such a gland becomes plugged with inspissated secretions or keratin, secretions build up and disrupt the gland, and these retained secretions become infected.

Culture of the contents of pilonidal abscesses and of sebaceous cysts shows a similar, more or less constant mixture of bacteria. Apart from propionibacteria and coagulase-negative staphylococci from the skin microflora, the predominant flora is anaerobic, a finding supported by the typically foul smell of the pus. Anaerobic gram-positive cocci are always present and may be able to break down the sebaceous secretions. They are usually found in mixed culture with bacteroides of the asaccharolytic group, mostly *Bact. asaccharolyticus* and *Bact. ureolyticus*.

Treatment is primarily by surgical drainage supported by the use of metronidazole to eliminate residual anaerobes and improve the rate of healing.

Synergic (often incorrectly called synergistic) gangrene

A rapidly spreading superficial gangrene of the abdominal wall with widespread destruction of skin and subcutaneous tissue was first described by Meleney. He associated the condition with a synergic infection by anaerobic or microaerophilic cocci and aerobic staphylococci (but the incorrect spelling 'synergistic' has become traditional in this connection). Initial superficial damage or a surgical incision provides the portal of entry for the infection, which then spreads rapidly across the abdomen. It can be especially severe in patients treated with steroids. Fournier described a similar condition initially affecting the scrotum and perineum and spreading to involve the thigh and the lower abdomen. It is now clear that these are essentially the same condition and that synergic bacterial (anaerobic) gangrene can occur at any site close to a source of the anaerobes, which is usually the gastro-intestinal or genito-urinary tract. Modern bacteriological studies have shown that the causative organisms of these destructive lesions are strict anaerobes; anaerobic gram-positive cocci are always present, usually with asaccharolytic bacteroides—*Bact. ureolyticus* and *Bact. asaccharolyticus* in particular—and a synergic boost given by various relatively non-pathogenic facultative species. Treatment of established infections requires extensive surgical debridement and anti-anaerobe chemotherapy (metronidazole).

Synergic gangrene is linked by similarities in bacteriology with the less dramatic conditions, also involving superficial necrosis and ulceration, to be described in the next few paragraphs.

Peripheral gangrene

A group of superficial ulcerative conditions initiated by vascular insufficiency and minor trauma are included under this heading. They include diabetic gangrene, varicose ulcers and decubitus ulcers (pressure sores). In diabetic gangrene, peripheral arterial disease reduces the blood supply to the distal parts of the feet, the hypoxic area gradually moving proximally as the disease progresses. Tissue damage is caused initially by the hypoxia (dry gangrene). This tissue is then susceptible to bacterial infection. The trauma to which the hypoxic toes and foot are subjected in the course of everyday activities and the peripheral neuropathy that is a complication of diabetes make infection more likely.

Hypoxia of the skin of the lower leg occurs when there is venous insufficiency (varicose veins) with pooling of blood in the peripheral veins and failure of adequate circulation. This skin is highly susceptible to trauma, which may cause bleeding and then ulcer formation. The hypoxic skin does not heal and becomes secondarily infected.

A similar situation arises in bed-ridden patients, especially those who are paralyzed and unable to move their position frequently. Pressure on the weight-bearing areas, particularly on the buttocks and over the sacrum, reduces the blood supply to the skin. If the pressure is not frequently relieved by turning the patient the skin becomes necrotic and sloughs, leaving an ulcer known as a decubitus ulcer or pressure sore.

All of these necrotic, ulcerative lesions become infected by a mixed bacterial flora. Superficial colonization by enterobacteria, particularly proteus, is common, but of little pathogenic significance. Infection with *Staph. aureus* or β-haemolytic streptococci (groups A, C or G) may also occur, giving the characteristic features listed above; infection with group A streptococci is potentially the most serious but is relatively uncommon. In general, the bacteria that cause significant damage with increasing tissue destruction at the advancing edge of the ulcers are anaerobes—mostly the same ones found in the classical synergic gangrene: anaerobic cocci and asaccharolytic bacteroides. Treatment with systemic metronidazole in addition to local cleansing improves the rate of healing.

Tropical ulcers

A particular type of acute ulcer, most often of the leg but sometimes occurring in the arm, is seen in tropical areas. It is not related to vascular insufficiency and the initiating cause is not clear.

Such a 'tropical ulcer' begins as a small papule (possibly the result of an insect bite) that soon breaks down to form a shallow, red and intensely painful ulcer. The ulcer then enlarges, the base and edges because necrotic and the pain subsides. There is characteristic hyperpigmentation around established ulcers. Large numbers of bacteria of many species can be found in such ulcers, but most interest has focused upon the presence of spirochaetes, fusobacteria and other anaerobic bacteria thought to be significant at least in the continuing tissue damage, if not in the initiation of the ulcer. The ulcers generally heal slowly after several months, leaving large scars. Unlike the ulcers described earlier, tropical ulcers affect young and healthy people and do not appear to be related to malnutrition or other debilitating diseases.

Infections of Accidental Wounds

Infections associated with surgical wounds are described in Chapter 27. In this section we consider accidental wounds and animal bites.

Accidental wounds are the most common reason for attendance at accident and emergency departments. As regards risk of infection they may be divided into clean and dirty wounds. Wounds made by a clean sharp instrument are unlikely to be infected at the time of the trauma. They may be infected with exogenous organisms subsequently, by touch or from the air. Infections in these wounds are almost always caused by *Staph. aureus* or group-A streptococci. These cause purulent exudate (*Staph. aureus*) or cellulitis (group-A streptococci) as described above.

Dirty wounds contaminated with dust, soil, clothing, etc. are at least equally susceptible to staphylococcal or streptococcal infection, but are also subject to infection with organisms from the contaminating material, of which the most important are *Clostridium tetani* and *Cl. perfringens*.

Tetanus

Cl. tetani (p. 88), like all other clostridia, is an intestinal commensal. It is found rarely in human faecal flora but is common in those of many species of domesticated and wild animals. Soil and the general environment are contaminated with spores by faecal contamination. The highest concentrations of spores are in places where there is a high population of animals and on cultivated land where animal manure is used. A patient is at risk of tetanus if a penetrating wound is contaminated with soil or dust that might contain spores. The spores germinate if conditions in the wound are appropriate (p. 88). The vegetative cells remain at the site of infection, causing little or no damage themselves; however other bacteria implanted during wounding may cause local infection and by using up oxygen may promote the growth of *Cl. tetani*. Clinical tetanus is the result of toxin (tetanospasmin) production by the vegetative *Cl. tetani* cells. The toxin blocks the γ-impulses that modify reflex arcs leading to extreme hyperreflexia. The mode of action is described on p. 88. Although heavily contaminated wounds are at obvious risk, many cases of tetanus occur after minor or unnoticed injuries (e.g. rose-thorn injuries) of which there is no residual sign when the tetanus develops (cryptogenic tetanus). Inadequately sterilized dressings or suture materials have often caused post-operative tetanus in the past, and **tetanus neonatorum**, due to infection of the umbilical stump, is common in communities in which animal dung is used as an umbilical dressing.

Tetanus is best prevented by prophylactic immunization with three doses of tetanus toxoid (p. 367). This gives good protection for 10 years or so; a booster dose should be given every 5–10 years. If an immunized person has a dirty wound, local cleansing and a booster dose of toxoid (if the last dose was less than 2 years before) is all that is required. In a non-immune person, however, passive immunity (p. 367) must be provided by an injection of human anti-tetanus globulin if available; anti-tetanus serum raised in horses, with consequent risks of hypersensitivity reactions, is no longer recommended. Anyone given passive protection should also be given a full course of toxoid to avoid the need for further doses of serum.

Once tetanus has become established, there is no specific treatment. Anti-tetanus serum is of doubtful value (p. 367). Penicillin is used to eliminate viable *Cl. tetani* and prevent more toxin production. Therapy, however, depends upon sedation to prevent the spasms (diazepam and phenobarbitone) and maintenance of respiration, which usually requires paralysis with a curare derivative and artificial ventilation. Antibiotics may be needed to control chest infections and the bacteraemia that may arise from the intravenous catheters and cannulae needed to maintain hydration and nutrition.

Recovery takes 3–5 weeks and the infection does not confer immunity, presumably because the small amounts of toxin that cause disease are insufficient to activate the immune system and are also sequestered from the immune system when attached to nerve cells. Active immunization with toxoid should be initiated during the convalescent period.

Clostridial cellulitis and gas gangrene

The organisms responsible for these conditions—*Cl. perfringens* (p. 89) or in a few cases other clostridia such as *Cl. novyi* and *Cl. septicum* (p. 89)—are intestinal commensals of many animals and of man, and their spores are widely distributed in soil and dust and on human skin. They are commonly implanted in dirty wounds contaminated with soil, etc., but those entering surgical wounds are usually the patient's own intestinal organisms, either from the skin or in cases of intestinal surgery direct from the intestinal contents. Mere contamination of wounds with these organisms is of no importance, but they may go on to cause significant infection. **Clostridial cellulitis** is a spreading infection of the subcutaneous tissues characterized by much gas-formation and consequent crepitation of the tissues when handled, and this often leads to confusion with true gas-gangrene. However, the cellulitis is relatively painless and there is as a rule not much general systemic disturbance. True clostridial gas-gangrene, perhaps better described as **clostridial myositis**, occurs when there is damaged muscle with a poor blood supply, providing excellent conditions for anaerobic growth—often enhanced by the presence of foreign material. The clostridia multiply rapidly, producing necrotizing toxins that increase the tissue destruction and the amount of culture medium available. The muscle compartments are tense with gas produced by the bacterial enzymes and are painful and tender. The gas pressure still further reduces the blood supply, so promoting further anaerobic growth. The patient is ill from absorption of toxic materials and may become very much more so from clostridial bacteraemia and septicaemia; without appropriate treatment this condition is rapidly fatal.

Treatment of gas-gangrene is by extensive surgical debridement to remove necrotic tissue and to relieve the tension in the tissues. Penicillin is given intravenously in high doses to eliminate the clostridia. Polyvalent *Cl. perfringens* antitoxin may be useful, but is not readily available. Some workers have advocated the use of hyperbaric oxygen therapy to combat the anaerobic infection. Penicillin treatment alone will usually deal adequately with clostridial cellulitis. A picture resembling clostridial gas-gangrene is sometimes produced by anaerobic gram-positive cocci.

Animal bites

Bites from domestic or wild animals commonly cause tissue damage. Their severity varies from a minor puncture wound to extensive lacerations. As with other wounds, they may provide openings for exogenous infection with staphylococci and streptococci. However, they raise the special problem that the mouth itself has a large microbial flora, some of which is able to cause wound infections. The most common special pathogen in animal bites is *Pasteurella multocida* (p. 101), which is carried in the mouths of many domestic and wild animals, including dogs, cats and cattle, as well as sometimes causing respiratory infections and septicaemia in those animals. When implanted in a bite, *Past. multocida* commonly causes a local abscess, a spreading cellulitis with a clearly defined edge, lymphangitis and lymphadenitis; in this it combines the features of staphylococcal and streptococcal infection. The diagnosis is confirmed by culture of wound exudate, and the infection is treated by local drainage and penicillin.

The other potential pathogens from the animal's oral flora are streptococci and gram-negative anaerobic bacilli. Streptococci are occasionally isolated from bites, but rarely cause severe infections. *Bacteroides* species and fusobacteria, however, can cause abscesses and local tissue necrosis, particularly if there is sufficient tissue damage from the original bite to provide an anaerobic nidus. Such infections are best treated by local drainage and administration of metronidazole.

Rabies, the most serious disease spread by animal bites, is described on p. 222.

Helminth Infections of Soft Tissues

Onchocerciasis (p. 158).

Onchocerca infection occurs in rural communities in equatorial Africa and Central and South America. It may be contracted at any age and by the time that symptoms develop the patient may be a long way from the breeding sites of the *Simulium* flies that transmit it. The worms can live for 15 years or more and the effects of infection are progressive. The most serious consequence is blindness which develops in a proportion of those infected (p. 297). Long-standing infection leads to skin thickening and over-growth, sometimes called 'lichen-skin', predisposing to secondary bacterial or fungus infection. In African patients the skin (particularly over the tibia) may become thin and depigmented, a condition known as 'leopard skin'. The skin loses its elasticity and becomes wrinkled and atrophic. Marked inguinal lymphadenopathy covered by atrophic, baggy skin is not uncommon in Africans and is known as 'hanging groin'. The sac may contain inguinal or femoral lymph nodes, and lymphatic fluid may exude from its surface.

Diagnosis of onchocerciasis is usually made clinically but is confirmed by taking shallow slices of epidermal tissue without blood (skin snips) from a number of sites in the body (the buttocks are preferred in Africa and the scapular region in Central America); the risk of eye involvement is assessed by taking snips from the outer acanthus. The skin is teased apart in a drop of saline on a slide and then examined with a microscope for microfilariae.

Treatment of onchocerciasis is based on two approaches. Diethylcarbamazine is effective against microfilariae but not against the adult parasites. Suramin is effective against both adults and larvae but is toxic. In some countries the fibrous nodules containing the adult females are removed surgically (especially in Central America, where they are more common on the head). Treatment to eradicate microfilariae is essential to prevent blindness but is complicated by allergic reactions to dead microfilariae in the skin, which may be severe enough to prevent completion of the course. Such allergy is so common that it is the basis of a diagnostic test, the Mazzotti reaction, in which a single dose of diethylcarbamazine is given and can be expected to cause a skin reaction in an infected patient.

Loaiasis (p. 158)

Loa loa, which causes 'eyeworm' infection or Calabar swelling, is confined to West Africa; it is a relatively mild infection, characterized by transient painful subcutaneous swellings which last for a few days and are caused by the wandering adult worms. Occasionally the worms wander through the subconjunctival tissues of the eye, where they are easily seen (hence the term 'eye worm'). There is no clear evidence that *Loa* causes any lasting damage to the host, although the adults may live for several years.

Diagnosis is made provisionally from the characteristic swellings and confirmed by the presence in the blood of typical microfilariae which are most numerous around mid-day (diurnal periodicity). Treatment with diethylcarbamazine is effective against adults and larvae, or the adults may be removed surgically.

Dracunculiasis (p. 159)

Dracunculus infection occurs in West Africa, the Middle East and in the Indian subcontinent. The adult *Dracunculus* lives between the muscle fascia and in the subcutaneous tissues, predominantly of the lower limb and causes a blister or ulcer over the tip of the worm, through which the larvae are discharged (p. 159). Attempts are often made to remove the female worm by traction as soon as the blister bursts and the uterus prolapses. This can be successful but has to be done carefully, as otherwise the worm breaks, leaving part behind as an irritant foreign body, and secondary infection may occur (even tetanus).

Diagnosis of the infection by observation is straightforward; if confirmation is required the larvae are easily recognised and can often be obtained by running a few drops of cold water over the ulcer. No really successful treatment is known. By the time the diagnosis is apparent it is too late to prevent the incapacitating effects of the leg lesions, and treatment shortens the life of the worm by a few days at most. Infection is easily prevented by filtering drinking water through a piece of cloth to remove any *Cyclops* which are the intermediate hosts of the larvae (p. 159).

Cysticercosis (p. 155)

The larvae of *T. solium* (*Cysticercus cellulosae*) most commonly encyst in the muscles of pigs, but they can develop in man in two sets of circumstances: (1) eggs of *T. solium* may occasionally be ingested accidentally as a result of faecal contamination of fingers, food, etc; or (2) reverse peristalsis may cause tapeworm segments containing eggs to be regurgitated into the stomach. The eggs then hatch in the stomach, and the embryos penetrate the wall of the intestine with the aid of their embryonic hooks and are carried in the blood to various parts of the body where they develop into small bladder-like larvae, the cysticerci. Ingestion of faecal material usually results in only a few cysticerci in the muscles, and these cause little harm. However, if reverse peristalsis is the source large numbers of cysticerci may result, mainly in muscles but sometimes a few in other sites where they may have serious consequences—e.g. in the brain and spinal cord (p. 226). The cysticerci are small (1–2 cm in diameter) and when calcified are easily seen in radiographs. Unless they are present in large numbers or in vital centres they are most often discovered incidentally during investigations for other diseases (particularly radiology). There is no treatment other than surgical removal of individual cysticerci from vital centres.

Trichinellosis

In the majority of human *Trichinella* infections (p. 158) only a few encysted larvae are ingested (e.g. in uncooked pork products such as sausage meat) and infection is light. Eating meat from heavily infected wild pig or polar bear has resulted in massive infection and wholesale destruction of muscle tissue, with fatal consequences. Unlike cysticerci, calcified *Trichinella* cysts are not detectable by radiology as they are very small (about 1 mm in diameter). Peri-orbital oedema is a characteristic sign of larval invasion, sometimes accompanied by conjunctival haemorrhage and splinter-like haemorrhages under the finger nails; eosinophilia may reach 90% in severe cases. Muscular tenderness varies in degree with the severity of infection. Tissue invasion may be preceded by intestinal disturbances including diarrhoea, occasionally lasting for several weeks. An outbreak of intestinal disturbance in a group of people 2–7 days after eating uncooked pork products suggests an outbreak of trichinellosis.

Various immunological diagnostic tests are available. The bentonite flocculation test is specific, rapid and simple; immunofluorescence and complement-fixation tests are also used. Infection can be confirmed by microscopic examination of a muscle biopsy—unfixed and unstained ('squash') preparation—for the characteristic lemon-shaped cysts, or by feeding the biopsy to a laboratory rat; the tissues of the rat are then examined in squash preparations several weeks later.

Treatment with corticosteroids during the severely symptomatic stages may be beneficial but may prolong the intestinal stage (and thus the production of larvae). Mebendazole has some beneficial effect on larvae in the tissues.

Visceral Larva Migrans

Non-human ascarid parasites such as *Toxocara canis* and *Toxocara cati* (which are parasites of dogs and cats respectively) can develop to a limited extent in man. The adults are similar to *Ascaris* in general appearance and in life cycle. If the ova passed in dog or cat faeces are swallowed by man they hatch, liberating larvae which can penetrate the intestinal mucosa and undergo a prolonged migration through the tissues, during which little or no development takes place (**visceral larva migrans**). The onset of this condition in a young child is often associated with recent acquisition of a family pet (usually a puppy) with which the child has played. *Tox. cati* can be acquired by children playing in sandpits in which cats have defaecated. Diagnosis is usually made on the basis of clinical and haematological findings—hepatomegaly or chronic non-specific pulmonary disease with chronic eosinophilia, especially in young children who have close contact with dogs or cats. Serological confirmation can be obtained by an ELISA test with larval antigen. Rarely, the larvae may invade the eye (p. 297). Usually the symptoms are not severe and do not require treatment; in severe cases corticosteroids and some anthelmintics may be helpful.

Flies and their Larvae

Myiasis (infection by larvae of various species of flies) is common in animals and in man in the tropics. Infection may be an obligatory stage in the life cycle of some species of fly, an optional or opportunist stage in others. The adult flies deposit eggs or first-stage larvae on the skin (most commonly), in the nose or ears, in open wounds, or on soiled linen from which they are transferred to the skin. Some are swallowed and develop in the intestine, others may infect the genito-urinary tract.

Flies for which a host is essential for the development of the larval stages include *Cordylobia* (the tumbu or mango fly of Africa), *Cochliomyia* and *Chrysomya* (the New World and Old World screw worms respectively) and *Dermatobia* (the human bot fly of Central and South America).

A female *Cordylobia* fly lays a batch of 200–300 eggs in shaded sandy soil contaminated with urine or faeces, or on soiled clothes. The hatched larvae attach to and penetrate human skin, or attach temporarily to laundry placed on the ground or on low bushes to dry, transferring to man when the clothes are worn. In the skin they develop in about 20 days into fat oval maggots 1.0–1.5 cm (0.4–0.6 inch) long, which provoke painful inflamed boil-like swellings. They eventually wriggle out of the swellings to complete their development. Such a swelling contains serous fluid but not pus, and the posterior of the maggot with its characteristic respiratory spiracles can usually be seen in the centre. Multiple infection is common; the most usual sites are on the areas covered by underclothes and on the head, which is infected from bed linen. The swelling should be sealed with liquid paraffin or petroleum jelly to prevent the maggot from breathing, so that it partly emerges and can then be extracted by gentle pressure. Infection can be prevented by not spreading clothes and bed linen on the ground to dry and by thorough ironing.

The New World screw worm *Cochliomya* is found from the southern United States to Argentina and the Old World screw worm *Chrysomya* occurs in tropical Africa and most of Asia. The flies lay eggs around the edges of wounds, scabs, sores or mucous membranes; larvae hatch within 24 hours and invade adjacent tissue where they feed, mature and then drop to the ground in 4–8 days. Larvae of both types, which are obligatory parasites of living tissue, can cause considerable damage and disfigurement, especially of the face.

Dermatobia occurs in Central and South America to an altitude of about 900 m (3000 ft). Adult females attach their eggs to the abdomens of other insects (mostly day-biting mosquitoes and other blood-sucking insects); larvae develop in the eggs but hatch only when the carrier insect alights on a suitable mammal (including man) to take a blood meal. The larvae then drop off the host insect and quickly burrow into the subdermal layers of the skin, where they grow in 5–12 weeks into large maggot-like larvae 19–25 mm (¾–1 inch) long in small pockets; then they drop off the host. The boil-like swellings may appear on almost any part of the body; they suppurate and cause considerable discomfort and pain. The shape of the larvae, their covering of large hooks, and the depth of invasion makes their extraction difficult; surgical removal is often necessary.

Various flies that lay their eggs in decaying animal or vegetable matter occasionally lay eggs on open sores or wounds. The larvae feed on pus and dead tissue but do not usually attack healthy tissues. Larvae of other species may invade the nasal or ear passages, causing obstruction, severe irritation, purulent discharge, facial oedema, fever and even death.

Eye Infections

Infections of the eye and the orbit are of particular concern to patients and their doctors alike because of the sensitive nature and essential function of the eye.

Superficial bacterial infections of the eyelids and surrounding tissue are common and usually harmless. They are mostly staphylococcal and present as styes (p. 285). Streptococcal cellulitis may affect the soft tissues around the eye and may involve the loose connective tissue of the orbit, causing orbital cellulitis. A similar clinical picture may be caused by various other bacterial infections—e.g. by *Haemophilus influenzae* type b (p. 234) in children or by mixed anaerobic infection of the orbit secondary to chronic infection of the adjacent sinuses. Orbital cellulitis and other septic foci around the eye, especially at the medial margin, are of particular concern because the venous drainage of this area leads centrally to the cavernous sinus, and such infections may spread via these channels to cause cavernous sinus thrombosis.

Conjunctivitis

Infection of the outer surface of the eyeball (the conjunctiva) is a common condition. In older children and adults it is most commonly caused by adenoviruses, and occasionally by other upper-respiratory tract viruses. Conjunctivitis is one of the early features of measles. Virus conjunctivitis causes intense hyperaemia of the conjunctival vessels ('pink eye'); there is excessive fluid production but the exudate is thin and watery. Specific treatment is unnecessary but bathing and anti-inflammatory eye drops or ointment may ease the irritation, which is described as 'like grains of sand in the eye'.

The conjunctiva may also be infected by bacteria. These are mostly derived from the pharynx and upper respiratory tract, to which the conjunctivae are connected by the nasolachrymal ducts. The bacteria most commonly implicated are pneumococci and *H. influenzae*. *Staph. aureus* is a rare cause. The symptoms are those of any conjunctivitis: hyperaemia of the conjunctiva, irritation and excessive fluid secretion. In contrast to virus infection the exudate in bacterial conjunctivitis is purulent—it is thick and creamy and contains many polymorphs; the eyelids are crusted and there is pus in the angles ('sticky eye'). A bacteriological diagnosis is made by culture of the exudate on media suitable for respiratory tract bacteria. Treatment is by bathing to remove exudate and to ease the irritation and by applying chloramphenicol or some other appropriate antibiotic in the form of eye-drops every 2–3 hours or ointment every 4–6 hours.

Neonatal conjunctivitis and ophthalmia neonatorum

Newborn babies are particularly susceptible to conjunctival infections, acquired in most cases from the mother's genital tract. Classically **ophthalmia neonatorum** is a gonococcal infection acquired in this way. There is a severe conjunctivitis, with copious purulent discharge. The eyelids and peri-orbital tissues are inflamed and swollen, and orbital cellulitis may push the eye forward, but even so it is liable to be hidden behind markedly swollen lids with pus exuding between their edges. Untreated gonococcal ophthalmia may cause irreparable damage to the eye and used to be a common cause of blindness. Diagnosis is confirmed by finding intracellular gram-negative diplococci in gram-stained smears and by isolating *Neisseria gonorrhoeae* in culture. Treatment is by parenteral penicillin and by local bathing. In places where penicillinase-producing gonococci are common, spectinomycin or a third-generation cephalosporin may be used. The condition can be prevented by a dual approach—screening of pregnant women at antenatal clinics so as to detect and treat those with gonorrhoea before delivery, and prophylactic treatment of all neonates by instillation of penicillin eye drops. The increasing incidence of penicillinase-producing *N. gonorrhoeae* in many parts of the world has entailed a re-assessment of this treatment. Chloramphenicol or tetracycline may be an appropriate substitute, or a return to an earlier form of antiseptic treatment with silver nitrate eye drops. This is very effective and its initial introduction caused a great reduction in cases of gonococcal ophthalmia, but antibiotics were subsequently preferred because silver nitrate causes a moderate amount of irritation—a chemical conjunctivitis.

The other common cause of neonatal conjunctivitis, and the commonest in areas where prophylaxis against gonococcal ophthalmia is used routinely, is *Chlamydia trachomatis* (p. 115). Like *N. gonorrhoeae*, this is acquired by babies from their mothers during birth. It causes inclusion conjunctivitis. There is inflammation and formation of granulation tissue over the conjunctiva, with an exudate that is less copious and less purulent than in gonococcal infection. The diagnosis is confirmed by microscopic examination of smears of conjunctival scrapings (not just exudate) for cells containing typical chlamydial inclusions. Cultures of swabs taken from the conjunctiva and sent to the laboratory in chlamydia transport medium are made in tissue cultures which are examined for chlamydial inclusions after 3–5 days. Alternatively, chlamydial antigen may be detected by ELISA. Treatment is by local application of erythromycin ointment or by systemic treatment with erythromycin.

Conjunctivitis in neonates can also be caused by bacteria that cause it in any age group—e.g. *H. influenzae*, pneumococcus and *Staph. aureus*.

Trachoma

This far more severe form of *Chl. trachomatis* eye infection occurs in tropical countries and in all age groups. It begins as a simple conjunctivitis but progresses to involve the cornea in an intense destructive inflammation that leaves dense corneal scarring. In many underprivileged countries it is the cause of widespread blindness. The infection is spread by direct contact—e.g. via fingers—or indirectly via flies that are attracted to the conjunctival exudate and can readily transmit infection. Acute infections can be treated by tetracycline or chloramphenicol eye drops or ointment, but the vast amount of morbidity left by old infections can be treated only by corneal surgery to remove the scarred cornea and replace it with an implant. Trachoma is one of the world's major health problems.

Herpetic conjunctivitis and associated infections

Both herpes simplex virus and the varicella-zoster virus can cause eye infections. Herpes simplex infection affects the conjunctiva and, most importantly, the cornea, where it causes corneal ulcers known as dendritic ulcers because of their irregular branching appearance. They are intensely painful and leave corneal scars that may seriously affect vision. In all herpes infections the initial manifestation is followed by a latent period with recurrences, and recurrence of herpetic corneal ulceration is common. The diagnosis is made by corneal (slit-lamp) examination to see the dendritic ulcers, and treatment is with acyclovir or with idoxuridine in dimethylsulphoxide drops. This may not prevent recurrences but it stops extension of the ulceration, speeds healing and reduces scarring. Badly scarred corneas can be treated only by corneal transplants.

Varicella-zoster virus may affect the tissues around the eye as part of chicken pox, but actual eye involvement occurs particularly in one type of zoster (p. 131). This is ophthalmic zoster, which occurs when the virus has been latent in the ophthalmic ganglion and which affects the forehead and forescalp as far as the mid line (with sharp demarcation) and the eyelids and peri-orbital tissues. As a rule the optic fibres of the ophthalmic nerve are not affected, the eye itself is not involved and there is no risk to sight. In the less common cases when the eye itself is affected, there is a high risk of permanent damage and specific treatment with systemic acyclovir should be given.

Bacterial Endophthalmitis

The interior of the eye is a well-protected site. The outer coverings of the eyeball protect it from direct infection and there is no blood supply to the vitreous and aqueous or to the lens;

therefore, haematogenous infection cannot occur. The eye is at risk of infection, however, after surgery or other penetrating injury, and the factor that normally protects it then makes it very susceptible to infection, as it is also cut off from the help of inflammatory and immune responses; infection therefore often leads to the loss of the eye. Infection after accidental injury is usually due to staphylococci or streptococci, but occasionally to *Cl. perfringens*, which is particularly destructive, or to *Past. multocida* if the injury was an animal bite. Staphylococci may also cause infections post-operatively, but gram-negative bacilli are of special concern in this context. *Ps. aeruginosa* in particular has been known to cause outbreaks of post-operative endophthalmitis because its ability to grow in dilute antiseptic solutions has led to eyes being irrigated with it in concentrated culture. Such dilute antiseptic solutions should always be freshly prepared or taken from freshly opened sterile single-use sachets or other containers (p. 341).

Choroidoretinitis

The choroid and retina differ from the rest of the eye in having a rich blood supply and are liable to destructive infections of several kinds.

Tuberculosis

Retinal granulomas (tubercles) are common in miliary tuberculosis and can be seen by ophthalmoscopy. They clear following anti-tuberculosis therapy, leaving no residual damage. The eye is not a common site for the secondary (adult) form of tuberculosis.

Toxoplasmosis

Toxoplasma gondi infections are usually asymptomatic (p. 245) and only rarely affect the eye. Choroidoretinitis occurs in a very few cases of adult infection, causing blindness, but it is much commoner in intra-uterine and neonatal infections (p. 321). In these the association with encephalomyelitis and hydrocephalus is common enough for this triad to be highly suggestive of toxoplasmosis. Recovery is rarely complete; blindness or severe visual impairment and mental retardation are the usual result.

Toxocariasis

Choroidoretinitis with visual difficulties and occasionally causing blindness, may occur in young children as a complication of **visceral larva migrans** (p. 294). It is usually unilateral. It may require treatment with thiabendazole or diethylcarbamazine.

Onchocerciasis

As mentioned on p. 292, blindness is the most serious consequence of *Onchocerca* infection. In the savanna areas of West Africa the disease is widespread, and although it causes blindness in only a proportion of cases this still means that in some areas it blinds 15% of the population. In Central America blindness occurs in an even higher proportion of cases, but in a smaller proportion of the population because infection is less widespread. Blindness develops progressively as a result of invasion of the cornea, anterior chamber and iris by microfilariae. When these die they cause a punctate keratitis consisting of numerous 'snowflake' opacities, initially in the lower part of the cornea but extending and coalescing to form an opaque film that spreads upwards to cover the whole pupil. Damage in the posterior chamber follows a less clear sequence. Microfilariae are rarely demonstrated there but the choroid and retina become inflamed and the optic nerve atrophies; these changes are similar to those of toxoplasmosis. The pathological changes in the eye may have an immunological basis.

The diagnosis is made by slit-lamp examination, which enables live microfilariae to be seen in the cornea or in the anterior chamber. Nodules containing the adult worms are often present on the head in Central America and more often on the body in West Africa. This difference of nodule formation is related to the proportion of cases of blindness in the two areas, as microfilariae from nodules on the head are more likely to reach the eye. If *Onchocerca* eye infection is suspected, the head and body should be examined for nodules and skin snips taken from the buttocks, the shoulder blades and from the outer acanthus of the eye. Treatment with diethylcarbamazine, though effective against microfilariae, does not reverse established eye damage.

Congenital Infections (see also pp. 321–2)

The developing eye can be damaged, with resultant congenital blindness, by various transplacental infections of the fetus. Syphilis and rubella can each damage the developing lens, causing congenital cataract, and retard the development of the eye as a whole, causing micro-ophthalmia. Intra-uterine toxoplasmosis has been mentioned already (p. 297).

22
Bone and Joint Infections

Bacterial infection of bones (**osteitis**—usually **osteomyelitis** i.e. predominantly of bone marrow) or of joints (**septic** or **infective arthritis**) may come via the blood-stream (**haematogenous route**) usually from a septic focus elsewhere in the body, or it may be **direct**, via a wound or from an immediately adjacent focus. If not promptly diagnosed and effectively treated such infections may cause permanent disability.

Osteomyelitis – Haematogenous Route

Haematogenous osteomyelitis may be acute or chronic. It may be the primary feature of the illness, and is then usually due to *Staph. aureus* or other pyogenic bacteria; or it may be merely a part, albeit often an important part, of a generalized infection—e.g. a staphylococcal septicaemia, or a more chronic infection such as brucellosis, syphilis or tuberculosis.

Staphylococcal Osteomyelitis

This condition can affect any age group, but is commoner in children because their bones are actively growing and have a rich blood supply. The source of the *Staph. aureus* may be a recognizable distant lesion, such as a boil or impetigo, but is commonly undetected. This may well be because some trauma or other disorder in the bone has provided a suitable niche for development of organisms from a transient low-grade staphylococcal bacteraemia, which is probably a quite common event. The bone sites most often affected are those most liable to trauma—notably the upper end of the tibia and the lower end of the femur. A haematoma, even a small one resulting from only minor trauma, in a long-bone metaphysis provides a highly appropriate niche for the circulating bacteria. As in *Staph. aureus* infection elsewhere, there is acute inflammation and suppuration, but because bone is inelastic the infection if confined and tension builds up rapidly, interrupting blood flow and causing avascular necrosis of bone. The pus is forced through the Haversian canals to the bone surface, where it raises the periosteum, rupturing the subperiosteal vessels, and may produce a subperiosteal abscess. Separation of dead bone as a **sequestrum** is a common feature if the infection is allowed to become chronic. New bone laid down by the raised periosteum forms a sheath (**involucrum**) around the sequestrum. Outward rupture of a subperiosteal abscess may cause a superficial soft-tissue abscess that may drain to the surface through a sinus. Extension of infection to epiphyses and joints is rare because the periosteum is much more firmly attached to the epiphyses.

The symptoms of acute osteomyelitis are local pain—sudden in onset and aggravated by movement of the affected limb—and the systemic signs of acute bacterial infection, such as pyrexia, tachycardia and polymorphonuclear leukocytosis. The initial local sign is bone tenderness; heat, redness and swelling follow, often accompanied by a 'sympathetic' serous effusion in a neighbouring (uninfected) joint. Radiography usually shows no abnormality in early acute osteomyelitis; the first sign of the infection is an area of rarefaction, 7–10 days after onset.

LABORATORY DIAGNOSIS Bacteraemia is usual in acute staphylococcal osteomyelitis, and blood culture is as a rule the best means of isolating the organism. Direct aspiration of pus should not be attempted unless the infection is sufficiently advanced to require drainage of an abscess to relieve pressure.

TREATMENT Acute staphylococcal osteomyelitis usually responds well to appropriate antibiotic treatment if this is started early enough. Because antibiotics penetrate poorly into healthy bone (though better into acutely inflamed areas) it is customary and wise to use two anti-staphylococcal drugs. Because *Staph. aureus* is commonly resistant to benzylpenicillin, flucloxacillin should be one of these, accompanied by erythromycin, fucidin or clindamycin (which penetrates bone better than most antibiotics). Subperiosteal abscesses require surgical drainage, and if bone necrosis has occurred any sequestra must be removed.

'Cold' staphylococcal osteomyelitis

An insidious form of silent staphylococcal osteomyelitis, called 'cold' because it lacks the classic signs of acute pyogenic infection, may develop in patients with rheumatoid arthritis under steroid treatment. Despite its lack of signs it can cause severe bone destruction.

Haemophilus Osteomyelitis

Although the commonest cause of acute osteomyelitis in all age groups is *Staph. aureus*, in children under 5 years old a similar clinical picture may well be due to type b *H. influenzae*—sometimes in association with other manifestations of infection with this organism, notably meningitis (p. 214) or epiglottitis (p. 197). Blood culture is again the best means of laboratory diagnosis. Treatment should be with chloramphenicol.

Salmonella Osteomyelitis

Localization of *S. typhi* in bone during the septicaemic phase of typhoid (p. 231) may add osteomyelitis, with signs as described above, to the general picture of enteric fever. A different type of salmonella osteomyelitis affects patients with sickle cell disease. During the crises of this disease blocking of small blood vessels leads to hypoxia and death of small areas of bone, which may then become infected by bacteria. These may be *Staph. aureus* or other common pyogenic bacteria, but osteomyelitis in sickle cell disease is often caused by non-typhoid salmonellae—which very rarely cause it in other people.

Other Forms of Acute Osteomyelitis

Although an area of damaged bone would appear to be liable to seeding with any bacteria that cause bacteraemia, few of these cause osteomyelitis. On occasions this is achieved by pneumococci, streptococci or enterobacteria.

Brucellosis

Bone-marrow infection is common in brucellosis, but it does not present the signs of acute osteomyelitis—or indeed as a rule any signs suggesting bone involvement. Small granulomas and micro-abscesses form in the bone-marrow, and may provide reservoirs for persistent re-seeding of the blood-stream. Bone-marrow biopsy and culture may be useful in the diagnosis of brucellosis, and may allow isolation of the organism when blood cultures have been negative.

Syphilis

Bone lesions occur in the tertiary stage of acquired syphilis, and in congenital syphilis. In the tertiary stage they are **gummata**, such as occur in many other organs (pp. 278–9). Characteristic lesions are seen in babies with congenital syphilis (p. 321); there is widespread periostitis which causes new bone formation, seen on radiographs as double contours to the bones, and there is osteochondritis at the ends of the long bones. The typical lesion in older children with congenital syphilis is **sabre tibia**—a fusiform cortical thickening of the shaft of the tibia as a result of new bone formation; it is very painful.

Tuberculosis

The haematogenous dissemination of tubercle bacilli in primary tuberculosis (p. 206) may lead to detectable lesions in bone at that stage. More often, however, small granulomata form and remain quiescent until reactivated as manifestations of the secondary (adult) form of tuberculosis. Such a lesion takes the form of a **cold abscess**, which develops insidiously and causes an aching pain but none of the signs of acute inflammation; it eventually discharges through the skin, leaving a persistent sinus. In the spine, which is the commonest site, the infection begins in a vertebral body close to the disc and spreads to involve the disc and the neighbouring vertebra. Bone is destroyed, and the expanding abscess may compress the spinal cord, causing acute neurological disturbances. Collapse of several affected vertebral bodies can cause an angular kyphosis; if this occurs in the thoracic spine, it may interfere with cardiorespiratory function. From the lumbar spine a cold abscess may track along the psoas muscle sheath to discharge below the inguinal ligament—a **psoas abscess**. A retropharyngeal abscess from the cervical spine may rupture into the pharynx.

Tuberculosis may also affect long bones, usually in association with joint involvement (see below).

In miliary tuberculosis (p. 207), as in primary tuberculosis, tubercle bacilli are distributed to sites throughout the body, including the bone-marrow; and culture of bone-marrow may be useful as a means of isolating them and confirming the diagnosis, particularly in elderly patients with suspected cryptic miliary tuberculosis and no obvious and more accessible sites of infection.

Osteomyelitis Following Direct Infection

Non-haematogenous osteomyelitis may result from infection through a wound (accidental or surgical) or by extension of infection from a focus in adjacent tissue.

Compound Factures

When a fracture involves breach of the skin surface, the organisms that predominate as causes of infection in accidental wounds (p. 290) can enter the bone and cause osteomyelitis. Staph. aureus is again the most important of these, but a wide range of bacteria may be implanted in dirty wounds. Clostridia (notably Cl. perfringens) are of particular concern in soil-contaminated compound fractures, because damaged hypoxic bone provides a suitable nidus for anaerobic growth.

Post-Operative Infections

Infection following orthopaedic surgery is an important cause of failure of the procedures, particularly when prosthetic implants have been used to replace joints or to reconstruct damaged bones. Such implants provide sites for colonization by organisms of low virulence, notably the skin commensal Staph. epidermidis (p. 76). This may be implanted at the time of operation, but the low-grade infection that it causes may not become evident for many months. Once established, such an infection around an implant is virtually untreatable, and it is usually necessary to remove the implant and replace it—either immediately or later. Such replacement is costly, and once the site has been infected it provides an unsatisfactory base for the new prosthesis; repeated operations are progressively less successful. Prevention of the initial infection is therefore of the utmost importance. The staphylococci or other organisms causing such infections either come from the patient's own skin around the operation site or are shed into his environment by the surgical team. Auto-infection can be virtually eliminated by applying some agent such as povidone–iodine in compressses to the skin around the operation area for two days or so before operation, by careful use of sterile drapes to mask the skin around the incision, and by discarding the scalpel used for the skin incision so that it cannot carry organisms into the deeper layers. Exogenous infection can be reduced by laminar air flow in the theatre, by keeping to a minimum the number of people around the patient and in appropriate cases by dressing the surgical team in exhaust-ventilated 'space-suits'. A further line of defence can be provided by giving the patient anti-staphylococcal antibiotics during and for a day or two after the operation and by incorporating gentamicin or some other stable and appropriate antibiotic in the cement used to fix a prosthesis in place.

Infections from Adjacent Sites

Intact bone is fairly resistant to infection from outside, but chronic infections alongside it can eventually penetrate the periosteum and enter the bone. Chronic sinusitis, otitis media and mastoiditis all occur in cavities surrounded by bone, and all may give rise to localized chronic osteomyelitis. These infections may then spread further, causing extradural empyemas, brain abscesses or, less commonly, meningitis. Similarly, infections around teeth (e.g. peri-apical abscesses) may lead to osteomyelitis of the jaw, usually with a mixed bacterial population including mouth anaerobes and streptococci (p. 189).

Actinomycosis (p. 286) starting in the mouth may involve the jaw; destruction of tooth sockets is not uncommon.

Chronic ulcerative and gangrenous foot lesions, particularly those occurring in diabetics, may give rise to chronic osteomyelitis of foot bones, with mixed bacterial populations including many anaerobes. Such bone involvement greatly reduces the prospects of controlling the progressive tissue destruction and achieving repair and healing; it usually necessitates amputation back to healthy bone.

Septic Arthritis

Like osteomyelitis, septic arthritis is usually blood-borne from an infection site elsewhere in the body, and is most likely to occur in a joint that has suffered trauma (often only minor) and so provided a nidus for infection. Again *Staph. aureus* is the commonest bacterial cause, with type b *H. influenzae* responsible for some cases in young children. Adults with gonorrhoea may develop gonococcal septic arthritis 2–4 weeks after urethritis; they are also liable to develop a reactive arthritis (see below).

Non-haematogenous septic arthritis can follow penetrating injuries of joints, or can be an extension of an infection of adjacent tissue. Except in infants, osteomyelitis seldom extends into a neighbouring joint because the metaphyses and epiphyses of bones have separate blood supplies after the age of 1 year and the firm attachment of periosteum to epiphyses prevents subperiosteal spread into joints. **Tuberculosis** is an exception to this generalization; the joints it attacks most commonly are the hips and the knees, and in each case both joint and bone are involved. In the case of the hip the initial focus is usually in the femoral neck or the acetabulum, with infection soon spreading to the joint, causing joint destruction, subluxation and deformity. Knee infection begins in the synovium but spreads to destroy cartilage and bone.

Immune-Mediated (Reactive) Arthritis

Inflammation of joints, with pain and swelling (due to fluid effusion) and restriction of movement, can occur in association with various bacterial and virus infections. Fluid aspirated from such a joint is often cloudy, and its leukocyte content is typical of an inflammatory exudate, but the fluid is sterile and the joint is not infected. These cases of arthritis result from immune (hypersensitivity) reactions triggered by organisms causing infections elsewhere in the body. Reactions of this kind are seen, as indicated below, in rheumatic fever, in certain intestinal infections, in parvovirus infection and rubella (probably) and in Reiter's syndrome.

Rheumatic Fever

The clinical features and pathogenesis of this sequel of infection with Lancefield group A β-haemolytic streptococci are described on p. 77. Its onset, 2–3 weeks after the streptococcal infection (usually of the throat) is heralded in most cases by painful swelling of one or more limb joints; this reactive arthritis characteristically 'flits' from joint to joint during the illness.

Reactive Arthritis Associated with Intestinal Infections

Salmonella, shigella, campylobacter or yersinia infection of the intestinal tract can be followed by reactive arthritis, coming on some 1–3 weeks after onset of the intestinal symptoms and often after these have resolved. One or more joints may become painful and swollen for a time, but the condition resolves spontaneously, leaving no lasting effects. The arthritis has the characteristics of an immune-mediated reaction and is commonest in people with the HLA-B27 histocompatibility antigen, but its mechanism is not known (p. 116).

Rubella Arthritis

Painful joints with restriction of their movements are common in rubella, especially in women, and may belong in this category of immune-mediated arthritis. However, they occur at the same time as the rash and may be caused directly by the rubella virus. A similar form of arthritis occurs in parvovirus infections.

Reiter's Syndrome

This triad of urethritis, arthritis and conjunctivitis (sometimes with the addition of uveitis, circinate balanitis, mouth ulcers, or keratoderma blenorrhagica) has the characteristics of an immune-mediated reaction; all three of the classical triad need not be present in every case. It is commonest in men (male : female ratio 15 : 1) with certain HLA-B antigens, notably HLA-B27 (p. 116). It is particularly associated with chlamydial urethritis (the commonest form of non-gonococcal urethritis—p. 116). Symptoms appear 8–28 days after the initial infection when the infecting agent may no longer be present, but serological tests can be used to confirm infection. Recurrences, particularly of arthritis, are common. The whole syndrome (not just arthritis) can also be triggered by salmonella, shigella, campylobacter or yersinia intestinal infection.

Bone and Joint Effects of Protozoan and Helminth Infections

Apart from *Leishmania donovani* (see below), protozoan and helminth parasites of man rarely affect bones and joints directly—though malaria has important indirect effects on the bone marrow.

Leishmaniasis

L. donovani (which causes visceral leishmaniasis—p. 270) invades and destroys cells of the reticulo-endothelial system, including those in the bone-marrow. The marrow responds by increased white-cell production but reduced erythrogenesis, with consequent anaemia. Parasitological diagnosis of visceral leishmaniasis depends on demonstrating the presence of the parasites in the reticulo–endothelial cells, and material for this purpose is most safely obtained by aspiration of marrow, since liver or spleen puncture is hazardous. Marrow smears show macrophages filled with *Leishmania* amastigotes; myelocytes and neutrophils may also contain parasites.

Malaria

The large-scale destruction of circulating red cells in malaria (pp. 241–2) makes big demands on the bone marrow. In acute infection, particularly in falciparum malaria, the marrow is red and hyperplastic, but in chronic infections it is pale and unresponsive and the patient is anaemic.

Hydatid cysts

On rare occasions hydatid cysts (p. 271) develop within the medullary canals of long bones. They are often asymptomatic and unsuspected, but they can be detected by radiography. Cyst enlargement is limited by the surrounding bone, but this may be weakened by the constant pressure and is then liable to fracture.

Dracunculiasis

Dracunculus infection (p. 292) not uncommonly causes fibrosis around joints and in associated muscles and tendons. Secondary bacterial infection of the tracks formed by female worms may cause chronic abscesses and, along with calcification of dead worms, may lead to synovitis, arthritis and limb contractions.

23

Skin Infections

Everyone carries a large resident microbial population on skin surfaces and in the openings of hair follicles, sweat glands and sebaceous glands. This population comprises mainly gram-positive cocci of the genera *Staphylococcus* (mostly *Staph. epidermidis*—p. 76) and *Micrococcus* (p. 74) and gram-positive rods of the genera *Propionibacterium* (p. 85) and *Corynebacterium* (diphtheroids or coryneforms—p. 83) together with the yeast *Pityrosporum* (p. 29). The skin is also host to a variable number of transient or contaminant bacteria. The resident bacteria produce antibacterial substances that provide some protection against colonization by potential pathogens. Some of these substances are antibiotic-like peptides; others, such as propionic acid produced from lipid components of skin secretions by propionibacteria, are simple metabolic products that reach high enough concentrations to inhibit other species.

Skin commensals are not entirely harmless. Coagulase-negative staphylococci and (less often) coryneforms from the skin cause opportunist infections of prosthetic implants (CSF shunts, p. 226; hip prostheses, p. 301; artificial heart valves, p. 237). They can also cause significant bacteraemia in patients on immunosuppressive therapy for malignant diseases or on intravenous nutrition through long intravenous catheters (p. 351), and in those with implanted prostheses (p. 319). Their role in acne vulgaris is described below.

Bacterial Infections

Staphylococcal infections of hair follicles, sweat glands and subcutaneous tissues (boils, folliculitis, carbuncles, etc.) are described in Chapter 21. Other superficial skin infections are considered here.

Impetigo

This is a superficial infection of the epidermis particularly common in children. It is highly infectious, spreading readily from child to child, especially in conditions of overcrowding and close contact. It used to be very common amongst schoolchildren in Britain but is less so now, probably as a result of improved living conditions but possibly also because of widespread use of antibiotics and a general reduction in the virulence of streptococcal infections (p. 77). Impetigo is caused by β-haemolytic streptococci of Lancefield's group A (*Str. pyogenes*—p. 77) or by *Staph. aureus* or by the two together forming a mixed infection. The staphylococcal type is the commoner in Britain now. The two types of impetigo are similar in appearance, and clinical distinction between streptococcal and staphylococcal infection is unreliable. Superficial blisters form under the horny layer of the epidermis. They contain clear or slightly cloudy fluid, and when they burst their contents coagulate, forming adherent yellow scabs over raw and painful deeper skin layers. Mild infections respond to superficial antiseptic treatment; antibiotic treatment speeds the healing of more widespread infection. The skin heals with no evidence of scarring.

Erysipelas

Certain strains of β-haemolytic streptococci of Lancefield's group A cause erysipelas, an acute spreading infection confined to the dermis. It is commonest in the elderly and debilitated. Acute inflammation spreads through the dermis via the superficial lymphatics from the site of initial infection to form an expanding red, hot and tender area, with a clearly defined edge where the infected and normal tissues meet. The indurated skin is raised above the normal skin and tethering of hair follicles and glands produces a dimpled appearance (like the skin of an orange—'peau d'orange'). The streptococcus produces exotoxins that cause fever and general prostration. Without antibiotic treatment, infection may spread to the blood-stream, producing a septicaemia with a high mortality; this is now rare in developed countries. Erysipelas responds rapidly to treatment with penicillin.

Erysipeloid

This condition was given its name because its appearance has a superficial resemblance to erysipelas; but it is usually a trivial infection and does not cause systemic effects. It is caused by *Erysipelothrix rhusiopathiae* (p. 85), and is an occupational disease of people who handle fish or, less commonly, meat and poultry. It occurs chiefly on the hands, commonly on a finger. The bacteria, which are present on fish scales, meat, etc., enter through scratches and minor skin abrasions from fish bones, wooden boxes, etc.; constant wet conditions make skin more susceptible to infection. The resulting lesion is a painful, purplish or dull red local swelling that expands peripherally and tends to heal from the centre. It is usually self-limiting. Penicillin is the drug of choice for treatment but antibiotics do not have much effect upon the progress of the infection.

Scalded Skin Syndrome

This condition of babies and young children, also known as **toxic epidermal necrolysis** or **Lyell's disease,** is caused by an epidermolytic toxin produced by *Staph. aureus* strains of particular phage-types. The staphylococcal infection is usually at a site remote from the skin lesions. The toxin causes splitting of the skin, probably at the level of the desmosomes between the epidermal cells of the granular layer. Large fluid-filled bullae develop at sites widely scattered over the whole body, and often coalesce to form extensive blistered areas. The fluid is clear or slightly turbid. Although it may contain some inflammatory cells, it is as a rule sterile; staphylococci cannot usually be isolated from the bullous lesions unless they become secondarily infected. The bullae are fragile and soon rupture, and the outer layer of skin easily rubs off, leaving a weeping surface. The raw areas, denuded of their outer layers, are very painful, and much fluid and protein is lost through them. There are also systemic toxic effects, and in its more severe form the disease has a considerable mortality. Fortunately this severe form is rare. The bacterial gene responsible for epidermolytic toxin production is chromosomal in some strains but plasmid-borne in others—notably in those that produce large amounts of toxin and cause severe disease.

Acne Vulgaris

Despite many years of debate, the role of bacteria in this very common skin condition is still not understood. The basic lesion is a sebaceous follicle packed with waxy material but not yet open to the surface—a closed **comedo.** This may initially be sterile, and even the open comedo or **blackhead** and the later inflamed comedones do not harbour evident pathogens. Hormonal levels undoubtedly play a part in the development of acne, but a complex immunological response is probably also involved and may be triggered in part by bacteria, notably *Propionibacterium acnes*. Prepubertal children do not carry this organism, but colonization begins at puberty and rises rapidly, in parallel with sebum secretion, to reach a peak around 16–17 years. Sebum production inside follicles blocked by excess keratin produces the comedones, in which bacteria multiply vigorously without at first provoking an inflammatory reaction. Production of proteases, hyaluronidase and other enzymes by *Prop. acnes* may cause the follicle wall to become 'leaky'. This allows bacteria and their products, including irritant short-chain fatty acids from sebum breakdown, to reach the surrounding tissue and to trigger an acute inflammatory response, probably with a major immunological component; a typical acne pustule results.

Treatment usually consists mainly of local measures to remove excess sebum and to prevent the development of keratin plugs. Antibiotics are of doubtful value, except in the severest forms of the disease, in which there may be secondary bacterial involvement. Milder forms may be improved by long-term tetracycline therapy, but the improvement is not always maintained, relapse may in any case follow cessation of the treatment, and prolonged antibiotic use brings its own problems (p. 373).

Cutaneous Anthrax

Spores of *B. anthracis* (p. 86) may be inoculated into the skin of farm workers and others in contact with infected animals. Animal anthrax is rare in countries such as Britain, and human infection occurs mostly in dockers or factory workers who handle imported hides, bristles, wool or bone meal. Bone meal is also a possible source of infection for gardeners, who on rare occasions develop either cutaneous anthrax or, as a result of inhaling the bone meal, the pulmonary form (p. 202). A papule forms at the site of a skin inoculation, develops into a blister, becomes purulent (malignant pustule) and then turns into a dark-centred necrotic lesion surrounded by oedema, induration and a ring of vesicles. In the absence of effective treatment a severe and commonly fatal septicaemia may follow. A gram-stained smear of material from the lesion may give a strong indication of the diagnosis (p. 86). The organism is usually sensitive to penicillin and various other antibiotics, but the slow response of the indurated skin lesion to antibiotic treatment is liable to cause anxiety.

Ground Itch

Irritant skin infections are often seen in parts of the world where infections with hookworms (p. 260) and *Strongyloides* (p. 260) are common. Penetration of the skin by the larvae of these worms may be followed by secondary bacterial infection; local hypersensitivity to the larvae themselves also contributes to the irritation.

Burns

The readily accessible damaged tissue and the nutrient-rich exudate of burns constitute an excellent bacterial culture medium, and large burned areas almost inevitably become infected. The outcome of the infection depends on the virulence of the bacteria, several species of which are likely to be present simultaneously in such an inviting situation. β-haemolytic streptococci and *Pseudomonas aeruginosa* are particularly liable to cause serious trouble. Group A streptococci used to be the major problem, since they prevented wound healing, almost invariably caused skin grafts to fail and could invade and cause septicaemia, often fatal. Streptococci of groups C and G can cause similar problems, though they are less aggressive invaders. The ease with which streptococcal infections can be controlled by antibiotics has left the far less tractable *Ps. aeruginosa* as the commonest cause of serious infection of burns. It too can retard healing and cause postponement or failure of skin-grafting, thus extending the patient's stay in hospital; and in the more seriously ill patients it can cause septicaemia, with a high mortality. Its resistance to antiseptic agents and to most antibiotics makes it difficult to keep out of burns and to eradicate when it gets there. *Staph. aureus* infection of burns is common, but in most cases it causes few problems. However, some particularly aggressive staphylococcal strains can cause invasive and destructive infections, and can be of major importance if they become resident in a burns unit and are resistant to most antibiotics.

The management of burned patients in special units with particular attention to prevention of infection is described on p. 350. Because *Ps. aeruginosa* has, despite such measures, caused serious outbreaks of infection in burns units, attempts have been made—with some success—to produce a vaccine with which burned patients can be actively immunized against the common virulent serotypes of *Ps. aeruginosa*.

Virus Infections

Herpes Simplex

Probably the commonest human virus infections are due to herpes simplex viruses (p. 130). These viruses remain latent in the sensory ganglia of the nervous system and cause recurrent skin lesions when the host–parasite relationship is disturbed (p. 131). The common sites affected are around the mouth (**herpes labialis**—p. 190) and on the genitalia (**herpes genitalis**—p. 282). The lesions of herpes simplex start with an itching or burning pain, then superficial vesicles appear which soon break down to form shallow painful ulcers that gradually crust over and heal after 5–10 days. Secondary bacterial infection may occur in herpetic ulcers. For further details of herpes simplex see pp. 190, 221 and 282.

Warts

Common human warts are transmissible to other humans, and electron microscopy shows that they contain viruses which are morphologically papovaviruses (p. 129); but these have not been grown in tissue culture. The viruses infect epidermal cells, replicate in their nuclei, and cause epidermal overgrowth. This produces either a papilloma or a flat firm plaque (**verruca plana**). Infection spreads to adjacent skin areas, producing multiple warts; such spread occurs most readily when two areas of skin come into frequent contact—e.g. between the opposed surfaces of two adjacent fingers. Inward growth of the wart, into the skin, occurs on the palms of hands and particularly on the soles of the feet (**plantar warts** or **common verrucas**); such lesions have thick horny surface layers and are painful on pressure, especially that caused by walking. On mucosal surfaces warts have no horny layer and grow outward rapidly to form soft papillomata.

Most warts disappear spontaneously after periods that range from a few weeks to 2–3 years. There is no specific antivirus drug active against warts. Treatment with chemicals that destroy the papillomatous tissue and inactivate the virus particles is effective, but it may be necessary to continue treatment for several weeks, or even for months in the case of established plantar warts.

Molluscum Contagiosum

This infectious skin disease is caused by a poxvirus (p. 234) that has not been grown in tissue culture. Infection of the epidermal cells causes small pearly flesh-coloured nodules, with central cores of white caseous material that can be expressed by gentle pressure. Virus particles can be seen by electron microscopy of this material. The nodules may occur in any skin area except the palms and soles. Transmission is by direct, including sexual, contact.

Contagious Pustular Dermatitis (Orf)

Orf is an infection of sheep and goats that occurs world-wide. Vesicles form around the lips, eyes, genitalia and teats, develop into pustules and then crust over. Human infection results from contact with infected animals and is an occupational disease of farm and abattoir workers and veterinary surgeons. The typical lesions, known as a 'milker's nodes', occur on the hands and arms. They begin as inflammatory papules that enlarge to form granulomatous blisters and then crust over and regress; vesicle formation is rare in man. Healing may take several weeks. Despite their angry appearance the lesions are more or less painless.

Fungal Infections

The fungi that can cause skin infections in man are the yeast-like fungi of the genus *Candida* (mainly *C. albicans*); dermatophytes of the genera *Trichophyton, Epidermophyton* and *Microsporum*; and *Malassezia furfur*. Their characteristics and the conditions that they cause have been outlined in Chapter 10.

Cutaneous Candidiasis

Normal skin is resistant to colonization and infection by *C. albicans*, but this organism can cause dermatitis in areas of skin that are persistently moist (p. 138). Common sites are the skin folds of the groin, axilla and submammary regions, especially in obese subjects; women are more commonly affected than men. People whose hands are constantly in water because of their work may develop chronic candida dermatitis of their hands. In babies, *C. albicans* frequently infects areas of skin already damaged by napkin dermatitis, causing further inflammation and irritation. Skin affected by candida dermatitis is red, raw and moist with serous exudate. Treatment involves removing any known precipitating factors, keeping the affected skin dry by exposure to air and by the use of dusting powder, and local application of antifungal agents.

Candida infection in those whose hands are continually in water may take the form of chronic paronychia; its features are redness and swelling of the nail-fold, loss of attachment of the cuticle and distortion of the nail plate. Removal of the precipitating factor may permit recovery.

LABORATORY DIAGNOSIS Yeast forms and pseudohyphae of *C. albicans* may be seen in gram-stained smears prepared from candida lesions. They are best grown on Sabouraud's agar (p. 24).

Ringworm (The Dermatomycoses)

Dermatophyte fungi (p. 138) cause superficial infections of keratinized tissue—skin, hair and nails. These infections are known as **ringworm** (or **tinea**) because they cause roughly circular skin lesions.

Body ringworm (tinea corporis) is usually caused by *Epidermophyton floccosum* or by one of the *Trichophyton* species, less commonly by *Microsporum* species. Person–to–person spread of infection may be by direct contact or via fomites; foot–floor–foot spread (as for tinea pedis) may be important. The fungal hyphae spread through the superficial keratinized epidermis causing a slowly expanding lesion. There is inflammation of the epidermis, dermis and hair follicles and the surface is dull red and scaling; the active edge of the lesions may be marked by small blisters. As the lesion spreads the centre heals, producing a 'ring' lesion. One of the commonest sites of infection is the groin, where infection is known as **tinea cruris**; this is usually caused by *Trichophyton* species and the interdigital clefts of the feet are usually also infected. Animal ringworm fungi—notably *T. verrucosum* from cattle—sometimes cause tinea corporis in man; they tend to provoke vigorous local skin reactions and to be self-limiting.

Tinea pedis (athlete's foot) is the commonest form of ringworm and is also usually caused by *Trichophyton* species. Its spread from person to person is often via wet floors around swimming pools and communal showers and in bathrooms, especially as moist skin is readily infected. Athlete's foot usually begins as peeling and cracking between the toes, spreading to cause blistering, scaling and peeling over the surface of the toes and on the soles.

In **tinea capitis (scalp ringworm)** the infection involves the hair as well as the scalp skin. Organisms that may cause it include *M. audouini* and *T. tonsurans*, acquired from other humans by direct contact or via combs or brushes, and *M. canis*, acquired as a rule from domestic pets. There is usually mild inflammation, redness and swelling of the scalp; and the hairs, weakened by infection, break off close to the skin, leaving bald patches. *M. audouini* infection, now much less common in Britain and other developed countries than it was only a few years ago, affects children and may persist for long periods but clears spontaneously at puberty. The clinical diagnosis of scalp ringworm can be confirmed by exposing the hair to long-wave ultraviolet light (Wood's light), which causes green-blue fluorescence of *Microsporum* infected hairs.

Infection of finger and toe nails (**tinea unguium**) is usually by *Trichophyton* species. Invasion by fungal hyphae causes thickening, discolouration and distortion of the nail.

LABORATORY DIAGNOSIS A suspected dermatophyte infection can be confirmed by low-power microscopy of suitably prepared material from an active part of a lesion. Scrapings from the edge of a skin lesion, clippings or scrapings from a thickened nail, or abnormal and apparently infected hairs should be collected, preferably on black paper which can be folded to ensure their safe transport to the laboratory and which allows them to be easily found on arrival there. Appropriate treatment with potassium hydroxide makes the specimens translucent, and allows hyphae, spores and other distinctive features to be seen under the microscope. It is not as a rule possible to identify the dermatophyte species at this stage; culture of the material (not KOH-treated) on Sabouraud's or other suitable media allows development of colonies with recognizable species characteristics, but this may take several weeks.

TREATMENT This is considered on p. 138.

Dermatophyte hypersensitivity reactions are not uncommon. They are often manifested as eczematous lesions on the palms and fingers, but may take the more severe form of a widespread and intensely irritant urticaria. The provoking condition is usually a *Trichophyton* foot infection. Systemic corticosteroid and antihistamine treatment may be necessary for control of the hypersensitivity while the infection is being treated with antifungal drugs.

Pityriasis Versicolor

This benign skin infection is caused by *Malassezia furfur* (p. 137). Small brown or fawn spots are produced, usually on the chest. They do not show scaling unless deliberately rubbed, when small powdery scales are produced. They are not irritant and are little more than a minor inconvenience.

Protozoan Infections

Leishmaniasis is the only protozoan infection to be considered here. The four forms of cutaneous leishmaniasis, caused by *L. tropica* and *L. major* in the Old World and by *L. braziliensis* and *L. mexicana* in the New World, are described on pp. 147–8. All are spread by sandflies. A small papule, which may itch intensely, develops at the site of the sandfly bite and grows into a lesion several centimetres in diameter; this may then ulcerate and become covered by serous exudate. Ulceration occurs early in *L. major* infection but may take several months when the organism is *L. tropica*. Most of the ulcers caused by these two species heal spontaneously, leaving the patient with protective immunity, but in some patients the infection extends to become diffuse cutaneous leishmaniasis. Although these patients have strong antibody responses, they are unable to eliminate the parasites and new peripheral lesions form as the original lesions heal. *L. braziliensis* infection also causes skin ulcers, but spread via the lymphatic system can occur. Its skin lesions develop as in *L. tropica* infection, but the ulcers may become much larger and are often multiple, and in some patients they include severely destructive and disfiguring ulcers of the oral and nasal mucosa and underlying tissues.

Ectoparasites

Several arthropods live on the skin surface or in the epidermis of man. They cause considerable discomfort but little serious damage themselves, but some transmit important bacterial infections.

Scabies

The mites (*Sarcoptes scabei*) that cause scabies are minute white disc-like arthropods just visible to the naked eye. They are world-wide in distribution. Female mites burrow into the superficial layers of the skin excavating winding burrows from a few mm up to several cm long through the stratum corneum, extending by about 2–3 mm per day. They prefer loose, thin, wrinkled skin such as that between the fingers and around the wrists, elbows, feet, buttocks, scrotum and axillae. In women, mites may invade the skin under and around the breasts and nipples, and in young children with soft skin they may be found on the face. Eggs are laid in the burrows and develop into adults in about 2–4 weeks. A punctate rash develops along the tracks of the burrows, especially at the sites where the eggs have been laid and the larvae are developing; it is intensely irritant and leads to vigorous scratching, frequently followed by secondary bacterial infection which may mask the original condition. Infection with only a few mites may provoke a severe rash. A follicular papular rash may also occur in areas of the body not affected by the mites; it is a hypersensitivity reaction that occurs 4–6 weeks after an initial infection or within a few days of reinfection in individuals sensitized by a previous infection.

The use of corticosteroids in patients with scabies may cause a rare but highly contagious form known as Norwegian or crusted scabies, in which infection with vast numbers of mites causes thick scaly crusts over the hands and feet, scaling eruptions on other parts of the body, but generally not much irritation. Mites usually spend the whole of their 1–2–month life on man but they can live for a few days away from the human body, e.g. in clothing or bedding. Infection is transmitted by close personal contact. All members of a family are usually affected, as the infection spreads amongst those living in close association, especially those sleeping in the same bed. Infection often increases in wartime and after disasters when people have to live in overcrowded conditions. Scabies is also spread by sexual contact, the mites invading the skin around the genitalia.

The clinical diagnosis of scabies is confirmed by gently scraping away the skin over a burrow with a needle and picking out a female mite which can be examined with a hand lens. When scabies is diagnosed in an individual, all members of the family/household should be treated, whether or not they appear to be infected, because the condition is readily transmitted between people in close contact and reinfection occurs readily. Topical application of benzylbenzoate or sulphur preparations (e.g. monosulfiram) is effective; two or three applications at intervals of a few days may be needed to kill mites that hatch after the original treatment. Prevention depends upon personal hygiene. Scabies is especially common in communities with inadequate water supplies; the rise in standards of hygiene with improved water supplies goes a long way towards preventing its transmission.

Lice

The head louse and the body louse are two varieties of *Pediculus humanus*; the pubic louse (*Phthirus pubis*) belongs to a separate genus. The head louse is 2–3 mm long and infects the scalp only; the female cements white eggs, which are visible to the naked eye, to individual hairs. The eggs hatch in about 10 days, producing young which are similar to the adults. The body louse lives on the rest of the body, but not on the head; it is similar in appearance and life cycle to the head louse. The larval and adult forms of both types feed by sucking blood through their piercing mouth parts. Body lice transmit epidemic typhus (p. 111), and louse-borne relapsing fever (p. 232), but head lice do not transmit these infections.

Pubic lice are usually found attached to the hairs in the genital area but may occur elsewhere, e.g. the axillae. They are commonly transmitted by sexual contact. The adult is up to 2 mm long—wider and shorter than the head and body lice. Eggs are attached to individual hairs and the life cycle takes about 40 days. Larvae and adults feed on blood but are not known to transmit bacterial infections.

People differ in their reactions to louse bites; in very sensitive individuals they provoke macular swellings and intense local pruritus but a sensitizing period of 3–8 months (during which there may be 10^4–10^5 bites) is usually required. Secondary sensitization may develop 18–30 months after infection. Lice are transferred by direct contact and infection is associated with poor standards of hygiene. Head lice are most often transmitted by head-to-head contact, less commonly via shared hair-brushes and combs. Head lice have become less common amongst working-class children in industrial cities in developed countries. However, they have become commoner amongst children in 'middle-class' and rural communities, probably as a result of increased social contact and less time spent on grooming. Fewer severe infections are seen now, and more light infections, but heavy infection may build up in socially isolated children. Body lice are associated with the use of one set of clothing for long periods. Washing clothes does not destroy lice but they are susceptible to heat (e.g. washing in hot water and ironing). Complete changes of clothes every 5–12 days breaks the cycle of transmission.

Infection with head lice is diagnosed by seeing the eggs attached to hairs or the lice themselves, though these move fairly quickly. Body lice are less easy to find, the best method being to examine clothing, particularly seams, pleats and folds. Pubic lice usually remain attached to the hairs and do not move so rapidly as head lice.

Infection is treated by thorough washing of the body and/or hair with soaps or lotions containing insecticide and dusting clothing with insecticide powder. Several treatments may be necessary

Fleas

Only a few of the several thousand species of fleas are important pests of man. Those of medical importance are *Xenopsylla*, *Nosopsyllus*, and *Tunga* (the chigoe or jigger flea).

Xenopsylla, Nosopsyllus, Pulex and Ctenocephalides

Adult fleas are small insects, flattened from side to side, varying in colour from light to dark brown, and capable of jumping considerable distances with their powerfully developed hind legs. Both sexes feed on blood. The female lays eggs after a blood meal, usually in cracks and crevices in the floor or in dust and debris. Larvae emerge in 2–14 days, feeding on organic debris (they are occasionally found on people with dirty clothes and unhygienic habits) for a period from 2–3 weeks to several months. When adult fleas develop they may remain dormant for up to a year, e.g. until living quarters are re-occupied, when newly emerged fleas thirsty for their first blood meal may make repeated attacks. Although flea bites are considered to be little more than a nuisance, they provoke considerable irritation and discomfort in some people. The irritation often becomes worse some time after the bite; intense itching in sensitized people may last for several days. Of greater medical importance is the role of *Xenopsylla* species in the transmission of plague (p. 95) and of *Nosopsyllus* and one species of *Xenopsylla* in the transmission of flea-borne endemic typhus (p. 112).

The cat and dog fleas (*Ctenocephalides*) commonly bite man and, like the human flea *Pulex*, cause widespread nuisance but do not transmit other diseases. During the day most bites are on the ankles and legs but at night any part of the body may be attacked

Fleas are notoriously difficult to catch and therefore to identify with certainty. Cat and dog fleas can be detected on the fur of the neck and belly of the animal. Chemical insecticides (usually in powder form) are used to treat not only the human or animal host but also the domestic environment where most of the fleas will be present. In many parts of the world *Xenopsylla* and *Pulex* fleas have developed resistance to some insecticides. Outbreaks of plague have followed the destruction of wild or semi-domestic rodent populations by natural processes or after active rodent control programmes. With the loss of their normal rodent hosts the fleas turn to man for blood meals, transferring the plague bacilli at the same time.

Tunga

These fleas occur in tropical Africa and its major offshore islands and in Central and South America. The female fleas burrow through the skin into the soft tissue, frequently between the toes and under toenails, leaving only the tips of their abdomens at the skin surface. They feed on blood and become enormously distended with eggs in 8–10 days. The surrounding area becomes itchy and inflamed, and secondary bacterial infection is common. Eggs are laid in batches during a period of 7–10 days, after which the female dies but remains embedded in the skin. Infection is usually contracted from dirty floors of houses, hotels, etc. or from the ground. Wearing shoes is a simple but effective control measure. Infection may also occur on the soles of the feet, the buttocks of people (e.g. children, beggars) who habitually sit on the ground, the elbows and sometimes the arms; heavy infections may occur in leprosy patients. Female fleas embedded in the skin should be removed with a fine needle (preferably within a few days of infection) and precautions should be taken to prevent secondary bacterial infection.

The Compromised Host

Many different external and internal factors can adversely affect the balanced relationship between man and micro-organisms, so that the host becomes excessively liable to infection by recognized pathogens and also susceptible to attack by less virulent organisms that seldom or never harm people in normal health (opportunist pathogens, p. 31). Conditions that compromise host resistance in this way include those that allow organisms to by-pass the host's external defences (burns and other major skin lesions; and the many invasive procedures of modern medical care); diseases that cause general debility; defects of the immune system; and diseases or treatments that cause immunosuppression. The major importance of the compromised host in modern medicine is largely the result of technological and therapeutic advances. These have led to increased use of catheterization and other invasive procedures, to greatly increased life-expectation of patients with leukaemias and other malignant and debilitating diseases, and to frequent employment of cytotoxic or other immunosuppressive drugs. It is of course common for patients to have combinations of such compromising factors—e.g. a debilitating disease treated with immunosuppressive drugs and various invasive procedures; and these patients are often located in intensive care units with their special infection hazards (p. 350). It is also sadly common for patients to be successfully treated for their underlying conditions and then to succumb to infections resulting from their compromised state. It is therefore important that clinicians and microbiologists should know what forms of infection to suspect in particular groups of compromised patients; and that any suspicion of a developing infection in such a patient should lead to careful clinical examination including common sites for 'silent' infections; to radiological examination of the chest and other appropriate regions; to the sending of blood, sputum (if any), urine and appropriate swabs to the microbiology laboratory; and to prompt appropriate treatment.

We shall deal in Chapter 27 (under Hospital Problems) with some of the factors that allow organisms to by-pass the patient's external defences—wounds (p. 349), burns (p. 350), peritoneal dialysis (p. 351), injections and infusions (p. 351), urinary catheters (p. 352) and endotracheal and tracheostomy tubes (p. 352)—and with other aspects of the prevention of infection of patients who, being in hospital, are likely to be some degree compromised. In this chapter we shall consider immune deficiency states and a few other common causes of compromised host resistance.

Immune Deficiency States

Under this heading we include defects in phagocytosis as well as those in the antigen-specific immunity dependent on B and T lymphocytes (see Chapter 6). Much of our understanding of the role of immunological mechanisms in defence against infection has been derived from study of what happens when they are defective.

Primary immune deficiency states, genetically determined in many cases and manifested either in infancy or later in life, are rare but have been

particularly instructive. They can be classified as follows:
(1) Deficiencies of neutrophils or monocytes, resulting in inadequate phagocytosis and killing of micro-organisms.
(2) Deficiency of B lymphocytes and so of immunoglobulins—hence the name **hypogammaglobulinaemia.**
(3) Deficiency of the stem-cells from which both B and T lymphocytes are derived, with resultant deficiency of both cell types.
(4) Complement deficiencies.
(5) Thymic hypoplasia and consequent T lymphocyte deficiency.

We have already indicated, when discussing the immunological mechanisms in Chapter 6, the types of infection to be expected in patients with these deficiencies. Fuller lists of pathogens to be expected are given in Table 24.1, and some of the more important infections are discussed later in this chapter.

Important though the primary deficiencies have been to our understanding, they are far less common than **secondary immune deficiency** states, due to malnutrition, underlying malignant or other diseases or to the use of immunosuppressive or cytotoxic drugs.

Neutrophil Polymorph Deficiencies

The importance of neutrophils in defence is made clear when they are deficient either in numbers (neutropenia) or in quality.

Neutropenia

This has many causes, of which the commonest are acute leukaemia, aplastic anaemia and bone-marrow suppression following marrow transplantation or after use of intensive cytotoxic drug treatment for malignant disease. The likelihood and severity of infection are inversely correlated with the number of neutrophils in the circulating blood, becoming particularly high when that number falls below about 500/mm^3. The types of organism commonly responsible for these infections are shown in the first column of Table 24.1, and are usually endogenous. Infections are liable to develop rapidly in neutropenic patients and to be hard to recognize because of the lack of normal inflammatory responses. Laboratory confirmation of infection is also made difficult by the absence of the cellular responses that characterize acute infection in other patients—a blood neutrophil rise and the presence of 'pus cells' in infected urine or other body fluids. Because severely neutropenic patients are liable to be rapidly overwhelmed by such uncharacteristic infections, their temperatures should be regularly monitored; and any rise should provoke thorough microbiological screening for possible infecting agents (including the whole range relevant to such situations) and prompt empirical broad-spectrum antibiotic therapy, using cidal drugs (p. 377) because of these patients' own restricted ability to kill organisms. Lack of a prompt response to antibacterial drugs may be due to a fungal infection, for which amphotericin or one of the imidazole compounds should be used. Granulocyte transfusions may be necessary, and severe cases may require protective isolation together with carefully selected antibacterial treatment to reduce their endogenous flora (p. 348). In general, however, the normal flora of patients who have defective cellular or humoral defence mechanisms should be treated with respect, since its protective value is enhanced in such circumstances; and antimicrobial drugs should be used with as much restraint and precision as are compatible with the points already made in this paragraph about the difficulty of recognizing infections and the need to treat them urgently.

An unusual opportunist pathogen that sometimes causes infections in neutropenic patients is *Capnocytophaga ochracea*. This is a CO_2-dependent gram-negative bacillus. Infection spreads from periodontal foci to cause mouth ulcers and bleeding from the gums, and may develop into potentially fatal septicaemia.

Functional deficiencies of neutrophils

Diseases in which neutrophil polymorphs are normal in numbers but defective in some of their phagocytic functions (chemotaxis, opsonization, digestion, killing) include:

(1) The **lazy leukocyte syndrome**, in which a failure of the cells to respond to chemotactic

TABLE 24.1 Important Pathogens in Patients with Different Kinds of Immunodeficiency

Neutropenia	B lymphocyte (antibody) defects	T lymphocyte defects
Staphylococci	Pneumococci	Measles virus
Faecal streptococci	Haemophilus influenzae	Herpes simplex virus
Pneumococci	Streptococci	Varicella-zoster virus
Pseudomonas	Meningococci	Cytomegalovirus
Klebsiella	Pseudomonas	Mycobacteria
Proteus	Hepatitis viruses	Listeria monocytogenes
Serratia	Enteroviruses	Legionella pneumophila
Candida	Mycoplasma pneumoniae	Candida
Aspergillus	Giardia intestinalis	Aspergillus
Cryptococcus neoformans		Mucor
Histoplasma capsulatum		Cryptococcus neoformans
Mucor		Pneumocystis carini
		Toxoplasma gondi
		Cryptosporidium
		Isospora belli
		Strongyloides stercoralis

A generic name by itself means various species in that genus.

stimuli results in recurrent bacterial infections, notably gingivitis, stomatitis and otitis media.

(2) The **Chediak–Higashi syndrome**, an autosomal recessive condition characterized by partial oculocutaneous albinism and by the presence in the blood of abnormal neutrophils with giant lysosomes. These cells are defective in most aspects of phagocytic function, and the patients suffer from frequent pyogenic infections.

(3) The **hyperimmunoglobulin E syndrome**, in which a high serum level of IgE is coupled with defective neutrophil chemotaxis. Recurrent bacterial infections result.

(4) **Chronic granulomatous disease** (CGD), an X-linked or autosomal recessive condition in which phagocytosis is normal but the neutrophils (and also the macrophages) have a metabolic defect that prevents them from killing certain bacterial and fungal species. These organisms can therefore survive intracellularly and cause the granulomas (p. 267) which give the condition its name. A paradox of this disease is that, although inability to generate H_2O_2 is the defect that prevents the cells from killing ingested micro-organisms, the species that can cause chronic indolent infections in these patients are those that produce the H_2O_2-destroying enzyme catalase (e.g. those of the genera *Staphylococcus, Klebsiella, Serratia, Candida, Aspergillus* and *Nocardia*). This is because catalase-negative organisms (e.g. streptococci) cannot inactivate the H_2O_2 which they themselves produce, and so they commit a form

of suicide inside the phagocytes. CGD presents in early childhood with superficial 'cold' abscesses and associated local lymphadenopathy, followed by the development of deep-seated granulomas throughout the body, generalized lymphadenopathy and hepatosplenomegaly. The diagnosis can be confirmed by the nitroblue–tetrazolium (NBT) test; normal neutrophils can reduce NBT to a blue dye, but those from CGD patients cannot do so. Treatment is long-term prophylactic administration of bactericidal antibiotics, but many patients die before reaching adult life.

Mononuclear Phagocyte Deficiencies

Macrophage defects in CGD have been mentioned above, but this is rare. Since much of the mononuclear phagocyte (reticuloendothelial) system is located in the liver and spleen, the common causes of deficiency of this system are diseases of those organs or removal of the spleen, either of which can severely compromise the body's ability to eliminate micro-organisms from the blood-stream. Patients with chronic liver disease are liable to develop low-grade infections with opportunist organisms; and these are often accompanied by hypergammaglobulinaemia, since the damaged liver fails to detoxify endotoxin absorbed from the gut and allows it to pass through to the spleen, where its non-specific stimulation of B lymphocytes leads to overproduction of unwanted immunoglobulins. A well-recognized complication of splenectomy is fulminant septicaemia, which is commonly fatal. Patients whose spleens have been removed for treatment of thalassaemia or congenital spherocytosis run the greatest risk of such infections; they are also fairly common in patients with lymphoreticular neoplasms. Pneumococci are the commonest causes of post-splenectomy septicaemia, with other streptococci, meningococci and *Haemophilus influenzae* (often type b) as the most frequent alternatives. It is therefore wise to give polyvalent pneumococcal vaccine and long-term penicillin prophylaxis to the high-risk patients. Infection with the protozoan *Babesia bovis*, transmitted from wild or domestic animals by ticks, is a possible sequel of splenectomy in areas where this organism is endemic (p. 151). Babesiosis presents with fever, haemolytic anaemia and haemoglobinuria, and can be diagnosed by finding the parasites inside red cells in blood smears; it is commonly fatal in splenectomized patients.

Antibody Deficiencies

By far the commonest form of primary antibody deficiency, with an incidence around 1 in 700 people, is **selective IgA deficiency.** This is usually asymptomatic, but in a few people it results in recurrent respiratory tract infections. The problem is lack of surface (secretory) IgA (p. 53); and the patients with recurrent infections do not benefit, as do those with other immunoglobulin deficiencies to be described below, from injections of normal immunoglobulins, since these are not transmitted from the serum to the secretions.

A more serious disease due to primary antibody deficiency is **X-linked hypogammaglobulinaemia (Bruton type).** Children with this condition usually have their first serious infections at about 4–6 months old—i.e. soon after losing the protection of maternal antibodies—and commonly develop either pneumonia or meningitis caused by *H. influenzae* or pneumococci. Their serum IgG levels are usually below 2 g/litre, and other immunoglobulin classes are either absent or at very low concentrations. Such children have no general increase in susceptibility to virus infections, but enteroviruses are exceptions to this—possibly because the children lack secretory IgA in the gut. In particular they are prone to develop hepatitis A or poliomyelitis or a dermatomyositis-like illness associated with echovirus infection of the CNS. They can be protected from bacterial infections by regular injections of pooled normal human immunoglobulin.

Similar infections occur in patients with **late-onset primary hypogammaglobulinaemia,** but usually not until they are young adults. Other infections that they are liable to develop include: persistent *Giardia intestinalis* infection with malabsorption; fulminating (untreatable) cryptosporidial infection with profuse watery diarrhoea, malabsorption, weight loss and in many cases death (cf. AIDS, p. 283, and see p. 254); acute or chronic rotavirus infections; and septic arthritis due to *H. influenzae*, pneumococci or *Mycoplasma pneumoniae*.

In **severe combined immune deficiency states** (e.g. **Swiss-type hypogammaglobulinaemia**) there is a lack of both antibody production and cell-mediated immunity—i.e. both B and T lymphocytes are deficient. Affected children are unable to mount any effective immune responses, suffer from frequent severe infections and usually die of overwhelming infection before reaching 1 year old. Transplantaion of HLA-matched bone marrow, preferably from a sibling, now offers some hope in this grim situation, but care has to be taken to prevent reaction of the grafted cells against host tissues (graft-v.-host disease).

Secondary **hypogammaglobulinaemia,** with consequent increased liability to infection, can result from unrestricted proliferation of one particular cell-line to the detriment of B lymphocytes in general (e.g. in chronic lymphatic leukaemia or in myeloma); from hypercatabolism (e.g. in myotonic dystrophy); or from excessive immunoglobulin loss (e.g. in protein-losing enteropathy or in nephrotic syndrome).

T Lymphocyte Deficiencies

T cell deficiency can lead to chronic severe infections, particularly with viruses, fungi or protozoa (Table 24.1), as a result of lack of cell-mediated immunity, and associated deficiency of T helper function (p. 54) adversely affects antibody production, notably that of IgA and IgG. Primary deficiency of T lymphocytes results from either of two rare forms of congenital thymic hypoplasia—**DiGeorge syndrome,** in which there is hypoplasia of both the thymus and the parathyroid glands, and **Nezelof's syndrome,** in which parathyroid development is normal. Both lead to recurrent pulmonary infections, candidiasis and in many cases severe varicella–zoster or cytomegalovirus infection. Serum immunoglobulin levels are normal.

Complement Deficiencies

Despite the important role of complement in microbial killing (p. 57), infections are rare features of congenital complement-deficiency diseases. Those which result from C3 deficiency and C5–C8 deficiency are indicated on p. 57.

Infection as a Cause of Immunosuppression

It has long been recognized that some virus infections, notably measles, and some heavy parasitic infections can cause non-specific suppression of the immune mechanisms and open the way to secondary invasion, in some cases by opportunist organisms. Detailed research into the mechanisms of AIDS has given greater precision to this concept; in this case at least, the virus achieves its immunosuppressive effect by attacking T helper (T4) lymphocytes, and so impairing cell-mediated immunity and antibody production (p. 58). The consequent infections are described on p. 283.

Other Factors That Cause Immune Deficiency

Malnutrition

In the world as a whole this is the commonest cause of secondary immune deficiencies. It operates not only where there is famine but even in such apparently improbable places as the wards of major hospitals, where patients with serious diseases or unable over long periods to take normal diets may develop unsuspected but important nutritional deficiencies. Vitamins and many other dietary components are important to the body's defences, and protein deficiency has an obvious relevance to defence mechanisms that depend on rapid production of immunoglobulins. Malnutrition, impaired immunity and infection can become an inextricable triad, each of which can exacerbate the others. Most immunological defects of nutritional origin can be corrected by providing an adequate diet; but fetal malnutrition may result in long-term immunosuppression. If vaccination of previously malnourished children is to be effective, it must be accompanied by adequate supplementation of their diet. In severe protein-energy malnutrition there is marked involution of the thymus, and consequent lack of T-cell-mediated immunity; this is reflected in the frequency with which children suffering from such malnutrition die of tuberculosis, herpes simplex or measles. Measles in particular is a scourge in these circumstances, and has special clinical features—notably absence of rash (since this is a T-cell-mediated manifestation of the infection) and frequent presence of giant-cell pneumonia caused by the virus.

Hormones

There is a close developmental and functional association between the immune system and the neuro-endocrine network. Even psychosocial factors can alter immunity. For example, it is suggested that stress stimulates the hypothalamic–pituitary axis, causing increased ACTH secretion and a consequent rise in circulating corticosteroid levels. Corticosteroids, produced in excess in this way or given therapeutically, may be immunosuppressive and increase susceptibility to some infections.

Immunosuppressive drugs

Iatrogenic immunosuppression by drugs is a common cause of secondary immune deficiency in modern medical practice. Cytotoxic drugs—e.g. vincristine, 6-mercaptopurine, cytosine arabinoside, methotrexate—and corticosteroids are used for deliberate suppression of immunity in patients undergoing organ transplantation to prevent rejection of the transplant by the recipient's immune system. The same drugs are used in chemotherapy of some cancers and of leukaemias because their main targets are rapidly dividing cells, i.e. the malignant cells. However, normal polymorphs and lymphocytes are produced by other rapidly dividing cells, and immunosuppression by reducing their production is an inevitable side-effect of cytotoxic therapy. Patients receiving cytotoxic therapy or corticosteroid drugs for any reason are susceptible to a wide range of opportunist infections. They readily acquire exogenous infections, often with low-grade pathogens such as *Klebsiella*, *Enterobacter* and *Candida* or with recognized pathogens that produce particularly severe infections in these high-risk patients. Furthermore, suppression of the normal immune mechanisms allows reactivation of latent infections with viruses (CMV and other herpesviruses), protozoa (*Pneumocystis*, *Toxoplasma*), fungi (*Cryptococcus*), bacteria (*Myco. tuberculosis*) or helminths (*Strongyloides*). Acute bacterial infections are due to polymorph suppression (neutropenia), whereas reactivation of latent infections is the result of reduced T cell immunity.

Other Compromising Factors

Failure of Superficial Defences

If the superficial defence systems described on pp. 46–7 are defective, potential pathogens have increased opportunities to reach the underlying tissues in infective doses. Patients with large areas of raw skin (e.g. from exfoliative dermatitis or burns) are liable to develop serious infections including septicaemia (p. 60), and their resistance may be additionally impaired by loss of protein through their skin lesions. Immotility of the respiratory-tract cilia in Kartagener's syndrome is responsible for the recurrent and chronic infections of the paranasal sinuses, middle ears and lower respiratory tract which are a prominent part of the syndrome; and a somewhat similar situation arises, far more commonly, when previously normal ciliate epithelium is damaged by virus infections (notably influenza or measles) or by cigarette smoke or other toxic factors, with resultant liability to secondary bacterial pneumonia. Failure to clear excessively viscid mucus from the lower airways predisposes to bacterial infection in cystic fibrosis (p. 204).

Metabolic Disorders

Diabetics are excessively liable to infection, and part of this susceptibility is attributable to excessive amounts of glucose in, for example, the vaginal secretions and the urine—in both of which sites *Candida albicans* is consequently able to thrive; but this is by no means the whole story. Rheumatoid arthritis and renal failure are among the other diseases in which there is an increased susceptibility to many sorts of infection, probably due to a combination of abnormal tissue metabolism and impaired immune responses. In acute renal failure there is an increased risk of septicaemia caused by enterobacteria. In chronic renal failure the underlying increased susceptibility to infection increases the hazards of invasive procedures used in treatment—e.g. arteriovenous shunts or fistulae, through which *Staph. aureus* can reach the blood-stream and cause septicaemia (with a mortality rate around 20%), and chronic ambulatory peritoneal dialysis (p. 351).

Implantation of Foreign Bodies

The increasingly frequent implantation of foreign materials, in the form of prostheses, into the human body has created a range of new infection problems. Patients with such prostheses can reasonably be classified as compromised hosts, since the prostheses provide unusual ecological niches with no capillary blood supply, in which micro-organisms can establish themselves virtually out of reach of the host's cellular and humoral defences. *Staph. epidermidis* is the species that most frequently colonizes such inserts, at least in part because it is commonly introduced from the skin during the process of insertion. When a Spitz–Holter, Hakim or other similar valve is used for draining CSF from the ventricles of a patient with hydrocephalus, it is liable to infection of this kind and the infecting organisms have ready access to the meninges, as well as to the blood-stream if the valve is part of a ventriculo-atrial shunt. They may therefore cause meningitis or septicaemia—relatively subacute in onset as a rule if the organism is a coagulase-negative staphylococcus, but liable to be far more dangerous if it is a gram-negative bacillus or *Staph. aureus*. Antibiotic treatment, lacking the support of body defences, is virtually always ineffective for eradication of such an infection, and the prosthesis usually has to be removed. The same is true of infected joint prostheses (p. 301) and artificial heart valves (p. 237). A closely similar type of infection can develop on intravenous catheters (p. 351).

Drug Addiction

In addition to any impairment of their resistance to infection that results from the action of the drugs themselves, drug addicts commonly expose themselves to serious infection hazards by using contaminated syringes and needles for their injections. In this way they may acquire hepatitis (B or non-A non-B, p. 283) or AIDS (p. 128), but they may also inject into their blood-streams infective doses of many kinds of potential pathogens, ranging from the common pyogens to unusual organisms that have very few other opportunities to invade the human body. One of the most serious forms of infection in drug addicts is endocarditis, frequently affecting the tricuspid valve and most often due to *Candida albicans* or *Staph. aureus*—the latter being in many cases readily available for injection because the patients have multiple infected puncture wounds.

Age

As we shall see more fully in the next chapter, the fetus and the neonate are to some degree compromised hosts, in that their defences are immature. At the other end of life the efficiency of these defences may begin to fall off, and various other factors contribute to the increased susceptibility of the elderly to infections of many kinds—notably poor circulation of blood (leading to tissue anoxia and to impaired access by protective cells and antibodies), degenerative changes in vital organs, underlying chronic or malignant disease, dietary deficiencies in many cases, and sometimes hypothermia.

Reactivation of Latent Infections

Anything that decreases host resistance may allow a latent infection to become reactivated. With herpes simplex this can follow quite minor changes in the host's state (p. 131). Tuberculosis, before it could be controlled by drugs, was commonly 'lit up' by other illnesses or a deterioration in general health. Use of immunosuppressive drugs to prevent graft rejection may allow activation of latent cytomegalovirus infection in host or donor tissue. Many other examples could be given.

25
Congenital and Neonatal Infections

During pregnancy the fetus develops in a sterile environment, where no infection can reach it except via the mother. Infections acquired by the mother, before or during pregnancy, are commonly kept away from the fetus by her defence mechanisms; and the placenta also provides an impassable barrier to many potential pathogens. When infection does reach the fetus, the outcome depends on the nature of the organism, the stage of development of the fetus and of its immune mechanisms, its genetic susceptibility and the level of relevant maternal antibodies in its circulation. Development of the fetal immune system depends on the general health and nutritional state of the mother, and may be seriously and permanently impaired by intra-uterine malnutrition.

At and immediately after birth the neonate is confronted by a wide range of micro-organisms acquired from its mother, attendants and the general environment. Many of these organisms become components of its own commensal flora, and may be of protective value (p. 29), but others are potential pathogens. Maternal IgG antibodies transmitted across the placenta are an important part of the infant's immunological defence against these unfamiliar enemies; but transfer of such antibodies takes place mostly in the last few weeks of pregnancy, and therefore premature infants are not well supplied with them. This problem of prematurity is compounded by the low level of the premature infant's own ability to synthesize immunoglobulins. Even the full-term infant, though its immune system contains all elements necessary for response to and defence against infection, responds slowly—in part because it lacks the priming effect of immunological memory (p. 52)—and its production of antibodies is often both qualitatively and quantitatively inadequate. As for cellular defences, neutrophil polymorphs are usually produced in adequate numbers but are functionally immature; in particular, their chemotactic response is poor and they have limited ability to phagocytose micro-organisms, probably because the serum of a neonate is lacking in opsonizing immunoglobulins and in complement. Circulating lymphocytes are also produced in numbers comparable to those in older children, but neonatal T lymphocytes are predominantly suppressors (p. 54); this results in weak cell-mediated responses and in limited IgG and IgA production, since this process requires help from T lymphocytes.

Congenital Infections

Under this heading we include all infections acquired *in utero*, whether they have been eliminated before birth or are still active then. Intra-uterine infections can result in abortion or still-birth; or in the birth of a live but damaged child; or leave no lasting effect, so that the child is born normal and healthy. Infection is responsible for only a small proportion of all congenital defects, but this minority is of particular importance because it would not occur if such infections were prevented. The range of clinical features of congenital infections includes potentially reversible abnormalities such as low birth-weight, rashes, jaundice, hepatosplenomegaly and thrombocytopenia; developmental defects that result in deafness, blindness or congenital heart disease; and microcephaly, hydrocephalus and intracranial calcification, with associated brain damage and mental handicap. Some of these features are immediately apparent at birth, others may be noticed in the first few days, and still others may not be detected for months or even years.

Congenital **syphilis**, once common, has been almost eliminated in countries with health services capable of carrying out routine serological screening of pregnant women during the first trimester and of providing for prompt detection and treatment of sexually transmitted diseases. The acronym TORCH is used as a reminder of the four infectious agents that now have to be considered first when investigating a case of suspected congenital infection—TOxoplasma, Rubella virus, Cytomegalovirus and Herpes simplex virus; though herpes simplex is strictly a neonatal infection, acquired at birth, far more often than a congenital infection. Appropriate investigations when a congenital infection is suspected include: full clinical examination; full blood picture, including platelet count; liver function tests; ophthalmoscopy; radiology of the skull and long bones; attempts to isolate the relevant viruses from the placenta and from the infant's faeces, urine, CSF and throat; and appropriate serological tests on maternal and cord blood.

The common congenital infections mentioned above are now considered one by one in this section; but herpes simplex is deferred to the section on neonatal infections.

Syphilis (pp. 278–9)

Treponema pallidum infection of a fetus occurs when the mother has untreated primary or secondary syphilis at the time of conception or acquires the infection during pregnancy; the spirochaetes can cross the placenta. The prognosis for the fetus is excellent if effective treatment with penicillin or another suitable antibiotic is given to the mother, and therefore also to the fetus, soon after the latter becomes infected. Delay or inadequacy of treatment increases the chances that the fetus will suffer the consequences that inevitably follow untreated intra-uterine syphilis—still-birth, or premature birth of a heavily infected baby. Such widespread congenital infection by treponemes causes hepatosplenomegaly and jaundice, lymphadenopathy, nasal discharge ('snuffles') and skin rashes in the early neonatal period, followed later (often years later) by characteristic abnormalities of development of bones, teeth, eyes and skin and often by deafness. Histological examination of fatal cases shows large numbers of *T. pallidum* in the tissues, especially in liver, lung and bone-marrow. The most informative laboratory finding when the infant survives is the presence of anti-treponemal IgM (detected by the FTA test—p. 182) in its serum at birth or in the next few weeks; this indicates that the infection has spread to the infant since maternal IgM does not cross the placenta.

Toxoplasmosis (p. 151)

Congenital toxoplasmosis complicates about 0.5% of pregnancies in Britain. It results from maternal infection acquired during pregnancy, with transplacental transmission to the fetus. If this occurs early in pregnancy, still-birth may result, or the birth of a live baby with a disseminated infection that can cause choroidoretinitis, microcephaly or hydrocephalus, intracerebral calcification, hepatosplenomegaly and thrombocytopenia. Maternal infection during the third trimester can also be transmitted to the fetus, but at this stage of development it usually causes no damage. Maternal infection is commonly subclinical, and unnoticed unless serological screening is carried out. A rise in the mother's toxoplasma antibody titre during pregnancy, or the finding that she has IgM antibodies (indicating recent infection), raises the question whether to treat the infection (see below), which does not guarantee that the infant will be unaffected, or to terminate the pregnancy, even though it is not certain that the fetus has been damaged. As with congenital syphilis, the sooner after fetal infection the treatment is started the better the prognosis for the fetus. Spiramycin is the drug of choice for treatment of the mother and her fetus. An infant with congenital infection should be given alternating courses of sulphadiazine+pyrimethamine+folinic acid and of spiramycin for the first year, with prednisolone added if there is choroidoretinitis.

Rubella (p. 123)

This disease is generally mild and unimportant when acquired during childhood; but when contracted by a pregnant woman it can have serious effects on the fetus, especially if the infection occurs during the first trimester. It may lead to abortion, miscarriage or still-birth, or to the birth of a live infant with multiple abnormalities. In Britain 10–15% of women of child-bearing age are still susceptible to rubella, despite the availability of an effective vaccine (p. 369). In the absence of a protective level of antibodies in the mother's blood the virus can reach the fetus and damage it, but its chances of doing these things depend on the stage of pregnancy. In the first month of pregnancy infection involves the placenta in 85% of cases and reaches the fetus in 50%; and interference with fetal tissue differentiation at this early stage can cause major abnormalities. The likelihood of the virus reaching the fetus, and its chances of doing serious damage if it does get there, decrease as pregnancy progresses, and infections after the 20th week rarely cause defects. The most severely affected infants (usually those infected early in pregnancy) have gross abnormalities that are apparent at birth—often a low birth weight (less than 2500 g), and such defects as cataracts, congenital heart disease, microcephaly, hepatosplenomegaly and thrombocytopenia. Infection at any time during the first 20 weeks of pregnancy may cause less obvious defects, which may not become apparent until some time after birth; these include nerve deafness, mental retardation, seizures and choroidoretinitis. Infected babies may be profuse excreters of virus during their early months of life and constitute a potential hazard to other non-immune pregnant women and their fetuses. About 30% of infants born with congenital rubella die within a few months, mostly from pneumonitis or myocarditis. Survivors often require to attend special schools because of deafness or mental handicap.

Since maternal rubella is often subclinical, or causes only a rash that is easily attributed to some other cause such as an allergic reaction, and since rashes from other causes can easily be misdiagnosed as rubella, serological investigation of all suspected cases is highly important (p. 123). Pre-natal screening identifies the large proportion of women who need have no anxiety about rubella and the minority who should be immunized. In the absence of pre-natal information, screening at the beginning of pregnancy can achieve similar purposes except that those who are susceptible must not be immunized while they are pregnant (p. 369); and it also provides a base-line for serological assessment of possible rubella infection later. Confirmation of rubella infection early in pregnancy, by demonstrating a rise in titre of rubella antibodies or the appearance of specific IgM in the mother's serum or by isolation of the virus from her throat (or, far less commonly, from urine, faeces or CSF), suggests a high risk of serious abnormality of the infant if the pregnancy proceeds to term.

Cytomegalic Inclusion Disease (CMV Infection) (p. 132)

Intra-uterine cytomegalovirus (CMV) infection has a frequency similar to that given for toxoplasmosis in Britain. As with toxoplasmosis and rubella, fetal infection may follow primary and commonly subclinical maternal infection during pregnancy; but it can also follow reactivation of a latent CMV infection acquired earlier. As with rubella, infections in early pregnancy are the most serious; but CMV infection at any time in the second or third trimester may cause congenital defects. Another important difference between CMV infection and the other congenital infections that we have described is that it is neither treatable nor preventable. Fortunately, only about 5% of infected fetuses develop abnormalities that are apparent at birth or soon after; another 10% have neurological sequelae that become apparent later in life, including sensorineural hearing loss in some cases. The major abnormalities are similar to those listed for other congenital infections— low birth weight, hepatosplenomegaly and jaundice, choroidoretinitis, microcephaly or hydrocephalus and intra-cerebral calcification; but the association of periventricular calcification with microcephaly or obstructive hydrocephalus is strongly suggestive of CMV infection. The more severely affected infants die early in the neonatal period or have severe mental and neurological handicaps. Detection of maternal CMV infection is of limited value because there is little that can be done about it. CMV can be isolated from the urine, saliva, throat or CSF of infected infants, but less than 30% of them have the typical 'owl's eye' inclusions in cells in their urine (p. 132). Serological evidence of infection consists of the demonstration of specific IgM in the blood at birth or subsequently, or of a persistent or rising level of CMV antibodies (indicating their production by the infant rather than their transplacental acquisition).

CMV infection can also be acquired from an infected mother **at or after birth,** as she may be excreting the virus from various sites during either primary or recurrent infection. Her cervical secretions and her milk are the most likely sources of infection of the infant. Most of the infected infants remain asymptomatic, and although some develop pneumonitis none have neurological or other late sequelae of their infections. This may be because maternal IgG in the infant's circulation limits dissemination of the infection. More serious neonatal infections sometimes occur in infants, most of them premature and of low birth weight, who are infected via contaminated blood or via milk from a human breast-milk bank and who have no maternal antibodies to protect them. Babies with perinatal CMV infections need to be isolated if they stay in hospital, as after 3–12 weeks they begin to excrete the virus and can be sources of infection for susceptible pregnant staff members and for other infants with no CMV antibodies, especially those in special care units.

Other Congenital Infections

Congenital abnormality due to intra-uterine infection with **varicella-zoster** (VZ) virus is extremely rare; it results from maternal infection in the first trimester, and its characteristic features are encephalitis, low birth weight, hypoplastic limbs, skin deformities and choroidoretinitis. Maternal infection in the last few days before or within two days after delivery is associated with transplacental transmission to the fetus, which can result in severe and sometimes fatal illness in the neonatal period. As it is not certain that the baby has been infected *in utero*, it should be separated from the mother; and as there has been no time for her to make and transfer protective antibodies, the baby should be given zoster immune globulin (ZIG).

Tuberculosis acquired *in utero*, or by inhalation of infected amniotic fluid during delivery if the mother has genital tuberculosis, remains a problem in many parts of the world where this disease has not been brought under control. Fetal infection early in pregnancy may cause abortion, still-birth or severe congenital disease, whereas at the other end of the scale infants who became infected during the third trimester may be apparently healthy at birth, but may develop tuberculous pneumonia or liver disease later, perhaps after several months. Tracheal and gastric aspirates, CSF, urine and if necessary a liver-biopsy specimen should be collected from an infant suspected of having tuberculosis, and should be examined for tubercle bacilli by microscopy and culture.

Listeriosis in pregnancy may cause only a mild influenza-like illness in the mother, but may result in abortion or still-birth, or in the birth of a baby with bacteraemia, meningitis and other evidence of widespread infection.

Among the protozoan diseases, **trypanosomiasis** is occasionally transmitted *in utero*, and may result in death of the infant within a few months of birth. Congenital transmission of **malaria**, of any of the four types, is less rare, though still uncommon. It is most likely to occur when a woman with no previous immunity acquires falciparum malaria during her pregnancy; in such circumstances the placenta is often heavily loaded with infected erythrocytes, and it seems that fetal infection is by transfer of these intra-erythrocytic plasmodia, presumably made possible by some damage to the placental barrier. Such infection may result in abortion, or in the premature birth of a small baby who may be afebrile and free from evidence of malaria for the first few weeks or months, probably because the infection is suppressed by maternal antibodies. Congenital or neonatal malaria should be treated with appropriate doses of the standard antimalarial drugs (p. 242). Even though the risk of placental transmission of malaria is low, all pregnant women in endemic areas should have their blood examined for malarial parasites and should be treated if any are present, since the physiological stresses of pregnancy may cause a flare-up of the infection and enough erythrocyte destruction to threaten the life of the mother or to provoke abortion.

Neonatal Infections

The last half-century has seen a series of changes in the relative prominence of different micro-organisms as recognized causes of neonatal sepsis, and these changes have taken place against a background of major advances in neonatal care and in laboratory diagnosis and of the development of new antimicrobial agents. Perinatal mortality has declined steeply during the period, and deaths are now mostly due to congenital abnormalities incompatible with life, to respiratory problems and to infections in very small (less than 1250 g) premature babies.

Before the antibiotic era group A β-haemolytic streptococci commonly caused puerperal fever in mothers and were the prominent cause of neonatal sepsis. Then, after the arrival of penicillin, gram-negative bacilli such as *Esch. coli* and klebsiellae took the lead. In the 1950s *Staph. aureus* was responsible for many outbreaks of infection in neonatal units (as well as other parts of hospitals), but by the 1960s the gram-negative bacilli had reasserted themselves and accounted for 70% of all bacterial isolates from the blood of neonates. The 1970s saw the emergence of group B β-haemolytic streptococci (previously regarded as relatively unimportant organisms) as neonatal pathogens comparable in importance with *Esch. coli* in many units. The use of broad-spectrum antibiotic treatment aimed at covering these various possibilities may have been a contributory factor in the trend towards increasing importance of *Staph. epidermidis* (coagulase-negative staphylococci, often resistant to many antibiotics) in neonatal sepsis.

Septicaemia and Meningitis

The incidence of neonatal **septicaemia**, often with associated meningitis, is of the order of 1 to 5 cases per thousand live births in many communities. It is highest among small premature infants with complex medical or surgical problems. The outcome depends on the nature of the organism responsible, the maturity of the infant and whether there is meningitis, but reported mortality rates are high (25–35%) and call for urgent attention to prophylaxis, early diagnosis and vigorous appropriate treatment.

The organisms now most commonly responsible for septicaemia in the first two days of life are group B streptococci and *Listeria monocytogenes*. After this immediate postnatal period *Esch. coli*, klebsiellae and other gram-negative bacilli predominate; but as we have already indicated, *Staph. epidermidis* is an increasingly common opportunist invader in these young infants, and its isolation from a blood-culture should not lead to its being discounted out of hand as a skin contaminant (though it often is so). It is particularly liable to invade small compromised infants, notably those who are on ventilators or receiving parenteral nutrition or who undergo extensive surgery. *Candida albicans* is also being increasingly recognized as causing invasive disease in such infants; some 10% of infants acquire this species during birth from their mother's genital tracts, but colonization rarely progresses to septicaemia except in those who are compromised.

Septicaemia developing during the first day or so of life is likely to be manifested as shock, apnoea or progressive respiratory failure that is easily mistaken in a premature infant for hyaline membrane disease. Because delay in treatment, especially of group B streptococcal infection, may be fatal, respiratory difficulty in such a young infant should immediately raise suspicion of septicaemia. Blood, CSF, urine and swabs from the throat, the umbilicus and the external auditory canal (which may contain infected amniotic fluid) should be sent immediately to the microbiology laboratory, and antibiotic treatment should be started at once and should include penicillin or some other agent reliably effective against group B streptococci. Septicaemia that develops 48 hours or more after birth is less likely to present as shock and respiratory distress; the clinician has to be on the look-out for subtle changes in muscle tone, colour or feeding pattern. Again antibiotic treatment should be started promptly, after appropriate specimens have been sent for bacteriological examination. It may be possible to base the choice of drugs on the findings of immediate microscopy or rapid antigen-detecting procedures; the age of the child and the known prevalence of pathogens in the unit and their antibiotic-sensitivity patterns will also influence the choice.

Meningitis occurs in up to 30% of septicaemic neonates, but may initially be difficult to diagnose because there may be no signs pointing to CNS involvement; the presenting features may be merely poor feeding, vomiting and loss of muscle tone. This is the reason for routine examination of CSF in all neonates with suspected septicaemia. *Esch. coli*, group B streptococci and *List. monocytogenes* are responsible for most cases of meningitis in the first week of life, and the latter two for most of those occuring later in the neonatal period. Neonatal meningitis is more fully discussed in Chapter 16 (p. 216).

Escherichia coli

This species (its K 1 serotype in most cases— p. 91) is responsible for the great majority of all cases of neonatal septicaemia and meningitis caused by gram-negative bacilli. The source of the septicaemia is usually a primary focus in the respiratory, gastro-intestinal or urinary tract.

Group B streptococci (p. 78)

These organisms are carried in the vagina and rectum by many women. Colonization may be persistent, intermittent or transient. Studies of the vaginal carriage rate during the third trimester of pregnancy have given widely varying results, but show that swabbing during this period is of little value in predicting which women will be carrying these organisms at term. In general some 15–20% of women are carriers at term, and about one-third of babies born to these carriers are colonized at birth, with a greater likelihood of colonization if maternal carriage is heavy. The risk of colonization leading to neonatal illness is increased by obstetric complications, notably by prolonged rupture of the membranes or by premature labour; and it is greatest for babies who weigh less than 2500 g at birth and whose mothers have no antibodies to the capsular polysaccharide of group B streptococci and therefore cannot confer passive immunity on their infants. Two distinct forms of neonatal group B infection are recognized. In the acute or early onset form the infection is acquired as above from the mother during birth, and the infant develops shock and pneumonia (but only rarely meningitis) during the first day or so of life. The late onset form is less fulminant, comes on 1–12 weeks after birth and ususally presents as meningitis; the infection is usually acquired not from the mother but from an attendant or by cross-infection from another neonate. The late onset form is far less likely than the early onset form to be fatal. The treatment of choice for either is intravenous benzylpenicillin.

Listeria monocytogenes (p. 85)

Asymptomatic vaginal carriage of this organism is common. As noted above, intra-uterine infection can occur, and may kill the fetus. Neonatal infections resemble those due to group B streptococci, both in being divisible into early onset and late onset forms and in their clinical features. Post-mortem examination of still-born fetuses or of neonatal fatalities shows micro-abscesses or granulomas in the liver, lungs and adrenals. Treatment of neonatal listeriosis should be with ampicillin or amoxycillin and an aminoglycoside (usually gentamicin).

Pneumonia

Septicaemia in neonates commonly results in pneumonia. Conversely, primary pneumonia may develop into septicaemia. Infection of the amniotic fluid (chorio-amnionitis), which is commonly secondary to prolonged rupture of the membranes, readily leads to pneumonia if the baby inhales the infected fluid during birth. The infecting organisms are then those from the mother's genital tract, with group B streptococci being the commonest in recent years. *Chlamydia trachomatis* is a common vaginal pathogen (p. 280), and infects about 30% of infants born to mothers who carry it. Conjunctivitis is the common consequence of such infection (p. 296), but about 15% of infected neonates develop pneumonia. In contrast to the pneumonia caused by group B streptococci, which presents within the first two days of life, chlamydial pneumonia is of insidious onset, most commonly presenting 6–12 weeks after birth as cough and tachypnoea. Chlamydiae may be detected in or isolated from material from the throat, nasopharynx or conjunctiva, and the diagnosis can be confirmed serologically. The treatment of choice for neonatal chlamydial infections is parenteral erythromycin.

Neonates frequently acquire respiratory tract viruses—e.g. respiratory syncytial virus, adenoviruses or influenza viruses—from nursery staff, parents or other visitors. Since these viruses may be causing only mild disease in the adults but are liable to cause life-threatening pneumonia in the neonates, staff with obvious (even though apparently trivial) respiratory tract infections should not look after neonates.

Urinary Tract Infection

This occurs in about 1% of all neonates but in a much larger proportion of those who are premature or have congenital abnormalities or neonatal illnesses. It is often associated with bacteraemia. The causative organisms are usually *Esch. coli* or related gram-negative bacilli. The near-impossibility of collecting a urine fit for culture from a neonate by any other method makes suprapubic aspiration a necessary investigational procedure.

Staphylococcus aureus Infections

Skin colonization with *Staph. aureus* in the first few weeks of life is very common, and usually causes no problems. However, it may result in umbilical stump infection, skin pustules or 'sticky eyes' (see below), or less commonly in serious infections such as pneumonia, osteomyelitis, septicaemia or the 'scalded skin syndrome' (toxic epidermal necrolysis, p. 306). Routine prophylactic application of triple dye or hexachlorophane powder to the umbilical stump minimizes the incidence of *Staph. aureus* colonization and subsequent infection. Serious *Staph. aureus* infections in neonates, as in all other patients, require parenteral administration of cloxacillin or some other effective anti-staphylococcal antibiotic, An outbreak of *Staph. aureus* infection in a neonatal unit is a serious problem, requiring isolation of infected infants and screening of all of the infants and staff for carriage of staphylococci of the phage type responsible for the outbreak (p. 349). Prevention of such outbreaks depends on avoiding overcrowding, on careful hand-washing and on the other usual procedures for avoiding cross-infection (pp. 346–8).

Ophthalmia and 'Sticky Eyes'

Acquisition of a gonococcus from the mother's genital tract during birth can lead to **ophthalmia neonatorum**—a severe conjunctivitis coming on within a few hours of birth and, if not treated effectively, liable to develop into an infection of the rest of the eye that may cause blindness. Far more common in Britain today is neonatal conjunctivitis due to *Chl. trachomatis,* also acquired from the mother during birth but causing a less severe infection that typically comes on several days after birth. These conditions are discussed in Chapter 21 (p. 295). Among other causes of purulent conjunctivitis in the neonate is *Staph. aureus* infection, usually manifested by the lids being stuck together by exudate and therefore known as 'sticky eye'.

Enteritis

Bacterial colonization of the bowel begins within hours of birth, and the bacteria rapidly increase in numbers and variety. In breast-fed infants lactobacilli and bifidobacteria predominate, whereas enterobacteria are more prominent in those fed on artificial foods. Infants usually acquire their first enterobacterial strains from their mothers; but spread of **enteropathogenic** strains of *Esch. coli* (p. 252) from child to child can cause serious outbreaks of enteritis in neonatal units. *Clostridium difficile*, the organism associated with pseudomembranous colitis in adults (p. 259), is to be found in the faeces of some 50% of healthy infants, but although many of the strains isolated are toxigenic they do not cause trouble in this age group.

Because neonates lack the protective effect of gastric acid and have immature immunological defences, they are particularly at risk from infection by **salmonellae** (p. 92), since in them even those serotypes that are usually confined to the intestine in older patients can spread via the blood-stream to other parts of the body, notably the meninges. Even if salmonella infection in the early days of life does not cause serious illness, it is liable to result in persistent carriage and excretion—the salmonella becoming, as it were, a 'founder member' of the intestinal flora.

Necrotizing enterocolitis is an ischaemic necrosis of the gut wall (most frequently of the terminal ileum), with sloughing of the mucosa. It affects neonates, particularly premature and high-risk babies already needing intensive care, and its prevalence in neonatal units appears to be increasing world-wide. Its presenting features are vomiting, abdominal distension and bloody stools, with gas bubbles within the gut wall (pneumatosis coli) often demonstrable radiologically. Reported mortality rates cover the range 30–70%. Its aetiology is obscure; but among the many suggested contributory factors is hypoxic damage to the gut wall following umbilical vein catherization, with consequent invasion of the necrotic wall by various bacteria. Clostridia may well be important in this situation, which in some respects resembles the disease of New Guinea highlanders known as 'pig bel' and caused by *Cl. perfringens* type C (p. 89). As soon as the diagnosis of necrotizing enterocolitis is suspected oral feeding should cease and be replaced by parenteral nutrition, and broad-spectrum antibiotic treatment should also be given parenterally; ideally, bacteriological specimens should be collected before this treatment is started, but the complex mixed growths derived from them are of limited usefulness. Surgical resection of damaged bowel may be necesssary, especially if there is any perforation. Because of the uncertainty about the precise role of micro-organisms, strict isolation of affected babies is indicated.

Tetanus and Botulism

Neonatal **tetanus** is similar to tetanus in adults (p. 290) except in the circumstances that cause it; it is almost exclusively a disease of communities in which animal dung is the traditional dressing for the umbilical stump. Spores of *Cl. tetani* are common in such dung, and find the decaying stump an excellent growth site.

Infant botulism, on the other hand, is strikingly different from the disease caused in adults by the same organism (p. 258). It occurs when infants are fed material that contains *Cl. botulinum* spores but no significant amounts of toxin (honey has been incriminated in the USA). The spores reach the large intestine and germinate there, producing amounts of toxin that would be lethal if ingested ready-made; but since absorption from the large intestine is poor and the great majority of the toxin is excreted in the faeces, the symptoms are usually mild— weakness and flaccidity and difficulty in feeding for a while, but with complete recovery in nearly all cases. However, infant botulism may have been responsible for some cases of 'cot death' or 'sudden infant death syndrome'.

Antibacterial Treatment for Neonates

Bacterial infections in the neonatal period are commonly fulminating and life-threatening, and are often difficult to diagnose rapidly because of the frequent lack of specificity of the presenting features. Treatment therefore has usually to be empirical and to cover a broad spectrum of possibilities. Prophylaxis may be indicated—e.g. administration of intramuscular penicillin to all low birth-weight babies within 2 hours of delivery, and to other babies who show signs of developing respiratory difficulty soon after birth, virtually eliminates deaths from early onset infection with group B streptococci (p. 78), and at the same time prevents gonococcal ophthalmia (p. 295). Suspected neonatal septicaemia has commonly been treated with injections of both a penicillin (benzylpenicillin or ampicillin) and an aminoglycoside, but newer cephalosporins such as cefotaxime and ceftazidime are used instead in some places and their good penetration into CSF makes them appropriate for treating neonatal meningitis caused by gram-negative bacilli. Chloramphenicol has a wide range of activity against pathogens of importance in the neonatal period, but must be used with special care and only when clearly indicated because of its special toxicity to neonates (p. 394); serum levels should be monitored.

Herpes Simplex

As mentioned on p. 321, the TORCH acronym includes herpes simplex among the possibilities to be thought of first in cases of possible congenital infection, but in fact infection with herpes simplex virus (HSV) is acquired during or shortly after birth far more often than *in utero*. Some 0.1–1.0% of pregnant women shed HSV from their genital tracts, but in many of these the vaginal infection is recurrent and their infants, even if they become infected, are protected by maternal antibodies and so are unlikely to develop severe neonatal herpes; **primary** genital herpes in the mother is more dangerous to the infant. Significant neonatal herpes follows one of every 7500 births or so. It can cause conjunctivitis and keratitis; generalized infection with fever, irritability, a vesicular rash and cyanotic attacks; meningo-encephalitis; and gastro-intestinal infection with bleeding. The diagnosis can be confirmed in the laboratory rapidly by immunofluorescence microscopy or electron-microscopy of material from skin lesions, and less rapidly by isolating the virus from such lesions or from the conjunctivae, throat or CSF. Early diagnosis and immediate treatment with appropriate drugs (p. 135) have reduced the mortality of neonatal systemic herpes from about 80% to about 40%. Infection localized to the central nervous system and treated promptly has a better prognosis, with about 90% of the infants recovering and many having no residual handicaps. Routine screening of all pregnant women for vaginal HSV excretion is not practicable, but cervical culture should be carried out when there is a history of previous genital herpes or of a sexual partner who has herpes. If the culture is positive or the mother has active genital herpes, the baby should be delivered by Caesarian section, should be managed so far as possible without skin puncture (e.g. by use of an external electrode for monitoring) in order to reduce the risk of systemic herpes infection, and should be isolated to prevent spread of infection, if it develops, to other susceptible infants.

Enterovirus Infections

Coxsackie B viruses and echoviruses have been responsible for nursery outbreaks of infection clinically indistinguishable from bacterial septicaemia. The illnesses, commonly severe and in some cases fatal, are characterized by fever, tachycardia, tachypnoea, poor feeding and irritability, and in the later stages may include seizures, apnoea, hepatic failure and circulatory collapse. Laboratory confirmation of the diagnosis is by culture of nose and throat swabs, faeces and CSF. Affected infants should be isolated, and it may be necessary to close the nursery to put an end to cross-infection.

Hepatitis B

The frequency of perinatal acquisition of hepatitis B virus varies widely from country to country. It depends not only on the frequency of hepatitis B carriage among mothers but also on the proportion of such carriers who have HBeAg (p. 135) in their blood, and this proportion differs markedly in different ethnic groups. In Europe hepatitis-B-carrier mothers are uncommon, few of them are HBeAg carriers and less than 5% of them transmit their infection to their babies. In South East Asia, in contrast, where hepatitis-B-carrier mothers are far commoner, 35–50% of them are HBeAg carriers and this group transmit the infection to 70% of their babies. HBsAg-positive mothers who have antibodies to HBeAg are highly unlikely to transmit the infection to their babies. The virus does not cross the intact placenta, and the rare cases of intra-uterine infection probably result from leakage of maternal blood into the fetal circulation. More commonly the babies acquire their infections during or very soon after delivery. They seldom develop clinically apparent hepatitis, but usually have subclinical infections that lead on to chronic carriage and may cause hepatic cirrhosis or hepatocellular carcinoma in early adult life. Presence of HBsAg in cord blood does not necessarily indicate infection; and conversely it is often not to be found at the time of birth in the blood of babies who are infected. The diagnosis is made either by its presence and persistence or by the appearance of antibodies to it, which may take 6 months. Babies born to mothers who are HBeAg-positive, or are HBsAg-positive with no antibodies to HBeAg, can be protected from developing chronic hepatitis by giving them injections of hepatitis B immunoglobulin (starting within a few hours of birth) or of hepatitis B vaccine, or probably most effectively by giving a single dose of the immunoglobulin followed by a course of vaccine injections.

PART IV

Prevention and Treatment of Infection

26

Principles of Infection Control; Sterilization and Disinfection

Principles

Everyone knows, and most victims of infectious diseases would heartily agree, that 'prevention is better than cure'; but medical teaching and practice have traditionally been preoccupied with treatment of sick individuals. Today we have therapeutic capabilities, both medical and surgical, far beyond the dreams of only a few years ago; but even so, when prevention of a disease is practicable it is usually 'better than cure' because it does more to reduce suffering and costs less money.

Microbial disease can be prevented by:
(1) eliminating sources of the responsible organisms;
(2) preventing transmission of the responsible organisms; or
(3) raising the resistance of potential hosts so that they are not susceptible to the attacks of the organisms.

The extent to which each of these lines can be followed varies greatly from one disease to another. In many cases it is possible and necessary to advance along two or all three of them at the same time.

Elimination of Sources

The great majority of human microbial infections are acquired from other human beings or from animals. Whether their sources can be eliminated depends upon the ease with which cases and carriers can be detected and then treated or destroyed.

That the sources of a disease can be totally eliminated is illustrated by the recent history of smallpox (p. 134). In 1967 this disease was still a major problem, endemic in 38 countries; but by 1980 the World Health Organization could declare the whole world free of it. Rapid eradication, despite the lack of any drug effective in treatment, was possible because this was a disease of humans only, with no animal carriers, and because patients either died or overcame the disease completely—they did not become carriers. Success was achieved by identifying and isolating all cases (the only sources of infection, though viruses shed by them into dust could remain infective for many weeks), and by immunization of all possible contacts, for their own protection and to break the chain of transmission. Provided that laboratory cultures of the virus are all destroyed or properly controlled, and that the organism does not prove to be capable of survival outside the body for longer than is currently believed, smallpox should never occur again. (But see p. 134, footnote.)

Tuberculosis (p. 206) provides a somewhat more complicated illustration of the possibility of eradication, since man may be infected from cattle as well as from humans—and indeed cattle may acquire their infections from badgers and other wild animals. In a highly developed community all cattle can be located, and those with latent or active tuberculosis infection can be identified by tuberculin testing. If those giving positive reactions are destroyed, the human population is freed from any risk of tuberculosis of bovine origin. Finding human sources of tuberculosis infection is a bigger problem, in that human beings are more numerous than cattle and less easy to round up for regular testing. Furthermore, the value of the tuberculin test as a means of detecting latent or active tuberculosis has been considerably reduced by widespread BCG vaccination, which also causes positive tuberculin reactions. However, since infection with the human type of tubercle bacillus usually involves the lungs, it can often be detected on chest X-ray films. Mass-radiography campaigns, involving the collection of such films of as many members of a community as possible, have been widely and effectively used in the search for infected human beings, but as the incidence of tuberculosis decreases as a result of such measures, further routine screening of the apparently healthy population ceases to be cost-effective. As for the active cases that are discovered by one means or another, treatment (with temporary isolation of those that are 'open'—i.e. discharging tubercle bacilli in their sputum) must be followed by prolonged surveillance, since it is never safe to regard any tuberculous infection as permanently cured. The contacts of all active cases must also be carefully followed up.

Brucellosis (p. 232) presents a simpler control problem than tuberculosis because domestic animals (cattle in Britain) are the only sources of human infection. Elimination of infected cattle in Britain has put an end to acquistition of the disease here.

Anthrax (pp. 202 and 307) is predominantly a disease of domestic animals. As Pasteur demonstrated, the most dangerous source of this disease, at any rate for other animals, is the dead body of one of its victims. Here the problem is not to find the source but to deal with it, since the bacillus is a spore-former and hard to eradicate. Bodies of animals that die from anthrax should either be destroyed by burning or be buried deep in the ground. These are simple procedures when small animals are involved but less simple, when, as in an English zoo some years ago, the victims are elephants! The main sources of anthrax in countries such as Britain are hides, hair and bone-meal imported from countries where anthrax is endemic and where skin and bone are the marketable components of sick or dead animals. The hazards from such materials are reduced by hypochlorite or other appropriate treatment, preferably before shipment but, failing that, on arrival in the receiving country.

It may be difficult or impossible to eliminate the sources of diseases carried by wild animals. A thickly populated island such as Britain has great advantages in this respect, for there are no large tracts of uncontrollable waste land, and land animals cannot enter the country without human help. The rabies virus is not carried by any native wild animals, and has not been endemic among domestic animals since the beginning of the century, although there have been a few localized outbreaks following importation of infected pets. The strict quarantine regulations governing importation of dogs or other potential carriers have been of great value to the country, but are defied—usually for sentimental or commercial reasons—with a frequency which is particularly disturbing at a time when rabies among the fox population has been spreading westward across Europe. Control is far more difficult on a continent than on an island, and particularly in a country such as Canada, where various wild animal species suffer from rabies—some of them roaming over vast uninhabited areas and others, such as squirrels, coming into close contact with man and his domestic animals.

Persistent hunting for cases and carriers, which is so valuable in controlling chronic widespread diseases such as tuberculosis, is less well rewarded in connection with more acute and less common infections. With some of these the best way to locate sources of infection is to co-ordinate information about new cases. This is the purpose of legislation which makes certain diseases notifiable. To search Britain for carriers of typhoid, for example, would be an enormous undertaking with little likelihood of reasonable rewards; but as soon as a new case is reported, public health authorities can institute an intensive local detective operation. This may be made easier if there are several new cases due to the same phage-type of typhoid bacillus, since the search for carriers can then be concentrated in spheres of contact that are common to the people involved. However, finding a typhoid carrier is only part of the problem; eradication of his infection may be a much slower process (pp. 93-4).

There is no possibility of eliminating the sources of some of man's commonest and most important pathogens—e.g. *Staph. aureus* and *Str. pyogenes*—since these are normally carried by a high proportion of healthy people. However, all possible steps must be taken to ensure that people in certain types of employment—e.g. operating theatre staff and food-handlers—are not disseminating strains of such organisms with known propensities for causing trouble (pp. 253 and 360).

Prevention of Transmission

Control of migration and local movement

We have already touched upon the importance of quarantine regulations in preventing transmission of disease on the international scale. Such regulations require animals or human beings, entering a country in circumstances in which they might be incubating one of certain specified diseases, to be kept in isolation for a period which exceeds the incubation period of that disease. Such an approach is of no use in excluding carriers or those with diseases that have long and unpredictable latent periods. There are other ways of dealing with these, such as insisting that intending immigrants undergo appropriate investigations—e.g. chest radiographs, examinations of their faeces for pathogens, or serological tests for syphilis—before leaving their countries of origin, or that they are effectively immunized against diseases for which this is possible.

Unwanted and undocumented animal immigrants can be a serious problem. Hence ships in port have shields on their hawsers which prevent 'hitch-hiking' by plague-carrying rats, and aeroplanes which pass through yellow-fever zones are sprayed with insecticides in case they pick up infected mosquitoes.

Quarantine regulations can be imposed on a local basis. It is no longer generally regarded as necessary to restrict the movements of children during the incubation stages of the common infectious fevers of childhood, but similar restrictions can be applied to members of small units, such as families or schools or military camps, who are possibly incubating more serious diseases.

Isolation of infected patients is discussed on p. 347.

Control of insect vectors

Destruction of insect vectors plays a large part in the control of many diseases—notably of rickettsial infections (pp. 111–3), yellow fever (p. 267) and malaria (p. 241). While it may be difficult or impossible to eradicate an insect species from an area permanently, it is often possible to reduce its numbers to a very low level for a time, and during that time to treat any remaining human cases of the disease and so abolish the reservoir from which the insects might otherwise become infected on their return.

When the relevant insects cannot be destroyed, it may be possible to prevent them from acting as vectors. For example, mosquito nets, fly screens and insect repellents can keep biting insects from attacking prospective hosts, and flies which cannot reach human faeces or human food cannot transmit dysentery bacilli from one to the other.

Communal hygiene

The spread of disease is made easier when human beings live closely packed together in homes into which little bactericidal sunlight penetrates; in which fleas, lice and bugs abound as vectors; in which accumulated refuse encourages the breeding of disease-carrying vermin; in which lack of water supply discourages personal and domestic cleanliness; and in which there is no adequate provision for sewage disposal. In other words, it is generally true that a well-housed community with a main water supply and proper provision for the disposal of refuse and sewage is also a community relatively free from microbial diseases, and the provision of such conditions must be the aim of those interested in the prevention of such diseases. But transmission of infection is not, of course, limited to the home. In schools, shops, meeting halls, places of amusement, public vehicles, swimming baths and wherever else human beings come together they exchange their microbial parasites. Some of the factors concerned in such transmission are mentioned below under 'Personal hygiene'; public responsibility is generally limited to preventing gross overcrowding and seeing that the individual has the necessary facilities for hygienic behaviour and is encouraged to use them. During serious epidemics, particularly of droplet-borne diseases, it may be wise to prevent people from congregating indoors.

The communal and personal aspects of the proper treatment of food and drink are discussed on pp. 360–1.

Personal hygiene

The individual has a double responsibility in relation to the transmission of disease—to do his best to avoid being either a recipient or a donor. Much of what he has to do about protecting himself relates to the maintenance of his general health and specific immunity, and so is not the concern of this section; but there are some ways in which he can reduce or eliminate his risks of acquiring certain infections—e.g. he can avoid promiscuous sexual intercourse (without which the serious problem of sexually transmitted diseases would cease to exist), and if he lives in an area where diseases are transmitted by biting insects he can protect himself from them as indicated above. In many everyday matters it is hard to decide how far hygienic precautions should be taken. Theoretical considerations suggest scrupulous care to keep all potentially infected objects away from the lips and mouth; but since this includes virtually everything except food which has just been cooked or has been kept in a sterile container since being cooked, some compromise with practical reality has to be reached which involves avoiding only those things that are most likely to be contaminated. From a purely microbiological standpoint kissing is a deplorable habit!

Precautions against transmitting pathogens to others are of course most important when one knows that one has an infection to transmit, and at such times it may be one's duty to stay at home or in some other way to reduce one's social contacts to a minimum. However, personal hygiene should also take into consideration the possibility of carrying and disseminating a pathogenic organism without knowing about it. Particular attention needs to be paid to the excretions of the respiratory and alimentary tracts. Airborne droplets, which are the means of transfer of many bacterial and virus infections, can be intercepted to some extent by following the advice of the slogan: 'Coughs and sneezes spread diseases; trap the germs in your handkerchief'. However, the handkerchief itself, replaced in a warm pocket for a period of incubation and then shaken out vigorously before its next use, can make a considerable contribution to the microbial population of the air. There is much to be said for using (and disposing of) disposable tissues. Spitting of sputum on to the ground, where it is allowed to dry, is a potentially dangerous practice at all times and a serious menace if the sputum contains tubercle bacilli. Faeces may contain pathogenic bacteria, viruses, protozoan cysts and helminth eggs, and so good personal hygiene includes proper disposal of faeces and the washing of hands after defaecation. Food-handlers must be particularly careful to avoid transfer of their intestinal bacteria to food—which may prove a good culture medium for pathogens and so an excellent way of transmitting them to large numbers of people.

Raising Host Resistance

General considerations

As discussed in Chapter 24 many factors, including malnutrition (p. 317) and fatigue and lowering of the body temperature by exposure to a cold environment (chilling), decrease resistance to infection. These probably operate more by allowing latent infections to develop into overt diseases than by increasing susceptibility to fresh infection. Potential hosts who are well fed, well rested, well housed and well clothed are therefore relatively resistant to microbial diseases—though excess of food, rest or warmth may be harmful. Treatment of non-microbial diseases—notably of diabetes—may be important in raising resistance to infection.

Prophylactic Medication

Indiscriminate use of antimicrobial drugs to prevent infection can be dangerous and costly (p. 373); but there are some clear indications for such prophylactic medication. Examples are the taking of suppressive drugs by those visiting or living in malarious areas (p. 242), and the use of metronidazole to prevent wound infection following bowel surgery (see pp. 375–6, where other examples of the correct prophylactic use of antibacterial drugs are given).

Immunization

This very important part of the raising of host resistance has a chapter to itself (Chapter 28). Here it is sufficient to stress (1) the duty of the individual, for his own sake and that of the community, to see that so far as possible he and his family are immunized against any disease which is a potential menace in his particular environment; (2) the responsibility of public health authorities to see that he has the necessary information and facilities to carry out that duty; and (3) the responsibility of the employing authorities to ensure that doctors, nurses, ambulance drivers, medical laboratory staff and others who are specially at risk because of the nature of their work are adequately immunized against such infections as tuberculosis and poliomyelitis.

Sterilization and Disinfection

Sterilization means the killing or removal of all micro-organisms, including bacterial spores. **Disinfection** has a less precise meaning. The word suggests freeing an object from potentially harmful micro-organisms, but no process differentiates between these and non-pathogens. In practice, disinfection means use of chemicals (disinfectants—p. 340) to eliminate most micro-organisms, but chemicals cannot be relied on to destroy bacterial spores, and some widely used disinfectants are ineffective against viruses.

Micro-organisms vary greatly in their resistance to adverse physical and chemical conditions. Some fail to survive minor environmental changes, whereas others are difficult to kill. Spore-forming bacterial species (*Bacillus* and *Clostridium* species) are the most durable. These spores have the physical protection of their thick coats, their water content is low, their metabolic activity is minimal and since they do not divide they avoid the increased susceptibility of dividing cells to various noxious agents. (By contrast, fungal spores are reproductive, not specially resistant stages, and are readily killed by disinfectants.) *Mycobacterium tuberculosis* and related bacteria do not form spores, but having waxy hydrophobic surfaces and low rates of metabolism they are more resistant than most bacteria to drying and to chemical agents. However, they do not share the heat-resistance of the spore-formers. Drying by exposure to ordinary atmospheric conditions rapidly kills many bacteria and such viruses as those of influenza, mumps and measles; but some important non-sporing pathogens, such as *Myco. tuberculosis*, *Staphylococcus aureus*, rickettsiae and *Coxiella burneti*, can survive in dust for long periods, and sporing organisms can do so almost indefinitely. Freezing kills some organisms, especially if they are in a liquid medium that is frozen slowly, but more commonly it can be a valuable means of preserving microbial viability. Even delicate organisms such as viruses and *Haemophilus influenzae* survive for many months at temperatures between -20 and $-70°C$ if they are frozen rapidly.

Sterilization

Items can be sterilized by using **heat, irradiation** or **filtration,** under carefully defined and observed conditions. The killing of a microbial population is a continuous process, not an instantaneous event. The time taken to complete it depends on the size of the population (among other things). Therefore if a sterilizing procedure kills 90% of a particular bacterial population every minute, 5 minutes of such treatment would be virtually certain to eliminate a population of 100 organisms (failing only once in a thousand times) but would be unreliable against a population of 10 000 (failing once in ten times) and could not be expected to kill all of 1 million organisms. Several hours of such treatment could be expected to deal with astronomical numbers of organisms of that strain; but in fact prolonged exposure to heat selects increasingly resistant individuals from the bacterial population, and the sterilizing process has to be restricted to that which gives a reasonable degree of safety without doing excessive damage to the article or material that is being sterilized. A reasonable degree of safety depends on the dimensions of the problem (in terms of bacterial populations and numbers of items to be sterilized) and upon the possible consequences of failure. For example, failure to sterilize 1% of bottles of culture media is of little consequence, but such a failure rate would be potentially disastrous among bottles of fluids for intravenous administration to patients or ampoules of vaccines meant to contain dead pathogenic organisms. Thorough cleaning of instruments and apparatus before submitting them to a sterilization process decreases the risk of sterilization failure, because it reduces the number of micro-organisms likely to be present and removes material that might protect them against the process.

When in the following paragraphs we say that items 'can be sterilized' by the temperature–time combinations named, we mean that these combinations are commonly used because they can be relied on to sterilize the types of item mentioned, yet are not excessively destructive to them.

Heat sterilization

Heat is the most effective agent for killing micro-organisms, and is the best means of sterilization for all articles to which it is applicable. Dry heat kills micro-organisms by oxidation, moist heat by denaturing their proteins. Most vegetative bacteria are killed in a few minutes at 60°C, but killing of spores by dry heat may take as much as an hour at 160°C. Heating to 60°C for 30 minutes will inactivate most viruses (except polioviruses and hepatitis B virus), and also rickettsiae and chlamydiae; but *Cox. burneti* is more resistant and can survive heat only a little less than that used in the pasteurization of milk (p. 358). Moisture increases the heat-susceptibility of micro-organisms, but even so some spores may survive prolonged boiling.

(a) Dry Heat

INCINERATION   Total destruction by burning in a furnace is a useful means of eliminating the microbial content and consequent infection hazard of such disposable items as dirty dressings, pathological specimens in destructible cartons, some forms of laboratory cultures and the bodies of small dead animals.

FLAMING   Bacteriological wire loops and various other metal instruments can be sterilized by heating them to redness in a flame, but such treatment blunts cutting instruments. A less destructive modification, which is not a fully reliable means of sterilization but is sometimes useful as a 'first-aid' measure when proper facilities are not available, is to dip the instruments in methylated spirit (or pour it on to them) and set fire to the spirit—taking care to keep one's fingers out of the resulting conflagration!

DRY HEAT IN AN OVEN   Glassware, surgical and laboratory apparatus and instruments of many kinds, some forms of dressing and many other solid heat-resistant items can be sterilized by heating them at 160°C for one hour in a hot-air oven. They are usually packed in containers or paper wrappings that are impenetrable to bacteria so that they remain sterile after removal from the oven. To ensure uniform heating of the contents of the oven, there must be room for free circulation of air between the items (overloading prevents this) and the circulation should be maintained by a fan. Time must be allowed for the entire load to reach 160°C before starting to measure the hour. The measures for ensuring proper use and for monitoring the performance of ovens are much the same as those for autoclaves (see below).

(b) Moist Heat

STEAM UNDER INCREASED PRESSURE   Many materials, instruments and pieces of apparatus which cannot tolerate being heated to 160°C can be sterilized by exposing them to pure steam at more than atmospheric pressure (and therefore at more than 100°C). The instrument used to achieve these conditions is the **autoclave**—essentially an enlarged and sophisticated version of the domestic pressure cooker (which itself can be used for the same purpose on a small scale). The efficacy of steam as a means of killing micro-organisms depends on the fact that on reaching the surface of an object that is cooler than itself the steam condenses, giving up latent heat and rapidly raising the temperature of the object. The sudden decrease in volume draws in more steam, and so penetration into porous structure is good. For steam to be effective, it must be free of air, since this reduces its partial pressure and therefore the temperature achieved. Upward displacement of the air by rising steam, as in the pressure cooker, is an inefficient process because air is heavier than steam. Downward-displacement autoclaves, with steam entering at the top and displacing the air downwards and out at the bottom, are adequate for some purposes, provided that the load is not packed in such a way as to retain the air (e.g. in upward-facing bowls). In more efficient autoclaves the air is evacuated by suction before the steam is allowed in. After the load has been exposed to pure steam for long enough to reach the appropriate temperature and has been at that temperature for the appropriate time, the steam is evacuated and the vacuum is maintained for long enough to dry the load, after which dry air is introduced and the load is allowed to cool. Adequate time/pressure combinations for sterilization of appropriate items are 15 minutes at 15 lb/in^2 (100 kPa) (temperature 121°C), or 3 minutes at 30 lb/in^2 (200 kPa) (134°C)—the latter being attainable in modern high-vacuum autoclaves.

Clearly an oven or an autoclave (or any other form of sterilizing apparatus) can sterilize its load only if it is correctly loaded and correctly controlled throughout the sterilizing process. Loading depends on human attendants, but subsequent control can be maintained by thermometers, thermostats, clocks and other devices. Automatic recorders can show whether these mechanisms are working or have worked correctly, but the human attendants must take due notice of the recordings in order to decide which loads have been properly processed. As protection against mechanical and human errors, **indicators** should be incorporated in the load. (It may be thought necessary to do this for every load, or only from time to time.) Chemical indicators which change colour when adequately heated are in common use; some are designed for testing hot-air ovens and some for testing autoclaves. The chemical indicator can be incorporated in adhesive tape as stripes that are the same colour as the tape until heated. The presence of coloured stripes on such a tape that has been wrapped round or strapped across a package is an indication that the article has been heated to an appropriate temperature. In the Bowie–Dick test for autoclaves, a diagonal cross of indicator tape is placed in the middle of a standard pack of towels to be sterilized; by the end of the cycle the stripes on the tape should have changed colour uniformly to the centre of the cross, indicating adequate steam penetration. Such methods indicate indirectly whether bacteria should have been killed, but a more direct approach is to use bacterial indicators (which can also be used for checking other means of sterilization, such as irradiation or gas sterilization). These consist of standard numbers of viable spores of suitable *Bacillus* species, commonly carried on filter-paper strips or threads, which are placed inside appropriate containers at selected points within the load, and which should fail to grow when placed in culture media after undergoing the sterilization process.

STEAM AT ATMOSPHERIC PRESSURE Many culture media and other aqueous liquids which would boil in an oven can be sterilized in an autoclave. However, some are damaged by temperatures more than a little above 100°C. Twenty minutes at this temperature kills all vegetative micro-organisms. Cotton-wool-stoppered bottles or tubes of liquid can be heated in a **steamer;** free steam is generated by boiling water and is retained under a conical lid which has a small escape vent at the top. Such treatment does not kill spores; but if these are present and the liquid is suitable, they will germinate after it has been removed from the steamer and allowed to cool. The vegetative cells can then be caught and destroyed by similar heating on the next day. Further heating on the third day is usually added for extra security. This process of intermittent steaming is called **Tyndallization** after its originator.

LOW-PRESSURE STEAM WITH FORMALDEHYDE A combined physical–chemical method of sterilization applicable to many heat-sensitive materials is exposure to steam and formaldehyde in a special chamber at sub-atmospheric pressure (temperature about 80°C).

BOILING WATER Boiling-water baths have been commonly used in hospital wards and operating theatres for the treatment of bowls, instruments, etc., and were often described as 'sterilizers'. They cannot sterilize consistently, since some sporing organisms can survive boiling. Such pieces of apparatus may have their uses, provided that their limitations are remembered and that they are not, for example, expected to sterilize instruments used in wounds with possible clostridial infections.

Radiation

Sunlight has an antimicrobial effect by virtue of its content of ultraviolet light, and artificial ultraviolet light is used in sterilization of air, of some forms of apparatus, and of materials such as plasma which would be rendered useless by heating or by chemical treatment. Although ultra-violet light does very little harm to the material treated, it has little power of penetration and can therefore sterilize only surface or thin layers of material. X-rays, gamma-rays and other penetrating ionizing radiations kill micro-organisms by damaging chromosomal DNA and are much more efficient sterilizing agents. Under controlled conditions gamma-rays can be relied upon to kill all micro-organisms and in consequence are used commercially for sterilizing disposable instruments and equipment. Nearly all viruses except the 'slow viruses' (p. 223) are inactivated by ultraviolet light and by other types of irradiation, though there are quite large variations in the dosage required by different viruses.

Filtration

Air and liquids can be sterilized by passing them through filters that remove bacteria and larger particles. Free virus particles are small enough to pass through the majority of filters in common use, but for many purposes this is of no importance. Many viruses are arrested by the filters because they are contained in droplets or in cells.

FILTRATION OF AIR If a glass tube is tightly plugged with cotton wool, bacteria and other particles contained in air which enters the tube become entangled in the fibres of the cotton wool. This must have been sterilized by heating before use as a filter, and must be non-absorbent because motile organisms can swim through it if it becomes wet. Such simple filters are used in microbiological laboratories to allow sterile air to enter tubes and flasks containing cultures. Larger and more elaborate filters are used to provide sterile air for operating theatres and other places where it is particularly important to avoid airborne contamination. These are 'absolute' filters; whereas those used to provide clean air for wards, etc. remove particles and most micro-organisms but do not sterilize.

FILTRATION OF LIQUIDS Liquids can be sterilized by passing them through cellulose membrane filters. The liquid to be sterilized is drawn through a funnel into a sterile bottle or flask which is connected via an air filter to a vacuum pump. For small volumes a small membrane filter in a plastic holder can be attached to a syringe. Filtration is more time-consuming than heat-sterilization because the filter assemblies have themselves to be heat-sterilized beforehand, and there is always a risk of accidental contamination of the material because of leakage of unsterile air into the apparatus. It is used only for materials such as serum and yeast extracts which would be adversely affected by heat. Cellulose membranes of known pore size can also be used to determine the sizes of microbial particles or to separate organisms of different sizes. **Seitz** filter pads (made of compressed asbestos) can be used to remove bacteria from liquids, but are not sterilizing filters, as viruses can pass through them.

Disinfection

Disinfectants are widely used in clinical and laboratory medicine to treat objects and materials which are potential sources of infection or contamination but for which sterilization by heating is impossible, inconvenient or unnecessary. They are substances with useful antimicrobial activity which do not have serious general destructive effects such as are possessed by strong acids and alkalis, but which are too toxic for systemic use in the treatment of microbial infections. They are referred to as **bactericidal** when they kill bacteria (cf. fungicidal, germicidal) and as **bacteristatic** when they only prevent multiplication. It means nothing however to describe a substance as bactericidal or bacteristatic without defining the concentration in which it is used, the identity and state of the organism and the conditions under which the two come into contact. Disinfectants vary in the speed with which they kill micro-organisms—e.g. alcohols and hypochlorites have lethal effects in minutes whereas glutaraldehyde is much slower. There is in general a range of sensitivity of different bacteria to disinfectants, from the gram-positive species (other than the spore-formers), which are usually the most sensitive, through the gram-negative species and the mycobacteria to the highly resistant spore-formers. Lipid-containing viruses tend to be sensitive to phenols, whereas their non-lipid-containing counterparts are resistant. Disinfectants are inactivated to varying degrees by organic matter such as pus, blood or faeces; therefore items to be disinfected should whenever possible be first thoroughly cleaned or flushed through with large volumes of clean water, a process which in itself will greatly reduce the microbial load. The pH of a disinfectant solution may be of critical importance—e.g. phenolics and hypochlorites are more active at acid pH, whereas quaternary ammonium compounds and glutaraldehyde have optimal activity at high pH. Unfortunately, hypochlorites and glutaraldehydes are unstable at low and at high pH respectively and both require to be adjusted to the correct pH just before use. Many disinfectants have to be freshly diluted for use because they are unstable at in-use dilutions and rapidly become inactive. Even when stability is not a problem, dilute disinfectant solutions should be used only from freshly opened sterile bottles or sachets, as otherwise they may become contaminated with relatively resistant organisms as *Pseudomonas* species, and such contaminated solutions have caused numerous outbreaks of hospital infection.

Standardization of disinfectants presents serious problems. Results obtained in a test-tube culture in a fluid medium may be irrelevant to the choice of a disinfectant for the treatment of floors, furniture, or wounds. **Use-dilution tests**—e.g. the Kelsey–Sykes test—serve to indicate the 'working' concentrations of a particular disinfectant for a range of applications. Normally one dilution is specified for 'clean' conditions and another for 'dirty' conditions where organic matter is present and inactivation may be a problem. **In-use tests** check that pathogens cannot survive or multiply in the dilutions of disinfection solution currently available in the hospital or laboratory. Some examples of well-known disinfectants and the properties which play a large part in determining their suitability for particular tasks are listed in Table 26.1 (p. 344).

Hospital infection control committees are responsible for the formualtion and dissemination of disinfection policies that specify the appropriate actions and agents (with concentrations) to be used for decontamination of skin, equipment or environmental surfaces throughout the hospital. Examples of such problems and recommendations are:

(1) Treatment of excreta that may contain pathogens, of discarded microbial cultures and preparations, etc.

Unusually dangerous items should be sterilized in an autoclave or by some other effective form of heating before disposal, but for routine use the clear phenolics are appropriate because of their powerful and rapid action even in the presence of organic matter. However, they are ineffective against most viruses. Formalin is often used for the treatment of faeces in chemical closets, but its usefulness for other purposes is restricted by its pungent smell.

(2) Treatment of surfaces of dressing trolleys, bedside lockers, laboratory benches, etc.

A detergent–hypochlorite mixture has the advantages of physically removing adherent and possibly contaminated matter from the surfaces and of being effective against viruses—including hepatitis viruses which have to be constantly remembered when blood may have been spilled. Hypochlorite solutions are widely used as disinfectants in virology laboratories. They are the most effective agents for virus disinfection but they have to be used in concentrations (10 000 ppm) higher than those which kill vegetative bacteria (1000 ppm). However, hypochlorite may cause corrosion of metal and is inactivated by organic matter. Clear phenolics are appropriate for surface disinfection when there is no significant virus hazard.

(3) Treatment of instruments and apparatus that would be damaged by heat-sterilization

Such items must be thoroughly cleaned before any attempt at chemical sterilization. This reduces the number of organisms present and deprives them of protection by blood clot, pus and other extraneous material. Hollow tubular instruments such as endoscopes and catheters, and more complex devices such as heart–lung machines, are difficult to clean, but prolonged flushing with running water immediately after use often decreases the problem. They can then be exposed to low-pressure steam with formaldehyde (p. 340) or, if that is not possible, to a 2% solution of glutaraldehyde. This is a powerful bactericidal agent given adequate time (20 minutes for vegetative bacteria, 30 minutes for *Myco. tuberculosis* and 3 hours for spores), but it is expensive and unstable in the alkaline pH range in which it is active; and instruments immersed in it need to be rinsed in distilled water to remove the disinfectant. The gas ethylene oxide is a sterilizing agent that penetrates into relatively inaccessible sites in complex pieces of apparatus, and is also used for sterilizing objects as different as bone grafts and disposable plastic syringes. It is used in a special chamber, and requires a high humidity to be effective. It is toxic, and when mixed with oxygen in a wide range of proportions it is highly explosive.

(4) Treatment of skin

The skin microflora consists of 'residents'—e.g. coagulase-negative staphylococci and diphtheroid bacilli—which grow on the skin and are not readily removed by washing or disinfection, and 'transients'—organisms, often gram-negative, that have been deposited on the skin and can usually be rapidly removed by washing or disinfection. Skin cannot be sterilized. There are however various sets of circumstances in which steps must be taken to reduce the bacterial population, and in particular to eliminate pathogens.

(a) Hygienic hand disinfection

When there has been known or possible contamination of the skin—usually of the hands—as a result of contact with patients or with laboratory cultures, the organisms in question ('transients') are on the surface of the skin and can be relatively easily removed with soap and water. For additional safety, the hands can be first immersed for a few minutes in a disinfectant such as a 5% aqueous or alcoholic solution of chlorhexidine. Alternatively, an alcoholic solution of a skin disinfectant can be rubbed over the skin after washing and allowed to evaporate; but this is not a means of decontaminating unwashed skin. An alcoholic disinfectant is required to reduce the risk of cross-infection with enteroviruses and rotaviruses, since chlorhexidine has no virucidal activity.

(b) Surgical hand disinfection

A more difficult problem is the treatment of the hands of doctors and nurses in order to minimize the chances of their passing on their own resident and transient flora, possibly including pathogens such as *Staph. aureus*, to their patients during surgical procedures. A satisfactory approach to the problem is first to wash the hands, wrists and fore-arms carefully with a **surgical scrub,** consisting of a detergent and either chlorhexidine (e.g. in 'Hibiscrub') or povidone–iodine—i.e. iodine carried on an iodophor (e.g. in 'Betadine'), which retains the efficacy of the long-established tincture of iodine but does not cause staining or sensitization of the patient's skin. 70% isopropyl alcohol or some other suitable alcoholic preparation should then be rubbed over the washed skin. A 2–3 minute surgical scrub is adequate. Vigorous prolonged 'scrubbing up' removes surface organisms but drives others up from the hair follicles and sweat glands; culture of the skin after this procedure may yield more bacteria than before.

(c) Pre-operative skin disinfection

A patient's skin can be satisfactorily prepared for surgical incision by first cleaning it with soap or detergent and then painting it with chlorhexidine or povidone–iodine in 70% alcohol, either of which acts rapidly and has a prolonged antibacterial effect. This is also appropriate in preparation for needle puncture for special purposes—notably for lumbar puncture or for setting up a blood culture (pp. 227 and 240)—but on most other occasions application of 70% isopropyl alcohol to the skin is sufficient. For efficacy and also for the patient's comfort, the alcohol should be given time to dry before the needle is inserted.

There are numerous techniques for assessing the quantitative and qualitative bacteriology of hands before and after disinfection, and after wearing surgical gloves for 2–3 hours; the finger-streak method, in which the subject places the fingers on the surface of an agar plate, is the simplest.

Suggestions for Further Reading

See end of Chapter 27, p. 361.

TABLE 26.1 Some Properties of Some Commonly Used Disinfectants.

Since the actions of disinfectants depend very largely on the conditions under which they are tested, entries in this table should be regarded as wide generalizations.

Class of compound	Examples and trade names	Activity against — Vegetative bacteria	Activity against — Viruses Rickettsia Chlamydia	Activity against — Fungi	Inactivation by organic matter	Toxicity to human tissues
Environmental disinfectants						
Clear phenolics	'Hycolin' 'Stericol' 'Clearsol'	All[1]	Some	Good	Slight	Moderate
Chloroxylenols	'Dettol'	Some	Nil	Nil	Moderate	Moderate
Hypochlorites and Dichloroisocyanurates	'Chloros' 'Domestos' 'Milton' 'Kirbychlor' 'Presept'	All[2]	Good	Good	Marked	Moderate
Quaternary ammonium compounds	Cetrimide[3] ('Cetavlon')	Most	Some	Good	Marked	Slight
Skin disinfectants						
Diguanides	Chlorhexidine ('Hibitane') Chlorhexidine + Cetrimide ('Savlon')	Most	Slight	Good	Marked	Slight
Alcohols[4]	Ethanol Isopropyl alcohol	All	Good	Good	Moderate	Moderate
Iodine and iodophors	'Betadine' 'Disadine'	All	Good	Good	Marked	Slight to moderate
Hexachlorophane	'Phisohex' 'Sterzac'	Most	Nil	Nil	Slight	Slight[5]
Triclosan	'Manusept' 'Phisomed'	Most	Nil	Nil	Slight	Nil
Instrument disinfectants						
Aldehydes	Glutaraldehyde ('Cidex')	All[1,2]	Good	Good	Moderate	Moderate
	Formaldehyde[6]	All[1,2]	Good	Good	Moderate	Marked

Notes (1) Active against tubercle bacilli.
(2) Active against bacterial spores.
(3) An important feature of cetrimide is its inefficacy against *Ps. aeruginosa*.
(4) Absolute alcohol is a relatively ineffective disinfectant; its activity is increased by dilution to 70% v/v in water.
(5) Hexachlorophane solutions may be absorbed through babies' skin, with possible toxic effects. Powders containing hexachlorophane used for dusting umbilical stumps of neonates do not carry this risk of toxicity.
(6) Formaldehyde is mainly used as a gas for fumigation; Formalin, a 40% aqueous solution of formaldehyde, is too irritant to be used as a general disinfectant.

27
Infection Control in Hospital and in the Community

Hospital Problems

The population of a hospital necessarily includes many disseminators of pathogenic organisms and many people whose illnesses or injuries make them particularly susceptible to infection. In the past, before the nature and modes of spread of micro-organisms were understood, hospitals were fearful places into which patients were loath to go. Highly infectious diseases such as cholera were liable to spread uncontrollably among the overcrowded patients and their attendants; childbirth was commonly followed by puerperal fever which might well be fatal; and wounds all too often became gangrenous and gave rise to fatal septicaemia. 'Let him bear in mind' wrote the early nineteenth-century surgeon, John Bell, concerning hospital gangrene, 'that this is a hospital disease; that without the circle of the infected walls the men are safe; let him therefore hurry them out of this house of death ... let him lay them in a schoolroom, a church, on a dunghill or in a stable ... let him carry them anywhere but to their graves.' This terrible position was transformed by the introduction of Lister's antiseptic techniques (p. 7) and then of aseptic surgery, and by the development of other ways of controlling infection. Antimicrobial drugs have played a part in this control, though early hopes of their dramatic success in this field have been disappointed by the emergence of drug-resistant strains, especially of *Staph. aureus* and of enterobacteria. But the fact remains that hospitals are by their nature places in which infection is a grave menace that can be reduced to a minimum only by constant care on the part of all concerned.

The term **hospital-acquired infection** (or, in American literature, **nosocomial infection**) is used for any infection which a patient acquires in hospital, whether it becomes apparent during his stay there or only after his discharge, and whether the organism came from another patient (**cross-infection**), or was one that he himself was formerly carrying in another site (**endogenous, self-** or **auto-infection**). In cross-infection the organism may have been brought into the hospital by the other patient, but commonly it is resident in the hospital or in a particular ward, to be found on the floors, walls, bedding, etc., and maintained by infection of successive generations of patients. Endogenous infection is illustrated by infection of a wound with a *Staph. aureus* which the patient was carrying in his nose when he was admitted to hospital, or by urinary infection with *Esch. coli* from his own intestine. While such infection is not due to hospital organisms, it may well have been made possible by operative or other procedures carried out in the hospital and hence be a direct consequence of hospital admission; and in practice it is often impossible to distinguish the two types of hospital-acquired infection because the sources of infecting organisms are not known.

The size of the problem of hospital-acquired infection is indicated by a study of 18 000 hospital in-patients in England and Wales in 1980. Of this total, 9.2% had hospital-acquired infections, the commonest being those of the urinary tract, which affected 2.8% of all hospital patients and accounted for 30% of all hospital-acquired infections. The comparable figures for wound infections were 1.7% and 18.9%, and those for lower respiratory tract infections were 1.5% and 16.8%. Such national figures are of only limited relevance to any local situation, which should be continuously monitored by a local Infection Control (IC) team—commonly comprising an IC nurse, an IC officer (often a microbiologist) and an IC committee (p. 353). They should place special emphasis on high-risk areas—notably surgical wards and intensive care, burns and special care baby units. They should be aware not only of the incidence of hospital-acquired infections but also of the local prevalence of antibiotic-resistant bacterial strains, since this also varies greatly from area to area.

Wound sepsis and other forms of hospital-acquired infections cause the death of some patients, and prolong the stay in hospital of many others. Such prolongation may be a serious matter for the patient and for his family; his maintenance in hospital and treatment are expensive; and meanwhile a bed is occupied which might otherwise be used for another patient.

Of course, problems similar to those encountered in hospital arise with patients nursed at home, and much of what is said in this chapter applies also to them. But they are less liable to cross-infection, and any bacteria that they do acquire are more likely to be sensitive to antibiotics. Multiple drug-resistance is a feature of 'sophisticated' organisms bred in the highly selective environment of a hospital (p. 383). However, antibiotic-resistant strains of *Staph. aureus* and other bacteria are increasing in frequency among the general population, presumably derived from patients who have been discharged from hospital.

Patient Isolation

Spread of infection requires a **source** of organisms, a **susceptible host** and **means of transmission.** Exogenous infection of a hospital patient may come from other patients, hospital staff or visitors (who may have acute infections or be incubational or asymptomatic carriers—p. 27) or from contaminated medications, equipment or other objects. Host susceptibility varies greatly, and is much affected by the general health or debility of the patient as well as by the virulence of the organism. Isolation procedures can be used either to prevent a patient with a dangerous infection from being a hazard to others or to protect specially susceptible patients from exogenous sources of infection: in either case its purpose is to block the routes of transmission.

It is obvious that patients with easily transmissible and serious diseases such as Lassa fever, diphtheria, typhoid or even open tuberculosis should not be nursed in open wards among patients suffering from other diseases. In accordance with the circumstances prevailing, they should be isolated either in special hospitals, in special wards for patients with the same condition or in separate rooms or cubicles. Similarly, children with such conditions as measles or whooping cough should not be nursed in general children's wards, though in many cases there is no reason why they should not be nursed at home. The desirability of isolating patients with *Staph. aureus* infections is less widely recognized. In most hospitals these are so numerous and the number of suitable cubicles is so small that there would be no possiblity of isolating them all, and in most cases this is not necesssary. But some, such as those with staphylococcal pneumonia or with large infected areas of dermatitis or burns, liberate very large numbers of staphylococci into their environment and certainly require to be isolated if possible. This is particularly important if the *Staph. aureus* in question is resistant to a number of antibiotics and in particular to cloxacillin. If such a patient is in a ward, the staphylococci become freely distributed throughout the ward, colonize and multiply in the noses of other patients, invade wounds and respiratory tracts, and may make it necessary for the ward to be closed; attempts, not always successful, must then be made to rid the ward of its staphylococci by extensive washing with disinfectants. *Staph. aureus* can be particularly troublesome in the nursery of an obstetric unit, and babies with even minor lesions should be isolated. Isolation is also desirable for babies with *Esch. coli* gastro-enteritis, and indeed for patients of all ages with diarrhoeal diseases; and for many patients with *Ps. aeruginosa* infections.

Three forms of isolation are described below. Every hospital should have an agreed policy as to which of these (perhaps with appropriate variations) should be applied to different categories of patients and infection problems; and this should include clear guidelines from the IC committee about collection, labelling and transport of dangerous pathological specimens (p. 162), about use of hot-water-soluble bags for sending contaminated linen to the laundry, and about the safe disposal of infected rubbish, especially of contaminated syringes and needles.

Standard (source) isolation

This is the form of isolation appropriate for most patients with significant transmissible infections. Its basis is the provision of an isolation cubicle or single room, of which the door should be kept closed and which should be so designed, equipped and managed that as far as possible no micro-organisms can pass from it to the adjoining ward. Since for administrative reasons it is usually close to a ward and the patient is usually cared for by the ward staff, the success of isolation depends on the thoughtfulness and scrupulous carefulness of everyone concerned. The ventilation of the cubicle must be so designed that all air from it passes to the outside of the building (negative-pressure ventilation), not into the ward or corridors. Washing facilities for the patient and attendants must be provided inside the cubicle. Attendants should put on gowns or disposable plastic aprons on entering the cubicle and remove them on leaving, remembering that this is not a mystic rite but an attempt to prevent contamination of themselves or their clothes and consequent carriage of pathogens out of the cubicle. Disposable gloves should be worn and non-touch techniques used when dealing with infected sites or contaminated materials. As always, hand washing before and after contact is the most important measure in preventing the spread of infection. Everything inside the cubicle should be regarded as contaminated: dressings should be discarded into paper bags in which they can be removed to an incinerator; bedding and clothes should be placed and sealed in distinctive bags before being sent to the laundry; excreta should be treated with appropriate disinfectants; and cutlery and crockery should be disposable and go into the incinerator bags, or (if that is not possible) should not be allowed to return to a kitchen without sterilization. When the patient finally leaves the cubicle, it should be thoroughly washed with a detergent-disinfectant preparation and all equipment should be sterilized so far as possible.

When isolation is necessary but no cubicle is available, **barrier nursing** of patients in open wards is commonly used. The patient's bed is surrounded by screens and a routine similar to that for true isolation is applied to the area inside the screens. While this procedure serves as a reminder to the patient's attendants to use special care, it is not a satisfactory alternative to the use of cubicles, particularly as airborne micro-organisms are not impressed by the 'barrier'.

Strict (special) isolation

This is the form of isolation appropriate for patients with highly transmissible and dangerous infections such as Lassa fever, viral haemorrhagic fevers, anthrax, rabies and diphtheria. The first two of these at least should whenever possible be nursed in designated hospitals equipped with plastic isolator tents and providing highly specialized nursing; but when strict isolation is carried out in other hospitals it requires, in addition to the standard procedures, the wearing of masks when in the patient's room (to be discarded on leaving) and of gloves when handling the patient or bedding.

Protective isolation

This is used for patients whose resistance to infection is seriously impaired by extensive burns, severe bone-marrow disease, treatment with immunosuppressive or cytotoxic drugs, heavy radiotherapy or any other cause of marked immunodeficiency (p. 313). Such patients can be protected to some extent by nursing them in cubicles with sterile air constantly supplied under sufficient pressure to ensure that air flow at all doors and windows is outwards from the cubicles at all times (**positive-pressure** or **plenum ventilation**); this prevents entry of airborne organisms. Far greater protection can be provided by enclosing patients and their beds in plastic isolator (**Trexler**) tents, again with sterile air supplied under pressure, and with access ports to allow feeding, nursing procedures, etc. If microbial contamination of all items entering the isolator is minimized, such immunodeficient patients are protected against virtually all microorganisms except their own. Even these need to be reduced in number in some cases, to lower the risk of opportunist infections (p. 27). The bacterial population of the intestinal tract can be substantially reduced by oral administration of non-absorbable antibiotics such as framycetin + colistin + neomycin (= FRACON); it seems to be best to aim these at the aerobic organisms and to leave the anaerobic flora intact as a protection against overgrowth by opportunist invaders of the intestine, and frequent monitoring of the faecal flora is advisable. Reduction of the normal skin flora by application of disinfectants may be helpful to some severely immunodeficient patients (but see p. 29, protective function).

General Ward Hygiene

Even the patients who do not require isolation must always be regarded as potential sources or recipients of infection with pathogenic organisms. Overcrowding increases the ease with which organisms can pass from one bed to another, and lack of ventilation allows a high concentration of micro-organisms to be built up in the ward air. Bedding, particularly blankets, can become heavily loaded with bacteria, which are thrown off into the air by vigorous bed-making. For this reason it is bacteriologically desirable, though not always administratively convenient, that dressings which have to be changed in the ward should be dealt with before beds are made, and indeed as early in the day as possible since any movement of a patient in bed adds to the bacterial content of the air. Blankets should be made of cotton, and laundered before being used for another patient. Ward dust is often rich in pathogenic organisms and should be removed by a vacuum cleaner rather than a brush; and the vacuum cleaner must be fitted with a suitable filter so that it does not fill the ward air with bacteria collected from the floor. Appropriate steps should be taken to prevent transmission of pathogens by bed-pans and other common utensils, and each patient should have his own thermometer, or disposable thermometers should be used.

Hand Washing

The hands of hospital staff members are the most important vehicles of cross-infection from patient to patient or from equipment to patient. Instruments and bedding can be sterilized, but not hands. Hand washing is the single most important means of preventing the spread of infection in hospital. Hands should be washed after contact with an infected patient's body, excretions or secretions, and before performing any invasive procedures or touching any open wound or any part of a high-risk patient. Hands should also be washed between all patient contacts in intensive care and similar units. Techniques of hand disinfection are discussed on pp. 342–3. Pathogens commonly transmitted by hand are *Staph. aureus* and the enterobacteria and other gram-negative bacilli that are particularly important in hospitals because they are often resistant to antibiotics. Viruses can also be transmitted by hand—particularly enteroviruses and rotaviruses, which can cause outbreaks of infection in special-care baby units.

Staff Carriers of Pathogens

Hand washing is primarily aimed at preventing transmission of 'transient' hand flora (p. 342). Many microbial species, however, can be carried and disseminated around a hospital by true carriers—those in or on whose bodies pathogenic organisms are multiplying. These organisms include *Str. pyogenes,* salmonellae and viruses of respiratory infections—but once again *Staph. aureus* is by far the most important species in this context. There is little point in routine swabbing of hospital staff aimed purely at detecting carriage of this species, since this is common (p. 189) and there is little that can be done about it or indeed needs to be done about it in most cases. There is some point in such routine swabbing of the staff of special units if the phage-types and antibiotic sensitivities of carried strains can be determined. In this way it may be possible to pick out an occasional member of the staff who is carrying an organism which could be really troublesome. Frequently, however, the search for carriers of such strains is carried out not as a routine but in an attempt to find the cause of an outbreak of staphylococcal sepsis. It may be that the number of cases of staphylococcal infection in a ward or unit has been abnormally high and that at least a proportion of them were due to strains of the same phage-type. Swabbing of the staff and of all other regular visitors to the ward may reveal one or more carriers of the appropriate phage-type. It does not follow that these were the source of infection for the patients—they may in fact have acquired their staphylococci from the patients or from a common source—but as carriers of organisms which have shown themselves capable of causing an outbreak of sepsis these people must be excluded from contact with patients, and may be allowed to return only when such carriage ceases. This may occur spontaneously after a few weeks, or may be assisted by application of antiseptic or antibiotic creams or sprays to the nose. The use of systemic antibiotics for such a purpose is seldom justified. Medical, nursing or other hospital staff who have staphylococcal lesions such as boils must be removed immediately from duties in which they may infect patients, since the nature of their lesions proves the pathogenicity of their staphylococci. Similarly, members of staff who develop diarrhoea must be removed from such duties until it can be established that they are not excreting pathogens, as should any who are found to have *Str. pyogenes* in their throats or noses or in skin lesions.

Surgical Wound Infections

Any breach of the skin surface, whether accidental or surgical, provides an open door for bacterial infection. Wound infection may be trivial, with simple local erythema, swelling and tenderness; or there may be pus formation, fever, wound dehiscence and delayed healing; or the infection may extend to cause local thrombophlebitis, lymphangitis and even septicaemia, shock and sometimes death. Four categories of surgical wounds are recognized, differing in their liability to develop infections.

(1) **Clean wounds** are due to elective surgery that does not involve entering the gastrointestinal, genito-urinary or respiratory tract. Infection rates of under 2% are the norm, with *Staph. aureus* as the commonest infecting organism. Such infection is often exogenous and airborne, in contrast to the endogenous infections that predominate in the other categories of wound.

(2) **Clean contaminated** or **potentially contaminated wounds** occur when a site which has a resident bacterial flora is entered without significant spillage, as in an uncomplicated appendectomy or an elective cholecystectomy. Infection rates of 5–10% are reported in this group.

(3) **Contaminated wounds** occur when there is significant spillage of bacterial flora—e.g. from operations on the intestinal tract. The degree of contamination during the procedure will influence the sepsis rate, which is often in the range 15–20%. The enterobacteria and anaerobes (mostly *Bacteroides fragilis*, p. 102) predominate in these infections.

(4) **Dirty** or **infected wounds** occur when surgery involves a perforated viscus, or is to drain an abscess, or when devitalised tissue must be removed after trauma. Infection rates of greater than 30% are common in this group and antibiotic therapy is an essential part of the treatment.

Aseptic surgical technique is aimed at preventing entry of bacteria from any source into wounds during operations, and dressings should be so designed and applied that they fulfil the protective function of the missing skin barrier. A drain in a wound makes it more difficult to exclude airborne organisms and also is liable to predispose to infection by damaging the tissues. When dressings are changed, great care must be taken to protect the wound or burn from infection and to dispose of dressings from already infected wounds in such a way that organisms from them are not transferred to other patients. While there are considerable advantages in changing dressings in special side-rooms rather than in open wards, careless technique which allows contamination of the air and equipment of such dressing rooms can be the means of extensive cross-infection.

Many factors influence the rate of infection in surgical wounds—the quality of the surgical technique (particularly the degree of trauma to the tissues and the adequacy of haemostasis), the duration of the operation (the longer it lasts the higher the risk of infection), the efficiency of the surgical scrub and wound preparation (see p. 343) and, in the case of bowel surgery, the quality of the bowel preparation and the appropriateness of the prophylactic antibiotic regimen (p. 376). The presence of a foreign body such as a hip prosthesis or an artificial heart valve will predispose to infection with low-grade pathogens; *Staph. epidermidis* in particular has emerged as important in such situations. Host factors are also important; increasing age, malnutrition, underlying disease (e.g. diabetes) or steroid therapy increase the risk of infection.

The importance of airborne infection in clean surgery has been particularly thoroughly investigated in relation to hip replacement. The low incidence of infection can be significantly reduced by ventilating the theatre with ultra-clean air, by placing the patient in a Trexler plastic isolator and by the operating team wearing whole-body exhaust suits which prevent their organisms from reaching the patient through the air (or in other ways). However, a similar reduction in wound infection following such operations—1.5% to about 0.6%—has been achieved by antibiotic prophylaxis (p. 376) without such care to prevent aerial spread.

The relative importance of different routes of spread for a particular potential wound pathogen may be determined by its own properties. For example, staphylococci and streptococci can persist in dried exudates and in dust far better than can enterobacteria and related organisms; whereas the latter survive well in moist conditions and can thrive in nebulizers and in ventilating machines, from which they may be sprayed into patients' wounds or respiratory tracts.

Antibiotic prophylaxis in surgery

The use of antibiotics to prevent wound infections is discussed on pp. 375–6.

Special Units and Procedures

Burns units

Patients with extensive burns provide bacteria with large areas of exudate and damaged tissue which constitute an excellent culture medium (p. 307). Infection may delay healing, lead to breakdown or rejection or skin grafts, or even—in such a compromised host—cause septicaemia and death. Such patients are best cared for in special burns units—isolation wards with individually ventilated air-locked rooms, or at least with plenum-ventilated dressing rooms. Changing of dressings on burns is doubly hazardous, because it exposes the lesions to the air and because enormous numbers of organisms may be dispersed into the air if the burns are already infected. Cross-infection can be avoided by using techniques that minimize such dispersal, by using antibiotics to eliminate or control such infections as occur in the unit, and by isolation procedures. Group A streptococci, once the chief threat to burned patients but readily controlled by antibiotics, are now far less important in this context than *Staph. aureus* and antibiotic-resistant gram-negative bacilli—klebsiellae and other enterobacteria and *Ps. aeruginosa*.

Intensive care units

Such units are necessary in modern hospitals; but they are liable to provide ideal places for cross-infection, because they contain concentrations of patients who are abnormally susceptible to infection because of major surgery or trauma or other factors, who are subjected to numerous invasive diagnostic procedures, and who are linked to various pieces of ancillary equipment that require repeat handling by staff and offer many routes for infection—e.g.

ventilators, dialysis machines, suction apparatus and intravascular, urethral or other catheters. The antibiotic-resistant gram-negative bacilli referred to in previous paragraphs are again important here, often first colonizing the patients and then causing significant infection in the same or other patients. The risk of outbreaks of infection in such units can be minimized by constant vigilance about hand washing, attention to proper technique when aspirating respiratory secretions or inserting catheters, and frequent bacteriological monitoring of the patients, their equipment and appropriate parts of their environment.

Continuous ambulatory peritoneal dialysis (CAPD)

CAPD, increasingly used in recent years in treatment of end-stage renal disease, involves the introduction of a special (Tenckhoff) in-dwelling catheter into the peritoneal cavity. Dialysis fluid is infused into the cavity, left there for 6–8 hours, drained out and replaced, in a continuous cycle. Patients are instructed in sterile technique for changing the dialysate, but since this is a procedure repeated very many times it is impossible to eliminate the possibility of organisms gaining entry to the peritoneal cavity via the tubing, connectors or catheter and so causing peritonitis. This may be heralded by mild pyrexia, but the diagnostic finding is cloudiness of the drained fluid, with its polymorph count exceeding $100/mm^3$. In around 40% of episodes the organism causing the peritonitis is a coagulase-negative staphylococcus; other episodes are due to *Staph. aureus*, enterobacteria, α-haemolytic streptococci, enterococci, pseudomonas and candida. Antibiotic treatment should of course be appropriate to the organism isolated, but cephalosporins or aminoglycosides administered in the dialysis fluid, alone or with vancomycin, are commonly effective.

Injections and infusions

Introduction of a needle and extraneous fluids into a patient's subcutaneous or muscular tissues or his blood stream may introduce pathogens. This risk may be minimized by adequate skin preparation (p. 343), aseptic technique including particular care not to contaminate the prepared skin with organisms from the operator's respiratory tract or hands, and use of properly sterilized syringes, needles, blood-giving sets and fluids.

Studies of the incidence of bacteraemia asssociated with intravenous devices have shown that this occurs in 0.2–0.5% of patients with peripheral intravenous drips and in 3.8–12% of those with the central 'long-line' catheters that are frequently used in the management of seriously ill patients. There are various routes by which organisms may gain entry to an intravenous infusion system, the commonest being the insertion site itself—up to 10% of patients with intravenous devices develop thrombophlebitis. The development of infection is often heralded by a spike of fever or a rigor. The drip should be discontinued, blood cultures taken and the catheter tip sent for microbiological investigation. When *Staph. aureus* or an intestinal gram-negative bacillus (e.g. a klebsiella, *Serratia marcescens* or a pseudomonas) is grown from such a tip, it is readily recognized as a pathogen; but it is important that coagulase-negative staphylococci should not be dismissed as skin contaminants as they are the commonest colonizers of silastic catheters, adhering to their surfaces and forming microcolonies with persistent shedding of bacteria into the blood and low fever. *Candida albicans* is particularly liable to cause trouble in long-line catheters used for parenteral nutrition.

Certain pathogens may be transmitted from person to person by transfusion of blood or blood products. Since there is little that can be done to sterilize blood, donors must be chosen with care to avoid any likelihood of their blood containing these pathogens. They must be in good health and give no history of hepatitis, syphilis, drug addiction or homosexuality, or of acquired immunodeficiency syndrome (AIDS) in themselves or their contacts. Serological screening of blood to exclude transmission of syphilis (p. 278), hepatitis B (p. 265) and AIDS (p. 283) is now routinely performed in blood transfusion centres.

Surgery in patients with protozoan or helminth infections

Such infections sometimes call for special care before, during or after surgery. The faeces of patients who might be infected with *Ascaris* should be carefully examined before intestinal surgery and infection should be treated, as otherwise the worms may penetrate and mechanically disrupt the wound. Amoebic abscesses (p. 269) can usually be treated by chemotherapy, sometimes assisted by aspiration, and open surgical drainage is rarely necessary; it must always be preceded by chemotherapy, since opening of an untreated amoebic abscess frequently leads to fatal generalized amoebiasis—a point that must be remembered when opening any abscess that might be amoebic. Opening of a hydatid cyst (p. 271) requires special care not to spill the contents into the peritoneal cavity or tissues, since this may result in an anaphylactic reaction to the spilt fluid or in dissemination and implantation of the immature scolices contained in the 'sand' in the fluid. Incomplete removal of viable germinal epithelium from the lining of a hydatid cyst results in the formation of multiple cysts.

Care of Particular Body Tracts

The urinary tract

The passage of a catheter or other instrument into the bladder can cause infection of the urinary tract, usually with *Esch. coli* or other gram-negative bacilli, and even bacteraemia ('catheter fever'). Such instrumentation should therefore be carried out only when essential, with full aseptic techniques, using pre-sterilized disposable catheters whenever possible. When the bladder is drained continuously through an in-dwelling catheter, a **closed drainage system** should be used, to prevent ascending infection from the air or environment. Such a system must provide some means of drawing off an uncontaminated specimen of urine, for bacteriological examination, at a point near to the end of the catheter; and it should end in a plastic collecting bag with a non-return valve and preferably with some means of emptying it without entry of organisms, as frequent changing of bags is a common means of introducing infection. Care must be taken to see that infection does not develop in the urethral orifice around the catheter.

One-third of all hospital-acquired infections are urinary tract infections (pp. 273 and 346). Patients at greatest risk are those with indwelling catheters, especially women and those over 40 years of age. The incidence of infection increases markedly with the duration of catheterization. The source of the infecting organism is often the patient's own faecal flora. Infected patients in turn become reservoirs of multi-resistant gram-negative bacilli, which can lead to direct cross-infection of others, especially in intensive care units, or can indirectly infect others via contamination of the hospital environment. Hospital-acquired urinary tract infection may lead to the serious complications of ascending infection, with pyelonephritis and renal failure and death or, more acutely, to septicaemia and shock—a well-recognized problem in patients undergoing genito-urinary instrumentation or surgery.

The respiratory tract

Patients who lie still for long periods as a result of unconsciousness, major operations, paralysis or other causes, and some who have respiratory tract virus infections or other predisposing conditions, are liable to acquire pneumococcal, staphylococcal or other infections of their lower respiratory tracts. Endotracheal and tracheostomy tubes increase their risks of infection by by-passing the normal defences of the respiratory tract. Apart from general hygienic measures there is little that can be done to prevent these infections. Attempts at antibiotic prophylaxis often do more harm than good (p. 376); antibiotic treatment of an established infection is a different matter and should begin promptly.

Many of the outbreaks of legionnaire's disease (p. 203) recognized in the USA, Britain and elsewhere since the late 1970s have occurred in hospital, where they have mainly affected debilitated or otherwise compromised patients. Means of spread of this disease and preventive measures are considered on p. 356.

Respiratory tract virus infections spread

readily in hospitals, as in other closed communities. Of particular importance are outbreaks of respiratory syncytial virus infection among young children and of influenza A or B in geriatric wards.

The alimentary tract

Outbreaks of *Esch. coli* and rotavirus gastro-enteritis among babies and of *Sh. sonnei* dysentery occur from time to time in hospitals. They can be prevented or controlled by measures already outlined in this chapter—isolation, general hygienic precautions and exclusion of carriers. However, the scale of catering required in hospitals makes them potential sites for large outbreaks of food-poisoning, often affecting patients and staff and usually due to salmonellae, *Cl. perfringens* or campylobacters. The high proportion of elderly, debilitated or otherwise vulnerable people in hospitals means that fatalities are relatively high in such outbreaks. It is important that all hospital food-handlers should be adequately trained, and that the precautions outlined on pp. 360–1 should be scrupulously observed. Nursing and other staff who develop gastro-enteritis and continue on duty during the acute phase of the illness can be highly effective transmitters of the causative organisms to their patients, since they are likely to be excreting them in large numbers and can easily contaminate their hands. Staff with diarrhoea should therefore be removed from duty and investigated microbiologically as soon as possible, and should not be allowed back until they have recovered clinically. Persistent excretion of pathogens by a staff member after diarrhoea has ceased should not lead to a serious risk of infection of patients if normal hygienic practices are observed, but even so it is unwise to allow a salmonella-excreter, for example, to work amongst such high-risk patients as neonates, the elderly or the immunocompromised.

Investigation of Outbreaks of Hospital Infection

We can deal only very briefly with this large and complex subject. Investigation of such an outbreak is a detective operation, and begins with the accumulation of evidence—the number of patients involved; their distribution in the hospital or ward; the times of onset of their symptoms and the probable times at which they were infected; whether all or the majority of cases followed operation and, if so, whether they were operated on in the same theatre or by the same team; and any other clues as to the way in which they became infected. If their infections are all due to apparently identical bacteria, a human carrier or other source of the appropriate organism must be sought, whereas an outbreak of infection due to various organisms suggests a breakdown in theatre or ward ventilation, in aseptic techniques or in the sterilization of dressings or instruments. Such work is best carried out by an IC team (p. 346), and can be valuable provided that there is some mechanism for ensuring that appropriate administrative action is taken when there is evidence of trouble. A small IC committee—including, for example, a surgeon and a bacteriologist, the IC officer (if he is not one of the first two), the IC nurse (who must be of adequate seniority to speak for the nursing administration) and a hospital administrator, with other members co-opted as the occasion demands—can be a suitable body for dealing with such episodes, as well as for defining the hospital's policies for preventing cross-infection. (It is, of course, easier to define policies than to ensure that they are carried out by all concerned!)

Protection of Health-Care Staff

Health authorities commonly have occupational health services that collaborate with IC teams to screen staff for evidence of infection or carriage before they start work, to co-ordinate their immunization programmes and to minimize their risks of acquiring infections from patients.

Tuberculosis

All hospital workers should have tuberculin tests and chest radiographs when they take up their employment, and those with negative tuberculin tests should be offered BCG immunization (p. 365). Those working with tuberculous patients or in microbiology laboratories should have annual chest radiographs. Only those who are known to be tuberculin-positive should be allowed to attend patients with known or suspected open tuberculosis (p. 207); and such patients should be isolated in single rooms or special wards, for the sake of staff as well as other patients, until they have received appropriate anti-tuberculous treatment for 2–3 weeks. Patients with tuberculosis but no tubercle bacilli visible in their sputum are essentially non-infectious. Sputum specimens from patients known or suspected to be tuberculous should be sent to the laboratory in sealed containers inside plastic bags, with the request forms not included in the bags, and both the specimens and the forms should be appropriately labelled so that the specimen container is opened in the laboratory only in a suitable safety cabinet. When an unsuspected open case of tuberculosis is discovered, the occupational health service should be informed and should follow up all exposed staff.

Hepatitis B

The risk of hepatitis B infection in health-care staff was highlighted in the 1960s by serious outbreaks, with fatalities, among the staff of renal dialysis units and a considerable number of infections among laboratory staff. Transmission to hospital patients has been less common. The major risk to staff is from the blood of acute hepatitis B patients or of HBeAg-positive carriers (p. 135). The common routes of transmission are accidental injury with a blood-contaminated needle or other sharp instrument, contamination of skin cuts or scratches or accidental splashing of infected blood on to the conjunctiva or a mucosal surface—e.g. of the mouth. Any such incidents should be immediately reported and recorded, and an appropriate senior member of staff should attempt to establish the source of the blood and the HBsAg and HBeAg status of the patient from whom it came. Hepatitis B immunoglobulin should be given within 48 hours if the patient is HBsAg-positive or if the source of the blood is not known. The risk of needle-stick and other injuries can be reduced by careful handling of all sharp instruments and by discarding them into puncture-resistant bins. Further trouble in renal dialysis units has been prevented by excluding from them all HBsAg-positive staff and patients; and strict guidelines for collection, transmission and laboratory handling of blood and other fluid samples from known hepatitis B carriers or those at high risk of being so have further reduced the hazards to health-care staff. Blood from such high-risk patients should be collected only by trained staff wearing disposable gloves, and should be transported in leak-proof bottles or tubes kept upright in sealed plastic bags and separate from the request forms. Specimens and request forms should be conspicuously marked 'Hepatitis Risk' or 'Danger of Infection', in accordance with local codes. However, these precautions applied to blood from known high-risk patients must not lull anyone into forgetting the possible hazards of blood from unsuspected carriers (p. 162). In Britain hepatitis B in health-care staff is recognized as an industrial disease for which they can claim compensation.

Acquired Immune Deficiency Syndrome

So far it appears that the risk to health-care staff of acquiring AIDS is less than that of acquiring hepatitis B. Needle-stick injury is again the most likely means of transmission.

Precautions in collecting and transmitting blood or other fluid samples should be as for hepatitis B.

Community (Public Health) Problems

The principal vectors for spread of infections through human communities are air, insects, water (including sewage), milk and food. We have discussed various aspects of the subject of airborne infection (notably on pp. 35, 301 and 346–50); the few measures to reduce it that can be taken by public health authorities include prevention of overcrowding, ensuring adequate ventilation of places where people congregate and, in special circumstances, discouraging them from doing so (p. 336). Control of insect vectors has been considered in relation to various insect-borne infections (notably on pp. 111–3, 241 and 267) and summarized on p. 335.

Water-borne Infection

Water, particularly drinking water, has enormous potential as a means of spreading microbial diseases (p. 254); and provision of safe drinking water is thus of great importance to public health. Bacteriological monitoring of a water supply is an essential part of the procedure for ensuring its continuing safety.

Many factors influence the total bacterial content of a water supply, including the following:

(1) Rivers fed by surface drainage contain many bacteria, some collected from the air by the water as it fell in the form of rain or snow and many collected from the soil or other surfaces on to which it fell or over which it passed.

(2) Water from deep wells or springs usually has a low bacterial content because it has undergone filtration as it percolated through the soil to reach underground lakes or rivers.

(3) Bacterial multiplication may occur in running water if it contains suitable organic nutrients and if the temperature and other conditions are appropriate.

(4) When water comes to rest in large lakes and reservoirs, its bacterial content is as a rule greatly reduced as a result of sedimentation and other factors.

(5) Supplies to most developed human communities are usually first stored, then further purified by filtration through sand-beds (which owe their efficiency as filters to a surface layer of protozoa and algae), and finally chlorinated. Free chlorine, in a concentration as low as 1:5 000 000, rapidly kills nearly all vegetative bacteria provided that there is very little organic matter present. However, the chlorine levels usually found in treated water are insufficient to kill the cysts of amoebae and *Giardia*.

It is not merely the total content, however, which is important in assessing the suitability of a water supply for human consumption, though a high total count certainly suggests unsuitability. What matters is the possibility that it contains potential intestinal pathogens. (Non-intestinal pathogens that can be carried in drinking or other water are considered separately below.) Important among these intestinal pathogens are: typhoid, paratyphoid, dysentery and cholera bacilli and campylobacters; viruses, notably those of poliomyelitis and hepatitis A; and the protozoa *Giardia intestinalis* and *Entamoeba histolytica*. Presence of such organisms results from contamination with human excreta, or in some cases with animal or bird droppings. There are a number of obvious precautions which can be taken to reduce the risk of such contamination—e.g. drawing supplies only from relatively uninhabited catchment areas; ensuring that one community is not discharging its untreated sewage into a river upstream from the point at which another is drawing off its water supply; or, in places where there is no main water supply or main drainage, seeing that wells are so situated and protected that there cannot be any leakage into them from pit-latrines and the like. A dangerous level of contamination is highly improbable in the case of water taken from a fast-running stream in a hill or mountain area above the level of human habitation. With this exception it is virtually always unwise to drink unboiled water derived from a source that has not been subjected to thorough and repeated testing, as described below, or to filtration and chlorination with adequate bacteriological control.

Routine examination of water supplies aimed directly at detection of bacterial pathogens is impracticable. With very rare exceptions (e.g. when urine from a renal carrier of *S. typhi* has entered the water), any bacterial hazard in drinking water is due to faecal contamination. Where this has occurred, pathogens are likely to be heavily outnumbered in the water by other faecal bacteria. Water found to contain substantial numbers of enterobacteria, and in particular of faecal-type *Esch. coli* (recognizable by their ability to grow at 44°C in the presence of bile salts) is presumptively contaminated with faeces and therefore possibly with faecal pathogens.

From what has been said above it is obvious that proper management and disposal of **sewage** is important for the safety of a water supply. Sewage collection and water distribution systems inevitably follow similar routes in residential areas, and it is therefore important to ensure that there can be no leakage from the one to the other, even in times of heavy rainfall or flooding; and the processing of sewage must be such as to minimize the number of potential intestinal pathogens contained in effluent that is discharged into rivers which provide the water supplies of communities downstream. Similarly, the indiscriminate spreading of slurry over agricultural land must be prevented because it may well lead to contamination of the environment and so of a drinking water supply (as well as to spread of pathogens by other means). Although the provision of a 'pure' water supply is the aim of the public health measures outlined above, water must also be supplied in adequate quantity for the drinking, washing, etc. needs of the population. The provision of sufficient 'pure' water would be prohibitively expensive for many less developed countries, but generally 'safe' supplies can be produced at much lower cost and be made more widely available by measures such as digging wells or constructing simple collection and filtration systems for drinking water, and in the absence of a sewage system, the construction of latrines to prevent contamination of water supplies.

Non-intestinal infections spread by water (For intestinal infections see pp. 254–5)

Legionella pneumophila is present in many natural water supplies and can establish itself in cooling towers, air-conditioning systems and both hot- and cold-water supplies—notably in large buildings such as hospitals (p. 203). Human infection results from inhaling droplets released from cooling towers, air-conditioning systems, shower heads, heated whirlpool baths etc. into the surrounding air. Heavy contamination of water is particularly liable to occur when water systems are inadequately cleaned and maintained or are turned on again after being out of use for some time. The organisms flourish at temperatures between 20 and 45°C, particularly when they can colonize rubber washers, blind loops of plumbing or rust and sludge in calorifiers, etc. They can be eliminated from a water system by thorough cleaning of tanks and pipes and by treatment of the whole system with chlorine or another suitable biocide. They can also be killed by heating above 60°C, and maintenance of such temperatures in hot-water systems keeps them safe from legionella colonization but introduces the risk of scalding users.

Outbreaks of **pseudomonas skin infections** (rashes and folliculitis) have been associated with use of heated whirlpool baths (hot tubs and jacuzzis). Such problems have arisen where the whirlpools have not been properly maintained, filters have been infrequently cleaned and replaced, and inadequate attention has been paid to disinfection with chlorine or bromine and to control of pH and temperature. The organism can be grown from the skin pustules, from the water in the pool (in high concentrations) and from the pool surrounds. *Pseudomonas* multiplies rapidly in the hot water if there is inadequate disinfection. *Ps. aeruginosa* is also recognized as a cause of otitis externa in swimmers and saturation divers who are working in extremely humid conditions.

Leptospirosis (p. 267) used to be common in sewer workers and miners who worked in water contaminated with leptospire-infected rat urine; rodent control and protective clothing have almost eliminated this form of infection. Another factor may have been the widespread use of detergents, which are toxic to leptospires. However, the popularity of water sports—swimming in fresh-water ponds and streams, canoeing and raft racing—and the use of canals for recreational boating has seen an increase of leptospirosis in these groups. It is also a common disease in the tropics, particularly associated with irrigation schemes.

Primary amoebic meningo-encephalitis caused by the free-living amoeba *Naegleria fowleri* (p. 220) is a rare but often fatal condition which can affect children and young adults who swim in warm fresh-water lakes, ponds or stagnant water.

'Monday morning fever' is a form of extrinsic

allergic alveolitis (p. 141) resulting from inhalation of 'organic soup' formed in contaminated recirculating water in a humidification or water-spray air-conditioning system and released into the air as an aerosol when the system is turned on after a period of inactivity. The probable offending agents are thermophilic actinomycetes, amoebae and enterobacterial endotoxin.

Our consideration of water-borne disease should include infections that are spread not by water itself but by vectors that depend on water. Malaria (p. 241) and yellow fever (p. 267) are spread by mosquitoes that have their larval stage in water. Onchocerciasis (p. 158) is spread by *Simulium* black flies whose habitat is along river banks. The three species of *Schistosoma* (p. 155) that have become widespread in the wake of so many irrigation schemes have an essential developmental stage in certain species of water snail, and can infect man only by penetrating skin immersed in water. Schistosomiasis shares with the diarrhoeal diseases a common preventive measure; if sewage contamination of the water is prevented, then the route from the human intestine or bladder to the snail intermediate host is blocked.

Milk-borne Infection

Human milk taken by the baby straight from the mother is seldom a vector of pathogens, though transmission of cytomegalovirus and other viruses to unprotected infants by this route is possible. In human milk banks the milk is pasteurized to prevent feeding of contaminant micro-organisms to the infants, though this process may deprive them of passive immunity by destroying maternal antibodies. Cow's milk, on the other hand, presents many important problems. It may contain pathogens derived from the cow; the circumstances of its collection, unless carefully controlled, permit it to become heavily contaminated with a wide variety of micro-organisms; it is a good culture medium for many of these; it may spend hours or days at temperatures suitable for bacterial multiplication before it is consumed; and, as the result of pooling, milk from a single cow may be distributed to a large number of human beings.

Organisms which may be present in cow's milk include the following:

(1) **Pathogens excreted in the milk or derived from the animal's udders.** The most important of these are *Myco. tuberculosis* and *Br. abortus* (*Br. melitensis* being transmitted similarly in goat's milk in some countries). Other pathogens which occasionally come under this heading are *Staph. aureus, Str. pyogenes* (including scarlet fever strains), and *Cox. burneti*.

(2) **Pathogens derived from the animal's faeces.** The risk of such faecal contamination, and of consequent transmission of salmonellae or campylobacters, is greatly increased during an outbreak of scours in the herd.

(3) **Pathogens derived from the hands or respiratory tracts of dairy workers during or after milking; from utensils or bottles washed inadequately or with water from a contaminated supply; or from airborne or other contamination.** These include *Staph. aureus, Str. pyogenes,* salmonellae and shigellae, campylobacters and possibly viruses such as those of poliomyelitis and hepatitis A.

(4) **Non-pathogens** of many varieties and from many sources, which may multiply in the milk and cause souring or other changes in it. Souring, which is due to the production of lactic acid from lactose by such organsims as *Str. lactis* and lactobacilli, is deliberately encouraged for such purposes as the production of butter and cheese, but is otherwise undesirable.

Regular tuberculin-testing of cattle and examination of milk for brucella antibodies (see p. 358) make possible the detection and slaughter of animals with these infections. Scrupulous attention to the general health of the cattle and to any local lesions of the udders are other important steps towards eliminating pathogens of the first category. Subsequent contamination of the milk can be reduced by a high standard of hygiene in cow-sheds and by efficient washing and sterilization of all utensils and containers. Multiplication of organisms can be kept to a minimum by cooling the milk as soon as it is collected, keeping it cool during transmission, and delivering it to the consumer early in the day while the atmospheric temperature is still low.

But even a combination of all these measures does not guarantee that the milk will be safe to drink. This can be achieved only by heating it, as described below.

Pasteurization

A process which Louis Pasteur devised to deal with a problem of the wine industry has been generally adopted as the most satisfactory treatment for milk. If milk is heated to 63–66°C and kept at this temperature for 30 minutes (the **Holder process**) or is heated to 71°C for at least 15 seconds (the **high-temperature short-time process**), all vegetative pathogens are killed; spore-forming pathogens are of no importance in this context. The milk must then be rapidly cooled to 10°C or less, so that surviving organisms do not multiply. The adequacy of the treatment which a specimen of milk has received can be tested by the **phosphatase test**. This depends upon the fact that the enzyme phosphatase is constantly present in fresh milk and is destroyed by heat treatment that just conforms with the above standard. Inability of treated milk to liberate phenol from disodium phenylphosphate is therefore evidence that it has been effectively pasteurized.

Pasteurization does not impair the taste of milk, and there is no evidence that it appreciably lowers its nutritional value. It does make the milk entirely safe to drink, provided that it is delivered into sterile bottles immediately after treatment and is kept sealed until it is consumed.

Sterilization and ultra-heat

Two forms of more vigorous heat treatment are used to destroy all bacteria in milk and so prevent it from being soured or otherwise spoiled by bacterial multiplication. Such milk has a long shelf life at room temperature. Sterilized milk is heated to 105–115°C in its container whereas ultra-heat treated (UHT) milk is heated to 135–150°C and then aseptically dispensed into sterile containers.

Bacteriological examination

The bacterial content of milk can be assessed by the **methylene blue reduction test**. This is a non-specific test for the presence of organisms producing enzymes that reduce and thereby decolourize methylene blue. Milk collected with reasonable care to avoid contamination, and therefore suitable for distribution or pasteurization, should not have a bacterial population that is able under standard conditions to decolourize methylene blue in less than 30 minutes. Examination for individual pathogenic species such as *Myco. tuberculosis* or *Br. abortus* is carried out by culture on suitable media. Milk from cows with brucellosis may contain brucella agglutinins. Even when such milk has been mixed with a large volume of milk from healthy cows, these agglutinins can be detected by the sensitive **brucella ring test**. In this test, haematoxylin-stained dead brucellae are added to a sample of the milk; if they are agglutinated, they rise up in the fat globules and form a blue ring in the cream layer.

Milk products, such as butter and cheese, and milk-containing foods, such as ice-cream and custards, are of course liable to contain pathogens similar to those found in milk, and are exposed to greater risks of contamination by handlers. It is, however, much more difficult to devise bacteriological standards and tests for these products, apart from cultures to exclude the presence of named pathogens.

Food-borne Infection

Many foods make good microbial culture media and 'go bad' unless protected from the deleterious effects of multiplying bacteria and fungi. This protection can take the form of cooking, refrigeration, drying or the addition of sugar, salt or other preservatives.

The role of food as a vector of pathogenic organisms is not closely associated with obvious deterioration. Food which has been made unpalatable by the activities of multiplying organisms is not necessarily harmful to eat, and on the other hand food may contain large numbers of pathogens and yet be normal in appearance and pleasant to eat; if this were not so, there would be few outbreaks of food-poisoning.

What we said above about the difficulties of devising bacteriological standards and tests for milk products applies even more strongly to other food substances, because of their great diversity.

Food-poisoning is a major health problem. 14 000 cases a year are notified in England and Wales and this is undoubtedly an underestimate of the true incidence, since many cases are not reported by patients to clinicians or by clinicians to the appropriate authorities. Early notification of an incident and the taking of specimens during the acute phase of the illness is extremely important if the causative agent is to be identified. The investigation of outbreaks is most efficiently carried out by a team comprising the consultant microbiologist, the community medicine specialist, the environmental health officers and the veterinary practitioners where there is animal involvement. Meat and poultry are the foods incriminated in the majority of food-poisoning incidents in Britain; dairy products, and in particular raw milk, are still important sources of salmonella and campylobacter infection in England and Wales, but not in Scotland since a ban on the sale of unpasteurized milk was introduced there in 1983. Food-poisoning incidents are most commonly caused by *Salmonella* species, followed in frequency by *Camp. jejuni, Cl. perfringens, Staph. aureus, Bacillus cereus, Vibrio parahaemolyticus*, enterotoxigenic *Esch. coli* and *Cl. botulinum*. The aetiology and clinical features of these types of bacterial food-poisoning are described on pp. 256–9.

Food (like water) can also be the vehicle for transmission of viruses that cause gastro-enteritis (e.g. rotaviruses—p. 252); such food-poisoning has a 24–72 hour incubation period, but the epidemiology is confused by the common secondary transmission of infection to close contacts who had not eaten the food responsible for the primary infection.

Shell-fish are commonly implicated in food-poisoning outbreaks. They have been the source of outbreaks of typhoid in the past and more recently of virus gastro-enteritis and of hepatitis A infection. Bivalve molluscs (oysters, cockles, mussels, clams and scallops) are the shell-fish most frequently involved, because they filter estuarine water and concentrate faecal bacteria and viruses which may be present as a result of sewage contamination. In some outbreaks of diarrhoea, small round viruses similar to those associated with winter vomiting have been demonstrated in the stools of those affected and in the shell-fish consumed. The problem of control and prevention is compounded by the fact that shell-fish are often eaten raw. Control of the infection risk depends on routine bacteriological examination of shell-fish and on effective purification by keeping them in clean water for an adequate time before they are marketed. *Vibrio parahaemolyticus* is a frequent contaminant of shell-fish and is responsible for 60% of food-poisoning outbreaks in Japan where uncooked seafood is very popular.

Travellers' diarrhoea

Diarrhoea in those who have recently arrived in foreign lands (often with questionable standards of hygiene) is an all too common experience (p. 251). Food and water are the likely sources of infection. The pathogens responsible are enterotoxin-producing strains of *Esch. coli*, viruses (especially rotaviruses and Norwalk virus), salmonellae, shigellae, campylobacters and *Giardia intestinalis* and *Entamoeba histolytica*. Many cases are self-limiting and settle in 3–4 days, although in infants a careful watch must be kept for signs of dehydration. Prevention depends, in part, on the areas to be visited but hinges on advice to avoid uncooked food such as salads and shell-fish; not to consume raw milk; and to drink bottled water or boil or chlorinate all drinking water. Fresh fruit is permissible if it can be washed and peeled. The prophylactic use of antibiotics has no proven value and should be discouraged.

Laboratory investigation of food poisoning

The nature of the organism responsible for a food-poisoning outbreak is often suggested by the pattern of the outbreak—whether it is confined to a household or involves large numbers of people, the time relationships, the nature and severity and duration of the symptoms, and so on. Except in salmonellosis the investigation of an outbreak has little bearing upon treatment and is principally concerned with finding out what went wrong and preventing further outbreaks. The laboratory's contribution is to try to isolate the responsible organism, by culture of faeces and vomit (if available) from a manageable proportion of the patients and by microscopy and culture of the offending food if it can be identified and if some of it is still available. Precise identification of the organism isolated—e.g. by phage typing (p. 75) or serotyping (p. 92)—may help to pinpoint the source of contamination of the food. Sometimes the history of the outbreak clearly incriminates a particular item or at least a meal, but in other cases detailed detective work is necessary. If the food is available and either a salmonella, a heat-resistant *Cl. perfringens* or a *B. cereus* is involved, there is not usually much difficulty in growing the organisms from the food and from at least some of the patients, and in showing that they are identical. *Cl. botulinum* is also likely to be recoverable from the food, and although it is unlikely to be recovered from the patients it produces a highly characteristic clinical picture. The case against a staphylococcus, however, may be difficult or impossible to prove. These organisms, being less heat-resistant than their enterotoxin, may have been killed by cooking. By using a highly sensitive enzyme-linked immuno-assay (p. 178) it is possible to demonstrate the presence of staphylococcal or *Cl. perfringens* enterotoxin in suspected foods and in the faeces from those involved in an outbreak.

Prevention of food poisoning

The responsibility for preventing food-poisoning is shared by many people, including public health authorities, wholesale and retail food distributors, caterers and their staffs, and housewives. The following are among the more obvious general precautions:

(1) All animals should be inspected before being slaughtered for human consumption, and their meat should be inspected afterwards, for evidence of relevant disease.

(2) All consignments of potentially dangerous food ingredients such as spray-dried or frozen eggs should be tested bacteriologically.

(3) All food should be protected from flies, rodents and other possible vectors of pathogens at all times—during distribution and storage and after being cooked.

(4) All food in or on which bacteria could multiply should be kept in a refrigerator, or at least cool, at all stages. This applies particularly to such excellent culture media as ice-cream mixtures and synthetic cream.

(5) Meat, poultry, etc. that have been stored uncooked in the frozen state must be allowed adequate time for complete thawing before being cooked, otherwise the heat may fail to penetrate them adequately.

(6) Cooking, especially of meat, should be thorough, and food which is not to be eaten immediately after cooking should be cooled rapidly. If reheating is necessary, temperatures of more than 60°C should be attained with a minimum of delay. These precautions are particularly important with large quantities, especially of meat. Cooling of large joints is accelerated if they are cut into several pieces immediately after cooking.

(7) Cooked food must be protected from the risk of contamination by contact with uncooked food, or with instruments or surfaces that have been used for uncooked food and not subsequently cleaned. In large kitchens the work-flow should ensure that uncooked and cooked foods are dealt with in separate areas.

(8) Known carriers of salmonellae, shigellae or *E. histolytica* should be excluded from work with foods. Appropriate laboratory tests for carriage of intestinal pathogens should be carried out on all those about to be engaged in the kitchens of institutions, restaurants, etc., or in the food-distribution trade at points at which contamination of the food could have serious consequences.

(9) A high standard of personal hygiene, especially the careful washing of hands after defaecation, should be observed by all food handlers (and indeed by everyone!).

Precautions that apply especially to particular types of food-poisoning are indicated in the following paragraphs.

Prevention of **salmonella food-poisoning** requires care in the production, handling and storage of food. Measures to reduce the amount of salmonella infection in domestic animals include improvement in feedstuff control, in accommodation for stock, in handling of stock at markets (where cross-infection can occur readily) and on journeys, and in procedures at poultry-packing works and abattoirs. Adequate thawing and thorough cooking of all poultry and meat and the avoidance of subsequent cross-contamination from raw meat are particularly important parts of food preparation.

Although typhoid is not strictly 'salmonella food-poisoning', international transmission of salmonellae is well illustrated by the 1964 Aberdeen typhoid outbreak, which was the largest of several that occurred in Britain within a few years in association with the distribution of corned beef imported in large cans from South America. These cans were in effect cultures of typhoid bacilli, which they had presumably acquired because river water that was contaminated with human excreta had been used to cool the cans after sterilization and had been sucked into them through faulty joints. The large scale of the Aberdeen outbreak—507 cases—was a result of contamination of a slicing machine and other utensils in the shop in which the corned beef was sold, and consequent transfer of typhoid bacilli to various other foods sold in the same shop. A fortunate feature of this series of outbreaks was the very low frequency of secondary cases—i.e. of patients who had not themselves eaten the contaminated food but had acquired their infection from those who had.

The source of **staphylococcal food-poisoning** is usually a food-handler with a staphylococcal finger infection; such a person should be excluded from work involving the handling of foods. As far as is possible, food that is to be eaten without further heating should not be touched by hand but by clean utensils and should be kept in a refrigerator.

Cl. perfringens **food-poisoning** is usually associated with mass catering. Contamination of meat with *Cl. perfringens* is impossible to prevent. Prevention of this type of food-poisoning depends upon adequate cooking with heat-penetration sufficient to kill spores (with sizes of joints and volumes of stews, etc., small enough to ensure this); not holding cooked foods for more than very short periods at temperatures suitable for bacterial growth; and thorough reheating of any food that has been held and needs to be hot.

Prevention of **botulism** depends upon careful supervision of commercial canning to ensure that canned vegetables, fish and meat items are heated to temperatures that are lethal to clostridial spores. If there is any ground for doubting the effective sterilization of home-preserved foods in a part of the world in which botulism occurs, they should be heated to 100°C for more than 10 minutes immediately before consumption to destroy *Cl. botulinum* toxin.

B. cereus **food-poisoning** can be prevented by not keeping rice in the warm moist conditions that favour growth of *B. cereus* and production of its toxins.

Suggestions for Further Reading

Maurer, I. 1978. *Hospital Hygiene*. 2nd edn. Edward Arnold, London. An excellent small book, full of good practical advice.

Hobbs, B.C. and Gilbert, R. 1978. *Food Poisoning and Food Hygiene*. 4th edn. Edward Arnold, London.

Benenson, A.S. (ed). 1981. *Control of Communicable Diseases of Man*. 13th edn. American Public Health Association, New York. Detailed information about the control of a wide range of infectious diseases.

Lowbury, E.J.L. et al. (eds). 1981. *Control of Hospital Infection: A Practical Handbook*. 2nd edn. Chapman & Hall, London.

28
Immunization

Immunological theory has been considered in Chapter 6. Here we are concerned with the artificial enhancement of immunity—i.e. immunization.

Active Immunization

This is the administration of an antigen which stimulates the recipient's own immunological mechanisms. The speed and vigour of his response depend on the dose of antigen, the route of administration, his age and whether he has met the antigen (or something closely similar) previously. The **primary response** to an antigen not previously encountered may take 1–3 weeks to manifest itself, and if antibodies are produced they are predominantly IgM (p. 53). The **secondary response** on meeting it again weeks, months or even in some cases many years later is more vigorous and much faster, and if it involves production of antibodies they are predominantly IgG; alternatively, or in addition, the T lymphocyte responses are enhanced and accelerated. These differences between primary and secondary responses, which are of central importance in the design of immunization regimens, are the result of immunological memory, as described on p. 52. For adequate and lasting immunity it is necessary, with many antigens, to give an initial course of three doses and a **booster dose** a few years later; and even then immunity may wane unless further doses are given every few years. Other antigenic stimuli—e.g. measles or mumps virus—may stimulate life-long immunity. There are some antigens that provoke adequate primary responses, but with no accompanying immunological memory that enhances the response to a second encounter. For example, immunization with pneumococcal or *Haemophilus influenzae* capsular polysaccharide results in a good and sustained antibody response (except in infants—pp. 55–6), but the response to a second injection is virtually indistinguishable from that to the first. The primary/secondary distinction is of course blurred if the antigen persists in the tissues, and particularly when it is a live organism that not only persists but proliferates. As in natural infections, there is then a period of continuous, and at least for a time increasing, stimulation, and the pattern of immunological response is modified accordingly. Substances such as alum are added to some injected immunizing agents, in order to delay their absorption and so prolong their effectiveness; the name **adjuvants** is given to such substances which, when administered with antigens, enhance their immunogenicity.

In general it is probably desirable that an immunization regimen should simulate as closely as possible the corresponding natural infection, and in particular the mode of entry of the pathogen into the host. Thus, while non-living antigens are most effective when injected, live attenuated vaccines are increasingly given orally or intranasally. For example, of the vaccines available for immunization against poliomyelitis (p. 368), the inactivated vaccine is given by injection and stimulates production of much antibody in the serum but virtually none in the intestinal secretions, whereas oral administration of live attenuated vaccine results in high levels of secretory antibody at the site at which natural infection would otherwise occur, as well as some serum antibody. Secretory antibody responses appear to be accompanied by little or no immunological memory.

Antibodies already in circulation, especially IgG, can suppress the response to active immunization. This is of practical relevance to the immunization of infants who still have placentally transmitted maternal IgG antibodies in their blood (pp. 54 and 55). Such serum antibodies can impair the response to parenteral immunization, but do not affect the response to oral vaccines, whereas the latter can be impaired by maternal antibodies in breast milk (p. 55).

Another complication of active immunization is illustrated by responses to influenza virus vaccines. One might expect that immunization with a vaccine containing a range of currently prevalent influenza A virus strains might result in protection against them all; but in fact the response is far from uniform, and is dominated by production of antibodies to the first influenza A virus strain the recipient ever encountered—perhaps many years before. The term **original antigenic sin** has been applied to such behaviour. It is characteristic of human responses to any virus with numerous cross-reacting types; and it has important implications for natural as well as artificially induced immunity, as exposure to a new variant of a virus may lead to production of antibodies which are not of the right specificity and give poor protection against the new strain. (See p. 371 for an example of a somewhat similar phenomenon in connection with the use of combined vaccines.)

Since active immunization of a subject who has no previous experience of the antigen takes several weeks to produce effective protection, it is of prophylactic but not of therapeutic value. Boosting of existing immunity may be of immediate value—e.g. in prevention of tetanus (p. 367).

Passive Immunization

This is the giving of ready-made antibodies, formed by another person or an animal in response to artificial immunization or to natural infection. If formed in an animal, they are to their human recipients foreign proteins and potential allergens. The antitoxin used in treatment of diphtheria is produced in horses, and causes few problems because nobody is likely to need repeated courses of it. This is not true of tetanus antitoxin, which may have to be given to the same individual prophylactically after a series of relevant injuries (unless he has been actively immunized). Repeated injection of horse proteins has led to hypersensitivity reactions to these foreign proteins—Type I reactions such as anaphylactic shock, which may be fatal, if the host produces IgE antibodies, or Type III reactions such as serum sickness if he produces IgG (pp. 61–2). Tetanus antitoxin is therefore now prepared by repeatedly immunizing human volunteers with toxoid (p. 367).

The γ-globulin fraction of pooled plasma from healthy blood-donors, known as **human normal immunoglobulin,** contains antibodies that reflect the prevalence of infectious diseases in the community. Injections of this material are used for such purposes as giving travellers a few months' protection against hepatitis A; reducing the severity of reactions to measles vaccine; or giving general temporary protection to patients with hypogammaglobulinaemia. Similarly, **human specific immunoglobulin,** derived from convalescents or from immunized volunteers (as mentioned above in relation to tetanus), can be used to confer temporary protection against a particular infection on exposed immunodeficient patients or those for whom active immunization is contra-indicated—e.g. by severe skin disease. Such specific immunoglobulin is also an important means of protecting normal people against rabies or hepatitis B if it is given soon after exposure.

Passive immunization begins to be effective as soon as the antiserum enters the recipient's circulation, but its effect lasts at most for only a few months. This is because the antibody molecules, like all protein molecules in the body, have only a short life, and they are not replaced as are those produced in response to active immunization. Consequently passive immunization is of use only for short-term prophylaxis or for treatment of existing infection.

The main differences between active and passive immunization can be summarized as in Table 28.1.

TABLE 28.1 Differences between Active and Passive Immunization

	Active	Passive
Immunizing agents	Live or dead organisms, toxoids	Sera from immunized animals or humans
Rapidity of protection	2 to 3 weeks' delay if no previous immunity	Immediate
Duration	Usually several years	At most a few months
Complications	Various (see later) but rarely serious	Anaphylaxis, serum sickness (animal sera)
Uses	Long-term prophylaxis Treatment only if previously immunized	Short-term prophylaxis Treatment

Vaccines

The words **vaccine** and **vaccination**, derived originally from the use of material from the cow (Latin *vacca*) for immunization against smallpox, have gradually extended in meaning to include all immunizing preparations of living or dead organisms or of materials derived from organisms. Vaccines can be classified as follows:

(1) **Live organisms of limited virulence**—usually attenuated derivatives of pathogens, but sometimes naturally occurring organisms closely related to pathogens (e.g. cowpox virus, or the vole tubercle bacillus which has been used for immunizing humans against tuberculosis). Because live cultures are used, it is important to ensure that they do not contain contaminant pathogenic organisms (see p. 365—the Lubeck disaster). A particular hazard of virus vaccines that have to be grown in primary tissue cultures—i.e. in cell populations directly derived from animal embryos or organs—is that they may contain wild viruses that were already in the tissues. Provided that a live vaccine 'takes' and multiplies in the recipient's tissues, a single dose usually gives a satisfactory degree of immunity, since it provides a prolonged antigenic stimulus.

(2) **Dead (or inactivated) organisms.** Suspensions of dead bacteria have on the whole been disappointing as inducers of immunity, though some are misleadingly successful in stimulating the production of measurable antibodies. The position may be improved by research aimed at finding for each organism the culture conditions and method of killing which give the best yield, not just of organisms but of those antigenic components which stimulate **protective** responses. Inactivated virus vaccines have been somewhat more successful, but disasters have resulted from failures of the inactivation processes and consequent inadvertent administration of live virulent viruses. Because a dead or inactivated organism does not multiply in the recipient's tissues, several suitably spaced and relatively large doses have to be given, usually by injection.

(3) **Purified microbial products.** The classical examples of these are **toxoids**, which are bacterial toxins—e.g. of diphtheria or tetanus—that have been made harmless by heat or formalin treatment, but are still effective as antigens and stimulate the production of antibodies which neutralize the corresponding toxins. Also in this category of products are the *B. anthracis* protein mentioned on p. 86 and the pneumococcal and meningococcal capsular polysaccharides. The ultimate aim in active immunization must be the maximum protective response with the minimum risk of unpleasant or harmful reactions. Administration of whole organisms, alive or dead, is a somewhat crude approach to this aim as compared with administration of refined preparations of 'protective' antigens, from which virtually all irrelevant and potentially 'reactogenic' material has been removed.

Immunization against Particular Diseases due to Bacteria

Tuberculosis

Immunization against tuberculosis, using the living attenuated bovine-type tubercle bacillus of Calmette and Guerin (BCG), has been in use since 1922, but its widespread adoption was greatly delayed by the 1930 Lubeck disaster, in which 72 of 251 BCG-vaccinated infants died of tuberculosis soon afterwards. There is little doubt that this was due to contamination of one batch of vaccine with virulent tubercle bacilli. Since the early 1950s BCG has been given to many millions of people with few complications; a very small number of generalized and fatal infections have been attributable to unsuspected T cell immunodeficiency.

The vaccine is given as a small intradermal injection of a suspension of the live bacilli—usually a reconstituted freeze-dried culture. For a variety of reasons the upper arm is the most satisfactory site for this and most other forms of vaccination. BCG vaccination in Britain (and in many other countries) is preceded by tuberculin testing (p. 184), and is confined to those giving negative results. This is partly because a positive result is presumptive (though not conclusive) evidence that the subject is already protected against tuberculosis; but mainly because BCG vaccination in the presence of tuberculin hypersensitivity may result in a destructive Type IV reaction (p. 64). However, the omission of preliminary tuberculin testing in large-scale BCG immunization programmes in developing countries speeds up the programmes, and hypersensitivity reactions have not been found to be a serious problem. In the absence of pre-existing hypersensitivity, a small tuberculous ulcer develops a few weeks after vaccination and persists for several months before it heals, leaving a small scar. The recipient usually gives a positive tuberculin reaction from the sixth week after vaccination.

While BCG vaccination does not give complete protection against tuberculosis, there is clear evidence from trials in Britain and many other countries that it decreases the chance of contracting the disease and virtually eliminates the danger of developing the more virulent primary form (p. 206). How long this protection lasts is not so clear. Reversion to tuberculin negativity after 3 or 4 years or less is quite common, and may be regarded as an indication for revaccination, but there is no certainty that protective immunity has also waned at the same time as the hypersensitivity, and revaccination may induce a vigorous local response. Some trials in other countries—notably the USA and India—have failed to show any protective effect following BCG vaccination. A possible explanation of this failure is that the populations concerned were already enjoying, as a result of natural infection with non-tuberculosis mycobacteria, a degree of protection against tuberculosis comparable to that conferred in other communities by BCG vaccination. As with any form of immunization, the cost-effectiveness of BCG immunization is decreasing as the risk of tuberculous infection decreases.

BCG vaccination is currently recommended in Britain for school children between the ages of 10 and 14, and also for contacts of cases of active respiratory tuberculosis and for high-risk groups such as immigrant populations and health service staff.

Whooping cough

This disease can be serious, even lethal, in the first year or so of life, and responds poorly to antibiotic treatment. Effective prophylaxis is therefore clearly desirable, and ideally would give protection even in the first months of life, since there is little transfer of maternal immunity to this disease. Vaccines consisting of killed suspensions of *Bordetella pertussis* have been available since the 1930s, and are given by intramuscular or subcutaneous injection. A full course of the vaccine currently available in the UK—a killed suspension of *Bord. pertussis* given by intramuscular or subcutaneous injection—confers a good level of protection in 80% of recipients, and in those who contract the disease after vaccination the severity of the illness is much reduced. There has been fierce controversy in Britain about the safety of pertussis immunization. There is a very small risk of serious neurological reactions;

encephalopathy has been estimated to occur in perhaps 1 in 300 000 injections but the true frequency of this complication is very difficult to assess, since comparable brain damage can occur in unvaccinated children, either from unrelated causes or following an attack of whooping cough. Unfortunately but understandably, public discussion of these doubts about whooping cough vaccination caused many parents to doubt the wisdom of having their children vaccinated at all—against any disease, not merely against whooping cough. The fall in pertussis vaccination rates in the mid-1970s was followed by widespread epidemics of whooping cough in Britain in 1978–1982, with a small number of deaths but a significant incidence of pulmonary and cerebral damage. On the basis of the available evidence it seems reasonable to recommend parents to have their children vaccinated against this disease, starting at 3 months old, in the absence of such specific contra-indications as a history of epilepsy in the family or of convulsions or cerebral irritability in the child during the neonatal period, central nervous system abnormalities or a current febrile illness (a contra-indication for any form of immunization).

Typhoid and paratyphoid

Immunization against enteric fever by injecting killed suspensions of the causative bacilli was introduced at the end of the last century and has been widely used, especially in war-time, ever since then; but for many years there was surprisingly little definite evidence that it gave any protection. This was in part because army units and other groups of people with high vaccination rates have usually also had high standards of general hygiene which could equally well have been responsible for their low incidence of enteric fever; and in part because the effectiveness of such a vaccine varies with its method of preparation. However, carefully controlled large-scale trials in Yugoslavia in 1954–55 established that recipients of a heat-killed phenol-preserved *S. typhi* vaccine developed typhoid significantly less often than did those who received either a *Sh. flexneri* vaccine (presumably irrelevant) or an alcohol-killed *S. typhi* vaccine (which, on the evidence of this trial, was of little or no value).

Subcutaneous injections of heat-killed suspensions of *S. typhi* and of *S. paratyphi* A and B or A, B and C (TAB or TABC) have been extensively used in the past. These commonly produce painful local swellings, often accompanied by fever and sometimes by rigors, and they provoke vigorous antibody responses to all of the species used, as measured in the Widal test (p. 180); but since there is no evidence that the paratyphoid components evoke any useful level of protection, and since they undoubtedly increase the adverse reactions, a monovalent *S. typhi* vaccine is now used. When given for the first time typhoid vaccine should be injected subcutaneously or intramuscularly but subsequent doses can be given intradermally since this further reduces the adverse reactions.

Diphtheria and Tetanus

Both active and passive immunization against these diseases date from the pioneer work of Behring and Kitasato at the end of the last century (p. 7), and from the time of their discovery there has never been any room for doubt about the efficacy of these procedures. In both diseases the damage is done by toxins and immunization is aimed at neutralizing these. Now that *Corynebacterium diphtheriae* has been virtually eliminated in Britain and many other countries there is little natural immunization from contact with the organism, and hence an increasing percentage of the population will be susceptible if not immunized.

ACTIVE IMMUNIZATION In each case toxoid for active immunization is made by heat- and formalin-inactivation of toxin obtained from a culture of the relevant organism. Elaborate steps are taken to purify this **formol toxoid** so as to eliminate unnecessary ingredients which might act as antigens and give rise to hypersensitivity reactions, or else might be directly toxic.

Diphtheria formol toxoid is an inadequate antigen when given on its own, but whooping cough vaccine given with it acts as an adjuvant (p. 362). Of various other means used to enhance the efficacy of the formol toxoid, adsorption onto aluminium phosphate or hydroxide is the most successful. Two intramuscular or deep subcutaneous injections of such an adsorbed vaccine, given 4–6 weeks apart, constitute an adequate primary course; adverse reactions are uncommon and virtually never serious in children up to 10 years old. Older children and adults are more likely to react to the vaccine and should be given a low-dose vaccine without prior Shick testing. The Shick test (p. 183) is now indicated only for assessing the immunity of staff likely to be exposed to diphtheria in the course of their work. Contacts of diphtheria cases should all receive erythromycin prophylaxis.

'Provocation poliomyelitis' is a possible complication of diphtheria immunization in a community in which polioviruses are circulating—as in Britain in the 1950s, before vaccination against poliomyelitis had been carried out here on a large scale. At that time it was found that paralysis of the injected limb occurred with a frequency of about 1 per 37 000 injections of diphtheria toxoid, and that it could follow use of formol toxoid with whooping cough vaccine or of alum-precipitated or aluminium-phosphate-adsorbed toxoid without whooping cough vaccine. It apparently depended on local tissue damage converting a non-paralytic poliovirus infection into a localized paralytic form of the disease.

Tetanus formol toxoid is itself a potent antigen, but an adsorbed vaccine is also available. For either, the recommended primary course consists of 3 intramuscular or subcutaneous injections, with 6 weeks between the first two and 4 months between the second and third. The aim of the infant immunization programme (p. 371) against tetanus is a high level of immunity in the general population; this level of protection should be maintained by booster doses of toxoid every 5–10 years. This is particularly important for high-risk occupational groups—e.g. farm workers, soldiers, bacteriology laboratory staff. A booster dose should also be given after any relevant injury occurring more than 2 years after the last previous dose (p. 290).

As we shall see later (p. 371), diphtheria and tetanus vaccines are commonly used in combination with one another and with other vaccines, and the timing of doses for routine immunization is therefore a matter of finding the best compromise between the optimal schedules for individual components of the mixture.

PASSIVE IMMUNIZATION The danger to life in **diphtheria** depends largely upon circulating toxin, and there is a good chance of neutralizing this by giving intravenous diphtheria antitoxin (usually horse serum) as soon as the diagnosis is made or suspected (p. 196). Since this is the only effective treatment, it must be given with the minimum of delay, but it must be preceded by a small subcutaneous dose of the serum to ensure that the patient is not sensitive to horse protein. If the test does cause any systemic reaction, desensitization must be attempted. There is virtually never any call for the prophylactic use of diphtheria antiserum.

Wounds and other lesions potentially infected with **tetanus** spores are common. Irrespective of the immunization status, careful surgical toilet, debridement of wounds including the removal of damaged and avascular tissue and the use of antibiotics will prevent multiplication of the bacilli. Treatment would be simpler and safer if everyone was actively immunized against tetanus in infancy and received regular booster injections. Any injured patient could then be safely and efficiently protected by giving a further dose of toxoid. As it is, all too often the patient does not know or is incapable of telling the doctor that he has been actively immunized. When the immunization status of a potentially infected patient is not known or when he is known not to have been recently immunized it is now agreed that human anti-tetanus serum (ATS) should be given in addition to the adsorbed vaccine if the wound, when seen, is more than 6 hours old, if there is contamination with soil or other matter likely to contain clostridial spores or if there is much devitalized tissue. Horse anti-tetanus serum should **not** be used.

It is customary to give intravenous ATS for therapy when tetanus has developed, but its value is doubtful; toxin already fixed in the nervous system cannot be neutralized even by a large amount of intravenous antitoxin. All that this might be expected to achieve is neutralization of further toxin released by the bacilli, and it is better to prevent this from happening by using antibiotics. It is important to remember that an attack of tetanus does not confer immunity (p. 291); therefore the patient who has recovered needs to be actively immunized to prevent further attacks.

Immunization against Particular Diseases due to Viruses

Smallpox

When, nearly 200 years ago, Jenner introduced a safe means of immunization against smallpox (p. 7), he began the first (by many years) and one of the greatest of medicine's success stories; but it was not until 1980 that the World Health Organization's campaign for the eradication of this disease culminated in the declaration that the world was now free of it (p. 333).

Poliomyelitis

Immunization against this disease, as introduced by Salk in 1953, involved repeated injections of formalin-inactivated suspensions of all three poliovirus types. **Salk-type** vaccines made a great contribution to the control of poliomyelitis, but **Sabin-type** vaccines, consisting of live attenuated viruses, are now in general use in the UK, the USA and many other countries, with the inactivated vaccine used only when live vaccines are contra-indicated—e.g. in pregnant women and in those with immunodeficiencies. Live vaccines are taken by mouth, and have other merits (pp. 362–3). The live viruses establish themselves in the recipient's intestinal wall, and in addition to providing him with an antigenic stimulus they are for some weeks excreted in his faeces and may be transmitted to other people around him. Immunity develops sooner and lasts longer after live-virus than after inactivated-virus vaccination, and it includes resistance to propagation of wild polioviruses in the intestine, whereas after inactivated-virus vaccination wild viruses are prevented from reaching the nervous system but are permitted to propagate in the intestine and so can be disseminated through the community. If live-virus vaccination is attempted in a community in which certain other enteroviruses are prevalent, it may be unsuccessful in some cases because of interference (p. 43). Similarly, immunization may also be unsuccessful in children with any form of infective gastro-enteritis because of this or some other, as yet undefined, type of interference. This is important in countries where infantile gastro-enteritis is common.

When vaccine strains of all three poliovirus types are administered simultaneously that of one type (commonly type 2) may establish itself in the intestine to the exclusion of the other two, and the resultant immunity is against only the one type. However, the strain that 'takes' on the first occasion is in consequence unable to do so if a further dose of the trivalent vaccine is given after a suitable interval, and immunization against all 3 types can be achieved by giving 3 doses of the trivalent vaccine at monthly intervals. It is important that the course should be completed, so as to achieve protection against polioviruses as a whole, not just some of them. The phenomenon of interference can be put to good use during an outbreak, since oral vaccine given to all contacts will protect them against the wild viruses long before they have developed protective levels of antibodies.

When, as in Britain at present, the circulation of polioviruses within the community is insufficient to maintain herd immunity (the importance of which is outlined on p. 59), it is essential that it should be maintained instead by a high level of vaccination—otherwise imported viruses can cause havoc. Furthermore, the community may need protection against the vaccine strains themselves. There has been no confirmation of early fears that the virulence of these might be rapidly enhanced by a few passages through human contacts of those who had been given the vaccine. However, a slower change is going on; some of the polioviruses circulating in Britain today, and capable of causing clinical illness with neurological symptoms, have genetic characters which indicate their derivation from vaccine strains that have been able to adapt to life in the community. A falling level of herd immunity might permit such strains to cause more serious trouble.

In Britain at present it is recommended that infants receive 3 doses of trivalent polio vaccine together with their triple vaccine (p. 371), starting at 3 months (p. 371). Non-immune parents should be offered vaccine at the same time as their children, to avoid any danger of the vaccine strain being transmitted to them with enhanced virulence, though the risk of this is very small. Vaccination is clearly indicated for previously unvaccinated individuals going from a country with a low prevalence of polioviruses to one where they are common.

Measles

Measles (p. 126) may be a severe infection, complicated by otitis media, bronchopneumonia and even in a few cases encephalitis (p. 221). Active immunization by means of a single injection of a live attenuated chick-embryo culture of an attenuated measles virus is commonly followed by fever and rash lasting 24–28 hours, sometimes by bacterial infection of the respiratory tract, and rarely by convulsions, encephalopathy or subacute sclerosing panencephalitis (SSPE, p. 221). However, the risk of these complications is tenfold less and they are far less serious than those of the natural measles which virtually every unvaccinated child acquires. Vaccination in the early months of life commonly fails to prevent subsequent development of measles, probably because maternal antibodies neutralize the vaccine. The beginning of the second year of life seems to be the best time for routine vaccination. American experience indicates that vaccination is highly effective, and that protection is still good after 16 years; but eradication of the virus from a community requires immunization of nearly 100% of its members.

It is important that children with histories of convulsions or epilepsy should be given measles vaccine because they are at high risk from the natural infection. In their case human normal immunoglobulin should be given simultaneously with the vaccine to minimize any reaction. Patients with congenital or acquired immunodeficiency should not be given measles vaccine. An outbreak of measles can be contained by immunizing susceptible contacts within 3 days of exposure, or by giving human normal immunoglobulin where the vaccine is contra-indicated.

Rubella

It is clearly desirable that all women of child-bearing age should be immune to rubella (p. 123), so as to avoid the hazards of intra-uterine infection (p. 322). In Britain at least 80% of such women are immune as a result of natural infection, which gives far more solid and lasting immunity than any vaccine at present available. There is thus a strong case for allowing rubella to persist in the community. This is the basis for the current British practice of making rubella immunization available to all girls between the ages of 11 and 13 years and to adult women whose serological tests indicate absence of rubella antibodies. A live attenuated vaccine is given subcutaneously. Since it is not certain whether these vaccine strains are themselves hazardous to fetuses, non-immune women should be warned against becoming pregnant within 3 months after immunization, and women who are already pregnant should not be actively immunized until immediately after delivery. There is at present little that can be done to prevent or treat rubella infection in an exposed non-immune pregnant woman; passive immunization of the mother with human gammaglobulin has been shown to be of very limited value as a protection to the fetus.

In the USA, the aim of the vaccination policy is the eradication of rubella, and all children (boys as well as girls) are immunized early in infancy and must have evidence of immunization before they are allowed to enter school.

Rabies

Immunization against rabies is described in Chapter 16 (p. 222).

Hepatitis B

Hepatitis B vaccine consists of a suspension of HBsAg particles (p. 135) purified from the plasma of hepatitis B carriers and inactivated. Studies in the USA suggest that the vaccine is 80–90% effective in preventing hepatitis B infection and does not carry a risk of transmitting AIDS (p. 283). The vaccine should be reserved for those who have special occupational risks of acquiring hepatitis B because they work with, or handle blood or other hazardous specimens from, patients who are likely sources of infection—those with liver disorders, drug addicts, mentally handicapped patients, those attending clinics for sexually transmitted diseases, etc.

Hepatitis B immunoglobulin is available for passive immunization of those who are accidentally inoculated by a needle-stick injury or who contaminate their eyes, mouths, cuts or scratches with blood from a known carrier of HBsAg. Its use for neonates born to infected mothers is considered in Chapter 25 (p. 329).

Influenza A

The composition of vaccine to protect against influenza A (p. 125) has to be reviewed annually and tailored to match the antigenic composition of currently circulating wild virus strains, since these undergo antigenic shift and drift (p. 125). In addition the phenomenon of 'original antigenic sin' (p. 363) may result in an ineffective response to even the most appropriate vaccine. Various forms of vaccine are available:
(1) Whole virus vaccine containing inactivated influenza virus.
(2) Split virus vaccine containing disrupted virus particles. These have fewer side effects than whole virus vaccines.
(3) Surface antigen vaccine which contains the haemagglutinin (H) and neuraminidase (N) antigens prepared from disrupted virus particles and absorbed on to aluminium hydroxide.
(4) Live attenuated vaccine which is administered intranasally and probably depends for its efficacy on stimulation of a local IgA response.

Influenza vaccine should be given in the autumn for those at special risk—i.e. patients with chronic disease of the heart, lungs or kidneys, diabetics, the immunosuppressed and, particularly, the elderly who are in long stay hospitals or homes. When an outbreak occurs it is too late for immunization of contacts to be effective, but chemoprophylaxis with amantadine soon after exposure may prevent influenza A.

Immunization Programmes

Active immunization of a large proportion of a population can be used either to bring a disease under control or to prevent its return to a community from which it has been eradicated. Such use of any immunizing procedure must be governed by the answers to the following questions:

(1) **Is it effective?** Smallpox vaccination had been widely used for over a century, and often with the backing of legal compulsion, before there was any formal proof of its efficacy. Today any prophylactic procedure must be tested in carefully designed trials, to ensure that any apparent benefit is not in fact due to other factors such as improved nutrition or hygiene. Part of the problem of assessing the value of whooping cough immunization (p. 365) is that the marked fall in the incidence of this disease in Britain following the introduction of general immunization occurred at a time of rapidly improving living standards.

(2) **Is it safe?** No form of vaccination is entirely free from risk—there are very few things in life which are! Safety in vaccination has to be assessed by comparison with the risk of being unvaccinated. Again, this point is illustrated by the pertussis controversy (p. 365). Poliomyelitis, on the other hand, is a disease in which the benefits from vaccination clearly outweigh (except in a community with negligible risk of infection) the small risk of complications.

(3) **How great is the need?** In general, the need for immunization against a disease depends on the risk of acquiring it, the likelihood that the illness will be serious, and the effectiveness of available means of treating it. Yet again, whooping cough provides a clear illustration: small children need protection against it, since it is potentially lethal to them; but there is no need to immunize adults, to whom the disease presents no serious hazard, though it may be an unpleasant experience.

(4) **Is it practicable?** This depends on many social and economic factors. Repeated small outbreaks of diphtheria, with quite high death

rates, have occurred in some countries which could, in theory, have prevented them, but which could not in fact afford to divert their available medical, administrative and financial resources from larger and more urgent problems.

(5) **Can it be made acceptable?** No programme of mass-immunization can succeed without popular support. This depends largely upon successful propaganda; and is much easier to secure when the disease is still prevalent and its dire effects are known. 'Diphtheria is deadly', an effective slogan in Britain of the early 1940s, is today liable to produce the reply, 'What is diphtheria? Surely it doesn't happen any more.' Popular support also depends upon absence of disfigurement (such as large vaccination scars) and absence of unpleasant (even if harmless) reactions. One reason for never immunizing anyone who is at all unwell is that any illness which he may be incubating will be attributed to the injection. A single death *following* (not caused by) immunization outweighs much expensive propaganda. Finally, immunization is made more acceptable by minimizing the number of visits to the clinic or surgery which are necessary. This means using potent and, where possible, combined vaccines.

Immunization of large numbers of people is made easier and safer by giving intradermal injections in the form of a high-pressure jet, from an instrument which does not touch the patient and therefore cannot transmit infections such as hepatitis B.

Combined vaccines

In Britain it is currently recommended that young children should be immunized against diphtheria, tetanus, whooping cough, poliomyelitis and measles. If immunization against each of these diseases was carried out separately, the primary courses alone would require at least 12 administrations of vaccine. However, three doses of **triple vaccine** (diphtheria and tetanus toxoids and a killed suspension of *Bord. pertussis*—DPT) constitutes an effective primary course for the first three diseases, the whooping cough vaccine acting as an adjuvant to enhance the efficacy of the toxoids as well as fulfilling its primary function. Three doses of oral live poliovirus vaccine can be given at the same times as the injections, so that after three visits to the doctor the child needs only to be immunized against measles (one visit). The increasing use of combined vaccines makes it necessary to consider the theoretical possibility of **antigenic competition**—impairment of response to one or other of two vaccines when the two are given together or sequentially. In practice, studies with combined vaccines have failed to show that this does occur. If it is necessary to administer two or more live virus vaccines they may be given simultaneously at different sites without any impairment of response; but if one live vaccine is given alone an interval of 3 weeks should be allowed before giving another.

Timetable for immunization of children

Attempts at immunization of a new-born baby are generally unsuccessful both because of the immaturity of its immune mechanisms and because maternal antibodies in the baby's circulation may interfere with the induction of an active immune response (pp. 54 and 363). BCG is

TABLE 28.2   Infant Immunization Schedule

Immunizations recommended	
At 3 months old	Triple vaccine (DPT; see above) and oral poliovirus vaccine
Again 6–8 weeks later	
Again 4–6 months later	
At 1–2 years	Measles vaccine
At 5 years (school entry)	Single booster doses of diphtheria/tetanus toxoids and oral poliovirus vaccine
At 10–13 years	BCG for tuberculin-negative children. Rubella vaccine for all girls
At 15–19 years (school leaving)	Single booster doses of tetanus toxoid and oral poliovirus vaccine

an exception to this, and is effective when given at birth. If advantage is to be taken of the triple vaccine, there is a problem about when to use it. Whooping cough immunization should start as early as possible, in order to protect the infant during the dangerous period; but diphtheria and tetanus immunization are more efficient if started when the child is at least 6 months old, so a compromise is necessary. The schedule of immunizations currently recommended for use in Britain is shown in Table 28.2, p.371.

Vaccination for international travel

To pass from one country to another a traveller may need to produce certificates of recent vaccination against one or more diseases. The details depend on his country of origin and his route, and vary from country to country and from time to time in accordance with the distribution of foci of active disease and the current anxieties of health authorities. The prospective traveller should make sure of the prevailing regulations by asking travel agents or official representatives of the countries to which he is going. Vaccination against **yellow fever** with live attenuated virus vaccine (obtainable only at specially designated centres) and an International Certificate of such vaccination are required for entry into some of the Central African countries in which this disease is endemic, and also for entry into many countries when travelling from the endemic zones of Africa and South America. The certificate becomes valid 10 days after primary vaccination or immediately after revaccination, and remains so for 10 years.

Vaccination against **cholera** with a vaccine consisting of a heat–killed suspension of *Vibrio cholerae*, though of limited efficacy, is required for entry into some countries and is indicated for those travelling to a country where cholera is endemic or epidemic. Such protection as may be gained lasts for only 3–6 months. The best protection is to avoid consuming food or water which may be contaminated.

Monovalent **typhoid** vaccine (p. 366) is recommended for travellers to endemic areas, which include the Mediterranean area as well as most tropical countries with poorly developed sanitation.

Travellers to any of the numerous countries where **hepatitis A** is common should be given human normal immunoglobulin, which protects against this disease for a few months. If a live vaccine is to be given the immunoglobulin should be given later, lest it impair the response to active immunization. Exceptions to this are oral poliovirus immunization, BCG and yellow fever vaccine. Frequent travellers to high-risk areas should be tested for possession of antibody to hepatitis A virus. If it is present, they do not require further immunoglobulin.

Chemoprophylaxis against **malaria** (p. 242), though it does not belong under the heading of vaccination, must be included in the provisions made for travel to any country where there is a risk of acquiring this disease.

Contra-indications to vaccination

Live vaccines should not be given to patients with defective immunity, whether primary or due to such causes as leukaemia, lymphoma, other widespread malignant disease or treatment with immunosuppressive drugs; and oral poliovirus vaccine should not be given to family contacts of such patients. Live virus vaccines should not be given to pregnant women, because of the risk (theoretical at least) of damage to the fetus.

Patients who are highly allergic to eggs may have serious hypersensitivity reactions to egg-grown vaccines such as those for influenza, measles and yellow fever. Some vaccines contain traces of antibiotics, notably penicillin, polymyxin or neomycin, but these do not usually cause any problems except in patients with existing extreme hypersensitivity.

Suggestions for Further Reading

Dick, G. 1978. *Immunisation*. Update Books, London and New Jersey.

29
Antibacterial Drugs*

Perspective

Up until 1935 no drugs were available for the treatment of systemic bacterial infections other than syphilis. Doctors could do no more than treat symptoms and, with the invaluable help of nurses, look after the patient's general condition while he overcame the infection himself, or failed to do so. Then came the sulphonamides and other drugs for systemic treatment of bacterial infections, notably the antibiotics. The situation was rapidly transformed, and the morbidity and mortality of bacterial infections were dramatically reduced (p. 9). It is hardly possible to overestimate the importance of antibacterial drugs.

However, this is far from meaning that today the correct answer to a bacterial infection is simply to give the right antibacterial drug. In many cases, as we shall see, choosing the right drug is by no means simple. Furthermore, choosing the right drug is only part of the right management of the patient. The doctor who treats pneumonia merely by giving a suitable antibiotic, and forgets the principles of general medical and nursing care and the importance, for example, of giving oxygen for hypoxia, may lose his patient even though he cures the infection. Similarly, the surgeon who ignores the rules of aseptic surgery and relies on antibiotics to prevent or cure wound sepsis is on the way to disaster. Antibacterial drugs must be seen in proper perspective. To use them is to intervene in the struggle between the host's defences and the invading organisms. If the weapons are rightly chosen and rightly used, such intervention is likely to be decisive; but at no time can we afford to neglect the defender's morale or supplies, or to allow the invaders to restore their numbers.

Hazards

Neither can we afford to forget that antibacterial drugs are foreign substances so far as the patient's body is concerned, and are potentially harmful to him. They vary in the frequency and severity of their adverse effects, but none of them is perfectly safe, and almost all of them have on occasions killed patients. Most of the troubles for which they are responsible fall into the following three categories:

(1) **Direct toxicity to host cells.** These drugs are necessarily toxic to living cells; but their clinical usefulness depends on **selective toxicity** (p. 389)—i.e. upon their being substantially more toxic to bacterial than to host cells. The β lactams, which act on bacteria by interfering with a mechanism of cell-wall synthesis not shared by mammalian cells (p. 389), are in general free from direct host toxicity, except at very high levels or by unrelated mechanisms (e.g. the nephrotoxicity of early cephalosporins). In contrast, trimethoprim interferes with a process common to bacterial and mammalian cells (folate synthesis—p. 388), but its greater affinity for the relevant bacterial enzymes than for their mammalian counterpart ensures that only when given in prolonged high dosage does it interfere

---

*Information about drugs for use against organisms of other kinds is to be found in Chapters 7–11, 15 and 17–21. (See also index, *Antimicrobial drugs*.)

with the process in the host. Cephalosporin nephrotoxicity, already mentioned, and aminoglycoside ototoxicity are examples of damage done to the host by mechanisms unrelated to the antimicrobial action of the drugs. In the very early days of chemotherapy Ehrlich introduced the concept of the **therapeutic index,** which is (the maximum tolerated dose) ÷ (the minimum curative dose). As patients vary in tolerance and bacteria in drug-sensitivity, this index can never by given a precise numerical value, but the underlying concept is important; and for every antibacterial drug that has been developed for clinical use there are many that have been tested and rejected because this index was too close to or even below unity. (This problem is even more serious in relation to antivirus drugs—see p. 135—since the metabolic processes of virus replication are in fact host-cell processes.)

(2) **Hypersensitivity reactions.** Patients may become hypersensitive to almost any antibacterial (and indeed, almost any other) drug, but the antibacterial drugs with which this type of problem is most commonly encountered are the penicillins, otherwise the least harmful of antibiotics. Penicillin hypersensitivity may be of any of the four types described in Chapter 6— rashes (Type I, III or IV), serum-sickness-like reactions with urticaria, fever and arthralgia (Type III), haemolytic anaemia (Type II) or anaphylactic shock (Type I). Hypersensitivity is not merely unpleasant or dangerous in itself; it also means that the patient is in future deprived of the possibility of being treated with any of the group of drugs to which he has become hypersensitive—and in practice this all too often means that the use of an antibiotic for a condition for which it was unnecessary or inappropriate debars its subsequent use in a more serious situation in which it might have been extremely valuable.

(3) **Alteration of the hosts's bacterial flora.** The doctor who prescribes an antibacterial drug is aiming it at a known or suspected pathogen, but the drug itself is by no means so selective! As we have indicated in earlier chapters, man's normal bacterial population is of great value to him, and any major interference with it can have unpleasant and sometimes serious consequences (pp. 30 and 259). The availability of powerful antibiotics has made such interference possible, and indeed easy to achieve. Freed from the restraining influence of their more numerous but antibiotic-sensitive neighbours, more resistant organisms that are normally present only in small numbers (e.g. *Candida albicans, Ps. aeruginosa* and some of the enterobacteria) or are unable to establish infection in the face of the body's normal flora (e.g. *Cl. difficile* p. 259) may be able to proliferate vigorously and may become 'opportunist' pathogens, particularly if the patient is debilitated or immunologically deficient. Furthermore, antibiotic-resistant variants of the pathogen for which the patient is being treated, or of any other pathogens that he may be carrying, may be present as a result of mutation or plasmid-transfer (pp. 18–20); and if so, the selective advantage conferred on them by the antibiotic treatment may enable them to multiply and so both to cause serious trouble to the patient and to be transmitted to other patients or potential patients in his vicinity. We shall return later in the chapter to the problem of bacterial resistance to drugs; here we would simply point out that a large part of the problem has been created by their use (wise or unwise).

Use and Abuse of Antibacterial Drugs

As far back as 1956 Professor Jawetz, one of America's leading authorities on this subject, hazarded a guess that not more than 5–10% of the vast output of antibiotics was employed on proper clinical indications. He went on to give a vivid description of the various pressures, notably from the manufacturers and from patients and their relatives, which cause doctors to misuse such drugs. The number of new antibacterial drugs introduced since 1956 has increased our capacity both for effective treatment and for misunderstanding and mistakes, and it may well be that Jawetz's estimate is not far from the truth today. To provide his patients with optimal antibacterial therapy the doctor needs to answer a number of questions, which we can group under three headings: *Why?, Which?* and *How?*.

WHY does this patient need antibacterial treatment?

This is the most important of the questions. Unless there is a valid reason—scientific, not just social or emotional—for giving an antibacterial drug, the patient would probably be better off without it. (For a placebo effect it is usually possible to choose something safer and cheaper.) Valid reasons include the following:

(1) **Treatment of a known or suspected bacterial infection** that is unlikely to undergo rapid and satisfactory spontaneous resolution, and that can be expected, on the available evidence, to respond to the drug given.

Patients in the early stages of acute upper respiratory tract infections are commonly given oral penicillin or some other antibiotic. This is often a clear example of the misuse of such drugs, since the infection is usually caused by a virus and is likely to be rapidly self-limiting; the antibiotics cannot therefore be expected to do good but may well do harm. However, if the patient has tonsillitis or pharyngitis and there are clinical, epidemiological or bacteriological reasons for suspecting that it is due to *Str. pyogenes*, penicillin treatment is indicated.

Sometimes a patient is known to have a bacterial infection due to an organism which is sensitive to antibiotics *in vitro*, but their clinical use is contraindicated by existing knowledge. For example, attempted antibiotic treatment of enteritis due to the 'food-poisoning' salmonellae may merely result in prolonged carriage and excretion of the offending organism (p. 257).

On many occasions, however, use of antibacterial drugs is clearly indicated. Patients with lobar pneumonia, purulent meningitis, serious post-operative wound infections, or specific bacterial infections such as typhoid, tuberculosis, syphilis or gonorrhoea—to name but a few of many possible examples—must receive prompt and effective antibacterial treatment. In many other cases the indications are less clear, but one or more antibacterial drugs should be given because of the probability that the patient has a bacterial infection, or of the possibility that he has such an infection which could be serious unless treated promptly. However, as we shall see more clearly when we come to our next question, if there is any doubt about the presence or nature of a bacterial infection it is the doctor's duty to ensure that so far as possible all necessary specimens for precise bacteriological diagnosis are collected before antibacterial treatment is given.

(2) **Prevention of bacterial infection.** There are a few definite indications for prophylactic administration of antibacterial drugs, such as the following:

(a) Patients who have had rheumatic fever need to be protected, especially during childhood, from *Str. pyogenes* infections which might precipitate further attacks. Fortunately this species is always highly susceptible to penicillin and nearly always to the sulphonamides, and either of these can safely be given in prolonged low dosage for prophylaxis.

(b) Patients with congenitally abnormal, diseased or prosthetic heart valves run the risk of developing bacterial endocarditis. This risk is increased by procedures that result in bacteraemia, notably major dental treatment—especially extraction—and tonsillectomy. In such cases the offending organism is usually an oropharyungeal streptococcus, often of the viridans group (p. 236). Similarly, enterococcal endocarditis may follow prostatectomy or other genito-urinary tract surgery or instrumentation in such patients. It is debatable whether they are at increased risk following gastro-intestinal tract instrumentation. Prophylaxis against such risks consists of ensuring that on entering the blood the bacteria encounter antibiotic concentrations that kill them before they reach and establish themselves on the heart valves. Such antibiotic cover should begin only just before the relevant procedure, so as to allow no time for proliferation of resistant bacteria; and probably does not need to be continued for more than 12 hours. One or two large oral doses of amoxycillin (a very well absorbed oral penicillin) may be all that is needed (p. 239).

(c) Operations involving the intestine, particularly the appendix, colon and rectum, almost inevitably result in some contamination of the peritoneal cavity and incised tissues with mixed intestinal flora. Various procedures for pre-operative 'gut sterilization' have been tried, without convincing evidence of success; but 'peri-operative prophylaxis', starting just before or during operation and discontinued after 24 or 48 hours, can reduce the frequency of wound infections. Its aim is not to alter the intestinal flora but to prevent intestinal organisms from colonizing and invading the incised tissues. Metronidazole is highly effective in such prophylaxis, because of the special importance of *Bacteroides* and other anaerobes in the establishment of these wound infections (p. 349), because of its selective activity against such organisms, and because it is well absorbed from the intestine and has little effect on anaerobes there but does achieve effective levels in the tissues. Antibiotic lavage of the wound and peritoneal cavity at the end of the operation with a solution containing tetracycline may have a similar effect.

(d) *Cl. perfringens* and other clostridia are predictably sensitive to penicillin (and to metronidazole, which is suitable for patients who are hypersensitive to penicillin). Patients at special risk of developing gas-gangrene (clostridial myositis) include: those who have suffered major trauma with soil contamination; those who have lower limb amputations for vascular disease; and those who have operations on their hip-joints or femoral heads. These three classes of patient have in common the presence of devitalized muscle or bone fragments and the likelihood of clostridial contamination of these—due, in the last two classes, to the frequency with which the patient's intestinal clostridia are to be found on the skin of the thigh and the difficulty of eradicating these sporing organisms by pre-operative skin preparation. For all such patients penicillin prophylaxis is indicated.

Antibacterial prophylaxis can be justified in a few more conditions—e.g. extensive burns, open heart surgery, leukaemic patients with severe marrow depression, and some patients with recurrent urinary tract infections—but in most others it is likely to do more harm than good. Antibiotics are frequently given to patients with virus infections, in order to prevent superinfection by bacteria. This is nearly always unwise; its most likely result is to ensure that the superinfection is by antibiotic-resistant bacteria and therefore more difficult to treat. Attempted prophylaxis against pneumonia in paralysed, unconscious or debilitated patients has much the same effect. In general it is better to wait until bacterial infection occurs and then treat it with an appropriate drug.

WHICH drug or drugs should this patient receive?

This question breaks down into a number of subsidiary questions:

(1) **What is the pathogen and to which drugs is it sensitive?** Proper use of antibacterial drugs requires that the problem may be defined as closely as possible, so that the best tool or tools for dealing with it can be selected. Sometimes the patient's clinical condition is characteristic of the activities of a particular pathogen—e.g. a typical staphylococcal abscess, typhoid or syphilis—and the drug sensitivities of that organism are predictable enough for a suitable drug to be chosen without waiting for laboratory help. Sometimes an illness could be due to any of a number of organisms but stained smears of pathological material give all the information that is needed; pneumococci may be recognized microscopically, for example, in pus from otitis media or in CSF from meningitis, and can be relied upon to be sensitive to penicillin (in most countries). But in many cases in which immediate drug treatment is necessary, it has at first to be based on informed guesses as to the organism and its sensitivities. Except in an emergency or when no laboratory facilities are available, **all specimens necessary for the**

isolation of the causative organism should be collected before treatment is started. If this is not done, treatment may obscure the diagnosis without being adequate to effect a cure. For example, a patient may have a streptococcal endocarditis which has not yet been diagnosed. If he is given bacteriostatic treatment before blood cultures have been set up, it may then be impossible to isolate the organism. His clinical picture may be temporarily improved, but it is highly unlikely that cure will result (p. 238). When treatment stops, he is likely to relapse, and his proper investigation and treatment will have been delayed by some weeks. Whenever initial treatment has been based on guesses, it needs to be reviewed in the light of subsequent laboratory reports about the nature and sensitivities of the organism isolated. However, the ultimate test of a drug's suitability is its therapeutic effect; laboratory sensitivity reports are at best only an indication of probabilities, and if the patient is responding well to the initial treatment, it is usually unwise to act upon a report which suggests that he ought not to be doing so! In many cases there is no urgent need to start treatment until the laboratory report is available.

(2) **Narrow or broad spectrum?** The ultimate in precision tools for dealing with identified pathogens—a dream in Ehrlich's time and still no more than a dream—would be an array of drugs, each one of which would with antibody-like precision attack one pathogenic species, leaving other micro-organisms and the host intact. At the other extreme would be a drug which would deal with all bacterial (and preferably other) pathogens. This might be a pharmaceutical manufacturer's dream, as he might hope that all doctors would give it to most of their patients, but to a bacteriologist it is more like a nightmare, because of its inevitable complex side-effects. Real-life antibacterial drugs come between these hypothetical extremes. Cloxacillin, benzyl penicillin, the macrolides and the polymyxins have useful levels of activity against only some parts of the 'spectrum' of bacterial genera, whereas some of the other penicillins, the aminoglycosides, chloramphenicol, the tetracyclines and others are effective against at least some members of most genera, and are often referred to as 'broad spectrum' antibiotics (particularly by manufacturers, who are naturally concerned to promote their widespread use). In practice, when the identity of the infecting organism is known or virtually certain the doctor's primary question about any particular antibiotic is not about the width of its spectrum but about its suitability for dealing with this pathogen; though when there is a choice between drugs that are equally appropriate in other respects, narrowness of spectrum is an asset, at least in theory. On the other hand, when the pathogen has not been identified or when there is more than one pathogen to be treated, it may be best to use a drug with a spectrum wide enough to cover all the probable or known organisms. The alternative approach is to use more than one drug—see below.

(3) **Cidal or static?** It appears to be true of the treatment of most bacterial infections that there is no need to use drugs that can kill the invading organisms; if they are prevented from multiplying, the patient's own defence mechanisms can eliminate them. However, there are exceptions to this general statement. It is essential to use bactericidal therapy when the patient's immunological defences are seriously impaired, or when the infection is overwhelming. Chronic bronchitics may have longer spells free of bronchial suppuration if the offending organisms, in the bronchial lumen and so apparently beyond the reach of the host's defences, are eradicated by treatment rather than merely suppressed (p. 201). But the classic, and in some ways the most surprising, example of a disease in which bactericidal treatment is required is bacterial endocarditis. It might seem that organisms in fibrinous vegetations in the heart and major blood vessels were well within the reach of blood cells and antibodies, but in fact all available evidence indicates that in such a site they are unusually well protected against both of these. Bacteristatic treatment, so long as it is continued, may bring about apparent cure; but the organisms persist inside the vegetations, ready to resume activity when the treatment stops. They require treatment with drugs that can kill them (p. 238).

(4) **One drug or drugs?** It is possible to demonstrate *in vitro* four types of results when antimicrobial drugs are mixed: **indifference,** the

combined effect being indistinguishable from that of the more powerful drug used alone; **addition,** the combined effect being the sum of the individual effects; **synergy,** the combined effect being greater than can be explained by simple addition; and **antagonism,** the combined effect being less than that of the more powerful drug used alone. The type of result depends upon drug concentrations, the microbial strain used and many other factors, and it is therefore meaningless to describe a drug combination as synergic or antagonistic without specifying the circumstances of such interaction. In general, synergy is likely to be observed only in a mixture of two bactericidal drugs (e.g. a β-lactam + an aminoglycoside) and antagonism only in a mixture of one bactericidal and one bacteristatic drug (e.g. a β-lactam + tetracycline). In at least some instances the mechanism of antagonism appears to be that the bacteristatic drug prevents the organisms from multiplying and so from entering the phase of growth in which they are susceptible to the bactericidal drug.

These various types of result also occur *in vivo*. Antagonism was clearly illustrated as far back as 1951 in one famous series of cases of pneumococcal meningitis, in which the mortality of patients treated with penicillin + tetracycline considerably exceeded that of comparable patients treated with penicillin alone. The possibility of such an interaction is a strong warning against the use of drug combinations without definite reasons. Furthermore, use of more than one drug increases the likelihood of adverse reactions; and when these occur in such circumstances it may be difficult or impossible to decide which of the drugs should be discontinued and might be dangerous if given to the patient on future occasions. Drug incompatibilities are also possible (see below), and the expense is of course increased—a factor which we have not mentioned so far in this chapter, but which needs to be considered in many decisions about antibacterial therapy. However, there are situations in which it is justifiable and may even be essential to use combinations of antibacterial drugs, including the following:

(a) When the patient is suffering from infection with two or more organisms and no single drug is likely to be effective against both or all.

(b) As a temporary measure in a severe acute illness which might be due to any of several organisms and when again no single drug is likely to be effective against them all—see p. 214 (meningitis).

(c) To prevent the emergence of resistant strains—see pp. 381-4 (drug resistance), p. 397 (*Staph. aureus* infections and tuberculosis).

(d) When synergy can be expected—see p. 288 (cotrimoxazole), p. 238 (bacterial endocarditis).

(e) When there is empirical evidence that a particular combination gives the best results—see p. 101 (treatment of brucellosis).

(5) **Is it compatible with other medication?** Even when an antibacterial drug is appropriate in all other respects for treating an infection, it may be contra-indicated or may have to be given with special precautions because of its possible interactions with other drugs that the patient is receiving. Some combinations of drugs are physically or chemically incompatible when mixed in high concentrations. Thus when a doctor prescribes two or more antibiotics which are to be mixed before injection or are to be given together in an intravenous infusion, he needs to be sure that they do not precipitate or inactivate one another. Similarly, when giving any antibiotic by slow intravenous infusion he must be sure that it is not adversely affected by other components of the infusion fluid (e.g. the aminoglycosides are incompatible with heparin, and ampicillin is fairly rapidly inactivated in the presence of 5% dextrose). This is clearly of great practical importance, since incompatibility may result in a patient never having an effective blood level of an antibiotic which is being given in apparently adequate doses. Also very important, and in general even more serious in their consequences, are the pharmacological incompatibilities of some drug combinations at the levels achieved in the patient. Renal damage may follow the giving of gentamicin with cephaloridine, for example, or of one of these with either of the diuretics frusemide or ethacrynic acid; and either nalidixic acid or

cotrimoxazole may enhance the effects of anticoagulant drugs and lead to severe bleeding. As more new drugs are introduced, and more undesirable interactions between older ones are recognized, this subject becomes increasingly complex and worrying. Information about known incompatibilities involving antibacterial drugs is to be found in the three books mentioned at the end of this chapter in Suggestions for Further Reading.

HOW should the drug or drugs be given?

This question overlaps with the previous one at many points. For example, however appropriate a drug may be for dealing with a particular pathogen, it is not the right one for the patient if it can be given only in a form inappropriate to his situation—e.g. orally to a patient who is vomiting, intramuscularly to one with a severe bleeding tendency, or intravenously to an out-patient. Nor is it sufficient to be able to give the drug to the patient; having been given, it must be capable of arriving at the site of the infection in adequate concentration. We therefore have to think about the route by which a drug can be delivered, not merely into the patient but to the place where it is needed; and also about dosage.

(1) **Route** For a superficial infection of the skin or an accessible mucous surface it may be possible to apply the drug in high concentration directly to the lesion; but unfortunately such topical application, notably of the penicillins, is particularly liable to provoke a hypersensitivity reaction and so to deprive the patient of subsequent systemic use of the group of antibiotics in question. A different form of local application that gives less trouble and is sometimes indicated is direct injection into a body cavity—intrapleural, intraperitoneal, intrathecal or intra-ocular, for example.

Oral administration is possible only if the drug can be produced in a palatable form; if it survives the action of the gastric secretions (or can be protected from it in capsules that dissolve in the small intestine); if it does not provoke significant gastro-intestinal upset; and above all if it is reliably absorbed into the blood (unless its site of action is to be the bowel lumen) and passes through the liver without being inactivated. Sometimes an ester or other derivative, itself not an effective antibacterial drug but a 'pro-drug', meets all of these requirements and is converted into the active form after absorption. Oral preparations have, in addition to their unsuitability for patients who are vomiting or cannot swallow, the disadvantage that their absorption may be unreliable in very ill patients, in whom it is particularly important to achieve good levels rapidly and consistently. Some antibacterials, notably metronidazole, are well absorbed when given as rectal suppositories, provided that the patient does not have diarrhoea.

Parenteral (i.e. non-alimentary) administration is usually intramuscular or intravenous. For intramuscular injection it is necessary to produce a strong solution (so that the volume is tolerable) which is of physiological pH and which does not cause excessive pain or damage the injected muscle; this is not possible for some antibiotics. Also, having been injected, the drug must be rapidly and reliably absorbed into the blood (unless it is deliberately given as a slow-release depot preparation, as is sometimes done with penicillin). Intravenous injection or infusion is in some ways the ideal way of getting a drug into the blood, but it may not be the most practicable or convenient. Furthermore, some drugs are difficult to give repeatedly by this route because they cause local phlebitis and thrombosis and a consequent shortage of accessible veins.

Once in the blood-stream, drugs vary in their distribution to tissues and body fluids and in their renal handling, and therefore a high blood level is no guarantee of good tissue levels. Indeed, a high and sustained blood level could well be due to the drug's inability to get out of the blood! Part of this variability in tissue penetration is related to differences in binding to plasma proteins. Virtually all antibacterials are bound to some degree, but some very much more than others—even others in the same group. There is a reversible equilibrium between bound and unbound drug, but it may be only the unbound portion that is free to diffuse out of the circulation—and even then its troubles are by no means over, as it may bind to tissue proteins. Bound drug may be inactive against bacteria, so two drugs may give comparable total levels but with one mostly bound and inactive and the other mostly free and active.

Rapid excretion of an antibacterial drug by the kidneys may make it highly suitable for dealing with urinary tract infections (p. 384), provided that it is not, like chloramphenicol, excreted mainly in an inactive form (p. 395). Such rapid excretion also necessarily means that blood and tissue levels are not well sustained—which may be an advantage, as we shall see below, under (2) (a). Conversely, persistent high levels will be achieved if a drug that is not excreted in bile or metabolized in the liver or elsewhere is only slowly excreted by the kidneys—either because of its nature or because of renal failure.

Antibacterial drugs also vary widely in their ability to pass into body fluids other than urine. For example, sulphadiazine or chloramphenicol levels in the CSF are usually 40–80% of the prevailing blood levels. On the other hand, only traces of the penicillins reach the CSF from the blood if the meninges are healthy; though much larger amounts go through and much higher CSF levels are achieved when the meninges are inflamed, and high concentrations can be achieved by direct intrathecal or intraventricular injection. Similarly, when ampicillin is used in treatment of suppurative chronic bronchitis it may pass fairly readily into the sputum at first, but as the infection is brought under control and the sputum ceases to be purulent its ampicillin content falls sharply. Antibiotics vary greatly in their ability to cross the placenta and reach the liquor. Penicillins do it particularly well, and liquor concentrations may exceed those in blood. These examples indicate something of the complexity of a subject which is of great clinical importance but is far from being thoroughly understood.

(2) **Dosage** Nowhere in this chapter do we go into details about dosage schedules for individual drugs; these are to be found in the books mentioned at the end of the chapter, as well as in the manufacturers' literature and in many other places. However, we do need to consider some general principles.

'Give enough, for long enough, and then stop.' This facile generalization embodies three important points:

(a) **'Enough'**. The aim of antibacterial treatment must be to ensure a drug level at the site or sites of infection which is sufficient to kill or inhibit the pathogen. The patient must therefore be given his drug in doses adequate to achieve this, with due allowance for his size and other factors that may affect the distribution of the drug. To give him less than this is to deny him the help of the antibiotic without necessarily sparing him the hazards; it may well encourage development of drug-resistance by the pathogen or by other potential pathogens; and it is a waste of the drug and of money. When bacteristasis is the aim, theoretical considerations suggest that an effective concentration should be maintained all the time. The same is not necessarily true for a bactericidal effect. For example, since the action of penicillins is on cell-wall formation, they can do nothing to a bacterium which is inhibited and not trying to to make cell-wall; and it is still debatable, after all the years for which we have had penicillins to use, whether the best mode of attack is by a sustained high level or by a transitory high level ('peak'), followed by a period ('trough') during which the level is low enough to allow any survivors to resume multiplication and so to be susceptible to the next peak. Sometimes toxicity also has to be considered. For example, with gentamicin it is possible to produce toxic effects (mainly on the 8th cranial nerve) by maintaining a blood level that is never high enough to be cidal to any but the most susceptible pathogens; it is therefore essential to achieve bactericidal peak levels but to ensure that for most of the time between doses there is a trough that is below the toxic level. With this and related drugs, laboratory monitoring of both peak and trough levels can be an important aid to treatment. This is particularly so when impaired renal function makes it difficult to predict the rate at which the drug will be excreted, and therefore the interval between peaks which is necessary to ensure adequate troughs.

(b) **'For long enough'.** The length of treatment necessary to eradicate a bacterial infection depends, obviously enough, on the nature and location of the pathogen. The tubercle bacillus, with its very long generation time by comparison with most bacteria, needs months or even years of treatment. A pathogen that is well 'dug in' in a fibrotic chronic lesion may call for considerably longer treatment than one that has just arrived and is causing an acute infection. The need for prolonged treatment in brucellosis is probably due to the relatively protected situation of the brucellae inside host cells. But so far as any generalization is permissible, it seems to be true that for many acute bacterial infections it is appropriate to give an antibacterial drug for 5–7 days. By then it will probably have done its job— or failed to do it, in which case a change of treatment is indicated and may well have taken place already. Unfortunately patients who are not closely supervised, including doctors themselves, commonly give up or forget their drugs as soon as symptoms are abating; relapse of infection following inadequate treatment is therefore all too common. Perhaps we need to learn more from the veterinary profession about the value of a single very large dose of an antibacterial. Venereologists, with their special problems of patient supervision, have made some progress in this direction (pp. 279 and 280); and there is now evidence that acute urinary tract infections can often be treated effectively in one day, or even by a single dose (p. 275).

(c) **'And then stop'.** No good purpose is likely to be served, and harm can be done, by continuing to give an antibacterial drug after it has had a proper chance to do its job. 'Tailing off'— i.e. continuing for a while with reduced doses—is even more deplorable.

Drug Resistance

Resistance of a given organism to an antibacterial drug is seldom absolute; it can usually be overcome by increasing the drug concentration. However, in the clinical context an organism is said to be resistant if it is not killed or inhibited by drug concentrations readily attainable in the patient. (This usually means blood and tissue concentrations; an organism resistant to these may of course be sensitive to the higher concentrations attainable in urine or by topical applications)

Even the broadest of broad-spectrum antibacterial drugs is ineffective against some bacterial genera, against some species of other genera, and usually against some strains of species that are in general sensitive to it. Bacteria may be resistant to a drug because of their intrinsic lack of susceptibility (the first four mechanisms described below) or because they can damage the drug (the fifth mechanism).

Mechanisms of Resistance

(1) Relative impermeability of the bacterial cell wall or cell membrane to the drug in question, so that it has difficulty in reaching its target site. For example, gram-negative bacteria in general are resistant to penicillin and some related drugs, to some extent at least, because these cannot readily penetrate the lipopolysaccharide/lipoprotein outer membranes of their cell walls and reach the target enzymes at the mucopeptide–synthesis sites within the cell walls.

(2) Lack of affinity between the drug and the appropriate bacterial receptors. Penicillin-binding proteins (PBPs) are bacterial membrane components that act as receptors for penicillins and other β-lactams. Inhibition of growth correlates with saturation of these sites—the lower the affinity the greater the amount of drug needed for inhibition. In gram-positive cocci and some other bacteria, resistance to particular

penicillins may depend on low affinity between them and the PBPs.

(3) Lack of suitable intracellular target. Examples are sulphonamide resistance due to possession of a dihydrofolic acid reductase that does not 'confuse' sulphonamides with PABA (p. 387); and streptomycin resistance due to lack of a ribosomal protein to which streptomycin can bind.

(4) Ability of the bacteria to switch to alternative metabolic pathways unaffected by the drug.

(5) Ability of the bacteria (whatever their intrinsic sensitivity to the drug in question) to produce enzymes that destroy or inactivate it. Such enzymes may be cell-bound (e.g. those that inactivate aminoglycosides) or released (e.g. β-lactamases, which open up the β-lactam rings of penicillins and cephalosporins, and the acetyltransferases that acetylate and inactivate chloramphenicol). Their production may be constitutive or inducible (p. 15).

Combinations of these mechanisms operating in the same bacteria can create special treatment problems. For example, some gonococcal strains that produce penicillinase (and might therefore be expected to be treatable with a penicillinase-resistant β-lactam) also have intrinsic resistance due to low affinity of their PBPs for penicillin.

Selection and Mutation

Despite the limited range of effectiveness of any one antibacterial drug even when it is first introduced, a small part of our present vast armoury of such drugs would have been sufficient, had they become available simultaneously in the 1930s, to deal with the great majority of pathogenic bacteria then in circulation. In fact, however, new drugs have appeared on the scene gradually; and in most cases the arrival of a newcomer and its widespread use have been followed by proliferation of bacterial strains, including pathogens, resistant to it. *Staph. aureus* has been notably successful in keeping pace with new discoveries. In relation to penicillin, evolution of new strains has played a relatively small part, the increase in frequency of penicillin-resistant *Staph. aureus* being largely due to a 'take-over' by existing resistant strains as their more sensitive colleagues were eliminated. But with most other drugs a more important mechanism is that sensitive strains give rise to occasional resistant mutants; these normally have no particular survival value and indeed may be at a disadvantage when in a non-selective environment, but in the presence of an appropriate concentration of the drug in question they alone are able to multiply, giving rise to a new strain with increased drug resistance (p. 19). In most cases each mutation involves only a small increase in resistance, and therefore such selection depends upon the drug concentration being not much above the minimum to which the original strain is sensitive; but with streptomycin in particular a marked increase in resistance may develop by a single mutational jump that renders the ribosome insusceptible, rather than by a series of short steps.

Prevention of the emergence of resistant mutants is one of the main indications for the clinical use of combinations of drugs (p. 278). As a result of spontaneous mutation, one cell in every thousand million (10^9) might be resistant to drug A and one in 10^{12} might be resistant to drug B. 10^9 and 10^{12} are not very large populations by bacteriological standards. But provided that the mechanisms of action of the two drugs are unrelated, the incidence of cells resistant to both should be one in $10^9 \times 10^{12}$, which is a very large population. Therefore if both drugs are given in adequate dosage, the risk of the emergence of a resistant strain is very much less than if either is used alone.

Resistance Transfer

Chromosomally determined resistance to one or more antibiotics can be transmitted from one bacterial strain to a related but previously sensitive strain by bacteriophage transduction (p. 19). This is one mechanism for transmission of antibiotic resistance between gram-positive cocci, and occurs in other bacterial groups.

Transferable or infective resistance among gram-negative and some gram-positive bacteria is a major threat to the antibiotic control of bacterial diseases. As explained on p. 20, it depends on bacterial conjugation and transfer of plasmids. It is important for the following reasons:

(1) The transferable plasmids commonly determine resistance to several unrelated drugs (multiple resistance), and organisms possessing them have a selective advantage in the presence of any one of these drugs. Thus, for example, the reduction in numbers of streptomycin-sensitive organisms in the intestines of a streptomycin-treated patient or animal may favour the growth of an enterobacterial strain resistant to streptomycin and also (incidentally) to other aminoglycosides, tetracyclines, chloramphenicol and sulphonamides, substances which the patient or animal has never received.

(2) Such plasmids are transferable not merely to related strains of the same species but to strains of other species and genera. Thus a patient under treatment with streptomycin for tuberculosis might have a large population of multiresistant but harmless *Esch. coli* in his intestine as a result of the mechanism described in the last paragraph; he might then ingest some shigellae or salmonellae to which the plasmids determining the multiple resistance could be transferred in his intestine; and these pathogens might then be unresponsive to treatment with any of the drugs concerned. Strains of *Sh. flexneri* resistant to chloramphenicol and various other antibiotics were current in Mexico before 1972, and in that year *S. typhi* strains with the same resistance pattern appeared there and were isolated in Britain and other countries from patients who had acquired them in Mexico. These *S. typhi* strains were resistant by virtue of plasmids which they had presumably acquired from the *Sh. flexneri* strains, either directly or via intestinal commensals, probably other enterobacteria. Furthermore, it is not only to one another that enterobacteria can transfer plasmids coding for resistance; *N. gonorrhoeae* and *H. influenzae* strains that produce β-lactamases by virtue of plasmids acquired from enterobacteria have become world-wide problems.

(3) It is possible for multiresistant enterobacteria to develop in farm animals and be transmitted to man. Some antibiotics are widely used as food supplements for young animals, since partial suppression of intestinal flora can accelerate their weight gain. Prophylactic and therapeutic administration of antibiotics to sick farm animals is also widespread, and often economically important, but it is unco-ordinated. Resistant organisms that have proliferated as a result of such antibiotic usage have excellent opportunities for dissemination because of the prevailing conditions for maintenance and marketing of stock and because of the transfer of young animals from farm to farm. Multiresistant strains of *S. typhimurium* arose among cattle in Britain as a result of sequential acquisition of resistance factors, and are now widespread among calves. There have been cases of infection with such strains among farmworkers and families who drank unpasteurized milk, and there have been several deaths—in part because of the difficulty of finding a suitable antibiotic for treatment of an invasive infection by such an organism.

(4) There is no reason to suppose that what happens among farm animals following widespread use of antibiotics does not also happen, in some measure at least, in human populations similarly treated. For some years development of resistance by man's bacterial flora was mainly a problem of hospitals, where antibacterial drugs are most heavily used and where they can be detected in the dust and even in the air! However, it now seems that widespread use of antibacterials in general practice encourages the spread through the community of multiresistant strains brought home by patients returning from hospital, and may be even responsible for adding others that are 'home-grown'.

Although the progressive sophistication of man's bacterial flora, including his pathogens, appears to be an inevitable consequence of the use of antibacterial drugs, we cannot afford to have a complacent or defeatist attitude to this growing problem of drug resistance. There are ways in which we can slow down or arrest its growth. As we have indicated repeatedly, antibacterial drugs should not be prescribed—to men or to animals—without valid indications for doing so. We have also indicated other precautions that should be observed so as to minimize the proliferation of resistant strains—

such as isolating patients who are distributing these strains into their environments; always giving antibacterial drugs in adequate doses; and using drug combinations when appropriate. Another approach to the problem that has had some successes is to introduce an **antibiotic policy** for a hospital or area. This usually involves designating certain antibiotics as available for general use but withdrawing others from circulation or permitting their use only on rare and special occasions. If all goes well (and in particular, if all relevant clinicians abide by the policy), the incidence of strains resistant to the reserved antibiotics falls considerably in the months following institution of the policy, and in due course it may be judged right to reintroduce these drugs for general use and to withdraw others.

Laboratory Procedures

Sensitivity Tests

We said on p. 376 that the proper use of antibacterial drugs requires that the problem be defined as closely as possible. Frequently an important part of that definition is to determine the antibacterial sensitivities of the pathogen. Various methods are available for routine testing of isolates. Many routine diagnostic laboratories use some form of disc–plate method. In essence, this involves inoculating the whole surface of a plate of suitable culture medium with the organism under test, and then placing at appropriate intervals on the surface of the plate a number of filter-paper discs impregnated with different antibacterials. The plate is then incubated, usually overnight. The antibacterials diffuse out of their discs into the medium, and bacterial growth is inhibited in a circular zone around any disc that contains a drug to which the organism is sensitive. 'Sensitive' is of course not an absolute term. The object of this type of test is to distinguish between strains susceptible to drug concentrations that are attainable in patients (and therefore designated **sensitive**) and significantly less susceptible strains (designated **resistant**)—with the option, thought by some authorities to be a useful one, of dividing the latter into those which might be treatable by achieving unusually high concentrations (designated **moderately resistant, moderately sensitive** or **intermediate**) and those that appear unquestionably resistant. For each drug the disc content should be such that a sensitive strain (as just defined) will give an inhibition zone large enough for any significant decrease to be readily detectable but not so large as to interfere with the zones around neighbouring discs. The amount that achieves this depends on the diffusibility of the drug through the culture medium and also on the levels of it that can be attained in the patient. (For most purposes this means attainable blood levels, but for tests on urinary tract pathogens it usually means the higher levels attainable in urine.) Zone sizes are at the mercy of many factors besides the amount of drug; these include the heaviness of the bacterial inoculum, the medium composition and pH, and the temperature and atmospheric conditions of incubation. Reproducibility of results depends on careful control of all of these factors. One method of standardization depends on faithful reproduction of precisely the right conditions and then the comparison of the inhibition-zone measurements with a table of the results to be expected from sensitive, moderately resistant and fully resistant strains of the species in question. The other approach is to compare the zone sizes given by the test organism with those given by a control-sensitive strain tested at the same time and under the same conditions (indeed, on the same plate in the widely used Stokes procedure).

Like any *in vitro* method, the disc–plate method can at best only give an indication of what might happen in the very different circumstances prevailing inside the patient's body. In its basic form it does not distinguish between bactericidal and bacteristatic action. This distinction can be made by the further procedure of transferring to fresh culture plates the bacteria that were originally inoculated on to the zones in which growth was inhibited, and so giving them a chance to show whether they are still alive. One anomaly of the disc–plate method is that resistance dependent on inducible enzyme (notably that of *Staph. aureus* to penicillin) is manifested not so much by a reduction in size of the zone of inhibition as by vigorous growth around the edge of the zone; bacteria near to the disc were overcome before they could produce enzyme, but those further out had time to defend themselves against the advancing drug, and then had the benefit of the additional nutrients diffusing from the nearby depopulated zone.

It is possible to use the disc–plate technique for **direct sensitivity determinations**—i.e. discs can be applied to plates inoculated with pus, sputum or other pathological material and inhibition can be observed in this primary culture. In such circumstances there is virtually no control of inoculum size, and difficulties of interpretation may arise if the specimen contains a mixture of bacteria; for example, it is impossible to assess the penicillin sensitivity of other strains in the presence of one which produces penicillinase. However, this method has the compensatory advantage of speed, since it may be able to provide a rough guide to appropriate therapy as soon as any visible growth is present—sometimes in a little as 6 hours from the time of the collection of the specimen, well before it would be possible to pick single colonies from the primary culture and set up a properly standardized test.

Sometimes it is desirable not just to classify an organism as sensitive or resistant but to determine more precisely the smallest drug concentration that will inhibit it—the **minimal inhibitory concentration (MIC)**. This is done by testing its ability to grow in the presence of the drug in a series of concentrations, either in tubes of broth or incorporated in solid media in plates. By appropriate subculture from such a series of tubes or plates it is possible to determine also the **minimal bactericidal concentration (MBC)**. Since MIC and MBC values are expressed in numbers (e.g. 0.5 μg/ml or mg/l) they appear more accurate than they are; with some organisms and some antibiotics in particular, they are markedly dependent on inoculum size and the precise techniques used. Some organisms are inhibited but not killed by usually bactericidal antibiotics; such **tolerance** results in a wide difference between the MIC and the MBC. This phenomenon may have important clinical implications—e.g. in the treatment of endocarditis.

The **break-point method,** used by some as an alternative to the disc–plate method for routine sensitivity testing, is a simpler form of the incorporation MIC method. It is usually carried out on plates, using for each drug either one plate containing a concentration regarded as the most relevant break-point between sensitivity and resistance, or two plates with a fourfold difference in concentration. 20–30 different strains can be tested on each plate by replicate spot-inoculation (which can be mechanized). Use of two plates allows strains to be called fully sensitive (i.e. to both concentrations), moderately sensitive (to the higher concentration only) or resistant (to both).

Assays of Drug Levels in Body Fluids

Chemical methods (e.g. high-performance liquid chromatography) are available for the assay of some antibacterial drugs but may give misleading results through failure to distinguish between biologically active drug and inactive derivatives. Immunological methods (e.g. RIA, p. 178) also fail to make this distinction, but they are highly sensitive and give rapid, accurate assays which are unaffected by the presence of other antibiotics. Biological methods (bio-assays) are also used. A simple bio-assay method is similar in general design to the disc–plate sensitivity test method. A plate is inoculated with an appropriate standard bacterial strain sensitive to the drug to be assayed. In place of the discs of the sensitivity test method, small cylindrical wells are cut in the medium; some of these are filled with various dilutions of the fluid under test, and others with fluid containing known concentrations of the drug in question. By comparison of the diameters of the zones of inhibition around the standard solutions after incubation and those around the dilutions of the fluid under test, the drug concentrations in the latter can be calculated (provided that any other antibacterial that might be present in the specimen has been inactivated or is without effect on the test organism). More rapid results can be obtained by means of sensitive systems for early detection of either metabolic inhibition or enzyme production; these are used to compare the effects of dilutions of the test fluid and of standard solutions of the drug on a test organism.

Most antibacterial treatment can be carried out satisfactorily without any monitoring of the levels achieved in blood and other body fluids. This is because, by the time that an antibacterial drug is on the market, the ranges of levels to be expected following recommended dosage schedules have been reliably established. Monitoring may be required, however, when there is wide individual variation in the levels resulting from a standard dose, so that some patients need unusually high dosage in order to have levels within the therapeutic range. This is true, for example, of gentamicin and related drugs, and for them there is the additional complication of a narrow margin between dosage which is adequate and that which may give toxic levels (pp. 380 and 393). For these and many other antibacterial drugs, impaired renal function is one of the most important factors that may invalidate deductions based on results obtained in healthy volunteers, so that it is necessary to check the levels actually achieved in the patient (p. 398). Other reasons for monitoring include attempts to deal with a moderately resistant organism, for which unusually high concentrations of the drug are required; determining whether oral administration, the least reliable means of administration for most drugs but in some cases the most convenient or most acceptable, is giving adequate levels in a particular patient; and the need to discover whether a drug is penetrating in adequate amounts into some body fluid other than blood. In general, assays as part of patient management are aimed at determining either that the patient is having enough of the drug for a therapeutic effect to be likely, or that he is not having too much and therefore exposed to unnecessary hazard.

As an alternative to assay of a drug by use of a standard bacterial strain, it is sometimes more satisfactory to carry out a direct test of the effectiveness of the patient's serum or other body fluid against a culture of his own pathogen, isolated before he started the treatment. This approach is particularly useful when the patient is receiving more than one drug, since assay of one in the presence of others may be difficult. It is common practice in the management of bacterial endocarditis (p. 238) to expose a culture of the offending streptococcus (or other bacterium) to serial dilutions of the patient's serum in nutrient broth, and to test for cidal action by subculturing these serum dilutions after overnight incubation. Treatment may well fail unless for at least part of each day the concentration of drugs in the serum is such that a 1 in 8 dilution of the serum is cidal to the streptococcus (under defined test conditions).

The Drugs Themselves

The contents of this section are subject to the following limitations:

(1) Dosages are not given. They are not necessary for a discussion of the principles of antibacterial therapy, and for its practice they would have to be considered in far greater detail than our present space permits.

(2) **Trade names** are also omitted. It is understandable that each manufacturer should want to promote the sales of his own product rather than those of identical or similar materials produced by other firms, but much confusion results from the consequent multiplicity of names. The marketing of mixtures of antibacterial drugs under names which suggest that they are single new compounds is another source of bewilderment. The wisest policy is to ignore trade names and to think entirely in terms of official names.

(3) **Antibacterial ranges** of drugs are impossible to define briefly in other than general terms. Strains of the same species may differ widely in their sensitivities, and strains resistant to a particular drug are likely to become far more common in any community in which that drug is widely used, as we have already noted. In this section we have merely indicated the important groups of organisms commonly sensitive to each drug or class of drugs.

(4) The drugs described here as bactericidal are those of which it is possible to achieve in the

body concentrations that are lethal to susceptible organisms. They may have only a bacteristatic action when used in lower concentrations or against less susceptible organisms. Some of the drugs described as bacteristatic are in fact bactericidal *in vitro* but not in concentrations that have any relevance to their therapeutic use.

Chemotherapeutic Agents

Synthetic chemicals active against microorganisms *in vitro* and of sufficiently low toxicity to be administered systemically are called chemotherapeutic agents. Scientific chemotherapy began in the first decade of this century (p. 9), when Ehrlich introduced the therapeutic use of organic arsenical compounds, starting with atoxyl for trypanosomiasis and arsphenamine for syphilis. Then in 1935 Domagk showed that prontosil (sulphonamidochrysoidin) could cure streptococcal infections in mice or human beings. Other therapeutically useful sulphonamides followed, differing from one another in solubility, toxicity and degree of absorption from the intestine, but all working in the same way against bacteria (see below). The discovery and development of this group made a great impact on medical practice because of their efficacy against a wide range of important bacterial pathogens, and also stimulated a great deal of activity in the search for other safe and effective antimicrobial drugs. Of the many chemotherapeutic agents in current clinical use some have already been mentioned in connection with leprosy (p. 105) and infections with viruses (pp. 135–6), fungi (p. 142), protozoa (pp. 152 and 242–3) and helminths (in Chapters 15 and 17–21). Others are described in the next few pages and on pp. 397–8 (tuberculosis).

Sulphonamides

These drugs are derivatives of *p*-aminobenzene sulphonamide and therefore they closely resemble *p*-aminobenzoic acid (PABA). Many bacteria need PABA for synthesis of tetrahydrofolic acid, which is an essential coenzyme for purine and pyramidine synthesis. If these organisms take up sulphonamide instead of PABA, conversion of PABA to dihydrofolic acid (the precursor of tetrahydrofolic acid) is blocked and bacterial metabolism is arrested. Such bacteristasis depends on a considerable excess of sulphonamide over PABA in the environment, and is reversible by addition of more PABA. There is no bactericidal action.

Sulphonamides are effective against many pathogenic bacteria, cheap to produce, usually harmless to the patient (though with occasional serious toxicity), and of such diverse pharmacological properties that there is likely to be at least one appropriate to any particular situation. Yet this once invaluable group of drugs is now of limited and decreasing medical importance—partly because of more widespread bacterial resistance but mainly because on almost all occasions when they might be used there are antibiotics available that can be expected to do a better job. Sulphonamides that are rapidly absorbed when given by mouth and rapidly excreted into the urine, achieving high concentrations there without the problem of insolubility described below in connection with sulphadiazine, are still commonly used for treating acute urinary tract infections, because they have roughly the same high success rate as more expensive drugs; but even here it is arguable that trimethoprim should be given, as well as or in place of sulphonamide (see below). **Sulphadiazine** is rapidly absorbed when given by mouth, or can be injected, but renal excretion is slow enough to give good blood levels and there is little protein binding; consequently it diffuses well into tissues, and also into the CSF, which suggests that it might be useful for treating meningitis due to sensitive meningococci (p. 214). However, like many of the early sulphonamides it is poorly soluble in acid or neutral urine, and the patient must be given alkalis and plentiful fluids to prevent renal tubular obstruction by sulphonamide crystals. **Sulphadimidine** is also well absorbed, and is preferable in treating bacterial meningitis because it is less prone to crystalluria. **Sulphacetamide,** one of the earliest sulphonamides, is not used systemically, but it is highly soluble and penetrates exceptionally well into the eye, so it is still widely used in the form of eye-drops and eye ointment. Sulphonamides that are virtually non-absorbable when given by mouth have been extensively used, together with antibiotics, for pre-operative suppression of bowel flora, but with little benefit. Some other sulphonamides with special uses are mentioned below.

The occasional serious toxic effects of sulphonamides mentioned above include (in addition to renal tubular obstruction) the rare but often fatal form of erythema multiforme known as the Stevens–Johnson syndrome; other allergies, including an illness resembling serum sickness; and bone-marrow depression, which has on a few occasions developed into a fatal agranulocytosis. Sulphonamides are contraindicated in late pregnancy and the neonatal period, since they may induce kernicterus in the infant by displacing bilirubin from its protein-binding sites.

Trimethoprim, cotrimoxazole, cotrimazine, pyrimethamine

Trimethoprim interrupts folate metabolism, as do sulphonamides, but at a later stage—the conversion of dihydrofolic acid to tetrahydrofolic acid. This stage also occurs in mammalian cells, but there it is far less susceptible to the action of trimethoprim; and the risk of host folate deficiency following prolonged use of the drug can be avoided by giving ready-made folate, which can be used by mammalian but not by bacterial cells. In other respects trimethoprim is less toxic than sulphonamides. It is a broad-spectrum antibacterial agent which is bacteristatic and slowly bactericidal. It can be useful in treatment of urinary tract infections, and is far better than sulphonamides at penetrating into sputum. Bacterial resistance and host hypersensitivity to it do not develop so readily as to sulphonamides. For a number of years trimethoprim was available for clinical use only in mixtures with sulphonamides (**cotrimoxazole** = trimethoprim + **sulphamethoxazole**; **cotrimazine** = trimethoprim + **sulphadiazine**); this was because of *in vitro* evidence of synergy against some bacteria, and the fear (not justified by events) that bacterial resistance to 'unprotected' trimethoprim would rapidly decrease its value. There is therefore much more information available about the use of the mixtures than about that of trimethoprim alone.

Cotrimoxazole has proved effective in treating typhoid and brucellosis, in which the ability of trimethoprim to penetrate into host cells and reach intracellular bacteria is important. It is also effective against some protozoa—for treatment of toxoplasmosis or pneumocystis infection and for prophylaxis against the latter in patients with leukaemia and other forms of compromised resistance (p. 315).

Pyrimethamine, a compound related to trimethoprim but with more effect on human folate metabolism, is not a useful antibacterial agent. However, it is used in weekly doses for malaria prophylaxis—either alone, with **dapsone** (as for leprosy) or with **sulfadoxine,** a long-acting sulphonamide (p. 242)—and also for treatment of toxoplasmosis, again with a long-acting sulphonamide.

Metronidazole

This substance has been in use since 1959, in the form of oral tablets, for the treatment of trichomonas infections (p. 152). Later it was discovered also to be effective in giardiasis and in amoebiasis. Some years passed, however, before clinical use was made of its inhibitory action on anaerobic bacteria. Metronidazole itself is inactive against micro-organisms, but on entering the cell of an anaerobic organism—protozoal or bacterial—it is reduced to a form that interacts with DNA. All strictly anaerobic bacteria are affected in this way. Its widespread use has not yet been followed by the emergence of resistant strains in significant numbers. It is of particular value against strains of *Bacteroides fragilis*, which are highly susceptible to it and are often resistant to most of the commonly used antibiotics. Patients who cannot take it by mouth can have it in the form of rectal suppositories or intravenously. It is widely and effectively used for prophylaxis in patients undergoing colonic and gynaecological surgery and for treatment of those with cerebral abscesses, wound infections, septicaemia and other conditions in which anaerobes might be or are known to be involved. It is also used for treatment of anaerobic vaginosis (in which anaerobes and *Gardnerella vaginalis* are implicated, p. 282), and remains the treatment of choice for trichomoniasis, giardiasis and amoebiasis. Its side-effects are an unpleasant metallic taste and sometimes, particularly in elderly patients, a dose-related CNS toxicity manifested as peripheral neuropathy, vertigo, ataxia or disorientation.

Nitrofurantoin and nalidixic acid

These two unrelated substances have in common that when given by mouth they are rapidly absorbed and are excreted in high concentration in the urine. They are of use only in treatment of urinary tract infections. Nitrofurantoin achieves urinary concentrations that are bactericidal to some members of most bacterial groups, especially if the urine is acid—a condition not likely to be fulfilled if the infecting organism is a *Proteus* or other urease-positive strain, converting urea into ammonia. Nalidixic acid is bacteristatic. Its range does not include gram-positive organisms or *Ps. aeruginosa*, and therefore its usefulness is restricted to enterobacterial infections. Thus although both of these drugs have been widely used in treatment of urinary tract infections they are not reliable for this purpose without laboratory evidence of a suitably sensitive pathogen, and even then resistance may develop during treatment.

Quinolones

This group of compounds related to nalidixic acid are recent introductions to medical practice. The important members are **ciprofloxacin, ofloxacin, norfloxacin** and **pefloxacin**. All can be given by mouth and ciprofloxacin also by injection. Their complex mode of action involves inhibition of bacterial topo-isomerases (gyrases)—enzymes concerned with coiling and uncoiling of DNA strains during DNA synthesis. Their antibacterial activity is far greater than that of nalidixic acid and covers a wide range of bacteria—many gram-positive and gram-negative aerobes, including *Pseudomonas* species, but few anaerobes; and it appears not to be susceptible to plasmid-mediated resistance (p. 383).

Antibiotics

The term antibiotic, coined in 1942 by Waksman to describe a newly discovered class of antimicrobial agents, is now a household word. According to his definition, an antibiotic is a substance, produced by micro-organisms, which can inhibit the growth of or even destroy other micro-organisms. He specified that dilute solutions of the substance must have these properties, thus excluding, for example, the lactic acid produced by lactobacilli (p. 29). The limits of his definition have been stretched by man's activities in synthesizing compounds similar to or identical with those made by micro-organisms. It is hard to deny the name antibiotic to a substance such as chloramphenicol, which is now man-made but is chemically identical with an antibiotic produced by a bacterium. With semisynthetic penicillins and other antimicrobial substances now produced by chemical alteration of microbial products there is no longer a clear dividing line between chemotherapeutic agents and antibiotics.

To be of clinical value an antibiotic must have selective toxicity (p. 373). The success of penicillin and the other β-lactams is due to their mode of action being interference with synthesis of cell-wall mucopeptide, a vital component of bacterial cells which is not shared by human or other eukaryotic cells (p. 10). In some circumstances and some bacteria this interference prevents formation of the cross-walls essential to bacterial division; and in others it causes formation of a faulty cell-wall with gaps through which osmotic pressure causes the cytoplasmic membrane to bulge out and rupture. Other antibiotics—the aminoglycosides, chloramphenicol, the tetracyclines, erythromycin and clindamycin—interfere at various points in the protein-synthesizing activities of bacterial ribosomes, and their clinical usefulness depends on structural differences between bacterial and mammalian ribosomes. Both of these forms of interference with the processes of growth are effective only against multiplying bacteria, as is the inhibition of bacterial RNA synthesis by the rifamycins; but the damage that the polymyxins inflict on bacterial cytoplasmic membranes, leading to leakage of cell contents, is lethal even to resting cells.

β-lactams

The basic building block of all β-lactams is a β-

lactam ring. Attachment of different adjoining rings and side chains produces a wide range of penicillins, cephalosporins, penems and monobactams. β-lactams vary in their susceptibility to β-lactamases, a range of enzymes which share the ability to split open the β-lactam ring and destroy the compound's antibacterial activity in doing so. β-lactams also differ in their affinity for the various penicillin-binding proteins (PBPs, p. 381) in the cell walls of bacteria, and such differences largely determine their antibacterial ranges.

(a) Penicillins

The name 'penicillin', used on its own, refers to the first-discovered member of the group, **benzylpenicillin** (penicillin G). This remains one of our most valuable antibacterial drugs. It is bactericidal to a range of bacteria that includes many streptococci, the neisseriae (though some *N. gonorrhoeae* strains produce penicillinase), most gram-positive bacilli, some gram-negative bacilli (especially when it is in high concentration), spirochaetes and actinomycetes. Most *Staph. aureus* were sensitive to it when it was first introduced, but penicillinase-producing and therefore resistant strains now predominate in many areas, especially in hospitals. It is a remarkable drug in that it is tolerated by the body in almost unlimited quantities; such adverse reactions as do occur are due to hypersensitivity and are virtually independent of dosage (apart from the toxic encephalopathy occasionally produced by giving very large doses to patients with poor renal function, in whom exceptionally high blood levels can be achieved, or by giving a large excess intrathecally). Two of its unsatisfactory features are that it has to be given parenterally because it is acid-labile and destroyed to an unpredictable degree in the stomach, and that high blood levels are hard to maintain because it is rapidly excreted in the urine. The latter feature is occasionally beneficial, since high urinary concentrations can be achieved; more often it is merely wasteful of a relatively inexpensive compound; but sometimes, when persistent high blood levels are required, it is necessary to give **probenecid**, which blocks renal tubular excretion of penicillin.

It is still an open question whether the most effective way of using penicillin is to maintain continuous high blood levels. It is possible that as good or better results can be obtained in most cases by intermittent high levels (p. 378). These can be achieved by suitably spaced intramuscular or intravenous injections. A steady blood level can be maintained by continuous intravenous administration, provided that incompatibilities are avoided (p. 379). However, it is often sufficient and more convenient to give a daily intramuscular injection of a slowly absorbed compound which will maintain an adequate blood level throughout the day, and in some cases there is a considerable advantage in being able to give a single dose which will maintain a moderate blood level for many days (see for example p. 279, treatment of syphilis). **Procaine penicillin, benethamide penicillin** and **benzathine penicillin** are slowly absorbed compounds of benzylpenicillin which respectively give useful blood levels for many hours, several days and several weeks after a single intramuscular injection.

By varying the substrate for growth of the moulds that make benzylpenicillin, the acid-resistant **phenoxymethyl penicillin** (penicillin V) can be produced. When given by mouth, salts of this compound are sufficiently well absorbed to give reliably useful blood levels.

In 1959 it was discovered that the 'nucleus' of all penicillin molecules, 6-aminopenicillanic acid, could be made by substrate variation or by enzymic removal of side chains from benzylpenicillin or other existing penicillins. By attaching various side chains to this nucleus it has been possible to make a vast range of new semi-synthetic penicillins, some of which have special properties that make them useful therapeutic agents. These include:

(i) **phenethicillin** and **propicillin,** acid-resistant and absorbed more efficiently than phenoxymethyl penicillin, but correspondingly less potent and therefore comparable with it in effectiveness.

(ii) **methicillin, cloxacillin** and **flucloxacillin,** penicillinase-resistant and therefore active against penicillinase-producing staphylococci. The cloxacillins are both acid-resistant, but the much more complete absorption of flucloxacillin makes it the obvious choice for oral administration.

(iii) **ampicillin,** and for oral administration, its much better absorbed esters **talampicillin, pivampicillin** and **bacampicillin** ('pro-drugs' which are converted to ampicillin after absorption); somewhat less potent than benzylpenicillin against many gram-positive bacteria, but markedly more effective against *Streptococcus faecalis,* some of the enterobacteria (including salmonellae and shigellae) and *H. influenzae;* liable to provoke rashes (distinct from those due to hypersensitivity to all penicillins), especially in patients with infectious mononucleosis.

(iv) **amoxycillin,** structurally related and similar in range to ampicillin; better absorbed than ampicillin when given orally (comparable to the esters); more effective for treating intrabronchial infections because of its better penetration into sputum (p. 201), and for eradicating some other infections (e.g. typhoid, p. 269—its biliary concentration and ability to penetrate host cells are probably important here).

(v) **mecillinam,** and its ester **pivmecillinam;** little activity against gram-positive organisms but highly active against some enterobacteria by a mode of action on cell-wall formation different from that of other penicillins; synergy with ampicillin in treatment of resistant enterobacteria.

(vi) the antipseudomonal penicillins **carbenicillin** and **ticarcillin** (the latter more potent and now more commonly used), needing to be injected, active against *Proteus* species and some other enterobacteria as well as *Pseudomonas* species but relatively ineffective against gram-positive organisms; and **carfecillin,** given by mouth, converted to carbenicillin (of which it is an ester) after absorption, and reaching the urine (only) in concentrations that may be useful in treating *Ps. aeruginosa* urinary infections.

(vii) the ureidopenicillins **mezlocillin** (more active than ticarcillin against various enterobacteria, and comparable to it against *Ps. aeruginosa*), **azlocillin** (highly active against *Ps. aeruginosa*) and **piperacillin** (possibly the most active of the three against many enterobacteria as well as against *Ps. aeruginosa*). All have to be given by injection.

The penicillins all share the lack of toxicity of benzylpenicillin, but also share its liability to provoke hypersensitivity reaction (p. 374). When hypersensitivity occurs, it applies to all members of the group, so that none of them can be given to a patient who has become sensitive to one of them.

(b) Cephalosporins

Nearly 20 years after the mould *Cephalosporium acremonium* was found to be a source of antibiotics, the first semi-synthetic derivatives of one of these were released for clinical use in 1964. Others followed, with increasing frequency, and the array of cephalosporins now available or due for general release in the near future is bewildering—especially to those who have to grapple with their trade names as well as their official names! Correlation between their times of introduction and their biochemical and biological properties led to their being classified in 'generations'.

FIRST GENERATION Those in general use before about 1975 are similar to ampicillin in their antibacterial range, though somewhat more resistant to staphylococcal and enterobacterial β-lactamases. **Cephaloridine** and **cephalothin,** both needing to be injected, are the most active against gram-positive cocci and many common enterobacteria; but they are now seldom used, since better drugs have become available and cephaloridine is nephrotoxic, especially in patients with existing renal impairment or on treatment with certain diuretics or an aminoglycoside. The first-generation cephalosporins that can be given orally—**cephalexin, cephradine, cefaclor** and **cefadroxil**—continue to be widely used, but have no obvious advantages over amoxycillin (and are far less potent than benzylpenicillin against pneumococci and other streptococci). They were thought at one time to be safe to give to patients with penicillin hypersensitivity, but in fact some 10% of such patients are also hypersensitive to cephalosporins.

SECOND GENERATION Following the widespread use of penicillins and the early cephalosporins, bacteria (notably enterobacteria) resistant to them became increasingly frequent in clinical specimens and as pathogens. Much of this resistance is due to production of β-lactamases. These enzymes differ from one another in the range of substrates that they hydrolyse and the rates at which they do so; some are penicillinases only and others attack cephalosporins. Ability to resist most enterobacterial β-lactamases is the distinctive feature of cephalosporins such as **cefuroxime, cefamandole** and **cefoxitin** (which is structurally a cephalosporin, but is also called a cephamycin because it was derived from cephamycin C, a streptomyces product). They all require injection. They are active against staphylococci (though less so than earlier cephalosporins) and a wide range of gram-negative bacteria including *N. gonorrhoeae* and *H. influenzae*, and cefoxitin in particular is active against anaerobes.

THIRD GENERATION Increased antibacterial potency against a wider range of organisms and still further resistance to β-lactamases are the features that characterize this generation—**cefotaxime, ceftizoxime, ceftriaxone, ceftazidime, cefoperazone** and **latamoxef** (moxalactam). They are up to 100-fold more active against many species that were regarded as sensitive to their predecessors, though less active against *Staph. aureus*, pneumococci and some other gram-positive species and ineffective against cloxacillin-resistant *Staph. aureus* and against enterococci. Ceftazidime in particular has useful activity against *Ps. aeruginosa*, and against various other gram-negative bacilli, such as *Acinetobacter* species, which are uncommon pathogens but of disproportionate importance when they do cause infection because they are virtually untreatable with earlier antibiotics. **Cefsulodin** differs from the members of the group already mentioned in that it has little action against enterobacteria but good antipseudomonal activity. All of these drugs have to be injected, but several of them—notably ceftriaxone—give persistent high serum levels after each injection and so can be given less frequently. The wide antibacterial spectrum of most of them makes them suitable for initial treatment of septicaemias and other severe infections, particularly in compromised hosts—in whom their bactericidal action can be important. Good penetration into CSF means that they can be used for treating meningitis, and they are particularly valuable in neonatal meningitis due to gram-negative bacilli (p. 217). One special problem that has arisen on some occasions when drugs of this group have been used is the emergence of resistance to them and other β-lactams, due to de-repression of chromosomal β-lactamase, in organisms of genera including *Pseudomonas, Enterobacter* and *Serratia*. Such induction of β-lactamase production probably accounts for observed examples of antagonism between these drugs and penicillins. Disturbance of the intestinal flora is a side-effect seen in particular with latamoxef, which is excreted in the bile and has more action than other cephalosporins on gram-negative anaerobes (though not enough to make it a first-line drug for use against them); its use may result in post-treatment diarrhoea—sometimes pseudomembranous colitis (p. 259). Latamoxef and cefoperazone have also been known to cause bleeding in some patients, associated with prothrombin–time elevation; the probable mechanism is vitamin K depletion due to interference with the intestinal flora (p. 18), together with an immune-mediated thrombocytopenia triggered by the antibiotic.

(c) Clavulanic acid, sulbactam

These β-lactams have no useful direct antibacterial action but are powerful β-lactamase inhibitors. The lack of any β-lactam drug with high resistance to β-lactamases and suitable for oral administration has meant that such inhibitors have a clinical role as protectors of the older orally administered β-lactams—e.g. as in the commercially available mixture of amoxycillin and clavulanic acid.

(d) New β-lactams

Recently introduced β-lactams with interesting prospects but as yet no adequate clinical assessment include: **temocillin,** with a high level of β-lactamase resistance and good activity against most gram-negative genera other than *Pseudomonas* and *Bacteroides*; **aztreonam,** the first monocyclic β-lactam to be evaluated, highly active but only against gram-negative aerobes; and **thienamycin,** the forerunner of another new group, with perhaps the widest antibacterial spectrum of any known β-lactam, but rapidly destroyed by renal dehydropeptidase enzymes and so needing concurrent administration of the enzyme inhibitor cilastatin.

Fucidin

This sodium salt of fusidic acid derived from the fungus *Fusidium coccineum* is unusual among antibiotics in having a steroid molecular structure. It is well absorbed when given by mouth, is apparently free from serious toxic effects, and is highly effective against most strains of *Staph. aureus*, including penicillinase-producers. Claims have been made that it is particularly good at eradicating staphylococcal infections of bones and joints. Resistance to it develops readily among staphylococci in culture, and sometimes occurs during treatment also. In order to preserve its value, it seems best to use it against staphylococci only, and always in conjunction with another antibiotic so as to reduce the risk of developing resistance (p. 378)—e.g. with cloxacillin or flucloxacillin.

Aminoglycosides

Most antibiotics of this group (and indeed of other groups yet to be discussed) are derived from *Streptomyces* species. Aminoglycosides of such derivation have names ending in 'mycin', whereas those derived from *Micromonospora* species have names ending in 'micin'.

Like the β-lactams, aminoglycosides are bactericidal. They are effective against many strains of *Staph. aureus* and of enterobacteria, though resistance among the latter is an increasingly common problem because they can produce an assortment of drug-degrading enzymes (cf. β-lactamases). Some are effective against *Ps. aeruginosa* (see below). They are not effective against anaerobes, or against streptococci (except when acting synergically with a penicillin—see pp. 238 and 378). They are not absorbed to a useful extent when given by mouth. Dosages have to be carefully controlled to avoid damage to the 8th cranial nerve, and these drugs may also be nephrotoxic (see below).

When introduced in 1944 **streptomycin** was the first effective antituberculous drug, and it remains important in this field (p. 398). For the various other purposes for which it was once used it has been almost entirely replaced—at first by **kanamycin,** which has a somewhat wider range of activity and perhaps less toxicity; and then by **gentamicin.** This drug is more active than other aminoglycosides against a wide range of bacteria, and was the first member of this group to be effective against *Ps. aeruginosa*. It is a most useful drug for treatment of proven or suspected septicaemia due to organisms presumably originating from the alimentary or urinary tracts—after surgery or as a complication of a malignant growth, for example; but for such purposes it needs the assistance of metronidazole or some other drug capable of dealing with *Clostridium* and *Bacteroides* species, and possibly also of a penicillin to deal with streptococci. **Tobramycin** is less active than gentamicin against many species but rather more active against most *Pseudomonas* strains; but there has been a marked increase in frequency of *Ps. aeruginosa* strains producing enzymes that destroy both gentamicin and tobramycin. **Amikacin,** a kanamycin derivative that is resistant to these enzymes, is useful for treating infections due to such organisms. **Netilmicin** is effective against some gentamicin-resistant gram-negative bacilli and possibly less toxic than gentamicin.

Monitoring of serum levels (p. 368) is of great importance when using parenteral aminoglycosides. This is because effective treatment requires adequate peak serum levels around 1 hour after each dose (8–15 μg/ml for gentamicin), but avoidance of toxicity requires low 'trough' levels between doses (less than 2μg/ml for gentamicin). Since the fall depends on renal excretion, the standard interval between doses (8 hours for gentamicin) has to be increased in the presence of renal failure. To carry out the monitoring tests satisfactorily the microbiologist needs to receive 'trough' and 'peak' serum samples and to know their time relationship to aminoglycoside doses, the patient's age (and weight and sex in difficult cases), the state of the kidneys and whether any other antibacterial drug has been given (since if he is using a bio-assay he will have to inactivate that drug or use a test organism resistant to it). The major toxic complications of aminoglycoside therapy are: ototoxicity, due to 8th cranial nerve damage and most likely to occur when the patient is elderly, has impaired renal function or is given prolonged treatment; and their nephrotoxicity, due to accumulation of the drug in renal tubular cells and their consequent necrosis and more likely to occur if the patient is concurrently receiving vancomycin, cephaloridine or certain diuretics (cf. p. 378).

Neomycin and **framycetin** are kanamycin-like drugs too toxic for systemic use. Neomycin is sometimes given by mouth as a means of attack on intestinal pathogens, but may do more harm than good (p. 257). It can be used to reduce ammonia production by bacteria in the gut of patients with liver failure. Each of these drugs has been used in antibiotic combinations for reducing bowel flora in immunocompromised patients (p. 348). Either can be applied topically in the form of powder, creams, ointment, eye or ear drops, etc., but these preparations must be used with great caution, since prolonged and heavy topical application can lead to absorption of toxic amounts. Gentamicin has also been extensively used in such topical forms, but this practice seems to have been responsible for the increasing frequency of gentamicin-resistant strains of *Ps. aeruginosa* and other species.

Spectinomycin (an aminocyclitol, not an aminoglycoside but closely related to that group) has been widely and successfully used for treatment of penicillin-resistant gonococci (p. 280). For this purpose it is given in a single large intramuscular dose. There are no established indications for its use in other ways or for other purposes.

Mupirocin

This naturally occurring substance produced by *Pseudomonas fluorescens* possesses broad-spectrum antibacterial activity and may be a useful topical agent for the treatment of skin infection.

Chloramphenicol

Originally derived from *Streptomyces venezuelae* in 1947, chloramphenicol is now made synthetically. It is well absorbed when given by mouth, and passes from the blood stream into the CSF more readily than any other antibiotic. It can also be given parenterally, but the intravenous route should be used, as absorption after intramuscular injection is slow and gives relatively poor blood levels. It is effective against a wide range of bacteria, and also against rickettsiae and chlamydiae. It is bactericidal to *H. influenzae*, and a valuable drug for treatment of the life-threatening infections of children caused by type b strains of this species (p. 215). Against other genera it is bacteristatic only. This may explain its failure to eradicate *S. typhi* infections (p. 93), even though it is highly effective in controlling the acute illness of typhoid (except when this is due to a chloramphenicol-resistant strain). In the years following its introduction it was widely used for infections of many kinds, but then came numerous reports of fatal bone-marrow depression following its use. Since then there has been a general tendency in North America, Britain and some other countries to regard it as highly dangerous and to use it almost exclusively for treatment of typhoid and paratyphoid (for which amoxycillin and co-trimoxazole are now valid alternatives—p. 94) and of severe infections due to *H. influenzae* or to bacteria resistant to all reputedly safer antibiotics. In other countries it has continued to be frequently prescribed or to be obtainable without prescription. On the available evidence it is impossible to be sure how dangerous chloramphenicol is, or indeed whether all batches are equally dangerous. Possibly because it has been 'in disgrace', it can be of great value in treatment of life-threatening infections because most strains of staphylococci and of other troublesome bacteria are still sensitive to it. Caution is necessary when giving it to new-born infants, since they inactivate and excrete it very slowly and therefore high dosage may lead to excessive blood levels, collapse and death ('the grey syndrome'—p. 217). Blood levels can be determined in a few hours by bio-assay and the dosage adjusted if necessary.

Trough levels of 5 µg/ml and peak levels below 20 µg/ml are required to ensure a therapeutic effect and avoid toxicity. It is of little value in urinary tract infection because the kidneys excrete it mainly in a form that is not active against bacteria. Chloramphenicol drops or ointment are extensively and effectively used for treating eye infections; but there is evidence that this practice is leading to the emergence of *H. influenzae* strains with transferable multi-resistance.

Tetracyclines

Chlortetracycline (first isolated in 1948) and its now numerous relations are all derived, directly or with subsequent chemical manipulations, from *Streptomyces* species. Like chloramphenicol, they are effective when given by mouth. They are purely bacteristatic. They have much the same antimicrobial range as chloramphenicol, including mycoplasmas and ureaplasmas, rickettsiae, chlamydiae and *Cox. burneti*. Bacterial resistance develops rather readily, and has become increasingly common even in species such as *Str. pyogenes* and *Str. pneumoniae* which do not readily become resistant to drugs of most other groups.

Chlortetracycline, oxytetracycline and **tetracycline** itself, the original members of this group, are all still in use. Their rather inefficient and irregular absorption when given by mouth necessitates high dosage, with frequent gastro-intestinal side-effects: nausea and vomiting are probably due to chemical irritation of the stomach, and diarrhoea results from derangement to the normal bowel flora by unabsorbed drug. In its extreme form this derangement may lead to intestinal candidiasis, or on rare occasions to staphylococcal enteritis (p. 259). The chief advantage of more recently introduced tetracyclines is that better absorption and slower excretion permits lower oral dosage. **Demeclocycline** is comparatively well absorbed and rather slowly excreted. Excretion of **doxycycline** is so slow that a daily dose is sufficent, as against 6-hourly doses for the older tetracyclines. As doxycyline is excreted in bile and also directly through the wall of the small intestine into the faeces, as well as in urine, it can be used (with caution) in renal failure without the danger of excessive serum levels. **Minocycline** is very well absorbed and slowly excreted, so that it too can be given as a daily dose, and it has the additional interesting property of being effective against many staphylococcal and other strains that are resistant to other tetracyclines; but unfortunately many patients develop light-headedness, dizziness and tinnitus.

Permanent yellow staining of teeth may follow administration of tetracyclines to children less than 12 months old or to their mothers during the later months of pregnancy. Other reasons for avoiding these drugs during pregnancy are reports of severe liver damage in pregnant women following high dosage and unconfirmed allegations of teratogenic effects on fetuses. Because of these various forms of toxicity, tetracyclines are less widely used than formerly. In Britain their main uses are in the treatment of non-specific urethritis, exacerbations of bronchitis, mycoplasma pneumonia, psittacosis and Q fever.

Erythromycin

This macrolide, produced by *Streptomyces erythreus*, has an antibacterial range similar to that of benzylpenicillin. In high concentrations *in vitro* it is bactericidal to susceptible organisms, but this probably has only limited relevance to its action in the body. It can be given orally as erythromycin base, which is rapidly destroyed by gastric acid and poorly absorbed; as stearate which is less acid labile and better absorbed; or as the esters succinate or estolate which are absorbed best of all. The estolate may cause acute but reversible liver damage. Intravenous erythromycin is useful in the treatment of severe infections, notably legionellosis (see below). Staphylococci in particular readily become resistant to erythromycin but this can be discouraged by giving another antibiotic at the same time (p. 397). Erythromycin is sometimes a valuable alternative to penicillin in patients hypersensitive to the latter; it has been widely used in paediatrics, largely because of its lack of toxicity and its availability in the form of an acceptable oral suspension; and it has recently acquired a new level of importance as the drug of choice for *Campylobacter* and *Legionella* infections, for the treatment and prophylaxis of diphtheria and whooping cough and for the treatment of chlamydial infection in pregnancy or in the neonatal period. It is an alternative to tetracycline for treatment of infections due to mycoplasmas and ureaplasmas, chlamydiae, rickettsiae and *Cox. burneti*.

Clindamycin

This is a synthetic derivative of another streptomyces antibiotic, **lincomycin**. It closely resembles the macrolide antibiotics in most of its properties, in spite of having a markedly different molecular structure. It is well absorbed when taken orally, and penetrates into bone better than most antibiotics; it is therefore a useful weapon for dealing with osteomyelitis. It is an effective alternative to penicillin for use against *Str. pyogenes*, and is also valuable for dealing with infections caused by anaerobes, particularly by penicillin-resistant strains such as *Bact. fragilis*. Being better absorbed from the alimentary tract than lincomycin, it does not cause the immediate gastro-intestinal upsets commonly associated with use of that drug. Pseudomembranous colitis (p. 259) is perhaps more often induced by one of these two drugs than by others.

Vancomycin

This is another streptomyces antibiotic. It has been associated with an assortment of toxic effects which may have been related to impurities in earlier formulations. It has to be given intravenously for systemic action, and is difficult to give repeatedly by that route because it causes thrombophlebitis. Nevertheless, its high activity against staphylococci (both *Staph. aureus* and *Staph. epidermidis*, including strains resistant to cloxacillin) and against streptococci means that it is sometimes useful—in treatment of severe infections due to organisms of these genera resistant to less toxic drugs; in *Staph. epidermidis* peritonitis complicating chronic ambulatory peritoneal dialysis; in staphylococcal infections associated with synthetic implants and intravenous catheters; in peri-operative protection of heart valves (p. 375) when the patients have already been on penicillin, with consequent modification of their resident bacterial populations; or in patients hypersensitive to β-lactams. Given by mouth, it is the treatment of choice for pseudo-membranous colitis (see above and p. 259) and for staphylococcal enteritis (p. 259). Virtually all of the drug is excreted unchanged by the kidney, and hence there is a risk of accumulation in patients with renal failure. Serum levels should be monitored to prevent side-effects, notably deafness.

Rifamycins

Although the members of this group of antibiotics derived from *Streptomyces mediterranei* have promising antibacterial activities *in vitro*, their clinical usefulness is impaired by their rapid excretion in the bile and the consequent impossibility of maintaining adequate blood levels. **Rifampicin**, a synthetic derivative of one of them, is less rapidly excreted, is well absorbed when given by mouth, and is bactericidal to staphylococci and non-faecal streptococci (at remarkably low concentrations), to *Myco. tuberculosis* and to many other bacteria. However, its most serious limitation is the speed with which resistant mutants emerge, and because of this it should probably never be used alone. Its considerable value in treatment of tuberculosis (p. 397) and leprosy (p. 105) is likely to be best preserved by using it only in exceptional circumstances for non-mycobacterial infections (e.g. in treatment of endocarditis due to a staphylococcus resistant to most other drugs). It has been used with some success to prevent spread and eliminate carriage of meningococci and of *H. influenzae* type b; but rifampicin-resistant meningococci have appeared following such use, the haemophili can be expected to follow suit, and there is at least a theoretical risk of a selective action on tubercle bacilli in populations so treated. When given in low dosage rifampicin is mostly excreted in the bile, but higher dosage saturates this excretory mechanism and it then appears in useful concentrations in the urine, which it colours red (as it may also do to the tears).

Polymyxins

These are polypeptides derived from bacteria of the genus *Bacillus* and first reported in 1947. **Colistin,** discovered in 1950 and at first thought to be a new antibiotic, is identical with polymyxin E. All are nephrotoxic but the toxicity of polymyxins B and E is not sufficient to preclude their systemic use when necessary. For this purpose they are administered intramuscularly. They are also used in the form of powders and ointments for the treatment of wounds and burns, and given by mouth for treatment of alimentary infections. They are not absorbed from the intestine. They are administered either as sulphates or as sulphomethyl derivatives (methane sulphonates), and the latter are preferred by some for parenteral use because they give less pain when injected and are less toxic. However, they are also less potent and more rapidly excreted by the kidneys—an advantage, perhaps, when treating urinary tract infections but not when sustained blood levels are needed. The polymyxins are bactericidal and are effective against *Ps. aeruginosa* and against gram-negative bacilli of most other genera except *Proteus*, but unfortunately their efficacy against *Ps. aeruginosa* is markedly reduced in the presence of calcium ions in concentrations found in blood and other body fluids.

Some Special Problems

Staphylococcus aureus Infections

This most adaptable species has continued to be a special problem throughout the antibiotic era. Apart from the methicillin-resistant strains described in the next paragraph. *Staph. aureus* strains are *intrinsically* penicillin sensitive, and benzylpenicillin remains the best antibiotic for dealing with those that cannot destroy it. However, such strains are now rarely encountered; most hospital strains and many of those from other sources produce penicillinase that destroys benzylpenicillin, ampicillin and related compounds rapidly—but is much less effective against the cloxacillin group and the cephalosporins. Cloxacillin or flucloxacillin is the first choice for dealing with penicillinase-producers (usually reported as 'penicillin resistant'), or with strains that are inaccessible for testing (e.g. because they are in deeply placed lesions) or that require urgent treatment before results of tests are available. Other antibiotics that can be used against *Staph. aureus* are erythromycin, fucidin, clindamycin and the aminoglycosides, with the more toxic vancomycin in reserve for special problems.

Some *Staph. aureus* strains, encountered with increasing frequency in hospitals, show genuine penicillin resistance that includes all penicillins, and usually the cephalosporins as well. They are known as 'methicillin-resistant *Staph. aureus*' (MRSA). Their resistance depends on unusual cell-wall composition, and is as a rule shown by only a small proportion of the cells of such a strain when incubated at 37°C, but this proportion is greatly increased at lower incubation temperatures (e.g. 30°C). Infections caused by such strains respond poorly to treatment with any of the penicillins. They are usually resistant also to several other anti-staphylococcal drugs but not to vancomycin.

Tuberculosis

Antibacterial treatment of tuberculosis has three unusual features: (1) the need to use combinations of drugs at all times, to avoid the emergence of resistance during therapy; (2) the need for prolonged treatment because of the very slow metabolism and multiplication rate of the tubercle bacillus; and (3) the nature of the drugs used.

To prevent emergence of resistance by selection of resistant mutants, it is necessary to use two effective drugs; and since the drug-sensitivity of the organism is not known until several weeks after starting treatment, it is standard practice to use three drugs initially, in case the infecting strain is resistant to one of them. The 'first line' triad in Britain and many other countries today, all of which can be given by mouth, are:

(1) **rifampicin** (p. 396), undoubtedly the most potent antituberculous agent avail-

able; costly; liable to cause liver damage (hepatitis).
(2) **isoniazid,** also highly effective, except that resistant strains are relatively common; cheap; hepatitis or peripheral neuritis may occur with higher doses.
(3) **ethambutol,** in low dosage a useful supporting drug and virtually non-toxic, but liable to cause optic neuritis when used in the higher dosage necessary if one of the others is contraindicated; should be avoided in young children and in the elderly.

Streptomycin (p. 393), the first drug to make an effective impact on the problem of tuberculosis, has the disadvantage that it has to be injected. It is the principal reserve drug after the first line triad, but it is also still widely used for initial treatment in countries where the high cost of rifampicin makes it unsuitable for this purpose. Other reserve drugs include **pyrazinamide** (particularly useful in tuberculous meningitis because of its good penetration into cerebrospinal fluid), **thiacetazone** and **cycloserine.**

The efficacy of modern antituberculous drugs—including the synergic bactericidal action of rifampicin and isoniazid against tubercle bacilli—means that it is no longer necessary to continue treatment for up to 2 years, as was the case not long ago. Three drugs for 2–3 months, with two of them (selected on the basis of laboratory tests) continued up to a total of 9 months, give good results in terms of clinical cure and low relapse rates—provided that the patients continue to take their drugs consistently. Failure of compliance with regimens is a major problem, especially where patients have little understanding of their disease. Continuing attemps to evaluate shorter courses with different dose combinations have indicated that shortening the period of treatment to 6 months may be feasible. Weekly or twice weekly supervised administration of high doses of drugs is more effective for some types of patient than unreliable self-administration of standard doses.

Patients with Impaired Renal Function

If a patient who needs antimicrobial therapy also has impaired renal function, additional factors influence selection of drugs and dosage schedules. The renal excretion rate is a major factor in determining the blood and tissue levels of many antimicrobial drugs. The aminoglycosides, for example, are normally excreted mainly by the kidneys and have serious toxic effects (including nephrotoxicity) at blood levels not far above those required for therapy; impairment of their renal excretion calls for carefully calculated dose-reduction and close attention to the levels actually achieved. On the other hand, the penicillins are also mainly excreted by the kidneys, but their lack of toxicity even at high levels means that they can safely be given to patients with impaired renal function, though it is wise to reduce the dosage. Blood levels of chloramphenicol, as determined by bio-assay, are little affected by renal failure, because this drug is conjugated by the liver to an inactive form. If chloramphenicol is to be used in the presence of renal failure it should be given in normal dosage, but it should be avoided if possible, because the conjugates accumulate in the blood and may be toxic. The tetracyclines too should be avoided in such circumstances, as they may precipitate severe and sometimes fatal uraemia—by direct action on the kidneys and also by an anti-anabolic effect which increases the amount of urea needing to be cleared from the blood. Fucidin and the rifamycins are among the few antimicrobial drugs that are excreted almost entirely by non-renal mechanisms.

Suggestions for Further Reading

Smith, H. 1977. *Antibiotics in Clinical Practice.* 3rd edn. Pitman Medical, Tunbridge Wells.

Garrod, L.P., Lambert, H.P. and O'Grady, F. 1981. *Antibiotic and Chemotherapy.* 5th edn. Churchill Livingstone, Edinburgh.

Greenwood, D. 1983. *Antimicrobial Chemotherapy.* Baillère Tindall, London.

APPENDICES
A Glossary of Technical Terms
B Meanings of Some Abbreviations
Index

APPENDIX

Appendix A
Glossary of Technical Terms

For the reader's convenience definitions of some terms in common use in medical microbiology and parasitology are brought together here. Definitions, or at least indications of the meanings, of many of them are also given in the main text on the pages cited. The meanings of many other terms not quoted here can be found by reference to the Index. Words in *italics* are defined elsewhere in the glossary.

Acid-fast (acid–alcohol-fast): Resistant to decolourization by acid (or by acid and by alcohol) after staining with hot carbol fuchsin, and so retaining a red colour when stained by the Ziehl–Neelsen method—p. 103.

Active (immunity, immunization): Dependent upon stimulation of the subject's own immunological mechanisms—p. 362. (Cf. *passive*.)

Adjuvant: A substance which, by delaying absorption of an antigen or by other means, enhances its antigenic efficiency—p. 362.

Aerobe: An organism which can live and multiply in the presence of atmospheric oxygen—p. 15.

Agglutination: Clumping together, e.g. of red blood cells or micro-organisms, on exposure to an appropriate antiserum—p. 176.

Aggressins: Extracellular enzymes that are produced by bacteria and damage tissue components—p. 17.

Allergy: Originally, any altered reactivity to a foreign substance; but now usually synonymous with *hypersensitivity*—p. 61.

Anaerobe: An organism which cannot multiply or survive for long in the presence of more than a trace of free oxygen—p. 15.

Anamnestic reaction: A rise of an existing antibody level in response to an irrelevant stimulus, such as an infection with an organism unrelated to that against which the antibody was originally formed—p. 180.

Antagonism (between antimicrobial drugs): Impairment of the efficacy of one or each drug in the presence of the other—p. 378. (Cf. *synergy*).

Antibiotic: A product of micro-organisms which, even when much diluted, is lethal or inhibitory to other micro-organisms—p. 389.

Antibody: A globulin which is formed by the human or animal body in response to contact with some foreign substance and which reacts specifically with that substance—p. 50 (Synonym *immunoglobulin*).

Antigen: A substance which provokes formation of antibodies—p. 49.

Antiseptic: Roughly synonymous with *disinfectant*—p. 340.

Antiserum: A serum containing antibodies for a given organism or toxin—p. 7.

Antitoxin: An antibody for a given toxin—p. 54.

Asepsis: Avoidance of infection—p. 7.

Attack rate: The proportion of people, within a group or population exposed to the risk of infection with a particular organism, who develop clinical illness as a result of that exposure.

Attenuated (organism): Reduced in virulence for a given host (but often retaining useful antigenicity for that host)—p. 38.

Bacillus: A 'little stick', a rod-shaped bacterium—p. 11.

Bacteraemia: Presence of bacteria in the blood-stream with or without resulting illness—p. 60. (Cf. *septicaemia*).

Bactericidal: Lethal to bacteria—p. 341.

Bacteristatic (bacteriostatic): Preventing multiplication of bacteria—p. 341.

Bacteriuria (bacilluria): Presence of bacteria (bacilli) in significant numbers in freshly voided uncontaminated urine—p. 272.

Brownian movement: Passive to-and-fro movement of small particles such as bacteria when suspended in a fluid medium, due to irregular bombardment by molecules of the fluid or its solutes—p. 164.

Capsid: The protein coat surrounding the *genome* of a virus—p. 12.

Capsomere: One of the units of which a virus capsid is composed—p. 12.

Capsule: A coating, commonly of polysaccharide, outside the cell-walls of some bacteria and fungi— pp. 11, 69, 141.

Carboxyphilic (organism): Needing for growth a higher concentration of carbon dioxide in the atmosphere than is found in air—p. 16. (Synonym CO_2-dependent)

Carrier: One who is harbouring but not currently suffering any ill effects from a pathogenic organism—p. 27.

Cell-line: An *in vitro* culture of mammalian cells of known origin, suitable for propagation of viruses— p. 169. (The word 'mammalian' in this definition is of course appropriate only in connection with the study of viruses that are parasites of mammals.)

Chemotherapeutic agent: A synthetic chemical suitable for systemic administration and effective in the treatment of microbial infections—p. 387.

Clone: A 'race' of cells derived from a single ancestral cell and sharing a single function, e.g. of producing a particular antibody—p. 51.

Coccus: A spherical or ovoid bacterium—p. 11.

Colony: A visible pile or mass of micro-organisms on the surface of a solid culture medium, resulting in most cases from the multiplication of a single organism or a very small number—p. 166. (A clump or chain of bacteria that together give rise to a single colony is called a **colony-forming unit**, and bacterial counts based on the numbers of colonies on plates are in fact counts of such units, not of total bacterial numbers.)

Commensal: Deriving nourishment from a host without being either beneficial or harmful to him— p. 26. (Cf. *pathogenic, symbiotic*.)

Complement: A heat-labile system with many components, present in the serum of man and of animals and playing a number of important parts in the mechanisms of immunity—p. 56.

Conjugation (bacterial): Exchange of genetic material between bacteria, a primitive form of sexuality— p. 19.

Constitutive (enzyme): Produced under nearly all circumstances, not dependent upon the presence of appropriate substrate—p. 15. (Cf. *inducible*.)

Cytopathic effect: Degenerative changes occurring in tissue-culture cells as a result of virus infection, the nature of the changes sometimes indicating the identity of the virus—p. 22.

Definitive host (in parasitology): The host in which the sexual reproductive phase of a parasite's life cycle occurs—p. 28. (Cf. *intermediate host*.)

Disinfectant: A substance, not an antibiotic, which has useful antimicrobial activity but is too toxic for systemic administration—p. 340.

Eclipse phase: The period after penetration of a virus into a host cell, before virus products can be detected in the infected cell—p. 22.

Elementary bodies: Single virus particles of some of the larger viruses, visible by ordinary light microscopy after appropriate staining—p. 12.

Endemic (disease): Persistently present in a given community—p. 31.

Endogenous (infection or disease): Originated by organisms or factors already present in the patient's body before onset of the condition—p. 32. (Cf. *exogenous*.)

Endotoxin: A toxic component of a micro-organism, largely dependent for its release on the death and disruption of the organism. In particular, lipopolysaccharide derived from the cell-walls of gram-negative bacteria—p. 40. (Cf. *exotoxin*.)

Enrichment medium: A medium used to encourage preliminary growth of an organism so as to enhance the chances of growing it on subsequent plate cultures—p. 93. (Cf. *selective medium*.)

Enterotoxin: A toxin that acts on the intestine—p. 40.

Epidemic (noun or adjective): A disease that temporarily has a high frequency in a given community—p. 31.

Eukaryote (adjective–**otic**): An organism with intracellular differentiation of nuclear and cytoplasmic compartments and organelles—p. 10. (Cf. *prokaryote*.)

Exogenous (infection or disease): Originated by organisms or factors from outside the patient's body—p. 32. (Cf. *endogenous*.)

Exotoxin: A toxin released by living micro-organisms into the surrounding medium or tissues—p. 40. (Cf. *endotoxin*).

Facultative (organism): Able to multiply in the presence of absence of oxygen—p. 15; or (in other contexts) adaptable in behaviour.

Fimbria (plural **fimbriae**): Hair-like protrusion from a bacterial cell, shorter than a *flagellum*—p. 11. (Synonym *pilus*.)

Flagellum (plural **flagella**): Whip-like organ of motion possessed by some bacteria and protozoa—pp. 11, 71.

Fomites (Latin, 3 syllables): Literally 'kindling wood', hence personal properties liable to convey agents that initiate diseases—p. 5.

Genome: The total genetic material of an organism; the nucleic acid core of a virus—p. 12.

Genotype: Genetic composition, whether manifest or not—p. 15. (Cf. *phenotype*.)

Gnotobiotic (animal): One delivered by Caesarian section and reared in an environment in which it is protected from microbial contamination or infection—i.e. it is 'germ-free' except for any viruses acquired *in utero*—p. 30.

Gram-negative: Staining red by Gram's method, through losing the primary stain during decolourization and taking up the counter-stain—p. 72.

Gram-positive: Staining violet or blue by Gram's method, through retention of the primary stain—p. 72.

Growth factor: An ingredient of which at least a small amount must be present in a culture medium in order that it may support the growth of a given organism or group of organisms—p. 16.

Haemolysis: Disruption of red blood cells. In connection with growth of streptococci on blood agar, destruction of all red cells around a colony and decolourization of the medium is called β-haemolysis, whereas destruction of most of the red cells and production of a green pigment is called α-haemolysis—p. 76.

Hapten: A substance which acts as an antigenic stimulus only when combined with a protein or other carrier, but which, even in the uncombined state, can react with the resultant antibody in the manner of a true antigen—p. 50.

Helminths: Worms—p. 14.

Heterologous: Related to a different kind of organism, a different disease, etc.—e.g. an *anamnestic reaction* is due to a heterologous stimulus.

Homologous: Related to the same kind of organism, the same disease, etc.—e.g. diphtheria requires treatment with homologous serum, serum containing diphtheria antitoxin.

Hypersensitivity: A form of immunity that may cause responses damaging to the host—p. 61. (Cf. *allergy*.)

Hyphae: Tubular fungal filaments—p. 13. (Cf. *mycelium*.)

Hypogammaglobulinaemia: Deficiency of circulating γ-globulins, resulting from inadequate production by B lymphocytes—p. 314.

Immunization: Artificial induction of immunity—p. 362. (Cf. *active, passive, vaccination*.)

Immunofluorescence: Use of fluorescent-dye-labelled antibodies to demonstrate antigen-antibody interactions—p. 178.

Immunoglobulin: See *antibody*.

In vitro: 'In glass', hence in laboratory apparatus.

In vivo: In a living animal or human being.

Inclusion bodies: Aggregates of virus particles, visible by light microscopy after appropriate staining, within the nuclei or the cytoplasm of infected cells—p. 12.

Inducible (enzyme): Produced only in the presence of an appropriate substrate—p. 15. (Cf. *constitutive*.)

Infection: The arrival or presence of potentially pathogenic organisms on the surface or in the tissues of an appropriate host—p. 31.

Infectious (disease): Transmissible from person to person by transfer of the causative organism—p. 31. (Synonym *communicable*.)

Infective (disease): Caused by an infecting organism (whatever its source of transmission)—p. 3.

Inoculation: (1) of man or animals: Introduction of material containing micro-organisms or their products into the tissues—usually for prophylactic purposes in the case of man.
(2) of culture media: Introduction into a fluid medium, or applications to the surface of solid medium, of material known or suspected of containing living organisms—e.g. Fig. 13.1—p. 166.

Inoculum: The particular portion of material used for a single inoculation.

Interference (by viruses): Modification of host cells infected with one type of virus so that other viruses are unable to multiply in them—p. 43.

Intermediate host (in parasitology): A host in which a parasite can undergo development and in some cases non-sexual reproduction, but not sexual reproduction—p. 28. (Cf. *definitive*.)

L-form: Cell-wall deficient mutant bacterium—p. 71. (Cf. *protoplast, spheroplast*.)

Lyophilization: Combined freezing and desiccation (freeze-drying), a means of long-term preservation of micro-organisms.

Lysis: Disruption (literally 'dissolving') of a microbial or other cell—p. 23 (Cf. *haemolysis*).

Lysogenic conversion: Alteration of the properties of a bacterium as a result of *lysogeny*—p. 19.

Lysogeny: A temporary stable relationship between a bacteriophage and its bacterial host, in which the phage is reproduced in step with the bacterium and thus handed on to succeeding generations of bacteria—p. 23.

Micro-aerophile: An organism which grows best in sub-atmospheric concentrations of oxygen—p. 15.

Monoclonal antibody: Immunoglobulin with a single antigenic specificity produced by a single *clone* of antibody-secreting hybrid cells—p. 56.

Monolayer: A sheet of tissue-culture cells one cell thick—p. 170.

Morbidity: The amount of illness caused by an infection (usually referring to its effect on a community rather than an individual)—p. 3.

Mortality (of an infection): The proportion of cases of the infection which are fatal (usually referring to a particular population or outbreak); *or* the contribution made by this infection to the overall death-rate in a particular population—p. 3.

Mutation: An alteration in genetic material—p. 18.

Mycelium: A mesh-work of fungal *hyphae*—p. 13.

Neutropenia: Lack of circulating neutrophil polymorphs, with consequent susceptibility to acute bacterial infection—p. 314.

Nosocomial (infection): Acquired in hospital—p. 345.

Nucleocapsid: The *genome* and *capsid* of a virus—p. 12.

Nucleoid (virus): Former synonym of *genome*.

Opportunist (pathogens): Not normally pathogenic but able to become so because of some deficiency of the host's defence mechanisms—p. 30.

Opsonization: Coating of micro-organisms with specific antibodies (**opsonins**), *complement*, etc., which facilitates phagocytosis—p. 53.

Pandemic (noun or adjective): World-wide *epidemic*—p. 31.

Parasite: Any organism that obtains nutrition and/or environmental protection from another. For the narrower special use of the word parasites in connection with protozoa, helminths and other complex (higher) organisms that infect man and animals, see p. 10.

Parasitology: The study of protozoa, helminths and other complex (higher) organisms that infect man or animals—p. 10.

Passage (French): Administration of a micro-organism to a host and its subsequent recovery from the host, usually carried out with a view to modifying the pathogenicity of the organism.

Passive (immunity, immunization): Depedent upon injection of ready-made antibodies and not upon the subject's own immunological mechanisms—p. 363. (Cf. *active*.)

Pathogenic: Actually producing or capable of producing disease—p. 26. (Cf. *commensal, symbiotic*.).

Petri dish: A shallow circular flat-bottomed glass or plastic dish used as a container for solid media—p. 166.

Phage type: The identity of a bacterial strain as indicated by its sensitivity or resistance to the lytic action of the members of a standard panel of bacteriophages (its 'phage pattern'); *or* a group of strains having identical or closely similar phage patterns—p. 75.

Phagocytosis: Engulfing of micro-organisms or other particles by host **phagocytes** (polymorphonuclear leukocytes and macrophages)—p. 48.

Phenotype: That part of the *genotype* of an organism which is expressed in a given situation—p. 20.

Pilus (plural **pili**): Synonym of *fimbria*—p. 11.

Plaque: A small roughly circular deficiency in the growth of a bacterial culture on a solid medium, resulting from local destruction of bacteria by bacteriophages—p. 23.

Plasmid: An extrachromosomal portion of genetic material (DNA)—p. 11.

Prokaryote (adjective–**otic**): A simple organism without intracellular differentiation of nuclear and cytoplasmic compartments—p. 10. (Cf. *eukaryote*.)

Prophage: Bacteriophage in a lysogenic relationship with its host—p. 23. (See *lysogeny*.)

Protoplast: A bacterium deprived of its cell wall and thus highly susceptible to osmotic distension and rupture—p. 71. (Cf. *L-form, spheroplast*.)

Purulent (discharge or exudate): Containing pus.

Pyogenic (infection or organism): Pus-forming.

Reagin(s): *Either* the serum component responsible for the Wassermann and related reactions—p. 181; *or* the IgE antibodies associated with certain types of hypersensitivity reactions—p. 61.

Replication: Virus reproduction, so called to emphasize that a virus does not produce **itself** but causes the host cell to make replicas of it—p. 12.

Reservoir host: A species of animal in which organisms infective for man may be maintained in nature—p. 28.

Saprophytic: Living on dead organic matter—p. 26.

Satellitism: Enhancement of bacterial growth on a solid medium around a source of a growth factor—p. 18.

Selective medium: A solid culture medium on which all but the desired microbial species are wholly or largely inhibited—p. 93. A **selective enrichment medium** is a fluid medium in which the desired species can multiply more rapidly than others likely to be present, so that a sample subsequently taken from it for inoculation of plate cultures is 'richer' than the original material in organisms of the desired species—p. 93. (Cf. *enrichment medium*.)

Septicaemia: A serious clinical condition (including shock) associated with the presence of pathogenic organisms in the blood—p. 60. (Cf. *bacteraemia*.)

Serology: Study of the antibody content of sera—p. 179—and also use of antisera in the antigenic analysis of micro-organisms and their products.

Serotype: The identity of a bacterial strain as indicated by antigenic analysis: *or* a group of strains shown by serological tests to be antigenically identical or closely similar—p. 92. (Synonyms: *antigenic type, serological type*.)

Specific: (1) Relating to a species. (2) Relating particularly to some other unit—e.g. type-specific. Hence commonly used in much the same sense as *homologous*.

Spheroplast: A bacterium similar to a *protoplast* except that the cell-wall damage is partial and reversible—p. 71. (Cf. *L-form*.)

Spirochaete: A member of one of several genera of spiral bacteria—p. 11.

Sterilization: The process of killing or removing all micro-organisms—p. 337.

Strain (of an organism): A culture all members of which are believed to be the progeny of a single organism—p. 67. (This is not an entirely satisfactory definition, and indeed the variations which inevitably accompany bacterial reproduction make the whole concept of a 'pure' strain fallacious.)

Symbiotic: Living in a mutually beneficial relationship with the host—p. 26. (Cf. *commensal, pathogenic*.)

Synergy: Combined action—e.g. of two or more micro-organisms in an infection—p. 289; or of two antimicrobial drugs that have a joint action against a target organism that exceeds the sum of their individual actions—p. 378.

Temperate phage: A phage capable of a lysogenic relationship with its bacterial host—p. 23. (See *lysogeny, prophage*.)

Therapeutic index (of a drug): The maximum tolerated dose ÷ the minimum curative dose—p. 374.

Titre: The highest dilution of a serum or an antigen preparation which gives a positive reaction under defined conditions—p. 180.

Toxin: A microbial product that damages the host—p. 40. (Cf. *endotoxin, enterotoxin, exotoxin*).

Toxoid: Toxin rendered harmless but still effective as an antigen—p. 364.

Transduction: Conveyance of genetic characters from one bacterial strain to another by means of a transfer of bacteriophage—p. 19.

Transformation: (1) Acquisition of genetic characters of one bacterial strain by a related strain grown in the presence of DNA from the first strain—p. 19. (2) Malignant change induced in a host cell by presence of a virus—p. 23.

Transport medium: A medium which increases the chances of survival of a micro-organism during transit from the patient to the laboratory—p. 163.

Vaccination: Originally, the use of cowpox material or of vaccinia virus in active immunization against smallpox; then all forms of active immunization using live organisms; now all forms of active immunization—p. 364.

Vaccine: Material used in vaccination; therefore this term also has an expanding meaning—p. 364.

Viraemia: Presence of viruses in the blood-stream—p. 60.

Virion: A virus particle, the virus unit corresponding to a single cell of a larger organism—p. 12.

Zoonosis (four syllables): A disease that man acquires from animals—p. 32.

Appendix B
Meaning of Some Abbreviations

Convenient though abbreviations may be, they are liable to cause confusion by meaning different things to specialists in different fields. The meanings given below are those which the appropriate abbreviations customarily have when they are encountered in the field of medical microbiology.

AAFB	acid–alcohol-fast bacilli—p. 103 and Appendix A.
AFB	acid-fast bacilli—p. 72 and Appendix A.
AHG	anti-human globulin (antibodies, serum)—p. 177.
AIDS	acquired immune deficiency syndrome—p. 283.
APT	alum precipitated toxoid.
ASO	antistreptolysin O—p. 77.
ATS	anti-tetanus serum—p. 367.
BALT	bronchus-associated lymphoid tissue—p. 55.
BCG	Bacille Calmette Guerin—p. 365.
C1, C2 etc.	complement components—p. 57.
cAMP	cyclic adenosine monophosphate—p. 252.
CAPD	continuous ambulatory peritoneal dialysis—p. 351.
CFT	complement-fixation test—p. 176.
CFU	colony-forming unit—Appendix A.
CIE	countercurrent immunoelectrophoresis—p. 175.
CMV	cytomegalovirus—p. 132.
CNS	central nervous system.
CPE	cytopathic effect—p. 170 and Appendix A.
CSF	cerebrospinal fluid.
CSU	catheter specimen of urine—p. 275.
DCA	deoxycholate citrate agar—p. 93.
DIC	disseminated intravascular coagulation—p. 40.
DNA	deoxyribonucleic acid—p. 11.
DPT	diphtheria, pertussis, tetanus combined (triple) vaccine—p. 371.
EB	Epstein–Barr (virus)—p. 133.
ECHO	enteric cytopathic human orphan (viruses)—p. 155.
ED	effective dose—p. 33.
ELISA	enzyme-linked immunosorbent assay—p. 178.
EMU (EMSU)	early morning (specimen of) urine—p. 272.
EPEC	enteropathogenic *Esch. coli*—p. 252.
ETEC	enterotoxigenic *Esch. coli*—p. 252.
Fab, Fc	antibody-binding (ab) and complement-binding (c) parts of immunoglobulin molecule—pp. 52–3.
FT	formol toxoid—p. 366.
FTA	fluorescent treponemal antibody (test)—p. 178.
GALT	gut-associated lymphoid tissue—p. 55.
GC	gonococcus—p. 82.
GCFT	gonococcal complement-fixation test—p. 82.
GLC	gas–liquid chromatography—p.73.
H	flagellar (antigens, antibodies)—from German *Hauch*, p. 91.

HAI	haemagglutination inhibition (test)—p. 183.	PMC	pseudomembranous colitis—p. 259.
HBcAg etc.	antigenic components of hepatitis B virus—p. 135.	PPD	purified protein derivative (of old tuberculin)—p. 184.
HIV	human immunodeficiency virus (AIDS associated)—p. 128.	PTAH	purified toxoid, aluminium hydroxide.
HSV	herpes simplex virus—p. 130.	PTAP	purified toxoid, aluminium phosphate.
HTLV–III	human T cell lymphotropic virus type III (now HIV)—p. 128.	PUO	pyrexia of unknown origin—p. 247.
HVS	high vaginal swab.	REO	respiratory enteric orphan (viruses)—p. 120.
ID	infective dose—p. 33.	RIA	radio-immuno-assay—p. 178.
IDU	idoxuridine—p. 135.	RNA	ribonucleic acid—p. 11.
IgA, IgE, etc.	immunoglobulins of classes, A, E, etc.—p. 53.	RPR	rapid plasma reagin (test)—p. 181.
INAH	isonicotinic acid hydrazide = isoniazid—p. 398.	RS (RSV)	respiratory syncytial (virus)—p. 127.
K	capsular or envelope (antigens, antibodies)—p. 91.	RTD	routine test dilution (of bacteriophage).
LD_{50}	dose lethal to 50% of a group of experimental animals—p. 38.	SSPE	subacute sclerosing panencephalitis—p. 127.
LF	lactose-fermenter—p. 90.	ST	heat-stable toxin (of *Esch. coli*)—p. 252.
LGV	lymphogranuloma venereum—p. 280.	TAB (TABC)	typhoid + paratyphoids A and B (and C) vaccine—p. 366.
LPS	lipopolysaccharide—p. 10.	TAF	toxoid–antitoxin floccules.
LT	heat-labile toxin (of *Esch. coli*)—p. 252.	TB	tubercle bacilli—p. 103—and hence loosely used to denote tuberculosis.
MBC	minimal bactericidal concentration—p. 285.	TCBS	thiosulphate citrate bile salts sucrose (agar)—p. 97.
MIC	minimal inhibitory concentration—p.285.	TPHA	*Treponema pallidum* haemagglutination (test)—p. 182.
MIF	macrophage-migration-inhibiting factor—p. 58.	TPI	*Treponema pallidum* immobilization (test)—p. 182.
MLD	minimum lethal dose (of a drug or microbial preparation).	TRIC	trachoma and inclusion conjunctivitis (agents)—p. 113.
MSU (MSSU)	mid-stream (specimen of) urine—p. 274.	TT	(1) tetanus toxoid—p. 367. (2) tuberculin tested (cattle)—p. 185.
NBT	nitroblue tetrazolium test—p. 316.	UHT	ultra-heat treated (milk)—p. 358.
NGU	non-gonococcal urethritis—p. 280.	UTI	urinary tract infection.
NLF	non-lactose-fermenter—p. 90.	UV	ultraviolet (light).
NK	natural killer (cell) p. 49.	VDRL	Venereal Diseases Research Laboratory (test)—p. 181.
NSU	non-specific urethritis—p. 280.	Vi	virulence (antigen of *S. typhi*, etc.)—p. 93.
NSV	non-specific vaginitis (anaerobic vaginosis)—p. 281.	VZ	varicella-zoster (virus)–p. 131.
O	somatic (antigens, antibodies)—from German *ohne Hauch*, p. 91.	WR	Wassermann reaction—p. 181.
OT	old tuberculin—p. 184.	XLD	xylose lysine deoxycholate (agar)—p. 262.
PABA	*p*-aminobenzoic acid—p. 387.	ZN	Ziehl–Neelsen (staining method)—p. 103.
PGU	postgonococcal urethritis—p. 280.		
PHA	passive haemagglutination—p.176.		

Index

Page numbers in bold type indicate major references

Abscesses 192, 233, 247, 248, 285, 287, 288, 291, 301
  see also Amoebiasis, Brain and Liver and Lung abscess
Absidia 141
Acanthamoeba 143–4, 220
Acetylcholine 87
Acid-fast bacteria 103–7
Acne 85, 306
Acquired immune deficiency syndrome (AIDS) **128**, 205, 245, 260, **283**, 319, 351, 354, 370
Actinobacillus 106
Actinomycetes, actinomycosis 11, **106**, 141, 142, 225, 286, 302
Acyclovir 135
Adansonian classification 68
Adenoviruses **129–30**, 192, 200, 245, 253, 295, 326
Adenylcyclase 97, 252, 255
Adhesins 39, 50
Adjuvants 362
Adsorption of viruses 21
Aedes mosquitoes 121, 122, 158
Aerobic respiration 15
  spore-bearers 87
Aflatoxins 24
Agar 7, 166
Age and susceptibility to infection 55, 319, 350
Agglutination 176
Aggressins 17, 41
Agranulocytosis 388
Airborne infections 34, 35, 301, 346–50
Air-conditioning systems and infection 102, 142, 203, 356, 357
Albert's stain 72, 84
Alcohols 341, 342, 343, 344
Alkaline peptone water 97, 262
Allergens, allergy 45, 61, 159, 223, 239, 259, 261, 271, 292, 363, 372, 388
  see also Hypersensitivity
Allergic respiratory diseases 61, 62, 141–2, 356
Alphaviruses 121
Alternaria 141
Amantidine 136
Amastigotes, protozoan 145, 146, 147, 148
Amies' medium 280
Amikacin 393

Amine test 282
Aminoglycosides 393–4, 397
Amodiaquine 242, 243
Amoebiasis **143**, 152, 225, **253**, 269, 355, 359
Amoebic meningo-encephalitis 143–4, 220, 356
Amoxycillin 391
Amphibia as sources of pathogens 106
Amphotericin B 142, 152
Ampicillin 391
Anaemia 45, 61, 129, 241, 260, 261, 270, 303
  see also Sickle-cell disease
Anaerobic bacteria 15, 106, 107, 110
  see also Anaerobic cocci, Bacteroides, Clostridial infection
  cocci 80, 197, 206, 225, 268, 288, 289, 291
  cultures 167
  fermentation 15
  pleuropulmonary infections 205
  respiration 15
  vaginosis 281
Anamnestic reactions 180, 181
Anaphylaxis 45, **61**, 271, 352, 363
Ancylostoma 157, 159, 209, 260
Animal bites 222, 291–2
Animals, as sources of pathogens 32
  see also Amphibia, Bats, Birds, Cats, Cattle, Dogs, Goats, Horses, Mice, Monkeys, Pigs, Poultry, Rabbits, Rats, Rodents, Sheep, Shell-fish, and many other references
  laboratory use 105, 109, 119, 169, 293; see also Guinea-pigs, Mice, Rabbits
Anopheles mosquitoes 148, 150, 158
Anthracoid bacilli 87
Anthrax 5, 6, 7, **86, 202**, 307, 334, 347
Antibacterial drugs 9, 373–98
  adverse effects 30, 259, 373–4
  antagonism 378
  bactericidal/static action 238, 377, 385, 386, 398
  hypersensitivity to 374, 390, 391
  incompatibilities 378
  in renal disease 381, 386, 393–4, 398
  levels in body fluids 238, 380, 385–6, 390, 393, 394, 398

  modes of action 387, 388, 389
  prophylactic use 208, 214, 239, 242–3, 295, 302, 327, 337, 350, 370, 372, **375–6**, 383
  resistance to 20, 346, 374, **381–4**, 397–8
  sensitivity tests 384–5
  spectra 377
  synergy 378, 398
  tolerance, by bacteria 385
  toxicity 373–4, 378, 380, 389, 398; see also *individual drugs* (387–98)
Antibiotic-associated diarrhoea 259
  policies 384
Antibiotics 389–98
Antibodies 8, 50, 52–6 (*main references*)
  see also Serodiagnosis
Antibody capture assays 178
  deficiencies 53, 316–7
Anti-DNAase B 78
Antigen-antibody interactions 174–8
Antigenic competition 371
  drift and shift (virus) 125, 370
  variation of borreliae 232
  of protozoa 59
  of salmonellae 92
Antigens 49, 50–1 (*main references*)
  see also Serological detection
Anti-human globulin (AHG) 177
Antimicrobial drugs, early 9
  for fungi 142
  helminths 246, 247, 260–2, 270–1, 277, 292, 293, 297
  protozoa 152, 242–3
  viruses 135–6
  see also Antibacterial drugs
Antimony drugs 152
Antiseptic technique 7, 345
Antisera 7, 8, 79, 85, 88, 89, 124, 196, 222, 291, 363, 367
  see also Human immunoglobulin
Antistreptolysin O 78, 195
Antitetanus serum (ATS) 367
Antitoxins 7, 51, 54, 363, 367
Aortic aneurysm 279
Aphthous stomatitis 130, 190
Apicomplexa 143
Aplastic crises 128–9
Appendicitis 287
Arboviruses 121–3, 223
Arenaviruses 123

408

Arsenical drugs 9, 108, 387
Arsphenamine 387
Arthritis, reactive 41, 116, **302–3**
  septic 234, 279, **302**
Arthrospores, fungal 24, 171, 172
Arthus 8, 62
Ascaris **156**, 159, 209, 271, 293, 351
Asepsis 7, 33, 345, 349, 353
Aseptic meningitis 219
  see also Meningitis, virus
Aspergillosis 140, 141–2, 205, 236, 240
Aspiration pneumonia 206
Astroviruses 121, 253
Athlete's foot 138, 309
Atoxyl 387
Attachment of bacteria 39
  viruses 21, 42, 43
  to phagocytes 48, 57
Attenuation of pathogens 7, **38**, 362, 364, 365, 368, 369, 372
Atypical mycobacteria 105–6, 208, 365
  pneumonia 110, 203
Auramine 104, 145, **165**
Australia antigen 265
Autoclave 339
Auto-infection 345
  –immunity 45, 47, 62–3
Automatic methods 241, 275
Autotrophs 16
Avery 19
Azlocillin 391
Aztreonam 393

B lymphocytes 49, 51, 52, 54, 55, 58, 62, 133, 313–4, 316–7
Babesiosis 151, 243, 316
Bacampicillin 391
Bacillary dysentery 249
  see also Shigellosis
Bacille Calmette-Guerin (BCG) 38
Bacilli 16, 69
  acid-fast 103–6
  gram-negative 90–103
  gram-positive 83–90
Bacillus anthracis 86–7
  see also Anthrax
  cereus food-poisoning 258, 361
  genus 86–7
  species as indicators of sterilization 87, 339
Bacteraemia 47, 57, 60, 230–5
Bacteraemic shock 40
  see also Shock
Bacteria, classification and nomenclature 67–73
  counting 21, 165; see also Quantitative bacteriology
  culture 165–8
  general features 10–11
  genetics 18–21, 382–3
  groups 73–4
  growth phases 20, 21
  isolation and identification 69–73

metabolism 15–18, 72–3
microscopy 164–5
pathogenicity 38–41
physiology 14–21
protective functions 16, 29
reproduction 18–21
staining 71, 164
taxonomy 67–9
Bacterial endocarditis 233, **235–9**, 247, 279, 319, 375, 377, 386
  infections, drug treatment see Antibacterial drugs
  laboratory investigation **164–8**, 210–1, 227–8, 231, 237, 240–1, 247–8, 262–3, 277–8, 286
Bactericidal action 341, 377, 385, 386, 398
Bacteriocines 30, 250
Bacteriolysis by host defences 53, 57
  by phages 13, 23
Bacteriophages 13, 19, 23
  see also Phage-typing
Bacteristatic action 341, 377, 385
Bacteriuria 272, 273
Bacteroides **102–3**, 191–2, 197, 206, 225, 226, **235**, 268, 280, 282, **288**, 289, 292, 349
Balanitis 282
Balantidium 151
Barrier nursing 347
Bartholinitis 279
Base-pair ratio 68
Bassi 5
Bats as sources of infection 139, 222
Baths and cross infection 138
  see also Swimming baths
Behring 7, 366
Beijerink 8
Bejel 109
Benzonidazole 152
Benzoylbenzoate 311
Bephenium 260
Betadine 343, 344
Beta- (β-)lactamases (penicillinases and cephalosporinases) 17, 167, 382, **390–2**, 397
  –lactams 389–93
Bifidobacterium 86
Binary fission 11, 14, 25
Bio-assay 385
Biotypes 92, 101
Bipolar staining 95
Birds as sources of pathogens 98, 106, 113, 114, 121, 125, 139, 141, 203, 223, 256, 355
  see also Poultry
Bithionol 209, 270
Black Death 95
Blackwater fever 45, 242, 278
Blankets 348
Blastomycosis 139, 220
Blindness 292, 295, 296, 297
Blocking tests 177
Blood agar 72, 166
  collection 240
  cultures 228, 231, 237, **240–1**

transfusion, splashes, etc. as sources of infection 33, 34, 133, 147, 150, 162, **265**, **266**, **283**, 351, 354, 370
Boils 285
Bone marrow 51, 299
Bordet 7
  –Gengou agar 100
Bordetella 100
  see also Whooping cough
Bornholm disease 119
Borrelia 110, 191, 232
Bot fly 294
Botulism 87, 258, 361
  infant 327
Boutonneuse fever 112
Brain abscess 197, **255**, **228**, 253, 302
Branching bacteria 106–7
Branhamella 82, 204, 219
Break-point sensitivity tests 385
Breast feeding and enteritis 55
Brill's disease 112
Bronchiectasis 201, 204
Bronchiolitis 127, 199
Bronchitis 200, 204
Bronchopneumonia 204, 205
Brucella, brucellosis **100–1**, **181**, **232**, 245, 248, 267, 300, 334, 357
Brucella ring test (milk) 358
Brugia 158, 246–7
Bubo 244, 280
Bubonic plague 244
  see also Plague
Bugs 146, 246, 335
Bunyaviruses 122, 223
Burkitt's lymphoma 23, 133
Burns 307, 313, 347, 350

Cagniard-Latour 6
Calabar swelling 292
Caliciviruses 120, 253
Calorimetry 275
Calymmobacterium (Donovania) 280
Campylobacter 97–8
  enteritis **251**, 303, 353, 355, 357, 359
  pyloridis 251
Candidiasis **137–40**, 142, 171, 192, 196, 205, 236, 240, 248, 259, 282, 283, 308, 318, 319, 324, 351
Canicola fever 218
Capnocytophaga 314
Capsid, virus 12, 21
Capsomeres, virus 12
Capsule swelling 79, 99, 176
Capsules, bacterial **11**, 41, 49, 50, **69**, 79, 86, 90, 91, 98, 99, 141, 362, 364
Carbenicillin 391
Carbohydrate fermentation 15, 73
Carbon dioxide and bacterial growth **15**, 80, 81, 82, 97, 100, 103, 107, 167
  radio-active 165, 241
Carboxyphilic bacteria 16
  see also Carbon dioxide
  streptococci 80
  see also *Streptococcus milleri*

Carbuncles 285
Cardiolipin 181
Carfecillin 391
Carriers, human, of pathogens **27**, **32**, 35, 83, 93, 94, 99, 189, 231, 255, 256 257, 258, 269, 281, 328, 334, 346, 348 360
Caseation in tuberculosis 207
Casoni's test 185, 271
Catalase 15, 17, 315
Cataract 322
Catheter fever 234, 352
Catheters as sources of infection 313, 351, 352
Cats as sources of pathogens 98, 101, 138, 147, 151, 159, 245, 293, 312
Cattle as sources of pathogens 86, 98, 100, 103, 109, 113, 114, 120, 134, 138 144, 151, 153, 155, 156, 208, 232, 251, 256, 266, 270, 333, 334, 357
Cavernous sinus thrombosis 295
Cefaclor and other antibiotics with names beginning with Cef- 91–2
Cell, cultures 116, 133
  –mediated hypersensitivity 62
  immunity 50, 51, 57–8, 59, 128
  membrane, bacterial 11, 381
  wall, bacterial 10, 381
Cellular defences 47–9
Cellulitis 60, 286, 290, 291, 295
Cephalosporins 391–2
Cephamycins 392
Cercariae 153, 155
Cerebrospinal fluid examination 214, 218, 219, 220, 221, 227
Cervicitis 279, 281
Cestodes 155–6
  see also Cysticerci, Echinococcus, Helminths, Hydatid, Taenia
Chagoma 146
Chagas' disease 146, 152, 246
Chain 9
Chancre 246, 278, 280
Chancroid 244, 280
Charcoal blood agar 100
Chediak-Higashi syndrome 315
Chemotaxis 49, 57, 58, 62, 314, 316
Chemotherapeutic agents 387–9, 398
  see also Antimicrobial drugs
Chemotrophs, chemo-organotrophs 15, 17
Chickenpox 113, 131–2, 134, 296
Chickens as sources of pathogens see Poultry
Chiclero's ulcer 148
Chitin 13
Chlamydiae 11, **113–6**, **203**, 204, 235, **280–1**, **296**, **326**
Chlamydospores, fungal 24, 171
Chloramphenicol 394
Chlorhexidine 342, 343
Chlorination of water 355, 356, 359
Chloroquine 242, 243
Chocolate agar 72, **81**, 99, 167
Cholangitis 268
Cholecystitis 268

Cholera 97, 254–5, 355, 372
Chorio-amnionitis 325
Choroidoretinitis 224, 297, 321, 322, 323
Chromosomes, bacterial 11
Chronic granulomatous disease 267, 315
Chrysoma flies 294
Chrysops flies 158
Chyluria 278
Ciliophora 143
Ciprofloxacin 389
Citrobacter 95
  see also Enterobacterial infections
Cladosporium 141
Clavulanic acid 392
CLED medium 262, 275
Clindamycin 396
Clofazimine 105
Clonal selection theory 51
Clone suppression 62
Clostridial food-poisoning **257**, 353, 361
  see also Botulism
  infections 252, 259, 291, 297, 301, 327; see also Pseudomembranous colitis, Tetanus
Clostridium 87–90
Clot culture 231
Clotrimazole 142
Cloxacillin 390, 397
Co-agglutination 76, 99, **176**
Coagulase, staphylococcal 17, 74, 75
Cocci 11, 69
  gram-negative 81–3
  gram-positive 73–81
Coccidioidin 139
Coccidioidomycosis 139
Cochlyoma flies 294
Cold abscesses 301, 315
  agglutinins 61, 110, 203
  enrichment 95
Colicines 94, 250
'Coliforms' 90
Colistin 397
Colonies, bacterial 166
  see also many places in Chapter 7
Colonization of hosts 31, 38
Colostrum 50
Commensals 26
Common cold (coryza) 119, 192
Communicable disease, definition 31
Complement 8, 48, 53, 54, **56–7**, 59
  alternative pathway 57
  classical pathway 57
  deficiency 57, 234, 317
  fixation tests 82, 113, 116, **176–7**, 181, 182
Compromised hosts 27, 139, 205, 220, 233, 251, 257, 267, **313–9**, 347, 351, 352, 370
  see also Drug addiction, Immune deficiency, Immunosuppression, and many other references
Concentration methods for tubercle bacilli 104

Condylomata 278
Conidia, conidiophores, fungal 24, 171, 172
Congenital infections 42, 320–4
  see also Placental and Vertical transmission
Conjugation, bacterial 19, 383
Conjunctiva as barrier to infection 46
Conjunctivitis 129, 131, 280, **295–6**, 326
Constitutive enzymes 15
Contagion 4, 33, 115
Contagious pustular dermatitis (orf) 308
Containers for specimens 162
Contamination, definition 31
Continuous ambulatory peritoneal dialysis (CAPD) 351
  culture 21
Cooked meat medium 167, 262
Coombs 61, 177
Corneal ulcer 296
Coronaviruses 127, 192, 200
Corynebacterium 83–5
Coryneform bacilli 83
  see also Diphtheroid bacilli
Cordylobia flies 294
Coryza (common cold) 119, 192
Cotrimoxazole, cotrimazine 388
Cough plate 201
Countercurrent immuno-electrophoresis (CIE) 175
Cowpox 7, 134
Coxiella 11, 113
  see also Q fever
Coxsackieviruses **119**, 190, 192, 219, 222, 239, 240, 328
Creeping eruption 159
Creutzfeld-Jacob encephalitis 223
Cross-infection, definition 345
Cross-over immuno-electrophoresis see Countercurrent
Croup 127, 199
  –associated virus 126, 199
Cryptococcosis 140–1, 220
Cryptosporidiosis 144–5, 254, 283, 317
Ctenocephalides fleas 312
Culex mosquitoes 158
Culture media for bacteria 72, 81, 84, 90, 93, 99, 104, 107, **166–7**, 210–1, 262
Curtis-FitzHugh syndrome 115
Cyclops crustacea 159, 293
Cycloserine 398
  cefoxitin fructose agar 262
Cystic fibrosis 204
Cysticerci 14, **155**, **226**, 261, **293**
Cystine-lactose-electrolyte-deficient (CLED) agar 262, 275
Cystitis 130, 272, 273
Cysts, helminth 155, 156, 158
  see also Cysticerci, Hydatid cysts, Trichinella
  protozoan 14, 35, 143, 144, 151, 173 174, 253, 254, 262, 269
Cytomegalic inclusion disease, cytomegalovirus (CMV) **132**, 193, 267, 283, 319, **322**

Cytopathic effect (CPE) 22, 170
Cytoplasm, bacterial 11
Cytoplasmic granules 48
  membrane 11, 381
Cytotoxicity, cell-mediated 49, 54, 58
  virus 42; see also Cytopathic effect

Dane particles 135
Dapsone 105, 107, 242, 388
Dark-ground microscopy 108, 164, 279
Davaine 5, 7
Deafness, congenital 321, 322
Decline phase of growth 21
Decubitus ulcers 289
Deep mycoses 44
Defence mechanisms of the body 46–63
Definitive hosts 28
Delayed-type hypersensitivity, mechanism 62
Delta virus 135, 266
Demeclocycline 395
Demyelinization 223
Dendritic ulcers 131, 296
Dengue 121, 245
Dental caries 191
  plaque 191; see also Dextran
Dento-alveolar abscesses 192
Deoxycholate citrate agar (DCA) 93, 94, 262
Deoxyribonucleases 75, 77
Deoxyribonucleic acid (DNA) 11, 12, 13, 18–19, 22, 23, 68, 134, 340, 389
  viruses 128–35
Dermatobia flies 294
Dermatomycoses 138
  see also Ringworm
Dermatophytes 138–9, 142, 171–2
Dextran and dental plaque 18, 39, 78, 191
Diabetes and susceptibility to infection 30, 39, 318, 350, 370
Diagnostic serology 8, 179–83
  see also Serodiagnosis
Dialysis and infection 351
Diarrhoea 119, 128, 144, 249–62, 293, 347, 349, 353, 359, 379
  see also many other references, notably in Chapter 18
Dick test 184
Dicrocoelium 270
DiGeorge syndrome 317
Diethylcarbamazine 244, 292, 297
Dihydrofolic acid reductase 382
Diloxanide furoate 152
Diphtheria 7, 19, **83**, **195**, 239, 346, 347, **366–7**, 371
Diphtheroid bacilli 83, 85, **305**
Diphyllobothrium 156, 261
Diplococci 69, 76, 79, 81
Dip-slides 275
Disinfectants 340–3, 344 (Table)
  standardization 341

Disinfection 337, 340–4
Disc-plate sensitivity tests 385
Disseminated intravascular coagulation (DIC) **40**, **57**, 111, 213, 234, 235
Dissemination of viruses 42
Dogs as sources of pathogens 98, 101, 109, 112, 138, 147, 148, 156, 159, 222, 251, 271, 293, 312, 334
Domagk 9, 387
Donovania (now *Calymmatobacterium*) 280
Double diffusion 175
Doxycycline 395
Dracunculus 159, 292, 303
Dressings, surgical 349, 350, 353
Droplets 34, 35, 113, 336
Drug addiction and infection 128, 236, 265, 266, 319, 351, 370
Dry socket 192
Durham 8
Dusch 6
Dust 34, 35, 113, 114, 123, 140, 291, 337, 348, 350
Dysentery, amoebic 143
  see also Amoebiasis
  bacillary 94, 249, 252; see also Shigellosis
  balantidial 150
  definition 250

Ebola virus 124
Echinococcus 155–6
  see also Hydatid cysts
Echoviruses 119, 192, 219, 222, 328
Eclipse of viruses 22
Ectoparasites 36, 283, 310
Ectothrix ringworm 171
Eczema herpeticum 130
Edelman 8
Effective dose (ED) 33, 37
Eggs, helminth see Ova
Ehrenberg 5
Ehrlich 8, 9, 108, 374
Electrical impedance monitoring 165, 275
Electron microscopy 8, 165, 168–9, 263
Elek plate 84, 175
Elementary bodies 8, 12
Elephantiasis 115, 158, 247, 280
ELISA 178
Eltor vibrio 97, 255
Emetine 152, 270
Empyema, extra- and subdural 226, 302
  gall-bladder 268
  thoracic 206
Encephalitis, meningo-encephalitis 121, 122, 126, 127, 129, 130, 131, 134, 143, 220–4, 280, 323
Endemic diseases 31
  myalgia 119
  typhus 112
Endocarditis 235–9
  see also Bacterial endocarditis

Endogenous infection 32, 38, 345
Endometritis 279
Endophthalmitis 296–7
Endothrix ringworm 171
Endotoxic shock 40
  see also Shock
Endotoxins 17, **40**, **57**, 63, 213, 215, 234, 235
End-product analysis 18, 73, 103
Energy sources of bacteria 15
Enrichment media 93, 262
Entamoeba 143
  see also Amoebiasis
Enteric fevers 92, 93, 180
  see also Paratyphoid, Typhoid
Enteritis, bacterial 249–52, 326
  see also Esch. coli enteritis, Gastro-enteritis
  candida 138, 259
  necroticans 252
  staphylococcal see Enterocolitis
Enterobacter 95
  see also next entry
Enterobacterial infections 90–5, 197, 205, 225, 234, 240, 268, 274, 324, 349, 350, 351
Enterobius 157, 261
Enterococci (faecal streptococci) 80, 236, 351
Enterocolitis, necrotizing 327
  staphylococcal 259, 262, 395
Enterotoxins 17, **40**, 97, **252**, **255**, 257, 259, 263, 359
Enteroviruses 117–9
Envelope, virus 12, 22
Enzyme immuno-assay 116
  –linked immunosorbent assay (ELISA) 178
Enzymes 15, 17, 21, 41, 77, 382, 389
  see also many other places
Epidemic diseases 31, 38, 214, 216
  keratoconjunctivitis 129
  myalgia 119
  typhus 111
Epidermophyton 138, 171, 172, 309
Epididymitis 279
Epiglottitis 197, 234
Epstein-Barr (EB) virus 23, 133, 193
Ergotism 24
Erysipelas 305
Erysipeloid, *Erysipelothrix* 86, 306
Erythema infectiosum 129
Erythritol and brucellae 39, 100
Erythrogenic toxin 19, 77, 194
Erythromycin 395, 397
Escherichia coli 91
  enteritis 251–2, 255, 327, 347, 353, 359
  in water supplies 355
  urinary tract infections 274, 352
  various infections 216, 268, 324, 325, 326, 327; see also Entero-bacterial infections
Espundia 148
Ethambutol 398
Ethylene oxide 342

Eukaryotes 10
Exogenous infection 32
Exotoxins 17, **40**, 53, 60, 194, 195, 290, 306, 327
 see also under various bacterial species
Eyeworm 159, 292
Extrinsic allergic alveolitis 62, 141, 356

Fab portion of antibody molecule 53
Facultative organisms 15
Faecal streptococci (enterococci) 80, 236, 351
Faeces, bacterial content 249
 laboratory examination 173, 262–3
 transmission of infection in 32, 33, 34, 117, 120, 121, 134, 143, 144, 151, 155, 156, 219, 221, 264, 266, 293, 355, 359, 360; see also many other places, notably in Chapter 18
Fansidar 242, 243
Farmer's lung 62, 141
Fasciola 153, 270
Fasciolopsis 153, 262
Favus 5
Fc portion of antibody molecule 53, 75
 receptors 53
Fermentation of carbohydrates 15, 73
Fever 63
Fifth disease 129
Filariasis **158**, 243, 245, **246–7**, 278, 292, 297
Filtration, to remove micro-organisms 340, 355
Fimbriae 11, 20, 39, 250
Fish as sources of pathogens 85, 106, 153, 154, 156, 259, 267, 271, 306
Flagella **11, 71**, 87, 88, 90, 95, 144, 145
Flagellar antigens 91, 92, 180
Flaviviruses 121–2
Fleas 16, 34, 36, 95, 111, 112, 244, 312, 335
Fleming 9
Flies as pathogens (myiasis) 294
 as vectors of infection 17, 253, 264, 296, 360
 see also Chrysops, Sandflies, Simulium, Tsetse flies
Florey 9
Flucloxacillin 390
Flucytosine 142
Flukes (Trematodes) 153–5, 209, 262, 270
 see also Helminths, Schistosomes
Fluorescence microscopy 164–5, 168
 see also next entry
Fluorescent antibodies, immunofluorescence 82, 102, 108, 113, 116, 148, 165, 170, **178**, 211
 treponemal antibody (FTA) test 182, 279
Folic acid metabolism 387, 388
Fomites 5, 32, 34, 250, 279
Food handlers 33, 258, 336, 353, 360, 361

–poisoning 256–9, 353, 358–61
 transmission of pathogens in 32, 34, 93, 94, 97, 121, 123, 144, 250, 254, 264, 293, 358–9; see also Fish, Meat, Poultry
Foot-and-mouth disease 8
Formaldehyde, formalin 341, 342
Formol toxoids 366
Fournier's gangrene 289
Fracastorius 5
Fractures, infected 301
FRACON antibiotics 348
Framboesia 108
Framycetin 394
Francisella 101
 see also Tularaemia
Friedländer's bacillus 92, 202
Frosch 8
Fucidin 393, 397
Fungal allergies 141–2
 infections 137–42
 laboratory investigations 171, 241
 serodiagnosis 183
 treatment see Antimicrobial drugs
Fungi, culture 171
 general features 13–14
 microscopy 171
 pathogenicity 44, 137–42
 physiology 24
 reproduction 24
Fungi imperfecti 24
Fusobacterium 102, 192, 196, 206, 225, 290, 292

G-6-PD deficiency 150
Gall-bladder infections 268
Gametocytes, plasmodial 148
Gangrene 289, 345
Gardnerella 282
Gas-gangrene 89, 291
–liquid chromatography (GLC) 18, **73**, 103, 228, 268
Gastric secretions 46
 washings 163, 210
Gastritis 251
Gastro-enteritis 120–1, 128, 130, 252–3, 359
 see also Enteritis
Gell 61
General paralysis of the insane (GPI) 225, 279
Genetics, bacterial 18–21, 382–3
Genital warts 283
Genome, virus 12
Genotype 15
Genotypic variation 18
Gentamicin 393
German measles 123
 see also Rubella
Germ-free animals 30
Ghon focus 207
Giant-cell pneumonia 127, 318
Giardiasis **144**, 152, **254**, 270, 316, 355, 359

Giemsa's stain 12, 173
Gingivitis 191, 196
Glandular fever 133, 245
 see also Infectious mononucleosis
Glomerulonephritis 41, 194, 237, 242
Glossina (tsetse) flies 145
Glucose phosphate broth 91
Glutaraldehyde 341, 342
Glycerol 104
Gnotobiotic animals 30
Goats as sources of pathogens 86, 100, 113, 156, 308, 357
Gonococci, gonorrhoea **82**, 234, 236, 244, **279–80, 295**, 302
Gordon 5
'Gram-negative shock' 40
 see also Shock
Gram's stain 10, **71–2**, 164
Granuloma venereum 280
Grey baby syndrome 217, 394
Griffith types of streptococci 78, 194
Griseofulvin 138, 142
Ground itch 307
Growth phases of bacteria 20, 21
Gruber 8
Guarneri bodies 22
Guillain-Barré syndrome 125
Guinea-pigs, laboratory use 38, 88, 101, 102, 104, 109
 worm (*Dracunculus*) 159, 292, 303
Gummata 278, 300

H agglutination 180
H antigens 91, 92, 180
Haemadsorption viruses 126, 170
Haemagglutination, virus 124
 see also Haemagglutinins
 –inhibition tests 177, 183
Haemagglutinins, antibody 50, 51, 176
 virus 124, 125, 170, 370
Haemogogus mosquitoes 122
Haemoglobinuria 278
Haemolysins 17
Haemolysis in culture media 72, **76**, 77, 78, 90
Haemophilus 98–100
 infections **197, 200, 201**, 202, 205, **214–6**, 219, 226, **234**, 240, 295, 300, 302
 see also Chancroid
Haemorrhagic fevers 121, 122, 124
Haematuria 155, 274, 276, 277
Hamsters, diagnostic use 109
Hand, foot and mouth disease 119, 190
–washing 342–3, 348, 351
Handkerchiefs and spread of infection 336
Hansen's bacillus 105
Hantaviruses 122
Haptens 50
Heaf test 184
Heat sterilization 338–40

Hedgehogs 105
Heidelberger 8
Helminth infections 153–9, 209, 245–7, 260–2, 270–1, 276–8, 292–4, 297–8, 303, 352
  see also Flukes, Roundworms, Tapeworms
  immunity 58–9
  laboratory investigations 172–4, 262, 277, 292
  routes 35
  serodiagnosis 183
  treatment see Antimicrobial drugs
Helminths, classification 154
  general features 14
  host-parasite relationships 27
  life-cycles 14, 27–28, 35
  metabolism 24
  microscopy 172–4
  nutrition 24
  pathogenicity 44–5
  physiology 24–5
  reproduction 25
Henle 5
Hen's eggs, laboratory use 169
Hepatic abscesses, granulomas 253, 267–8
Hepatitis 113, 132, 134–5, **264–8**
  A **119**, 134, **264**, 355, 357, 359, 372
  amoebic 269
  B **135**, 162, **265**, 319, 328, 351, 354, 369
  non-A non-B 135, 266, 319, 351
  other 267–8
Hepatoma 266
Herd immunity 59, 368
Herelle, d' 13
Herpangina 119
Herpes simplex **130**, 135, **190**, **221**, 244, 267, **282–3**, 296, 308, 319, 328
Herpesviruses 130–1
Heterophil antibodies 133, 193
Heterophyes 153
Heterotrophs 16
Hexachlorophane 344
Hibiscrub 342
High Temperature Short Time process (milk) 358
Higher bacteria 11, 106
Histoplasmin 139, 185
Histoplasmosis **139**, 240, 245, 248, 267
HLA antigens 50, 213, 303, 317
Holder process (milk) 358
Homogenization of sputum 210
Hookworms 157, 159, 209, 260, 307
Hormones and immunity 318
Horse serum in immunization 290, 363, 367
Horses as sources of pathogens 86, 112, 121, 151, 156, 223
Hospital-acquired infections 345–54
Host defences 46–63
Hot air ovens 338
Human immunoglobulin for immunization or treatment 127, 132, 222, 265, 290, 329, 354, **363**, 369, 370, 372

immunodeficiency virus (HIV, formerly HTLV III) 128, 283
  see also Acquired immune deficiency syndrome
Humoral defences 51
Hyaluronidase 17, 77, 306
Hydatid cysts 14, 45, **156**, **226**, **271**, 303, 352
Hydrocephalus 218, 224, 321, 323
Hydrogen ion concentration and bacterial growth 16
Hydrophobia 222
Hydroxystilbamidine 139
Hygiene, communal 335
  personal 336
  ward 348
Hymenolepis 156
Hyperbaric oxygen 291
Hypergammaglobulinaemia 59, 128, 316
Hyperimmunoglobulin E syndrome 315
Hypersensitivity 39, 41, 43, 44, 45, 58, **61–2**, 141–2, 194, 207, 288, 302, 310
  see also Allergy, Antimicrobial drugs
  skin tests for 139, 148, **184–5**, 207, 288
Hyphae, fungal 13, 171, 172
Hypnozoites, plasmodial 149
Hypochlorite 341, 342, 344
Hypogammaglobulinaemia 53, 314, 316–7

Idoxuridine 135
IgA, IgE, IgG, IgM see Immunoglobulins
Imidazole drugs 142
Immediate-type hypersensitivity 61
Immune adherence 57
  complexes 41, 48, **61**, 178, 179, 227, 237
  deficiency 131, 156, 180, **313–8**, 347, 363, 365, 368, 369, 372; see also Acquired immune deficiency, Compromised hosts
  response genes 50
Immunity 46–63
  age and 55
  genetic control 50
  herd 59, 368
Immunization 6, 7, 337, 362–72; see also Antisera, Immunoglobulins, Vaccines
  active/passive, differences 50, 362–4
  programmes 59, 370–1
Immunocompetence 50
Immunodeficiency see Immune deficiency
Immunodiagnosis 174, 185
  see also Serodiagnosis, Skin tests
Immunodiffusion 175
Immuno-electrophoresis 175
  see also Counter current immuno-electrophoresis

Immunofluorescence 165, 178
  see also Fluorescent antibodies
Immunogenicity 50
Immunoglobulins 8, 52
  see also Antibodies, Human immunoglobulins
Immunological memory 52, 320, 362, 363
  specificity 51
  tolerance 62
Immunology, beginnings and scope 7, 49
Immunosuppression 38, 44, 131, 132, 144, 151, 180, 216, 317, 318
  see also Compromised hosts
Impetigo 305
Inclusion blenorrhoea 115
  bodies 8, 12, 22, 130, 132, 222, 281
  conjunctivitis 115, 296
Incompatibilities, drug 378
Incubation periods of virus infections 42
Incubators, baby 96
Indicators of sterilization 87, 339
Indirect haemagglutination 378
Indole production 91
Inducible enzymes 15, 21, 384
Infection, definition 31
  Control committees, officers, teams 341, 346, 347, 353
  hazards from specimens 162
  some possible results 60–1
Infectious hepatitis 264
  mononucleosis (glandular fever) 132, 133, 151, **193**, 244, 245, 267
Infective dose (ID) 32
  drug resistance 20, 382–4
  hepatitis 264
Influenza, influenza virus infections 99, **125–6**, 135, 192, 204, 326, 352, 370
Infusions and injections, infections from 128, 147, 151, 351
  see also Blood, Drug addiction, Needle-stick
Inhibition tests 177
Inoculation of culture plates 166
Injections see Infusions
Insects 32, 34, 35, 121, 145, 223, 253, 335
  see also Bugs, Fleas, Flies, Lice, Mites, Mosquitoes, Ticks
Intensive Care units 313, 350
Interference, interferons **43**, 48, 49, 58, 170, 368
Interleukins 48, 49, 58, 63
Intermediate hosts 14, 28
Intestines, normal flora 29
Intracellular survival of pathogens 38, 41, 42, **49**, **58**, 101, 207, 267, 268
Intracerebral calcification 224, 321, 323
Intussusception and adenoviruses 130
Irradiation see Radiation
Isolation of patients 124, 335, 346–8
Isoniazid 398
Ivanovsky 8
Ixodidae (ticks) 122, 151

Janeway lesions   237
Jawetz   374
Jenner   7, 134, 368
Jigger fleas   312

K antigens   39, 91
Kabat   8
Kala azar   147, 152
Kanamycin   393
Kaposi's sarcoma in AIDS   283
Kelsey-Sykes test   341
Keratitis   130, 131
Keratoconjunctivitis   129, 131
Kerion   138
Ketaconazole   142
Killer cells   49, 54, 58
Kircher   5
Kitasato   7, 366
Klebsiella   **92**, 202, 324, 350
  see also Enterobacterial infections
Koch   5, 7
Koch's postulates   5, 37
Koch-Weeks bacillus   99
Koplik's spots   127
Kuru   223

L-forms   11, 71, 110
Labelling techniques   177–8
Lactic acid production   16, 357, 389
Lactobacillus   16, 29, **85**, 357, 389
Lactoferrin   46, 47
Lag phase of growth   21
Lancefield grouping of streptococci   76, 77, 80, 194
Larva migrans   157, 159, 260, 293, 297
Larvae, helminth   14, 25, 35, 153, 154, 156, 157, 158, 159, 209, 260, 262, 271, 292, 293, 297
  see also Cysticerci
Laryngitis   193
Laryngotracheitis, laryngotracheobronchitis   127, 199
Lassa fever   123–4, 346, 347
Latamoxef   392
Latent infection   22, **31**, 130–1, 132, 133, 245, **319**, 322
Latex agglutination   141, 171, **176**
Laundry   347
Lazy leukocyte syndrome   314
Lecithinase see Phospholipase C
Leeuwenhoek   5
Leishmaniasis   **147–8**, 152, 270, 303, 310
Legionella, legionellosis (legionnaire's disease)   **101–2**, 203, 352, 356
Lepromin test   288
Leprosy   105, 287–8
Leptospira, leptospirosis   **109**, **218**, 245, **267**, 356
Leptotrichia   102, 110, 191
Lethal dose (LD)   37–8
Leukocidins   17, 49
Leukotrienes   48
Levinthal agar   99
Lice   34, 36, 110, 111, 112, 232, 283, 311
Liebig   6
Limulus lysate test   17

Lincomycin   396
Linnaean taxonomy   67, 68
Lipopolysaccharides   10, 17, 40
Lipoteichoic acid   39
Liquoid   241
Lister   6, 345
Listeria, listeriosis   **85**, **217**, 248, 267, 323, 324, **325**
Liver abscesses, granulomas   253, 267–8
  biopsy   268
Loa, loaiasis   158, 292
Lobar pneumonia   202
Loeffler   8
Loeffler's serum   84
Logarithmic phase of growth   21
Loop, bacteriological   166
Louping ill   122
Löwenstein-Jensen medium   104
Lübeck disaster   365
Lumbar puncture   227
Luminescence biometry, luminometry   165, 275
Lung abscess   202, 204, **206**, 225, 253
Lutzomyia sandflies   147
Lyell's disease   306
Lymphadenitis, lymphadenopathy   **243–7**, 283, 286, 291
Lymphangitis   **243**, 246, 286, 291
Lymphocytes and immunological responses   8, 51, 57–8
Lymphocytic choriomeningitis   123, 220
Lymphogranuloma venereum (LGV)   115, 244, 280
Lymphokines   48, 58
Lysis, bacterial see Bacteriolysis
Lysogenic conversion, lysogeny   19, 23, 84
Lysosomes   48, 49, 315
Lysozyme   46, 47, 71
Lytic phages   23

M proteins, streptococcal   50, 78
MacConkey's agar   72, 80, **90**, 93, 167, 210, 262
Macrophage chemotactic factor   58
  migration-inhibiting factor   58
Macrophages   48, 49, 51, 58
Madura foot   107
Malaria   9, **148–51**, 224, 241–3, 248, 269, 278, 303, 323, 372
Malassezia furfur   137, 310
Malignant pustule   307
Maloprim   242, 243
Malnutrition   156, **317**, 336, 350
Malt worker's lung   142
Malta fever   100
Mango fly   294
Mannitol salt agar   262
Mantoux test   184
Marburg virus   124
Mast cells   61
Mastoiditis   194, 225, 226, 302

Maternal antibodies   50, 54, 55, 120, 362
Mazzotti reaction   292
McCarty   19
McFadyean reaction   86
McLeod   19
Measles   **126–7**, 204, 221, 245, 260, 317, 318, 346, **369**, 371
Meat, transmission of pathogens in   155, 306, 360
  see also Food
Mebendazole   261, 271, 293
Mecillinam   391
Melarsoprol   152
Meleney's gangrene   289
Memory cells   52
Meningitis   212–20
  bacterial (pyogenic)   212–9, 247, 302, 325
  fungal   220
  leptospiral   218
  otogenic   219
  protozoan (amoebic)   220
  tuberculous   218
  virus (aseptic)   117, 119, 123, 126, 130, 219–20
Meningococci, meningococcal infections   81, 213–4, 233
Meningo-encephalitis see Encephalitis
Merozoites, plasmodial   148
Mesenteric adenitis   245, 252
Mesophiles   16
Mesosomes   11
Messenger RNA   21
Metabolic disorders and susceptibility to infection   318–9
Metacercariae   153
Metachromatic (volutin) granules   72, 84
Metagonimus   153
Metchnikoff   8
Methicillin   390
Methylene blue reduction test (milk)   358
Methyl red test   91
Metrifonate   277
Metronidazole   152, 388
Mezlocillin   391
Mice as sources of pathogens   94, 109, 112, 123; see also Rodents
  laboratory use   88, 105, 119, 122, 123, 222, 232
Miconazole   142
Micro-aerophilic bacteria   15, 97, 98, 106, 107
  streptococci   197, 225, 289
Microbial ecology   26–30
Microbiology, scope   10
Microcephaly   224
Microfilariae   158
Micropolyspora faeni   141
Microscopy, methods   5, 7, 8, 71, **164–5**, **168–9**, 171, **172–4**, 262
Microsporum   138, 171, 172, 309
Milk antibodies   50, 55, 120
  bacteriological examination   358
  pasteurization   208, 232, 358
  sterilization   358

transmission of pathogens in   33, 34, 98, 100, 103, 113, 144, 208, 232, 251, 323, **357**
  ultra-heat treatment   358
Miller   8
Minimal bactericidal concentration (MBC)   385
  inhibitory concentration (MIC)   385
Minocycline   395
Mites   34, 36, 111, 112, 283, 310
Mobiluncus   98, 282
Molluscum contagiosum   308
Monday morning fever   356
Monkey B virus   130
  vacuolating virus   129
Monkeys as sources of pathogens   122, 124, 130, 223
Monoclonal antibodies   56, 58, 175
Monokines   48
Monolayers   116, 170
Mononuclear phagocytes   48, 61
  see also Reticulo-endothelial system
  deficiencies   316
Monosulfiram   311
Morganella   95
Mosquitoes   34, 246, 335
  see also Aedes, Anopheles, Culex
Motility of bacteria   **71**, 85, 87, 88, 90, 108, 109, 110, **164**
Moulds   13
Mouth, normal flora   29, 189
Moxalactam   392
Mucociliary escalator   47
Mucopeptide   10
Mucormycosis   141
Mucosal antibodies   55
Mucous membranes as barriers to infection   46, 47
Mucus   47, 200
Mumps   **126**, 220, 221
Mupirocin   394
Muramic acid   10, 11
Mutation of bacteria   18, 382–3
Mycelium   13
Mycetoma   44, 107, 287
Mycobacteria, atypical   105–6, 208, 365
  resistance to drying etc.   337
Mycobacterial infections   208, 243, 283
  see also Leprosy, Tuberculosis
Mycobacterium   103–6
Mycology   137–42
Mycoplasma   11, **110**, 200, **203**, 280
Mycoses   44, 137–42
Myiasis   294
Myocarditis   119, 125, 239
Myositis, clostridial   291
Myxoviruses   123–7

Naegleria fowleri   143–4, 220, 356
Nagler reaction   89
Naked viruses   13
Nalidixic acid   389
Natural killer cells   49, 58
Necator   157, 209, 260

Necrobacillosis   196, 235
Necrotizing enterocolitis   327
Needham   6
Needle-stick injuries   354, 370
Negative staining   69, 169
Negri bodies   22, 222
Neisser's stain   72, 84
Neisseria   81–2
Nematodes (roundworms)   14, **156–9**, 209, 246–7, 260–1, 267, 271, 292–3, 297, 303, 307
  see also Helminths
Neomycin   394
Neonatal antibody production   55, 120
  infections   115, 204, 216, 295, 320, **324–9**
Nephelometry   275
Nephrotic syndrome   278
Netilmicin   393
Neuraminidase   50, 124, 125, 370
Neurotoxins   45, 60, 87, 88
Neutralization tests   177, 182
Neutralizing antibodies   53–4, 177
Neutrophil deficiencies   314–6
Neutrophils   47, 48, 320
Nezelof's syndrome   317
Niacin production   104
Niclosamide   261
Nifurtimox   152
Nitroblue-tetrazolium (NBT) test   48, 316
Nitrofurantoin   389
Nocardia, nocardiosis   107, 225, 287
Non-gonococcal urethritis (NGU)   115, 280
Non-specific urethritis (NSU)   280
  vaginitis (NSV)   281
Norfloxacin   389
Normal microbial flora of man   29, 189, 249, 305, 320, 374
Norwalk virus   121, 253, 359
Nose, normal flora   29, 189
Nosocomial (hospital-acquired) infections   345–54
Nosopsyllus fleas   312
Novy-MacNeal-Nicolle medium   173
Nuclear bodies, bacterial   11
Nucleic acids (DNA, RNA)   11, 12
  see also Deoxyribonucleic *and* Ribonucleic acids
Nucleocapsid, virus   12
Numerical taxonomy   68
Nutrient agar   72, 166
Nystatin   142

O agglutination   180
O antigens   91, 92, 180
Oesophagitis   137
Ofloxacin   389
Old tuberculin (OT)   184
Onchocerca   158, 292, 297
Oocysts, toxoplasma   151
Operating theatres   35, 301, 340
Ophthalmia   115, 281, 295–6, 326
Opisthorcis   153, 270

Opportunist pathogens   27, **30**, 31, 44, **139–41**, 205, 313, **315** (table), 348, 374
Opsonins, opsonization   41, 53, 59, 79
Optimal proportions, antigen-antibody   175
Optochin   79
Orchitis   126, 279
Orf   308
Organ culture   170
Oriental sore   146–7
Original antigenic sin   363, 370
Ornithosis   114, 203
Orthomyxoviruses   125–6
Osler's nodes   237
Osteitis, osteomyelitis   226, 231, 247, **299–302**
Otitis   **196–7**, 215, 225, 226, 302, 356
Otomycosis   140
Ova, helminth   14, 25, 35, 45, 153, 154, 155, 156, 157, 159, 173, 174, 260, 261, 262, 276–7
Ouchterlony procedure   175
Oxidase (test)   81, 90, 96, 97, 101
Oxidation, bacterial   15
Oxygen requirements of micro-organisms   15, 72

Paludrine   243
Pandemic diseases   31, 125, 255
Papilloma viruses   129
Papovaviruses   129, 224, 283, 308
Para-aminobenzoic acid (PABA)   387
Paracoccidioidomycosis   139
Paragonimus   154, 209
Parainfluenza viruses   126, 199, 200, 204
Paramyxoviruses   126–7, 192, 199
Parasitism, types   26, 27
Parasitology, definition   10
Paratyphoid   94, 231, 355, 366
Paronychia   138
Parotitis   126
Parvobacteria   98–102
Parvoviruses   128–9, 303
Passive agglutination   176
  haemagglutination (PHA)   176
Pasteur   4, 6, 86, 222, 334, 358
Pasteurella   101
  multocida infections   222, 243, 291, 297
Pasteurization   103, 358
Pathogenicity   26, 32–3, 37–45
  see also many other places
Paul-Bunnell test   138, 193
Pebrine   6
Pediculosis   311
Penetration by viruses   22, 42
Penicillinases see Beta-lactamases
Penicillin-binding proteins   381, 390
Penicillins   9, 381, 390–1
Penicillium   9, 141, 142
Pentamidine   152
Peptic ulcers   251
Peptidoglycan   10
Peptococcus   80

Peptostreptococcus 80
Peptone water 90, 167, 262
Perfloxacin 389
Peri-anal abscess 287
Pericarditis 125, 239
Peri-odontal disease 191, 225
Peritonitis 138, 233, 279, 351
Peritonsillar abscess 193
Pernasal swabs 201
Peroxides 15
Pertussis (whooping cough) **100, 201,** 346, **365–6,** 371
Petri dishes 166
Pfeiffer 99
pH and microbial growth 16
Phage-typing 23, 75, 93, 94, 96, 360
 Staph. aureus 75, 258, 349
Phages 13, 19, 23
Phagocytes, phagocytosis 8, 41, 42, 47 **48,** 53, 79
Phagosomes 48, 49
Pharyngitis 193, 279
Pharyngo-conjunctival fever 129
Pharynx, normal flora 29, 190
Phase contrast microscopy 71, 164
 variation, flagellar 92
Phenethicillin 390
Phenolics 340, 341, 342, 344
Phenotypic variation 20
Phlebotomus sandflies 122, 147
Phosphatase test (milk) 358
Phospholipase C 17, 41, 89
Photochromogens 106
Phototrophs 15, 17
Picornaviruses 117–20
Pigments, bacterial 17–18, 74, 96, 102, 106
 fungal 171
Pig-bel 252, 327
Pigs as sources of pathogens 85, 100, 109, 120, 151, 153, 155, 158, 223, 293
Pili 11, 39, 250
Pilonidal abscesses 288
Pinta 108
Pinworms (*Enterobius*) 157, 261
Piperacillin 391
Piperazine 261
Pityriasis versicolor 137, 310
Pivampicillin 391
Pivmecillinam 391
Placental transmission of antibodies 50 54, 55, 362
 infection 29, 34, 35, 42, 132, 151, 224, 320–4
Plague 5, 95, 202, 244
Plaque, dental 18, 39, 78, **191**
Plaques, phage 23
Plasma cells and antibodies 52
Plasmids 11, 20, 383
Plasmodium 148–51
 see also Malaria
Plate cultures, inoculation 166
Plaut 191
Pleural effusion 207
Pleurisy 202
Pneumococci, pneumococcal infections **79–80,** 200, **202, 204,** 205, **216,** 219, 225, 226, **233,** 236, 240, 295, 316, 352
Pneumocystis 132, **151,** 152, **205,** 283
Pneumonia 202–6, 283, 325–6
Pneumonic plague 202
Pneumonitis 209, 260
Poliomyelitis and other poliovirus
 infections **219, 221,** 355, 357, 367, **368–9,** 371
Polioviruses 117–9
Polyanethol sulphonate (Liquoid) 241
Polymyxins 397
Polyoma virus 129
Polysaccharides, capsular or cell-wall 41, 49, 50, 69, 77, 78, 79, 81, 99, 176, 362, 364
Pontiac fever 203
Population screening for bacteriuria 276
Porter 8
Post-gonococcal urethritis (PGU) 116, 280
Post-infectious encephalitis 223
Post-operative infections 301, 349–50, 352
Poultry as sources of pathogens 34, 85, 98, 114, 147, 223, 251, 256, 306, 360
 see also Birds
Povidone-iodine 343, 344
Poxviruses 133–4
Praziquantel 262, 278
Precipitation, antigen-antibody 175–6
Pressure sores 289
Primary atypical pneumonia 110, 203
Probenicid 390
Proctitis 279
Proglottides, tapeworm 14, 155, 156, 262
Progressive multifocal leuko-
 encephalopathy 129, 224
Proguanil 243
Prokaryotes 10
Promastigotes, protozoan 147, 148
Prontosil 387
Prophage 23
Prophylactic immunization 362–72
 see also Antisera, Human immuno-
 globulins and Vaccines
 use of antimicrobial drugs 208, 214, 239, 242–3, 295, 302, 327, 337, 350, 370, 372, **375–6,** 383
Propicillin 390
Propionobacterium 85, 306
Prostaglandins 48, 63
Prostheses and infection 113, **226–7,** 233, **235,** 236, 301, **319,** 350, 375
Protein A, staphylococcal 75
 see also Co-agglutination
 synthesis, bacterial, and antibiotics 389
 human, and diphtheria toxin 83
Proteus 94, 274
 see also Enterobacterial infections
 OX strains in Weil-Felix test 113
Protoplasts 10, 11, 71
Protozoa, classification 144

general features 14
host-parasite relationships 27
life cycles 27–8, 35
metabolism 24
microscopy 172–4
nutrition 24
pathogenicity 44–5
physiology 24–5
reproduction 25
Protozoan infections **143–52,** 220, 224–5, 241–3, 245, 253–4, 269, 278, 281, 297, 303, 310, 323, 352
 immunity 58–9
 laboratory investigations 172–4, 262
 routes 35
 serodiagnosis 183
 treatment see Antimicrobial drugs
Providencia 95
Pseudohyphae 14, 282
Pseudomembranous colitis **259,** 263, 327
Pseudomonas **95–6,** 201, 204, 205, 234, 274, 297, 307, 347, 350, 351, 356
Psittacosis 114, 203
Psoas abscess 301
Psychrophiles 16
Puerperal fever, puerperal sepsis 5
Pulex fleas 312
Pure cultures or strains of bacteria 67, 72
Purified protein derivative (PPD) 184
Pus, composition 47
 specimens 163
Pustules 60, 285
Pyaemia 230
Pyelonephritis 272
Pyocine typing 96
Pyocyanin 18
Pyrantel 260, 261
Pyrazinamide 398
Pyrexia of unknown origin (PUO) 183, 247–8
Pyrimethamine 242, 388
Pyuria 272, 275

Q fever **113,** 204, 236, 238, 267, 357
Quantitative bacteriology 210, 273, 275
 egg counts (helminths) 277
Quarantine 223, 334, 335
Quaternary ammonium compounds 341, 344
Quellung reaction 176
 see also Capsule swelling
Quinolones 389
Quinine 9, 242
Quinsy 194

Rabbits as sources of pathogens 101
 laboratory use 38, 85, 108, 182, 239
Rabies 6, **127, 222–3,** 334, 347
Radiation, UV or X, effects on micro-
 organisms 19, 137, 138, 340

Radio-active CO_2 165, 241
  –immunoassay (RIA)   178
Rapid plasma reagin (RPR) test   181, 279
Rats as sources of pathogens   95, 109, 112, 148, 267, 335, 356
  see also Rodents
Rayer   5
Reaction of identity   84, 175
Reactive arthritis   41, 116, **302–3**
Reagin tests for syphilis   181
Reaginic antibodies   61
Redi   6
Reiter's syndrome   303
Relapsing fever   232, 248
Renal dialysis units   351, 354
  disease   318–9, 351, 398
Reoviruses   120
Replication of viruses   12
Reservoir hosts   28
Resistance to antimicrobial drugs   20, 346, 374, **381–4**, 397–8
Respiratory syncytial (RS) virus   **127**, 192, 199, 204, 326, 352
Reticulate bodies, chlamydial   12
Reticulo-endothelial system   48, 61, 230, 303, 316
Retroviruses   128
Reye's syndrome   125
Rhabdoviruses   127
Rheumatic fever   **194**, 302, 375
Rhinoviruses   119, 192, 193
Rhizopus   141
Ribonucleic acid (RNA)   11, 12, 21, 389
  viruses   117–28
Ribosomes   11, 43
Rice and food-poisoning   258
Ricketts   11
Rickettsiae   13
Rickettsialpox   113
Rifampicin   396, 397
Rifamycins   396
Ringworm (tinea)   44, 138–9, 142, 171–2, 309
Rocky Mountain spotted fever   112
Rodents as sources of pathogens   95, 101, 110, 112, 113, 124, 148, 151, 158, 232, 243, 312, 360
  see also Mice, Rats
Romanowsky stains   173
Rostellum, tapeworm   156
Rotaviruses   120, 252, 353, 359
Roundworms   14, **156-9**, 209, 246–7, 260–1, 267, 271, 292–3, 297, 303, 307
  see also Helminths
Roux   7
Rubella   **123**, 244, 245, 298, 303, **322**, 369

Sabin-type poliomyelitis vaccine   222, 368
Sabouraud's media   16, 24, 141, 171
Sabre tibia   300

Safe collection and transmission of specimens   162–3, 354
Saliva   34, 189, 193, 222
Salk-type poliomyelitis vaccine   222, 368
Salmonella   92–4
  food-poisoning   **256**, 353, 357, 359, 360
  infections   180, 300, 303, 327; *see also* Typhoid, Paratyphoid
Salpingitis   279, 281
Sanarelli   8
Sandflies   122, 147, 148
Sandfly fever   122
Saprophytes   26
Sarcomastigophora   143
Sarcoptes mites (scabies)   310
Satellitism   18, 98, 99
Scabies   283, 310
Scalded skin syndrome   306
Scarlet fever   194, 357
Schick test   85, 183–4, 367
Schistosomes   **155**, 209, **261**, 267, **270**, **276**
Schizonts, plasmodial   148
Schoenlein   5
Schroeder   6
Schwann   6
Scolex, tapeworm   14, 155, 156
Scombroid poisoning   259
Scotochromogens   106
Scrapie   224
Screw worm   294
Scrub typhus   112
Sebaceous cysts   288
Secretory IgA   53, 55
Selective culture media   93, 167, 262
  enrichment culture   93
Selenite broth   93, 262
Self-antigens   62
  –infection   345
Semmelweiss   5
Semple's vaccine   223
Sendai virus   126
Sensitivity tests (bacteria to drugs)   384–5
Septic shock   40
  see also Shock
Septicaemia   60, 140, 213, 217, **230**, 233, 234, 286, 306, 307, 324, 325, 352
Serodiagnosis   8, 113, 116, 123, **179–83**, 248, 277, 322, 323
Serological detection of microbial antigens   73, 169, 171, 178–9, 228
Serological tests for syphilis (STS)   181–2, 279
Serotyping   92, 98, 99, 109, 360
Serratia   95
  see also Enterobacterial infections
Serum hepatitis   264, 265
  sickness   62, 363
Sewage   254, 255, 267, 335, 355, 356, 357, 359
Sex fimbriae   11, 20
Sexually transmitted diseases   278–84
Shedders of staphylococci   35

Sheep as sources of pathogens   86, 113, 114, 151, 153, 156, 223, 270, 308
Shell-fish as sources of pathogens   97, 121, 154, 258, 264, 359
Shigellosis   **94, 249–51**, 303, 353, 355, 357, 359
Shingles   131
Shock (bacteraemic, endotoxic, 'gram-negative', septic)   40, 57, 234, 268, 352
  see also Disseminated intravascular coagulation, Waterhouse-Friderichsen syndrome
  viraemic   235
Shwartzman   8
Sickle-cell disease and trait   128, 150, 231, 300
Silver nitrate   296
Simulium blackflies   158, 292
Sinusitis (paranasal)   194, **197**, 225, 226, 302
Skin as a barrier to infection   42, 46
  disinfection   227, 240, 301, **342–3**
  normal flora   29, 305, 342
  scales and infection   35
  tests   139, 148, 183–5
Sleeping sickness   145, 152, 224, 246
Slow virus infections   23, 223
Small round structured viruses   121
Smallpox   7, 134, 333, 368
Snails   14, 153, 155
Soft chancre   280
Soil as a source of pathogens   32, 87, 88, 89, 139, 258, 287, 290, 291, 301
Somatic (O) antigens   91, 92, 180
Sore throat   193
Sources of pathogens   32, 34
  see also Air-conditioning systems, Animals, Baths, Blankets, Blood, Carriers, Catheters, Droplets, Dust, Food, Handkerchiefs, Infusions, Milk, Specimens, Soil, Swimming baths, Ventilators, Vertical transmission, Water, Wounds
Souring of milk   357
Spallanzani   6
Specimens, collection, transmission and hazards   162–3, 354
Spectinomycin   394
Spheroplasts   71
Spider's web clot   218
Spirochaetes   11, 69, 107–10
Splenectomy   233, 316
Spontaneous generation   5, 6
Spores, bacterial   11, 71, 83, 86, 87, 88, 257, 258, 259, 290, 291, 307, 327, 337, 361, 367
  fungal   24, 171, 172
Sporotrichosis   243
Sporozoites, plasmodial   148
Spotted fevers   112
Sputum, examination   210
  transmission of pathogens in   336
Staining of micro-organisms   71–2, 164–5, 173
  see also Albert, Giemsa, Gram,

Neisser, Romanowsky, Ziehl-Neelsen, Negative staining
Staphylococci, coagulase-negative 76, 233, **305**, 319, 350
see also many other places
Staphylococcus 74–6
  aureus, drug-resistant 397
    enterocolitis 259, 262, 395
    food-poisoning 257–8, 361
    infections 204, 225, **233**, 238, 240, **285**, 290, 295, **299–300**, 301, 302, 305, 306, 324, **326**, 346, 348; *see also many other places*
    drug treatment 397
    in hospital 349
Stationary phase of growth 21
Steam sterilization 338–40
Stevens-Johnson syndrome 388
Sterilization 337–40
  indicators 87, 339
  of milk 358
Steroids 193, 207, 282, 289, 293, 294, 310, 311, 318, 350
Stibogluconate 152
Sticky eye 326
Stokes procedure 384
Streptococcus 76–81
  group A infections **193**, 236, 239, 240, 243, 244, 286, 289, 290, 295, 297, 302, 305, 307, 349, 350, 357
  B infections 217, 324, 325, 326
  milleri 80, 197, 225, 268
  see also Anaerobic cocci, Micro-aerophilic and Viridans streptococci, Enterococci, Pneumococci
Streptokinase 17, 77
Streptolysins 77
Streptomyces 106
Streptomycin 393, 398
Strobila, tapeworm 14, 155, 156
Strongyloides 157, 159, 209, 260, 307
Stuart's medium 280
Styes 285, 295
Subacute sclerosing panencephalitis 127, 221, 369
Subclinical infection 31
Subdural effusions 215
Sulbactam 392
Sulphonamides 9, 242, 382, 387–8
Sulphur granules 107, 286
  in treatment of scabies 311
Superoxide dismutase 15
Superoxides 15
Suprapubic aspiration of urine 326
Suramin 152, 292
Swabs 163
Swimming baths 96, 106, 138, 356
Symbionts 26
Symmetry, virus 12
Synergic gangrene 289
Synergy between antimicrobial drugs 238, **378**, 388, 398
  bacteria 288, 289
Syphilis **108**, 225, 244, 245, 267, **278–9**, 298, 300, 321, 351
  serology 181–2

TAB and TABC vaccines 366
T helper cells 54, 58, 62, 128, 317
T lymphocytes 48, 49, **51–9**, 62, 128, 133, 313–4, **317**, 318, 320
T suppressor cells 54, 58
Tabes dorsalis 225, 279
Tache noire 111
Taenia 155, 261
  see also Cysticerci
Talampicillin 391
Tapeworms 155–6
  see also Cysticerci, Echinococcus, Helminths, Hydatid, Taenia
Teichoic acids 10
Tellurite medium 72, 84
Temocillin 392
Temperate phages 23
Temperature and microbial growth 16, 24, 72, 95, 105, 106, 119, 166
Tenosynovitis 279
Tetanus 7, **88**, 222, **290**, **327**, **366–7**, 371
Tetracyclines 395
Tetrathionate broth 262
Therapeutic index 374
Thermoactinomyces 141
Thermophiles 16, 97, 141, 142
Thiabendazole 260, 297
Thiacetazone 398
Thienamycin 393
Thioglycollate media 167
Thiosulphate citrate bile-salts sucrose (TCBS) agar 97, 262
Threadworms 157, 261
Throat, normal flora 29, 190
Thromboplastin 48
Thrush 137, 282
  see also Candidiasis
Thymus and immunological responses 8, 51
Ticarcillin 391
Ticks 34, 110, 111, 112, 113, 121, 122, 232, 243
Tine test 184
Tinea 138
  see also Ringworm
Tiselius 8
Tissue culture 8, 169–70, 364
Titre of a serum, definition 180
Tobramycin 393
Togaviruses 121–2, 223
Tonsillitis 193, 196
TORCH neonatal screen 321
Toxic epidermal necrolysis 306
Toxigenicity 40
  tests for *C. diphtheriae* 84, 175
Toxins 7, 17, 19, **40**, 45, 53–4, 75, 77, 83, 84, 89, 259
  see also Endotoxins, Enterotoxins, Neurotoxins
Toxocara 159, 267, 293, 297
Toxoids 7, 364, 366
Toxoplasma, toxoplasmosis **151**, 152, 193, 224, **245**, 283, **297**, 321
Trachoma 115, 296
Transduction, bacterial 19
Transfer factors 20
Transferable drug resistance 20, 382–4

Transformation, bacterial 19
  of host cells by viruses 23
Transmission of pathogens 32–3 (main reference)
  of specimens 163
Transport media 163, 169, 280, 281, 283, 296
Transtracheal aspiration 210
Traveller's diarrhoea 120, 249, 251, 252, **359**
Trematodes (flukes) 153–5, 209, 262, 270
  see also Helminths, Schistosomes
Treponema 108–9
  pallidum haemagglutination (TPHA) test 182, 279
  immobilization (TPI) test 182
  infections 225, 278–81; *see also* Syphilis
Trexler isolator 30, 348, 350
TRIC agent 113
Trichinella 158, 293
Trichomonas **145**, 152, 280, **281**
Trichophyton 138, 171, 172, 309
Trichuris 156, 260
Trimethoprim 152, 388
Triple concentration (trypanosomes) 146, 173
  vaccine 371
Trophozoites, protozoan 143, 144, 148, 151, 174, 245, 253, 254, 262, 269
Tropical ulcer 289
Trypanosomiasis **145-7**, 152, 224, 239, **245-6**, 323
Trypomastigotes, protozoan **145–7**, 152, 224, 239, **245–6**, 323
Tsetse flies 145, 224, 246
Tsutsugamushi fever 112
Tubercle bacilli 103–5
Tuberculin tests **184–5**, 207, 333–4, 354, 357, 365
Tuberculosis 38, **103**, **206–8**, **218**, 239, 244, 274, **275**, 297, **301**, 323, 333–4, 353, 357, 365, 397–8
  see also elsewhere
Tularaemia 101, 244
Tumbu fly 294
Tumour viruses 23, 128, 129, 131, 133, 266, 282
Tunga fleas 312
Turbidimetry 165
Twort 13
Tyndallization 339
Typhoid **93**, **231**, 248, **255**, **269**, 300, 346, 355, 359, 361, **366**, 372
Typhus 111–2, 248
Tzank cells 132

Ulcers 106, 139, 190, 231, 245, 253, 278, 280, 282, 283, 289–90, 292, 296, 302, 308, 310
Ultra-heat treatment (milk) 358
  –violet light 19, 137, 138, 340; *see also* Fluorescence
Uncoating of viruses 22

Undulant fever   232
  see also Brucellosis
Ureaplasma urealyticum   110, 280
Urease test   91
Urethral syndrome   115, 276
Urethritis   115, 116, 279, 280
Urinary tract infections   **272–8**, 326, 346, 352
Urine, collection and examination   174, 273, 274–5
Urticaria   61

V factor   98, 99
Vaccination   7, 317, 364–72
  contra-indications   372
  for international travel   372
Vaccines   78, 80, 81, 85, 86, 88, 93, 96, 99, 101, 105, 109, 111, 119, 122, 126, 133, 196, 222, 233, 290, 307, 316, 322, 328, 334, **364–72**
Vaccinia   134
Vacuum cleaners   348
Vagina, normal flora   29
Vaginitis   279, 281, 282
Vaginosis, anaerobic   281
Vancomycin   396, 397
Varicella (chickenpox)   131–2, 134
  –zoster virus   **131–2**, 152, 267, 296, 323
Varicose ulcers   289
Variola (smallpox)   7, 134, 333, 368
Variolation   7
Veillonella   81
Venepuncture   240
Venereal Disease Research Laboratory (VDRL) test   181, 279
Ventilation of theatres, wards, etc.   35, 301, 340, 347, 348, 350, 353, 354
Ventilators, patient   96, 324, 348, 350, 351
Ventricular tap   227
Ventriculitis   227
Verrucas   308
Vertical transmission of infection   29, 35, 132
  see also Congenital *and* Neonatal infections, Placental transmission
Vesiculitis   279

Vi agglutination   180
  antigen   93, 269
Vibrio parahaemolyticus food poisoning   258, 359
Vibrios   11, 96–7
  see also Cholera
Vidarabine   135
Vincent's angina   196
  organisms   110, 191
Viraemia   60, 235
Viral haemorrhagic fevers   124, 235, 347
Viridans (alpha-haemolytic) streptococci   **78**, 191, 192, 225, 226, **236**, **237**, 238, 351
Virion   12
Virology, beginnings   8
Virulence   37, 39–41
Virulent phages   23
Virus infections, chemotherapy   135–6
  incubation periods   42
  laboratory investigations   **168–71**, 193, 211, 227–8, 263, 269
  latent   **22**, 130–1, 132, 133, **319**, 322
  pathogenesis   41–3
  persistent   22, 42
  serodiagnosis   182–3
  slow   23, 223, 340
  treatment *see* Antimicrobial drugs
Viruses, classification and nomenclature   117, 118
  culture   169–71
  cytotoxicity   42
  defective   128
  general features   12–13
  haemagglutination by   124, 170; *see also* Haemagglutinins
  isolation and identification   169–71
  microscopy   168–9
  neutralization   54
  pathogenicity   41–4
  physiology   21–3
  purification   13
  replication   12
  tumour-associated   23, 128, 129, 131, 133, 266, 282
Vitamins, bacterial synthesis   18
Voges-Proskauer reaction   91
Volutin granules   72, 84
Vomiting   249, 253, 256–9, 379
von Pirquet   8
Vulvovaginitis   279, 282

Waksman   389
Warts   129, 308
Wassermann reaction (WR)   181
Water, examination   355
  purification   355
  transmission of pathogens in   32, 34, 93, 97, 98, 102, 121, 138, 142, 143, 144, 155, 159, 203, 253, **254–5**, 264, 266, 267, 277, 355, 360; *see also* Swimming baths
Waterhouse-Friderichsen syndrome   40, 57, 213, **234**, 235
Weil-Felix test   112, 113
Weil's disease   218
West Nile fever   121
Whitlow   130
Whooping cough (pertussis)   **100**, **201**, 346, **365–6**, 371
Widal test   8, 93, **180**, 231, 269
Wire loop, bacteriological   166
Wool-sorter's disease   202
Worms   14
  see also Helminths
Wound infections   290–2, 349–50
Wounds as sources of pathogens   349–50
Wuchereria   158, 246–7

X factor   98, 99
Xenodiagnosis   172, 246
Xenopsylla fleas   16, 95, 312
Xylose lysine deoxycholate agar   262

Yaws   108
Yeasts   13
  see also Candidiasis
Yellow fever   122, 267, 372
Yersin   7
Yersinia enterocolitica and *pseudotuberculosis*   95, 245, 252, 303
  pestis   95,
  see also Plague

Ziehl-Neelsen staining   72, **103**, 145, 164
Zoonoses   32, 223
Zoster (shingles)   131, 296